CATASTROPHE THEORY

E.C. Zeeman, F.R.S.　　Mathematics Institute
　　　　　　　　　　　University of Warwick
　　　　　　　　　　　Coventry

CATASTROPHE THEORY

SELECTED PAPERS, 1972 – 1977

1977

Addison-Wesley Publishing Company
Advanced Book Program
Reading, Massachusetts

London ● Amsterdam ● Don Mills, Ontario ● Sydney ● Tokyo

Library of Congress Cataloging in Publication Data

Zeeman, E   C
    Catastrophe theory.

    Includes index.
    1. Catastrophes (Mathematics)--Addresses, essays,
lectures. 2. Social sciences--Mathematical models--
Addresses, essays, lectures. 3. Science--Mathematical
models--Addresses, essays, lectures. I. Title.
QA614.58.Z43   514'.7   77-21459
ISBN 0-201-09014-7
ISBN 0-201-09015-5 pbk.

Original text reproduced by Addison-Wesley Publishing Company, Inc., Advanced Book Program, Reading, Massachusetts, from camera-ready copy prepared by the author. Copyright © by Addison-Wesley Publishing Company, Inc.

Published simultaneously in Canada.

American Mathematical Society (MOS) Subject Classification Scheme (1970): 34C99, 57D45, 58F99, 70K20, 73H05, 90A15, 92A05, 92A20, 92A25

All rights reserved. No part of this publication may be reproduced, stored in a retrieval system, or transmitted in any form or by any means, electronic, mechanical, photocopying, recording, or otherwise, without the prior written permission of the publisher, Addison-Wesley Publishing Company, Inc., Advanced Book Program, Reading, Massachusetts 01867.

Printed in the United States of America

ABCDEFGHIJ-HC-7987

To Rosemary

CONTENTS

	Page
PREFACE	ix
ACKNOWLEDGEMENTS	x

GENERAL INTRODUCTORY PAPERS

1. Catastrophe theory : Draft for a Scientific American article — 1
2. Levels of structure in catastrophe theory — 65

BIOLOGICAL SCIENCES 80

3. Differential equations for the heartbeat and nerve impulse — 81
4. Primary and secondary waves in developmental biology — 141
5. A clock and wavefront model for the control of repeated structures during animal morphogenesis (with J.Cooke) — 235
6. Gastrulation and formation of somites in amphibia and birds   (Addendum by R.Bellairs.) — 257
7. Dialogue between a Biologist and a Mathematician — 267
8. Brain modelling — 287
9. Duffing's equation in brain modelling — 293

SOCIAL SCIENCES 302

10. Some models in the social sciences (with C.A.Isnard) — 303
11. On the unstable behaviour of stock exchanges — 361
12. Conflicting judgements caused by stress — 373
13. A model for institutional disturbances (with C.S.Hall, P.J.Harrison, G.H.Marriage, P.H.Shapland) — 387
14. Prison disturbances — 403

PHYSICAL SCIENCES 408

15. A catastrophe machine — 409
16. Euler buckling — 417
17. Stability of ships — 441

MATHEMATICS 496

18. The classification of elementary catastrophes of codimension ≤ 5 (with D.J.A.Trotman) — 497
19. The umbilic bracelet and the double-cusp catastrophe — 563

DISCUSSION 604

20. Research ancient and modern — 605
21. Catastrophe theory : its present state and future perspectives (with R.Thom) — 615
22. Afterthought — 651

INDEX	659
PUBLICATION DETAILS	675

PREFACE

This book is a collection of papers that are variations on a common theme: the use of elementary catastrophes for modelling in the sciences. Catastrophe theory has many aspects, ranging from theorems in pure mathematics and applications in classical applied mathematics, to the more novel types of modelling in the biological and behavioural sciences. Each aspect demands its own language; and each of the papers in the book was written with one particular aspect in mind and first published in a journal or proceedings oriented to that aspect. In order to preserve those particularities, the papers are reprinted here unaltered.

The first paper is the original draft for an article that appeared in shortened form in "Scientific American", April 1976, and may be read as a general introduction. Alternative introductions are provided by Papers 2 and 10. The difference between these three introductory papers is one of emphasis, Paper 1 being oriented more towards psychology and physics, Paper 2 towards mathematics, and Paper 10 towards the social sciences. The rest of the papers may be read in any order, because each is independent.

The specialist papers have been grouped together in four sections: biological sciences, social sciences, physical sciences, and mathematics. Broadly speaking the book has been arranged so that the mathematical content increases steadily, culminating in the proof of the classification theorem in Paper 18. The final section contains three discussion papers.

Historically catastrophe theory began with the ideas of René Thom in the early 1960's. Both the mathematics and the applications were present from the beginning, each stimulating the other, as can be seen in Thom's classic book* on structural stability and morphogenesis, which he wrote during the late 1960's. By the end of the decade many of Thom's mathematical conjectures concerning the singularities of maps had been proved, with notable contributions from Arnold, Malgrange, Mather, and Thom himself, and these results have subsequently generated a great deal of mathematical research in neighbouring fields.

Meanwhile, during the 1970's, there has been a growing interest in using this method of modelling in a variety of scientific fields, and the motivation of most of the papers in the book has been to bring this possibility to the attention of scientists. Firstly, the geometry of the elementary catastrophes may be used to <u>describe</u> phenomena, particularly those in which gradually changing forces produce sudden effects. The philosophical justification for using the elementary catastrophes as models is that mathematically they are the higher dimensional analogues of the simple concepts of maxima, minima and thresholds. Already such a model may be scientific in the sense of reducing the arbitrariness of description, by providing a coherent synthesis of otherwise unrelated observations. Secondly, it may be possible to <u>explain</u> a model by deducing it from more fundamental hypotheses, as described in Paper 7; examples are given in Papers 4, 10, 11, 12, and 13. This type of explanation is a familiar

---

* see bibliographies of Thom on pages 63, 643-644, 647-648.

process in physics, as illustrated in Papers 16 and 17, but is a new paradigm in the biological and social sciences, due to the necessity of having to use sophisticated mathematics. Thirdly, the use of models to design experiments and <u>predict</u> new results is discussed in several papers.

Catastrophe theory is now moving into the experimental stage, as many experimental scientists are beginning to use the ideas to design their own experiments and test the predictions arising from the models. Some good examples are given in the Dialogue in Paper 7, which describes some of the successful predictions in embryology that have arisen from the primary and secondary wave model introduced in Paper 4. The experiments are concerned with somitogenesis, verifying the existence of a hidden primary wave of determination and a secondary wave of change of adhesiveness, and providing evidence for a clock that regulates the segmentation. It is not only in the biological sciences, but also in the physical and social sciences, that experimental work is being carried out and data fitted. Much of the work is as yet exploratory; nevertheless, I would venture to suggest that by the 1980's this type of modelling, which today is loosely called "catastrophe theory", will have proved itself a useful tool in many fields of science.

E.C.Z.
July 1977.

ACKNOWLEDGEMENTS

I am primarily indebted to the many mathematicians and scientists with whom I have discussed the various aspects of catastrophe theory during the last ten years, especially René Thom. I am also indebted to the many students, colleagues and visitors at Warwick University who have patiently listened to my lectures on the subject, and added their own ideas.

I particular I would like to thank my co-authors in papers 5, 10, 13 and 21 for allowing me to include our joint papers; David Trotman for allowing me to include paper 18, his lecture notes for a course I gave in 1973; Ruth Bellairs and Michael Berry for allowing me to print their photographs; and the Editors of the various Journals and Proceedings in which the papers were first published, for allowing me to reprint them.

I am very grateful to the Institut des Hautes Etudes Scientifiques and the Science Research Council for their support over many years. Finally my thanks to Elaine Shiels, who patiently typed many versions of all the papers.

GENERAL INTRODUCTORY PAPERS

1 CATASTROPHE THEORY

INTRODUCTION

Catastrophe theory is a new mathematical method for describing the evolution of forms in nature. It was created by René Thom who wrote a revolutionary book "Structural stability and morphogenesis" in 1972 expanding the philosophy behind the ideas. It is particularly applicable where gradually changing forces produce sudden effects. We often call such effects catastrophes, because our intuition about the underlying continuity of the forces makes the very discontinuity of the effects so unexpected, and this has given rise to the name. The theory depends upon some new and deep theorems in the geometry of many dimensions, which classify the way that discontinuities can occur in terms of a few archetypal forms; Thom calls these forms the <u>elementary catastrophes.</u> The remarkable thing about the results is that, although the proofs are sophisticated, the elementary catastrophes themselves are both surprising and relatively easy to understand, and can be profitably used by scientists who are not expert mathematicians.

In physics many classical examples can now be seen to be special cases of low dimensional catastrophes, and as a result the higher dimensional catastrophes are beginning to suggest new experiments and offer understanding of more complicated phenomena. However in the long run the more spectacular applications may well be in biology, providing models for the developing embryo, for evolution and behaviour. Much of Thom's book concerns embryology. Models in psychology and sociology suggest new insight into the complexity of human emotions and human relationships, and offer new designs for experiments.

This article is in three parts. In the first part we introduce the reader gently into the ideas by describing some simple applications to

---

Published (in a shortened form) in the Scientific American, April 1976, Volume 234, part 4, pages 65-83.

elasticity, aggression, emotions, war and economics; the objective is to lead up to a precise mathematical statement of one of the classification theorems, together with Thom's list of the 7 elementary catastrophes having a control space of dimension less than or equal to 4. In the second part we go more deeply into one particular application, namely a model for the nervous disorder anorexia nervosa, in order to illustrate the profundity and qualitative power of the new language in the human sciences. In the third part we select some familiar examples from classical physics, in order to illustrate the generality and quantitative power of the new language in the physical sciences.

CONTENTS.

PART ONE   Page

1. The cusp catastrophe   3
    Example 1   Aggression   3
2. The dynamic   8
    Example 2   A catastrophe machine   8
3. The mechanism of aggression   12
    Example 3   Territorial fish   13
    Example 4   Argument   15
    Example 5   Catharsis of self-pity   15
    Example 6   War   16
4. Divergence   17
    Example 7   Stock exchanges   21
5. The classification theorem   22
6. Thom's list of 7 elementary catastrophes   25
7. The butterfly catastrophe   29
    Example 8   Compromise   29

PART TWO
    Example 9   Anorexia nervosa   33

PART THREE
    Illustrations in physics   53
    Example 10 Phase transition   53
    Example 11 Light caustics   54
    Example 12 Euler buckling   58
    Example 13 Shock waves   61
    Example 14 Nonlinear oscillations   62

    Suggested further reading   63

PART ONE

1. THE CUSP-CATASTROPHE

The first elementary catastrophe is the <u>fold-catastrophe</u> but that is too simple to see what is going on, and so we shall start with the next one, the <u>cusp-catastrophe</u>, which contains many subtleties. The cusp-catastrophe is the 3-dimensional graph illustrated in Figure 4. But let us approach it by first considering the simpler 3-dimensional graph shown in Figure 1, which illustrates how profit x depends upon income α and costs β, by means of the simple formula, $x = \alpha - \beta$.

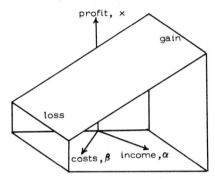

Figure 1. The graph of profit as a function of income and costs.

Here we represent α, β by axes in the horizontal plane C, and x by the vertical axis, and since the formula is linear the graph is a sloping plane in 3-dimensions. In particular :

    an increase in income causes an increase in profit;
    an increase in costs causes a decrease in profit;
    an increase in both causes no change in profit.

We might summarise this situation by saying "income and costs are conflicting factors influencing profit".

Example 1. Aggression.

Now let us pick a similar sentence from psychology. In Konrad Lorenz's book "On Aggression" he says that rage and fear are conflicting factors influencing aggression. The question is: can we represent this similar sentence by a similar graph? To make the question more specific first think of a dog, and then later we shall apply it to fish and humans. A preliminary problem arises: can we measure the rage and fear drives in a dog at any moment? Lorenz suggests that we can, and Figure 2 shows illustrations from his book; he proposes that rage can be measured by how much the mouth is open, and fear by how much the ears lay back. So let

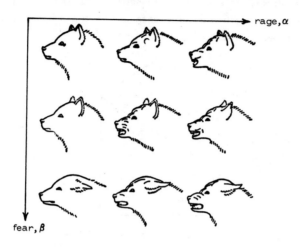

Figure 2. Rage and fear can be measured by facial expressions (after Konrad Lorenz).

us assume that rage and fear can be plotted as two horizontal axes, α and β. Meanwhile let us also assume we can devise some vertical scale x representing the resulting behaviour of the dog running from fight to flight, through intermediary behaviour such as growling, neutral and avoiding. We want to plot the graph x as a function of α and β. It is true, as before, that

an increase in rage causes an increase in aggression;

an increase in fear causes a decrease in aggression.

But what if we increase both rage and fear together? The least likely behaviour is for the dog to remain neutral, and the most likely behaviour is fight or flight, although which of the two he will choose may be unpredictable. Therefore one thing is sure : there is no simple formula like $x = \alpha - \beta$, and the graph cannot look like Figure 1.

How do we analyse the situation? One answer is to look at the likelihoods. So let us imagine a likelihood distribution for the behaviour x in each of the following four cases.

	Drives	Most likely behaviour
1	Rage only	Fight
2	Fear only	Flight
3	Neither	Neutral
4	Both rage and fear	Fight or flight

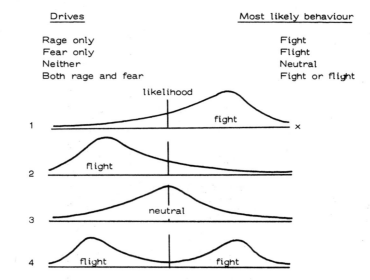

Figure 3. Likelihood of aggressive behaviour.

The interesting case is Case 4, where the distribution has gone bimodal. It should be possible to design experiments, in which the three variables are monitored by three observers for several minutes, and simultaneously fed into a computer, which could be programmed to draw the curves. From the curves the computer could then extract the 3-dimensional graph of the behaviour as a function of rage and fear. Above each point (α, β) of the horizontal plane C, representing given coordinates of rage and fear, is marked the point (or points) representing the most likely behaviour.

6   1. Catastrophe Theory

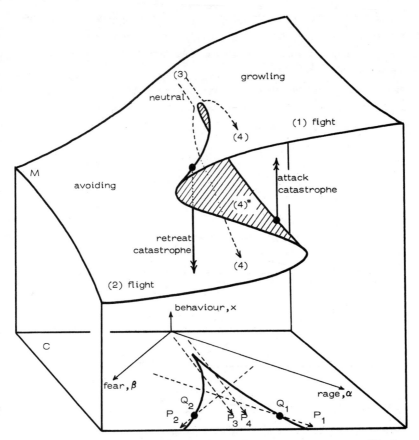

Figure 4. The cusp-catastrophe illustrating fear and rage as conflicting factors influencing aggression.

What catastrophe theory tells us is that if the likelihood distributions look like Figure 3, then the graph will look like the cusp-catastrophe surface M pictured in Figure 4. This follows from the main classification theorem, which we state later. In our experiment the graph could then be used for quantitative prediction of the dog's behaviour in the subsequent few minutes, predicting the sudden changes of mood in terms of the gradually changing facial expressions.

Already the graph has some surprising qualities, so let us analyse them. If there is rage only, then on the surface M above there is a single point marked (1), representing a fighting frame of mind, because the distribution of Case (1) of Figure 3 is unimodal. Similarly with cases (2) and (3). However in the interesting case (4) we obtain two points, marked (4) on the graph above, because the distribution has gone bimodal. Moreover there is another point, marked (4)*, in between these two, indicating the least-likely neutral behaviour. One of the reasons for including least-likely points on the graph, as well as the most-likely, is that it makes the graph M into a complete smooth surface; this is one of the consequences of the theorem. (Another reason for including these points is that it is sometimes useful to mark the threshold between the two modes of behaviour in the bimodal case). But it is important to remember when using the cusp-catastrophe that generally the middle sheet (shown shaded) represents least-likely behaviour, and only the upper and lower sheets represent most-likely behaviour.

The curve on the surface where the upper and lower sheets fold over into the middle sheet is called the fold-curve, and the projection of this down into the horizontal plane C is called the bifurcation set. Although the fold curve is a smooth curve, the bifurcation set has a sharp point, forming a cusp, and this is the reason for the name cusp-catastrophe. The cusp lines form the main thresholds for sudden behavioural change as we shall now explain.

The surface gives us a new insight into the dog's aggression mechanism. For, as his drives vary over the horizontal plane C, so his mood and behaviour will follow suit over the surface M above (except for the middle sheet). More specifically let us see what happens as his drives follow the dotted paths in C. Path P_1 begins with the dog frightened, cowering in a corner say, in a fleeing frame of mind, with his ears back. If we increase his rage, for example by approaching him too close and "invading his territory", then his mouth will begin to open, but he will remain cowering until the point Q_1 is reached. At that moment he reaches the fold curve at the edge of the lower sheet, and so the stability of his fleeing frame of mind breaks down, and he will suddenly catastrophically jump up onto

the upper sheet into a fighting frame of mind (indicated by the double
headed arrow). Consequently he may suddenly attack. Conversely suppose
he is in a fighting frame of mind, and we cause him to follow path P_2 by
increasing his fear in some way, then he will nevertheless remain in a
fighting frame of mind until the point Q_2 is reached, when he will suddenly
and catastrophically jump down onto the lower sheet into a fleeing frame of
mind. Consequently he may suddenly retreat. What actually causes these
sudden changes of mind? Why should the mood jump from one surface to
another? To answer these questions let us digress for a moment to a
simpler mechanical example, and later return to the problem of the dog.

2. THE DYNAMIC.

Example 2. A catastrophe machine.

To understand how continuous forces can cause catastrophic jumps the
reader is strongly recommended to make and play with the little toy
illustrated in Figure 5a.

The materials needed are 2 elastic bands, 2 drawing pins, half a
matchstick, a piece of cardboard and a piece of wood. Taking the unstretched
length of an elastic band as our unit of length, cut out a cardboard disk of
diameter about 1 unit. Attach the two elastic bands to a point H near the
edge of the disk - the easiest way to do this is to pierce a small hole at
H, push little loops of the elastic bands through the hole, and secure them
by slipping the matchstick through the loops and pulling tight, as in
Figure 5b. Now pin the centre of the disk to the piece of wood with drawing
pin A (with the elastic bands on the top and the matchstick underneath), and
make sure that it spins freely. Fix the other drawing pin B into the wood,
so that AB is about 2 units, and the hook one of the elastic bands over B.
The machine is now ready to go.

Hold the other end of the other elastic band : where you hold it is
the control point c. Therefore the control space C is the surface of the
wood. Meanwhile the state of the machine is the position of the disk, and
this is measured by the angle $x = B\hat{A}H$. When the control c is moved
smoothly, the state x will follow suit, except that sometimes instead of

1. Catastrophe Theory   9

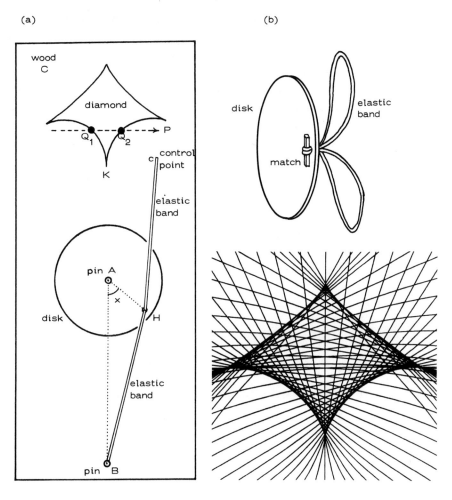

Figure 5. (a) A catastrophe machine; (b) how to attach the elastic bands; and (c) a computer-drawing by T. Poston and A.E.R. Woodcock of the diamond-shaped curve.

moving smoothly it will suddenly jump. Every time it jumps mark the corresponding control point c with a pencil dot. Soon you will build up sufficient dots to be able to join them up in a concave diamond-shaped curve with four cusps, as shown in Figure 5a. The computer-drawing of this curve in Figure 5c shows it as the envelope of a family of lines (moving the control along any one of these lines keeps the disk stationary). The diamond-shaped curve is the bifurcation set, because if c lies outside the curve there is a unique stable equilibrium position of the disk, whereas if c lies inside there are two stable equilibria. If we restrict ourselves to control points in the neighbourhood of the lowest cusp point K, then the stable equilibria will have small angle x (this is convenient because we can then measure x along a line, whereas the full state space is in fact a circle). Let f denote the energy of the elastic bands. Then by Hooke's Law we can compute f as a function of x, for a fixed control point c. Examples of the graph of f for control points outside and inside the diamond are shown in Figure 6.

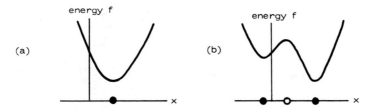

Figure 6. *Graphs of the energy in the elastic bands for control points (a) outside and (b) inside the diamond.*

The minima of f determine the stable equilibria (indicated by a solid circle) and the maximum of f determines the unstable equilibrium (indicated by an open circle). Figure 6 is analogous to Figure 3 with maxima and minima reversed. By the main theorem, the graph of equilibria as a function of c is equivalent to the cusp-catastrophe surface M shown in Figure 4. The upper and lower sheets represent the stable equilibria, and the middle sheet the unstable equilibria separating them. The cusp in Figure 4 is exactly the same as the cusp we have drawn on the wood in Figure 5, in the neighbourhood of the point K.

In this example it is easy to understand the dynamic : by Newton's law of motion the disk rotates so as to reduce f, and any oscillations are swiftly damped out by the friction at the drawing pin. Therefore x swiftly seeks a local minimum of f. As the control is moved slowly, then f changes slowly, but the dynamic keeps x at the local minimum of f, or in other words keeps x on the surface M.

Let us now see what happens if the control is moved slowly from left to right along the dotted path P shown in Figure 5a. The sequence of energy functions is shown in Figure 7. The state x starts in the unique

Figure 7. Changes in the energy graphs explaining the jump at Q_2.

minimum; nothing happens when a second minimum appears at Q_1. The second minimum eventually becomes deeper, but the state stays in the first minimum, held there stably by the dynamic. Eventually at Q_2 the stability of the first minimum breaks down as it coalesces with the maximum, and the state has to jump (or more precisely is swiftly carried by the dynamic) into the second minimum, which has now become the unique minimum. On the reverse journey the roles of Q_1 and Q_2 are reversed : the state stays in the right-hand minimum until Q_1, where it has to jump into the left-hand minimum again. This is called a <u>hysteresis cycle</u>, and is illustrated in Figure 8 (which is just the front section of Figure 4). The hysteresis of magnetism is in fact the same phenomenon happening to all the little magnets inside. Returning to the catastrophe machine, experiment

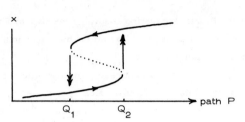

Figure 8. Going back and forth along P produces a hysteresis cycle.

will confirm that the cusp opposite K is similar, but the other two cusps are dual cusp-catastrophes, in the sense that the roles of maxima and minima are reversed : the upper and lower sheets are now unstable, and the middle sheet now represents a narrow pocket of stable equilibria. The difference can be observed by executing small circles of control around each cusp in turn.

3. THE MECHANISM OF AGGRESSION.

We now return to the rage and fear example and ask what is the dynamic? What mechanism can we add to the likelihood distributions of Figure 3 that will explain why the dog actively seeks and adopts the most-likely frame of mind? The answer lies in the underlying neurological activity of the brain. The brain may be regarded as a number of large coupled oscillators, each comprising millions of neurons. It is well known that non-linear oscillators can possess attractors (stable limit cycles), and that these attractors can typically bifurcate according to the cusp-catastrophe or higher dimensional catastrophes (See Figure 34). Therefore we may expect the elementary catastrophes to be typical models of brain activity, especially of activity in those parts of the brain such as the limbic system where the organs are more highly interconnected and consequently may tend to oscillate more as whole units (as opposed to the neocortex, whose different parts can oscillate differently at the same time, and whose activity can therefore be much more complicated).

According to Paul MacLean it is in the limbic system that emotions and moods are generated (while the neocortex determines the more complicated choice of behaviour within that mood). Therefore we might expect catastrophe theory to be the mathematical language with which to describe emotion and mood; and indeed it is striking that moods tend to persist, tend to delay before changing, and then tend to change suddenly, all of which qualities are typical of catastrophe models (see for example the hysteresis of Figure 8). Therefore it is not unreasonable to assume that Figure 4 is not only a model of the observed behaviour of aggression, but also a model of the underlying neural mechanism. Each point of the surface M represents an attractor of some huge dynamical system modelling the limbic activity of the brain, and the jumps occur when the stability of an attractor breaks down. The dynamical system remains implicit in the background; the only part that we need to make explicit for experimental predictions is the catastrophe model of Figure 4.

Moreover from the point of view of evolution one would expect this type of aggression mechanism to be very old, and therefore situated in a phylogenetically older part of the brain such as the limbic system. Consequently we should expect Figure 4 to be applicable not only to dogs, but to many species, under widely varying circumstances.

Example 3. Territorial fish.

Consider for example a territorial species such as some types of tropical fish. If we postulate that "size of invader and closeness to nest are conflicting factors influencing aggression", then the same argument leads again to the cusp-catastrophe. The resulting cusp in the control space is shown in Figure 9. Although Figure 9 is a somewhat simple picture, the argument deriving it has used a deep theorem, and we can deduce several consequences, leading to predictions which could probably be tested by fairly simple experiments. We might expect the cusp point to occur near (r_1, s_1) where s_1 is the size of the fish and r_1 the radius of his territory. If our fish meets a smaller adversary ($s < s_1$) then he

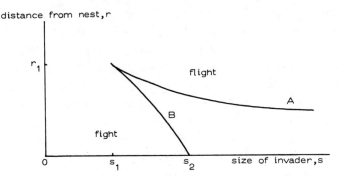

Figure 9. Size of invader and closeness to nest are conflicting factors influencing the agesssion of territorial fish.

will chase him away. However if he meets a larger invader ($s > s_1$) near the nest, then he will chase the invader up to cusp line A, where his frame of mind will then jump from valour to discretion, and he will return to the nest. Therefore the radius of the territory, measured by A, should be a decreasing function of s. Now suppose our fish goes foraging and meets a large adversary far from the nest : then he will flee home, and only jump into a fighting frame of mind and turn to defend the nest when he reaches line B. Therefore the hysteresis phenomenon of Figure 8 here predicts that radius of the territory as measured by B will be noticeably smaller that that measured by A. Moreover if the invader is very large ($s > s_2$) then our fish will continue to flee; he will never turn to defend the nest, which would then be lost, were it not for his mate, who, if she has been cruising near the nest, will automatically chase the invader out to line A. Therefore the model offers a measurable test of a simple explanation of why in territorial species the partner who happens to be nearer the nest displays the more vigorous defence. Similarly one could devise many qualitative experiments of this nature on other species to test the general hypothesis that aggression mechanism is stored as a cusp-catastrophe. However now let us be a little more adventurous and make some applications of the model to man.

Example 4. Argument.

In the first application suppose that the model is describing the emotional mood of an opponent in an argument. The behaviour axis runs as follows

```
        fight  ↑  hysteria
                  abuse
                  irrational argument
                  rational discussion
                  concessions
                  apologies
        flight |  tears
```

If we begin to make him angry and frightened, then we will first deny him access to rational thought, and force him to jump between irrational argument and concessions. If we make him a little more so, then we will next deny him access to those behaviour modes, and force him to make bigger jumps between abuse and apology. Finally if we make him very angry and frightened, then we will limit his possible behaviour to only hysteria or tears, and the very verbal usage of the phrase "hysterical tears" is confirmation of the catastrophic jump from one end of the spectrum to the other. If our purpose is to persuade him of something that we know will make him both angry and frightened, then the best policy is to state the case and go away; for then our absence will allow his anger and fear to subside, and give him access to rational thought again, and so enable him to see our point.

Example 5. Catharsis of self-pity.

In the second application, we modify the coordinates of rage and fear to frustration and anxiety, to suit the emotions of modern civilised man. One of the behaviour patterns of a fleeing frame of mind that we learn to adopt as children is the mood of self-pity. And when a child, or an adult, falls into a persistent mood of self-pity, it often seems as if sympathy is of no avail. But a sarcastic remark may suddenly induce a loss of temper, which in turn releases the tension, and seems to act as a catharsis for the mood of self-pity. This is a fairly common pattern of events, although

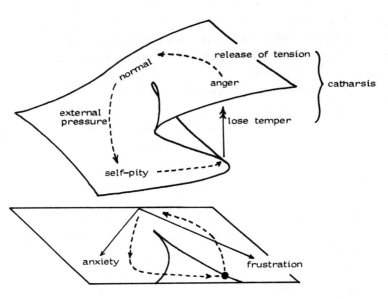

Figure 10. Anxiety can cause self-pity. Increasing frustration can cause catharsis of self-pity by loss of temper and release of tension.

it always seems a shame that sarcasm should suceed where sympathy has failed, even though the proverb tells us that it is necessary to be cruel in order to be kind. The model suggests that the pattern may in fact be just an automatic byproduct of the evolutionarily useful built-in aggression mechanism.

Example 6. War.

In the third application we go even further, passing from psychology to sociology, and apply the same model to whole nations instead of individuals. In place of rage and fear we substitute threat and cost. The behaviour axis represents the war policy of the nation, running from strong military action, through moderate or weak action to neutrality, withdrawal and surrender. The distributions of Figure 3 will represent the support amongst the population for the various policies. The two modes of case (4) are called doves and hawks. The dynamic in this example is the sensitivity of the government to its electors, continuously adapting its policy so as to increase its support. The path P_1 on Figure 4 represents a nation feeling increasingly threatened, but first pursuing a policy of appeasement until it reaches point Q_1 when it suddenly declares war. The path P_2 represents a nation suffering increasing costs, but first escalating the war, until it reaches point Q_2 when it suddenly surrenders. Paths P_3 and P_4 represent two nations both experiencing similar escalating threat and cost, but one finishing up in an aggressive mood and the other in an appeasing mood, very reminiscent of the Cuban Missile Crisis in 1962. America followed path P_3 to the right of the cusp, because her leaders first felt threatened by the presence of the missiles, and then increasingly appalled by the rising likelihood of a nuclear war. Russia on the other hand followed path P_4 to the left of the cusp, because her main feeling of threat came later as the crisis escalated. The art of diplomacy is to leave your adversary a clear avenue of retreat so that he can safely follow P_4, while you yourself follow P_3, otherwise if you overthreaten him too soon, you may force him to follow P_3 as well, with disastrous results for both of you.

Admittedly these applications may be gross oversimplifications, but the very fact that our minds indulge in simplification, analogy, abstraction and synthesis may in fact reveal that the model is telling us not so much about wars, but more about the way our minds work. The oscillations of our brains can bifurcate according to the elementary catastrophes, and so our minds automatically employ the latter for the subconscious organisation of our thoughts.

4. DIVERGENCE.

Let us re-examine paths P_3 and P_4 in Figure 4 in the context of our original application to the dog. Both paths begin at the same point and end at the same point in the control space, C, but induce divergent behaviour. Following P_3 the dog first gets angry and then frightened, but persists in a fighting frame of mind; conversely following P_4 he experiences the same emotions but in the reverse order, and persists in a fleeing frame of mind. Therefore although, as we have said, the behaviour in case (4) is unpredictable, if we happen to know the recent past history then it is predictable. Usually the psychologist's only defence against unpredictability is statistics, but the use of statistics in conjunction with a model of this nature is a much stronger weapon.

Notice that the change of behaviour under paths P_3 and P_4 was quite smooth without any catastrophes involved. Notice also that in the plane C the difference between the two paths may be only very marginal : all that matters is that they pass on either side of the cusp point. This phenomenon of a marginal change of path causing a major change in the behaviour we call <u>divergence</u>, and it is very common in biology and the social sciences. By contrast physics is generally non-divergent, because usually a small change in the initial data causes only a small change in the ensuing motion. It has long been a folk-lore that divergent phenomena in the "inexact" sciences could not be modelled by mathematics; but it is now realised that divergence is a characteristic property of stable systems, which can be both modelled and predicted, and the natural mathematical tool to use is catastrophe theory.

In fact the cusp-catastrophe shows that the five qualitative features of <u>bimodality</u>, <u>inaccessibility</u>, <u>sudden jumps</u>, <u>hysteresis</u> and <u>divergence</u>, are all interrelated (see Figure 11). And the deep classification theorem of catastrophe theory (stated below) permits us to enunciate the general principle that whenever we observe one of these five qualities in nature, then we should look for the other four, and if we find them then we should check whether or not the process can be modelled by the cusp-catastrophe. Indeed our verbal usage in ordinary language of pairs of opposites frequently indicates a bimodality that has grown smoothly out of some unimodality, and which may be modelled by the upper and lower sheets of a cusp-catastrophe surface. In Table 1 we give an assorted list of pairs, to indicate the scope of possible applications. In each case a fairly elaborate model can be developed, but we do not have space to develop them here, because each one needs careful analysis of what the axes should be, how they could be measured, what is the dynamic, and how to design experiments.

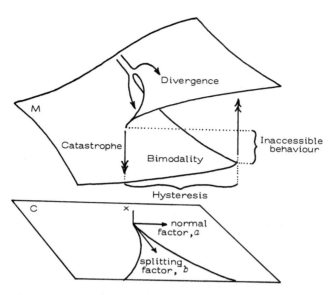

Figure 11. Five characteristic properties of the cusp-catastrophe are bimodality, inaccessibility, catastrophe, hysteresis and divergence. (The unstable middle sheet has been removed.)

In some cases the two control factors (or parameters) in C lie on either side of the cusp, such as α, β in Figure 4; in this case we call them <u>conflicting factors</u> as in all our previous examples so far. In other cases one of the control factors is perpendicular to the cusp axis, and the other lies along it, such as a, b in Figure 11 (see also Figures 12,17,30,31). In this case we call a the <u>normal factor</u>, because if $b < 0$ then x increases continuously with a, and we call b the <u>splitting factor</u>, because if $b > 0$ then M is split into two sheets. The equation of the standard cusp-catastrophe surface M illustrated in Figure 11 (with origin taken at the point on M above the cusp point) is
$$x^3 = a + bx.$$
By differentiating and elminating x, one obtains the equation of the cusp: $27a^2 = 4b^3$. To get the standard equation in terms of conflicting factors put $a = \alpha - \beta$, $b = \alpha + \beta$. We illustrate normal and splitting factors by Example 7, which is an application in economics.

Table 1.

	Phenomenon.	Bimodality.	Catastrophes.	Conflicting factors.
	Economic policy.	Deflation, reflation.	Stop-go.	Balance of payments, unemployment.
4	Embryology.	Ectoderm, mesoderm.	Cell determination.	Space, time.
2	More haste less speed.	Fast, slow.	Master, fumble.	Haste, skill.
				Normal/splitting factors.
13	Civil unrest.	Disorder, quiet.	Riot, restore order.	Tension/alienation.
*	Delinquency.	Harassed, lonely.	Crime, quarrel.	Approach–avoidance/anxiety.
20	Education.	Exciting, dreary.	Inspiration, disillusion.	Reward/punishment.
*	Emotional response.	Antagonism, sympathy.	Lose temper, reconcile.	Tiresomeness/demandingness of other person.
3	Heartbeat.	Systole, distole.	Contract, relax.	Pacemaker/blood pressure.

Numbering of relevant papers in this book. (For * see *Proc. Roy. Inst. Gt. Br.* 49 (1976) 77-92).

Example 7. Stock markets.

First we observe the bimodality in the common usage of the terms "bull market" and "bear market", which are situations that diverge smoothly from the more "normal market". Therefore we shall represent bull and bear by the upper and lower sheets of the cusp-catastrophe illustrated in Figure 12. Next we verify the presence of catastrophes ; a "crash" is a

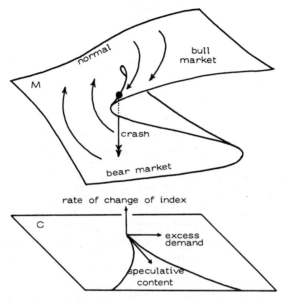

Figure 12. Excess demand is a normal factor, and speculative content a splitting factor influencing stock market behaviour.

sudden jump from bull to bear. This immediately prompts the question : why is this catastrophe more common than the opposite sudden jump from bear to bull? We shall explain the answer in a moment. First consider what is the behaviour axis? In a bull market the index is rising, while in a bear market it is falling, and so we choose rate of change of index to measure behaviour. What is the most likely reason for the index to rise in a normal market? Normally the main reason is excess demand for stock by investors, and so we choose excess demand as the normal factor. What causes an abnormal market? One of the main reasons is probably the speculative content, although this may be more difficult to measure; perhaps one might be able to measure the percentage of the market held by "chartists" (investors who base their investment policy on charts of previous performance) as opposed to "fundamentalists" (who base their policy upon research into the industrial health and growth-potential of the firms involved).

If the excess demand is zero, but the speculative content high, then it is unlikely that the index will remain constant; a random event may spark off a wave of confidence that induces a stable bull market, or equally a lack of confidence that induces a stable bear market. The main feature of a stock market is its centralised sensitivity, and this is what is responsible for the rapid dynamic that causes the behaviour to go bimodal, and follow the surface M shown in Figure 12.

We can now go further and introduce a slow feedback of the index upon the two controls, representing this by the arrows shown on the surface. A bull market encourages speculation, but overvaluation causes fundamentalists to sell and invest their profits elsewhere; therefore the arrows on the upper sheet come forward and bend to the left. Conversely a bear market discourages speculation, but an undervalued recovering market presents good opportunities for fundamentalists to reinvest; therefore the arrows on the lower sheet go back and bend to the right. We can now deduce from the model the characteristic overall stock-market cycle of growth, boom, recession and recovery. The anthropologist Michael Thompson has shown that similar economic cycles happen, and are even planned, amongst primitive New Guinea tribesman.

5. CLASSIFICATION THEOREM.

We are now ready to appreciate Thom's classification of elementary catastrophes, and to go on to the higher ones. In order to keep things as understandable as possible we begin by stating a simplified version of part of the theorem, and then elaborate upon it.

Theorem. Let C be a 2-dimensional control (or parameter) space, let X be a 1-dimensional behaviour (or state) space, and let f be a smooth generic* function on X parametrised by C. Let M be the set of stationary values of f (given by $\frac{\partial f}{\partial x} = 0$, where x is a coordinate for X). Then M is a smooth surface in C × X, and the only singularities of the projection of M onto C are fold curves and cusp-catastrophes.

Remark about singularities. Here a singularity means a point where a vertical line touches M. When we say that a singularity of M is a cusp-catastrophe we mean that near that point M is equivalent** to the standard surface shown in Figure 4. Equivalence preserves all qualitative features such as the foldcurve, the cusp, the bimodality, catastrophes, hysteresis, divergence and inaccessibility. Therefore what the theorem really says is that qualitatively Figure 4 is locally the most complicated thing that can happen to a graph. That is why the cusp-catastrophe can be used with such confidence in so many different fields, whenever a process involves 2 causes and 1 effect.

Remark about the function f. In the rage and fear example the function f is the likelihood function on X parametrised by C, shown in Figure 3. The stationary values of f are the maxima and minima, representing the most-likely and least-likely behaviour respectively. In Example 6 the function f measured the support for different policies. In the catastrophe machine f was an energy function, with minima representing stable equilibria, and maxima representing unstable equilibria. In economics f might be a cost function, in evolution a fitness function, in

---

Footnotes for the mathematically minded :

* Smooth means differentiable to all orders.
  Generic means that the map from C to the space of functions on X is transverse to the natural stratification. Almost all smooth functions are generic. Small perturbations of generic functions remain generic.

** Equivalence means there is a diffeomorphism from a neighbourhood N in the given C × X onto the standard C × X, throwing vertical lines to vertical lines, and throwing M ∩ N onto the standard surface.

engineering a Lyapunov function, in light caustics a geodesic distance, etc. In most applications there is, in addition to f, a gradient-like (dissipative) dynamic that maximises or minimises f, and is ultimately responsible for the sudden jumps. However the dynamic is not involved in the statement of the theorem, and so the theorem is also applicable to phenomena such as light caustics (Example 11 below), where there are discontinuities, but no dissipative dynamic and therefore no sudden temporal jumps.

Remark about dynamics. In many applications there are dynamics on X that are not gradient-like, and consequently there is no function f that is minimised. In this situation the elementary catastrophes do not describe all the possibilities, because non-elementary catastrophes can occur. Some non-elementary catastrophes are known (such as the Hopf bifurcation) but as yet the general classification problem is unsolved.

Remark about the dimension of X. The theorem remains true, word for word, if we increase the dimension of the behaviour space X from 1 to n; that is why it is both remarkable and difficult to prove. But the beauty of this result is that we can now use the theorem *implicitly*, in situations that would be far too complicated to measure or put on a computer. For example we can implicitly assume X is large enough to describe the states of a cell in the embryo, with at least 10,000 dimensions for representing the concentrations of the various chemicals involved. Or we could implicitly assume X is large enough to describe the states of the brain, with at least 10,000,000,000 dimensions for representing the rates of firing of all neurons. In physics we may wish to have X infinite dimensional (as in Example 12 below). But in each case the theorem explicitly hands us back the same simple surface of Figure 4, upon which to base a model. We feed in implicit complexity, and get out explicit simplicity; all the hard work of digesting the complexity has gone on behind the scenes in the proof of the theorem.

Remark about the fold catastrophe. If we reduce the dimension of C from 2 to 1, then the analogous theorem, 1 dimension lower, says that M is a smooth curve, and the only singularities are folds, such as the two that occur in Figure 8. Thus the fold-catastrophe appears in sections of the cusp-catastrophe, and the latter is made up of folds together with one new singularity at the origin. Similarly any higher dimensional catastrophe is always made up of lower dimensional ones, together with one new singularity at the origin.

6. THE SEVEN ELEMENTARY CATASTROPHES.

If we increase the dimension of C in the theorem, then this is quite a different kettle of fish. For example suppose C was 3-dimensional and X was 1-dimensional. Then C × X would be 4-dimensional, and M would be a 3-dimensional manifold (or hypersurface) in C × X. What is more, instead of being folded along curves M would now be folded along whole surfaces, and the bifuraction set instead of consisting of curves with cusp points in 2-dimensions would now consist of surfaces with cusped edges in 3-dimensions (see Figure 13). Moreover a new type of singular point would appear, called the swallowtail-catastrophe. It is impossible to draw the complete picture for M in the case of the swallowtail because this would require 4-dimensions. However just as the geometry of the cusp catastrophe can be described by drawing a cusp in 2-dimensions (and knowing that M is bimodal over the inside), so also we can derive some geometric intuition about the swallowtail by drawing its bifurcation set in 3-dimensions (see Figure 13). It is called the swallowtail or dovetail because it looks a bit like one : the name queue d'aronde was suggested by the blind French mathematician, Bernard Morin.

Now suppose we keep C 3-dimensional and allow X to be n-dimensional, n ≥ 2. Then two more singular points become possible, the hyperbolic-umbilic-catastrophe and the elliptic-umbilic-catastrophe, which we can again describe geometrically by drawing their bifurcation sets in 3-dimensions (see Figures 13,27,32). This completes the list of elementary catastrophes in 3-dimensions. If we now allow C to be 4-dimensional then we obtain two new singularities, the butterfly-catastrophe and the parabolic-umbilic-catastrophe (see Figures 14,16). And so on, until eventually the list becomes infinite. However the infinite list is more the concern of the

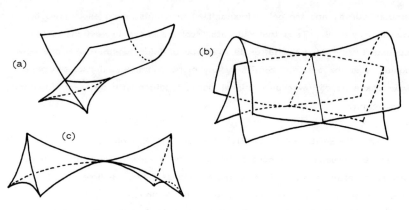

Figure 13. Bifurcation sets of (a) the swallowtail, (b) the hyperbolic umbilic and (c) the elliptic umbilic catastrophes.

mathematician, whereas here we are primarily interested in the lower dimensions because they are more relevant to applications. In particular we are interested in up to 4-dimensions, because the parameter space C often plays the role of space-time. Therefore the 7 elementary catastrophes up to 4-dimensions have all been given special names, whereas those above have not.

Table 2.

Dimension of C	1	2	3	4	5	...
Number of catastrophes	1	1	3	2	4	...
Names	fold	cusp	swallowtail hyp. umbilic ell. umbilic	butterfly par. umbilic – – –	–	...

Note that in Table 2 we have put each catastrophe in the lowest dimension where it first appears, although it also appears in all higher dimensions. For example fold points first occur when C is dimension 1 (in Figure 8) and fold curves also appear when C is dimension 2 (in Figure 4). Therefore all 7 catastrophes appear when C is dimension 4.

Each of the elementary catastrophes has a standard model, and in Table 3 we list some standard formulae for f, from which the standard models can be derived. In each case a,b,c,d are parameters for C, and x,y are variables for X. For the cuspoids M is given by $\frac{\partial f}{\partial x} = 0$, and for the umbilics M is given by $\frac{\partial f}{\partial x} = \frac{\partial f}{\partial y} = 0$. Note that the "standard" formulae are by no means unique, but are chosen for convenience in applications. In particular the fractions are there only so that they disappear when we differentiate to get the equation for M. For instance the cusp-catastrophe is given by $\frac{\partial f}{\partial x} = x^3 - a - bx = 0$ (see Figure 11). The minus signs are there in order to fit in with the notions of normal and splitting factors, etc.

Table 3.

		dim X	dim C	Function f
Cuspoids	Fold	1	1	$\frac{1}{3}x^3 - ax$
	Cusp	1	2	$\frac{1}{4}x^4 - ax - \frac{1}{2}bx^2$
	Swallowtail	1	3	$\frac{1}{5}x^5 - ax - \frac{1}{2}bx^2 - \frac{1}{3}cx^3$
	Butterfly	1	4	$\frac{1}{6}x^6 - ax - \frac{1}{2}bx^2 - \frac{1}{3}cx^3 - \frac{1}{4}dx^4$
Umbilics	Hyperbolic	2	3	$x^3 + y^3 + ax + by + cxy$
	Elliptic	2	3	$x^3 - xy^2 + ax + by + c(x^2 + y^2)$
	Parabolic	2	4	$x^2y + y^4 + ax + by + cx^2 + dy^2$

Each catastrophe has its own individual and surprising geometry. We do not have space to describe them fully here, and the reader is referred to Thom's book for more details. The fold (Figure 8) and cusp (Figure 4) are easy to understand, and the bifurcation set of the swallowtail and hyperbolic and elliptic umbilics are easily visualised (Figure 13). However the two 4-dimensional ones have to be approached more obliquely by drawing sections. Figure 14 shows sections of the bifurcation set of the parabolic umbilic in the (a,b)-plane for various fixed values of c,d on the unit circle in the (c,d)-plane; these sections are taken from Fowler's translation of Thom's book, and based on drawings by Chenciner and Jänich.

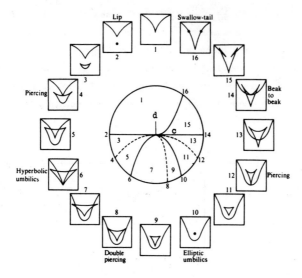

Figure 14. Sections of the 4-dimensional bifurcation set of the parabolic umbilic.

7. THE BUTTERFLY CATASTROPHE

Finally we come to the butterfly, which deserves more attention, because after the cusp it is the most important catastrophe for the behavioural sciences. We shall illustrate it by giving a fairly elaborate application to a nervous disorder, anorexia nervosa.

What bimodality is to the cusp, so trimodality is to the butterfly. We have seen how any evolution from unimodal to bimodal behaviour determines (by the classification theorem) the unique 3-dimensional geometry of the cusp-catastrophe, with its associated jumps, hysteresis and divergence, etc. Similarly any evolution from unimodal to trimodal behaviour determines the unique and much richer 5-dimensional* geometry of the butterfly-catastrophe. Since trimodality often emerges out of bimodality, the natural way to analyse the butterfly is to regard it as an extension of the cusp, as illustrated by the following example.

Example 8. Compromise opinion.

Suppose the cusp represents the polarisation over some issue in society; then the butterfly represents the emergence of a compromise opinion. For instance in the example above of nation at war, public support might be distributed as in Figure 15 (compare with Figure 3).

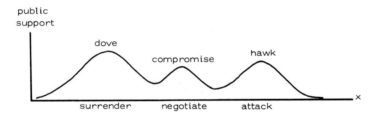

Figure 15. The emergence of compromise opinion.

---

* Each extra mode involves two more control factors, because it requires one more maximum and one more minimum (see Figure 15) and so the power of the leading term in the formula goes up by 2.

<u>The geometry of the butterfly.</u> There is one state variable x, and four control factors as follows

> a : normal factor
>
> b : splitting factor
>
> c : bias factor
>
> d : butterfly factor.

The behaviour lies on the "surface" M in 5-dimensions given by the equation
$$x^5 = a+bx+cx^2+dx^3,$$
which is obtained by differentiating the formula in Table 3 above. Since it is impossible to draw 5-dimensional pictures we have to make do with 2- and 3-dimensional sections. The bifurcation set lies in the 4-dimensional control space, and the top six pictures of Figure 16 show 2-dimensional sections of it parallel to the (a,b)-plane for different values of c,d. The top three pictures refer to d < 0. When c = 0 the section reduces to the cusp that we know already. The effect of the bias factor c is to bias the position of the cusp : when c < 0 the main body of the cusp swings to the right while the tip of the cusp moves up and bends over the left; when c > 0 the opposite happens. Meanwhile the effect of the bias upon the behaviour surface M is to move it up and down; this can be seen in the bottom pictures (vii) and (viii) of Figure 16 which show 2-dimensional sections of M drawn with control factor, a, horizontal and x vertical, for different values of b,c,d. Meanwhile the effect of bias upon the 3-dimentional sections of M is shown in Figure 19(i) and (ii), which are drawn with (a,b) horizontal and x vertical, for different values of c,d.

Now consider the effect of the butterfly factor d. Keeping c = 0, as d goes positive the cusp evolves into three cusps, which form a triangular "pocket", as shown in Figure 16(v). (Turning this picture upside down looks like a butterfly, which originally suggested the name.) Meanwhile the behaviour surface develops two new folds above the sides of the pocket as illustrated in Figures 16(ix) and 20. Above the pocket itself is a new triangular sheet of stable behaviour, which has grown continuously forward from the back, in between the upper and lower sheets at the front. This new sheet represents the third behaviour mode, for instance the emergence

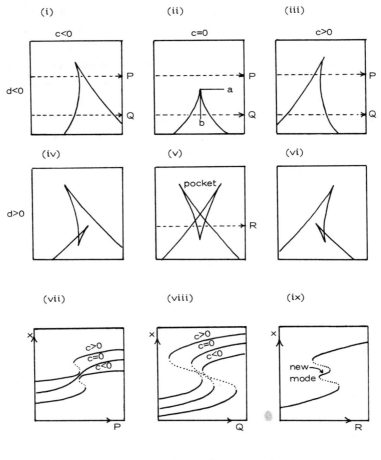

Figure 16. *The butterfly catastrophe.*

(i) - (vi) Sections of the bifurcation set in the (a,b)-plane for different values of c,d. The effect of the bias factor c is to swing the cusp to and fro. The effect of the butterfly factor d is to create the pocket.

(vii) - (ix) Sections of the behaviour surface over the dashed paths P,Q,R. The effect of c is move the surface up and down. The effect of d is to create a new sheet above the pocket, representing the third mode.

of the compromise opinion in Example 8 above. Notice that in Figure 16(iv) and (vi) the effect of the bias on the pocket is to reduce one side or the other until it disappears (by means of a swallowtail). Therefore the effect of bias is to destroy a compromise. In applications concerning the emergence of compromise, the butterfly factor will increase with time; at first the compromise is fragile, in the sense that its stability is broken by any perturbation across the nearby sides of the pocket; but as the pocket grows in size the compromise becomes stronger, in the sense of being stable under increasingly large perturbations.

PART TWO

ANOREXIA NERVOSA

Anorexia is a nervous disorder suffered mainly by adolescent girls and young women, in whom dieting has degenerated into obsessive fasting. It generally begins between the ages of 11 and 17, although it can start as early as 9 or as late as 30. It can lead to severe malnutrition, withdrawal and even death.

The proposed model is the joint work of the author and J. Hevesi, who is a psychotherapist specialising in anorexia. Hevesi has spent some 5000 hours during the last 5 years talking to over 150 anorexics and the model is based on his close observations. Of these 150 over 60 agreed to undertake his course of treatment, and of those treated he has achieved an 80% success rate of complete cure. His innovation is the use of trance-therapy. The Anorexic Aid Society in Britain recently conducted a survey of over 1000 anorexics, and the secretary of the society, Mrs. P. Hartley, who is a psychologist, writes : "I first read of Mr. Hevesi in several letters from patients who responded to my appeal for information about anorexia nervosa, and their experience re. treatment. These patients are the only ones who claim that they have recovered completely – i.e. those whose <u>attitude</u> to life has changed since undergoing Hevesi's treatment. They are not just eating properly (only the awful surface problem anyway) but living a full life as a <u>complete</u> personality." (her underlining).

The advantage of using mathematical language for a model is that it is psychologically neutral; it permits a coherent synthesis of a large number of observations that would otherwise appear disconnected, and in particular enables us to place the trance states in relation to other behavioural modes. As yet the model is only qualitative, in the sense that the predictions that have been verified by observation have been qualitative rather than quantitative. Nevertheless it does provide a conceptual framework within which the theory could also be tested quantitatively by monitoring patients. Meanwhile we hope that it may not only give a better understanding of anorexia and its cure, but also provide a prototype for understanding other types of behavioural disorder.

A striking feature of anorexia is that it sometimes develops a second phase after about two years, in which the victim finds herself alternately fasting and secretly gorging; the medical name for this is bulimia, and anorexics often call it stuffing or bingeing. If we regard the normal person's rhythm of eating and satiety as a continuous smooth cycle of unimodal behaviour, then we can interpret this second phase of bimodal behaviour, as a catastrophic jumping between two abnormal extremes. Therefore by the main theorem we can model the anorexic's behaviour by a cusp-catastrophe, in which she is trapped in a hysteresis cycle, as in Figure 17.

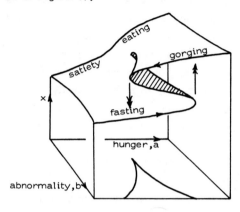

Figure 17. *Initial behaviour model for anorexia.*

Before we begin to analyse the model, we can immediately draw one important conclusion : <u>the victim will be denied access to the normal modes in between</u>. This denial of access to normal modes occurs already during the first phase of only fasting. Thus the main thrust of our approach will be to explain anorexia not as the complicated behaviour of a perverse neurotic, but as the logical outcome of a simple bifurcation in the underlying brain dynamics. If this is the case then catastrophe theory at once indicates a theoretical cure : if we can induce a further bifurcation according to the butterfly catastrophe, then this should open-up a new pathway back to normality. The practical problem is how to devise a therapy that will induce such a bifurcation, and this is what Hevesi's treatment achieves.

In Figure 17 we have chosen hunger and abnormality as the two control factors (a,b). Hunger is the normal factor because hunger normally governs the rhythmic cycle between eating and satiety; there are various known methods for measuring hunger, but we do not yet know which will be best to use for quantitative testing. We postpone the discussion on the measurement of abnormality until later. Meanwhile to measure the behaviour, x, it would be necessary to find some psychological index that correlates with the scale of wakeful states shown in Figure 18.

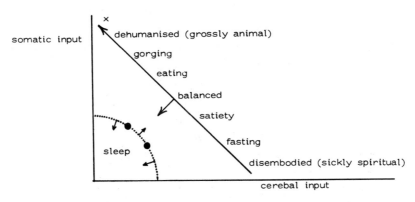

Figure 18. The x-axis measures both wakeful behaviour, and the relative weight given to the cerebal and somatic inputs to the limbic brain. The dotted line shows the boundary of the sleep basin, and the arrows show the movement of this boundary due to anorexia, leaving fixed the two nodal points.

What actually governs the behaviour is the underlying brain state, and if it is true, as MacLean suggests, that emotion and mood are generated in the limbic brain, then it is likely that x is measuring some property of limbic states. Since the limbic brain receives both cerebal inputs from the neo-cortex, and somatic inputs from the body, we might conjecture that x is some measure of the relative weight given to those inputs as shown in Figure 18. Of course such a conjecture must remain speculative until it is confirmed or rejected by future brain research. Nevertheless the conjecture has already proved useful in explaining many of the symptoms of anorexia, and, what is perhaps more important, enabled us to identify what may be the key operative suggestions in the therapy, as we shall see. Meanwhile the conjecture implies that the main neurological feature of anorexia is that during wakefulness the limbic brain is dominated either by cerebal inputs or by somatic inputs, while the balanced states have become unstable, and therefore inaccessible.

Before leaving Figure 18 notice that it is 2-dimensional. From the psychological point of view the natural axes to use are x and y, which are inclined at 45° to the neurological axes, cerebal and somatic. Here x measures the different wakeful states, while y measures the difference between wakefulness and sleep, and y exhibits the familiar healthy catastrophes of falling asleep and waking up. For a more complete model we ought really to use both the behaviour variables x and y, 5 controls, and a 7-dimensional catastrophe called E_6. However this is beyond the scope of this article, and so for simplicity of presentation we shall sacrifice y and use only x.

We now introduce a third control factor, c, which will play the role of the bias factor in the butterfly catastrophe. Define c to be <u>loss of self-control</u>, measured by loss of weight. Geometrically the effect of bias is to swing the cusp to and fro as in Figure 16(i) and (iii). The resulting effect on the behaviour surface is shown in Figure 19.

During the first phase of the disorder the anorexic is firmly in control of herself, and so c < 0 as in Figure 19(i). The normal person has learnt to perform the regular smooth cycle at the back, socially structured by mealtimes. The anorexic however finds herself trapped on the lower sheet at the front by the abnormality; in other words the limbic brain oscillates continuously in states underlying a fasting frame of mind all the time she is awake, even when she goes through the motions of eating. The frame of mind is predominantly cerebal, and the victims often speak in terms of "purity"; it tends to smother instincts

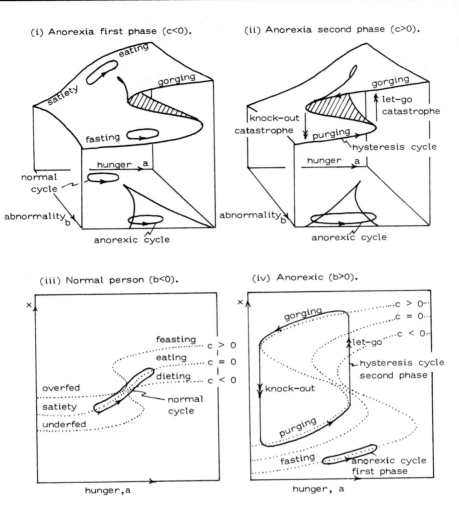

Figure 19. *The effect of the bias factor, c.*

(i) *Anorexia first phase (c<0). Strong self-control swings the cusp to the right, and abnormality displaces the normal cycle into fasting.*

(ii) *Anorexia second phase (c>0). Loss of self-control swings the cusp to the left, causing the anorexic to jump into the catastrophic hysteresis cycle of alternately gorging and purging.*

(iii) *Normal person (b<0). Changes in bias modify the behaviour slightly.*

(iv) *Anorexic (b>0). Changes in bias modify the behaviour dramatically.*

and produce excessive verbalisation. During the first phase victims often deny being ill, and refuse treatment.

Then as the anorexic gradually loses weight, she gradually loses control of herself; the bias factor c gradually increases, causing the cusp to swing gradually to the left, as in Figure 16(iii) and 19(ii). How far the cusp will eventually swing in relation to the cycle will depend upon the individual. If it swings sufficiently far for the right-hand side of the cusp to cross the right-hand end of the cycle, then this will cause the sudden onset of the second phase. For now, instead of being trapped in the smooth fasting cycle on the lower sheet, the victim finds herself trapped in the hysteresis cycle, jumping between the upper and lower sheets. The catastrophic jump from fasting to gorging occurs when she "lets go" : in the victim's own language, she watches helplessly as the apparent "monster inside herself" takes over, and devours food for several hours. Some victims vomit and gorge again, repeatedly. The catastrophic jump back occurs when exhaustion, disgust and humiliation sweep over her, and she returns to fasting for a day or several days. Some anorexics refer to this as the "knock-out". At each of the two catastrophes the limbic brain jumps from one set of states to the other, denying the victim access to the normal states between. Some anorexics even ritualise the catastrophes. The hysteresis cycle can be much longer than the previous cycle, because the after-effects of the gorge tend to prolong the fasting period.

Figures 19(iii) and (iv) show how the different cycles fit onto the sections of the surface that were illustrated in Figures 16(vii) and (viii). Notice that we have labelled the fasting period of the hysteresis cycle as "purging"; this is because it occurs at a different value of x to the "pure" fasting of the first phase. Indeed the two limbic states underlie quite different frames of mind; fasting is cerebally dominated, not allowing food to enter, while purging has the somatic element of getting rid of bodily contagion.

It is not known what proportion of anorexics switch into the second phase. Sometimes the switch occurs after a hospital treatment with drugs that are used to persuade the starving first phase anorexic to eat. If the effect of such drugs is to reduce cerebal inputs to the limbic brain in favour of somatic inputs, which is consistent with observed side-effects, then the drugs would be reinforcing the bias and therefore <u>causing</u> the switch. Thus the long-term harm caused by the use of such drugs may be greater than the short-term benefits.

Now we come to the cure. The strategic problem is how to persuade the anorexic to relinquish her abnormal attitudes, but this cannot be done directly. Therefore the practical problem is how to break the vicious circle :

The idea is to break in at the behavioural corner, by creating a third abnormal behaviour mode, during which the insecurity can be treated with reassurance. We will later show how this in turn causes a catastrophic collapse of the abnormal attitudes.

The new behaviour mode must lie between the abnormal extremes if it is going to provide a context within which reassurance can be effective. Therefore the butterfly catastrophe in Figure 20 tells us the geometric relationship that this mode must have (i.e. the dynamic relationship that the underlying brain states must have) in relationship to the existing modes.

Meanwhile Figure 18 shows that we must look for such a mode in the twilight zone between waking and sleeping for the following reasons. In the huge dynamical system modelling the limbic states of a healthy person, sleep is an attractor (i.e. a stable oscillation) with a stable boundary to its basin of attraction, separating it from wakefulness. In Figure 18 we have symbolically indicated the boundary by a dotted line. In the anorexic the boundary becomes fuzzy because the basin is being shifted, as indicated by the arrows; periferally the basin is being eroded by the increasing stability of the abnormal extremes, while in between it is being enlarged by the decreasing stability of the balanced states. These changes cause the sleeping patterns to be disturbed : sleep is fragmented, shifted around and edges of the fragments become fuzzy; the anorexic goes to bed late, wakes at night, sleeps little, find herself lounging about in her night clothes. Moreover, for exactly the same mathematical reason that temporary lakes sometimes appear on the boundaries of river basins near the nodal points in between erosion and growth, so fragile attractors may appear at the boundary of the sleep basin, particularly near the two nodal points marked in Figure 18. Therefore the anorexic finds herself spontaneously falling into fragile trance-like states, in the twilight zone between waking and sleeping, between dreaming and perceiving. At the somatic node these trance-like states are filled with thoughts about food, and lists of food, while at the cerebal node they are shot through with schemes and plans how to get through the day, how to manage social occasions and avoid set mealtimes, their preparation and aftermaths, shopping, cooking and washing up.

It is these confused trance-like states that are utilised by the therapist; <u>therapy builds upon naturally occuring processes</u>. Hevesi's treatment consists of about 20 sessions of trance-therapy over a period of 6 to 8 weeks, each lasting 2 to 3 hours. When the sufferer asks for help, the therapist begins by pushing aside the inconclusive and confusing contents of these states, pushing them away in their respective directions so as to create a new more balanced trance. Because of the state of the sufferer quite casual remarks can carry the force of suggestions, and thus the operative suggestions are actually made quite marginally, almost incidentally. Firstly a casual but firm announcement is made at the beginning (and adhered to throughout the treatment) such as "I don't care what you eat - we are not going to talk about eating or food", because this reduces the somatic input. Secondly after the formal step of going into the trance, a suggestion is made such as "Let you mind drift - don't think - look", because this reduces the cerebal input.

Thus the patient's mind is cleared of both food and scheming, and is free to look at itself. By contrast when she is fasting she is looking all the time at the outer world with anxiety, and when she is gorging she is overwhelmed by this same world, but during trance she is cut-off and isolated. By suspending the threats, the rules, the resistance and the hunger the trance gives temporary freedom from anxiety. She is able to look at the products of her own mind, and contemplate its images and memories. In this state she is open to reassurance, and, more importantly, <u>able to work out her own reassurance</u>.

The more the patient practises trance, the easier it becomes; reinforcement causes an increase in stability of the new attractor, and an enlargement of its basin of attraction. The trance states begin to emerge as the new middle sheet of Figure 20(i). Therefore we introduce the last control factor, d, as <u>reassurance</u>, measured by time under trance.

Summarising the four control factors :

    a : normal factor : hunger.
    b : splitting factor : abnormality, (measurement discussed below).
    c : bias factor : loss of self-control, measured by loss of weight.
    d : butterfly factor : reassurance, measured by time under trance.

Going into trance is a catastrophic jump from the lower sheet (because therapy usually takes place during the fasting part of the cycle) onto the middle sheet. Therefore the patient tends to <u>fall</u> into trance. What causes this jump? In fact the jump has two components, a relatively small one in the x-direction, and a larger one in the y-direction towards sleep, which is the second behavioural variable of Figure 18 that we have omitted from Figure 20 for simplicity. And it is not caused by a reduction in the abnormality, b, but by an increase in drowsiness, which is the fifth control factor, again omitted for the same reason; this is the only point where the simplicification has caused a slight geometrical inaccuracy in our pictures.

Coming out of the trance is another catastrophe, and causes the reverse jump back onto the lower or upper sheet, depending upon whether the left or right side of the pocket is crossed, as shown in Figure 20(ii). The patients confirm that when they awake from the first few trance

1. Catastrophe Theory

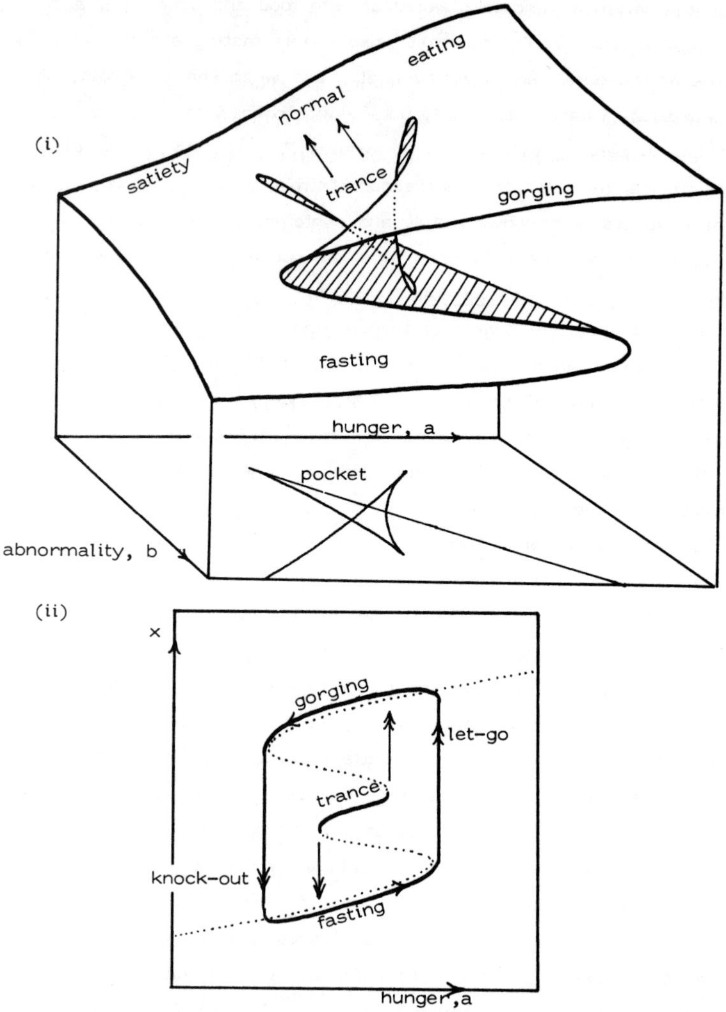

Figure 20. The effect of the butterfly factor, d>0.

(i) When therapy starts the trance states appear as a new triangular sheet of stable behaviour over the pocket (see Figure 16(v)) in between the upper and lower sheets. The new sheet opens up a pathway back to normality.

(ii) The trance states sit inside the hysteresis cycle; initially they are fragile, and coming out of trance is a catastrophic jump into either a fasting or a gorging frame of mind. (See Figure 16(ix)).

sessions they find themselves sometimes in a fasting and sometimes in a gorging frame of mind.

We now come to what happens during the trance. As the therapy progresses Hevesi's patients report that they experience three phenomena, of which the third is observable from the outside. Of course what the mind sees in trance it has put there, and interpreted in its own fashion, even though the images will naturally be made according to past experience, and the feelings will be such as are stored up from the past. The experience in trance may be compared to the steps in which an actor approaches a role. The first step is to envisage the part in a few simple strokes or characteristics; the second is to hear the lines the character is allotted to speak (say in a first reading through of the script); the third is to get into the part and play it, to act in front of an audience.

The first phenomenon is an experience of herself as a double personaility; one personality is usually described as the "real self" and the other is called various names by different patients such as "the little one, the imp, the demon, the powers, the spirit, the voice" or merely "it". Possibly the suggestion by the therapist to look rather than to think may prepare the way for the appearance of "persons", but usually the latter appear by themselves, and we shall argue below that the patient is in fact giving a logical description of herself. It is the voice, or however she describes it, who is apparently issuing the prohibitions over food : "The little one says I musn't eat". Typically the first appearance may occur about the third session : "I've got a voice", and then perhaps a couple of sessions later "This is the first time the voice has spoken in public".

The second phenomenon is an apparent transfer of important messages between the two personalities, such as the real self promising to "pay attention" to the little one, reassuring the little one that she "will not be forgotten", while the latter in return agrees to relax the prohibitions. Sometimes the little one is symbolically given a gift, such as a teddy-bear that she once longed for and never got.

The third phenomenon is a "reconciliation" or "union" or "fusion" of the two personalities, a "welcome possession" as opposed to the earlier malignant possession. Typically "She is coming out", or "She is very near the front", and then "The voice just seems to be part of myself". This third phenomenon is accompanied by a manifestation that can be witnessed by the therapist, such as speaking with a strange voice, and usually happens after about two weeks in around the seventh session. (depending of course upon the individual). When the patient awakens from this particular trance, she discovers that she has regained access to normal states, and is able to eat again without fear of gorging; she speaks of this moment as a "rebirth". Therefore during this trance the cure has taken place, a catastrophic drop in the abnormality, b, which we shall explain in a moment. Thus the trance states have opened up a path in the dynamics of the brain back to normality, indicated by the arrows in Figure 20(i). Subsequent trance sessions re-enact the reconcilitation in order to reinforce normal states and buffer them against the stresses of everyday life. At the same time the trance technique is itself reinforced, so as to provide a reliable method of self-cure, should the patient ever need to use it again at a later date.

Having dealt with the behavioural point of view, we now turn to the heart of the problem : What causes anorexia? Why can most slimmers diet without becoming anorexic? Why is there such a slow insidious apparently irreversible escalation of the disorder? How can we measure the abnormality, b? Why is the resulting neurosis/psychosis so rigid? How can there possibly be such a dramatically sudden cure?

We shall add one more cusp catastrophe to the model that will answer all these questions except the first. The reason that it cannot answer the first is that the model refers to what can be observed, whereas the original causes are probably hidden much earlier in childhood. We can offer an analogy, which may give some insight, but is not strictly part of the model. The metaphor is to describe anorexia as an "allergy to food"; of course it is *not* an allergy, but it does show some qualities similar to those of an immune system being set-up, switched-on, and inducing exposure-sensitivity. The origin of anorexia may occur in early childhood, when, perhaps for want of love or due to the inability to obtain the attention that it needs, the child retires into its shell; in other words the personality sets up an immunity against disappointment by turning inwards, and leaving the shell to act out the game of life. This immunity works well enough until the shell begins to grow and get out of hand, when the encapsulated core of the personality finds that it can no longer manage. It is then that the anorexia is switched-on, instinctively identifying food as the cause of growth. This may be why anorexia so often begins at the onset of puberty, or after a period of obesity. From now on the victim is exposure-sensitive to food, and being presented with food raises deep-seated anxieties. The logical reaction is to avoid stressful situations, and so the core begins to issue prohibitions to the shell concerning food. Consequently the victim begins to feel an urge to avoid food, which she cannot explain; when she attempts to explain it she tries to capture in words some quality of the urge : e.g. "the little one" is a recognition of its origins in childhood, "the imp" describes its bad quality, "the voice" its unidentifiableness. Usually such attempts are met with disbelief, and she soon stops trying to explain.

Our metaphor breaks down when the anorexia begins to escalate. This can be observed, and so can be put into the model, as follows. Increasing insecurity is observed, associated with a gradual escalation of abnormalities over food. A typical escalation might include the following stages, but of course each anorexic will differ in the details of her own particular escalation.

escalating stages of abnormality over food

| tummy-aches at school
| give up carbohydrates
| give up cooked meat
| elaboration of diet (weight watching, calorie counting)
| excessive diet (e.g. only cheese, peanuts, black tea)
| excessive activity
| deception (e.g. secret purging after each meal)
| manipulation (of rest of family)
| nocturnal eating only
b ↓ "I don't need to eat"

The important observations for our purpose are (i) there is an escalation of stages; (ii) at each stage a bimodal attitude is possible, normal or abnormal; (iii) when the anorexic reaches each stage, she will already have adopted abnormal attitudes towards all the previous stages, but as yet maintains normal attitudes towards the subsequent stages; (iv) increasing insecurity is associated with the escalation. Interpreting these facts geometrically gives the graph in Figure 21(i) showing the normality of attitude as a function of insecurity level, i, and abnormality stage, b. Abnormal attitudes begin at stage b_0, when insecurity has reached level, i_0. By the time insecurity has reached level i_1 the anorexic will have adopted abnormal attitudes towards stages up to b_1, but will so far have maintained normal attitudes towards stages beyond b_1. Thus b_1 measures the level of abnormality. Then, as the insecurity increases, so does the abnormality, following the curve in the horizontal plane, confirming that the onset of anorexia is a continual escalation by a succession of little catastrophes, little changes of attitude. Moreover as the disorder deepens, the individual catastrophes become bigger, and the attitudes towards earlier stages more abnormal (carbohydrates are at first avoided, and later feared).

We now appeal to the main theorem. The stability of memory and habit implies the existence of an implicit dynamic that holds the attitudes stably on the graph. The existence of a dynamic allows us to deduce that the graph is part of a cusp catastrophe. The right branch of the cusp marks the points where the stability of normal attitudes breaks down, and the attitude switches to abnormal, causing the gradual escalation of the disorder, while the left branch marks the points where the stability of abnormal attitudes breaks down, and the attitude switches back to normal. Thus the left branch (whose existence is a consequence of the theorem) predicts the possibility of a complete cure.

The left branch also explains why the disorder is rigid and seemingly irreversible, as follows. Suppose that after reaching i_1 the insecurity level drops again. Then the abnormality will <u>not</u> drop, but will stay fixed at b_1 until the insecurity has dropped to i_2 (where i_2 is given by the intersection of the line $b = b_1$ with the left branch), because all the abnormal

1. Catastrophe Theory   47

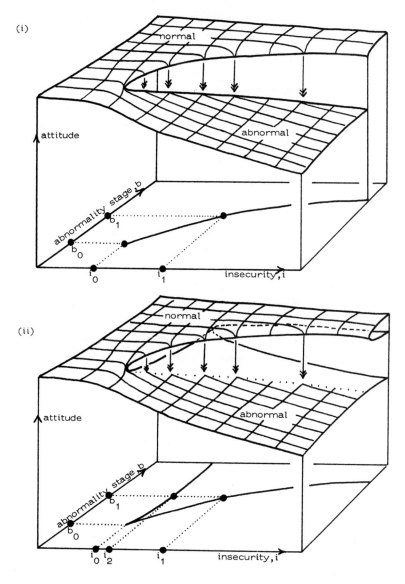

Figure 21. Abnormal anorexic attitudes.

(i) The graph showing the escalation of anorexia by a succession of little individual catastrophes, as abnormal attitudes are adopted towards each stage.

(ii) The graph embedded in a cusp-catastrophe. The left branch of the cusp indicates how far the insecurity must be reduced in order to effect a cure.

attitudes will be held stably on the lower sheet, up to the boundary of that sheet. Then, as the insecurity drops from i_2 to i_0, there is a sudden rush of attitudes switching back to normal, along the left branch of the cusp.

Notice that the worse the anorexia is, the more rigid and irreversible it is, because as b_1 increases, so does the length of the interval i_1-i_2, and hence the greater the reduction in insecurity that must be achieved before any improvement can take place. The model also explains why reasoning with the victim about her eating habits may be worse then useless, because it can only reinforce her insecurity and cannot change her attitudes; what is needed is the more fundamental reassurance about the source of insecurity. But the anorexic is not open to such reassurance while she is obsessed with, and transparently aware of, her abnormal behaviour. Hence the utilisation of the trance states, in order to give a temporary freedom from that obsession and awareness. Figure 22, which is deduced from Figure 21, shows how the therapist under these conditions can, by gently reducing the insecurity, trigger a dramatically swift catastrophic cure. Figure 22 also illustrates the difference between slimming and anorexia, showing how a quantitative difference in the initial insecurity can lead to a qualitative difference in the eventual outcome that will enable the slimmer to achieve her slimness without danger, but prevents the anorexic from escaping from her prison without help.

We have used the word "cure" in the sense of the fundamental change of attitude to life, referred to in the very revealing testimony of Mrs. Hartley quoted at the beginning; the actual physical recovery from the accompanying malnutrition and amenorrhoea will then follow naturally over the next few months. It is doubtful if this type of cure could be achieved while administering drugs that disrupt cerebal activity, because the recapturing of the whole delicate network of normal attitudes must depend not only upon reassurance, but also upon harnessing the full power of the cerebal faculties rather than suppressing them.

One of the most interesting points made by Paul MacLean is that the limbic brain is non-verbal, being phylogenetically equivalent to the brain

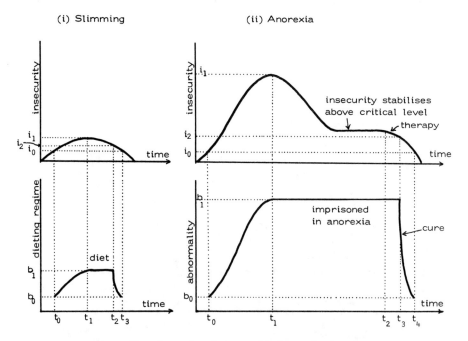

Figure 22. Comparison between slimming and anorexia.

(i) The <u>slimmer</u>, anxious about her size, reaches insecurity threshold i_0 at time t_0, and therefore begins to diet with regime b_0; reaches maximum insecurity i_1 at time t_1, and therefore stabilises dieting regime at strictness level b_1; finds, as dieting succeeds, that insecurity drops to the critical level i_2 by time t_2, and therefore rapidly relaxes her dieting regime; finds that insecurity drops to threshold i_0 by t_3, and therefore gives up dieting.

(ii) The <u>anorexic</u> begins the same, except that due to deep-seated anxieties reaches a much higher maximum insecurity i_1, causing her abnormality to develop and stabilise at level b_1; is prevented from reducing her insecurity to the critical level i_2 by the feedback from the abnormal behaviour forced on her by the anorexia, and therefore remains locked in the disorder; begins therapy at time t_2, which, by reassurance during trance, reduces the insecurity to the critical level i_2 by t_3, thereby effecting the catastrophic cure by t_4.

of a lower mammal. Therefore the problem of describing its activity in ordinary language is like trying to describe the conversation of a horse; no wonder anorexics have difficulty in explaining their symptoms. The patient can perceive that certain subsets of states are connected, and have boundaries, and so for her the most logical approach is to identify those

Figure 23. A painting by an anorexic of the tasks she saw ahead of her in life (reproduced with permission of A.H. Crisp).

subsets as "dissociated subpersonalities", and give them names. In the model these subsets are represented by the different sheets, and the structural relation between them is defined by the unique geometry of the catastrophe surfaces. Therefore we may identify those sheets with the patient's descriptions of her subpersonalities. For example the upper sheet of Figure 17 is often called the "monster within", and the lower sheet the "thin beautiful self". When she goes into trance the reduction of sensory input causes a shift in focus, from the close-up to the long-distance, from the immediacy of mood and behaviour to the long-term perspective of personality and insight. In terms of the model there is a shift from the perception of the states represented by the sheets of

Figure 17 to those of Figure 21. Therefore the "monster" and "thin self" recede in importance, and are replaced by the "real self" and the "little one" corresponding to the normal and abnormal sheets of Figure 21. More precisely it is the dynamic holding the attitudes stably on the abnormal sheet that is dimly perceived and interpreted as "prohibitions" by the voice or as "malignant possession" by the little one. The "reconciliation" refers to the left branch of the cusp, which marks the boundary of the abnormal sheet, where the stability breaks down and the catastrophic cure takes place. Thus the apparent nonsense spoken by some patients makes perfectly good sense within the framework of a complete model.

Finally comes the question of how the model can be tested scientifically. It already satisfies Thom's criterion for science, because by its coherent synthesis it reduces the arbitrariness of description. Furthermore it has survived a number of qualitative experiments between myself and Hevesi of the following nature. From the mathematics I would make some prediction, or get depressed about some failing of the model, and then when we next met Hevesi could confirm the prediction, or confirm that what I had thought to be a failing was in fact another correct prediction. Let me give some examples. The mathematics predicted the location of the trance state as the middle sheet of the butterfly. However at one stage I thought the model had failed because bias destroys the middle sheet, as can be seen from Figures 16(iv) and (vi), meaning that for those patients the trance was not accessible, but to my surprise Hevesi revealed that he found very confirmed fasters or very confirmed bingers more difficult to cure. Another prediction of the mathematics was the qualitative difference between the "fasting" and "purging" frames of mind, illustrated in Figure 19(iv); the correctness of this prediction concerning the operation of the bias factor gave further evidence in favour of using the butterfly catastrophe.

Perhaps our most striking experiment concerned the finding of the operative suggestions. Hevesi says that the trance is not like hypnosis, because the therapist does not attempt to control the patient. I was curious to know what he actually did during the trance, but when I asked him he maintained that he did not do very much. Meanwhile the mathematics was insisting that we ought to look at the underlying neurology as well as the psychology, even if only implicitly, in order to locate the dynamic; consequently we formulated the conjecture about inputs to the limbic brain in terms of MacLean's theories. It was only then, after watching himself with new eyes, that Hevesi was able to lay his finger on the operative suggestions that were reducing those inputs. Thus the model facilitated the communication of the therapeutic technique.

To test the model quantitatively would require monitoring a patient in different states; the prediction would be that the psychological data would give catastrophe surfaces diffeomorphic to those extracted from the accompanying neurological and physiological data, with the same bifurcation set. Different patients would have diffeomorphic bifurcation sets, or parts of them diffeomorphic.

PART THREE

ILLUSTRATIONS IN PHYSICS.

Example 10. Liquid/gas phase transition.

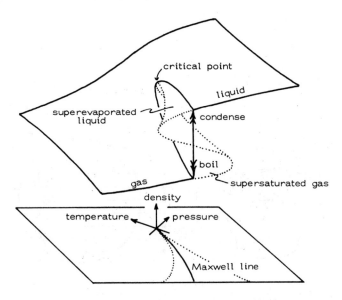

Figure 24. Temperature and pressure are conflicting factors controlling density.

Van der Waals' equation for phase transition is a cusp catastrophe surface, with temperature and pressure as conflicting factors controlling density. The upper and lower sheets represent the two phases of liquid and gas, while the two catastrophes represent the transitions of boiling and condensation. By going round the top of the cusp one can go from liquid to gas continuously. Since the controls are averaging devices, both catastrophes normally occur at the same temperature, on a line in the middle of the cusp given by Maxwell's convention. However both catastrophes can be delayed in the metastable states of superevaporated liquid or supersaturated gas. For instance clean water at atmospheric pressure can be gently heated beyond 200° before boiling, and when it does boil it explodes catastrophically with a sound like a pistol shot. The metastable states are exploited by high energy physicists in the bubble and cloud chambers. Since density is an averaging device, the shape of the surface very near the critical point is slightly distorted.

Example 11. Light caustics.

Light caustics are the bright geometric patterns created by reflected or refracted light. A familiar example is the cusp appearing on a cup of coffee in bright sunlight, caused by reflection of the sun's rays off the inside of the cup, as in Figure 25(i).

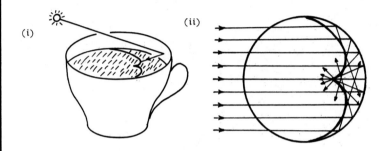

Figure 25. The caustic on a cup of coffee, caused by the sun's rays reflected off the inside of the cup, is a cusp-catastrophe.

Looking down on the cup from above, Figure 25(ii) shows that the vertical planes containing the sun's rays, after they have been reflected, envelop a vertical surface with cusp-shaped horizontal cross-section, which is called the caustic. Since the planes all touch the caustic there is a concentration of photons near the caustic (on its convex side), but of course these photons are invisible to the eye, because each is travelling along its appointed route. We can only see them if we place a screen in the way - in this case the screen is the surface of the coffee; then the concentration of photons hitting the screen will be scattered into our eye, causing the cusp-shaped section of the caustic to appear bright, with a sharp edge on the concave side and soft edge on the convex side.

Another familiar caustic is the rainbow. Here each water droplet refracts and reflects the suns rays to form a caustic surface approximately cone-shaped, with radial angle between $40°$ and $42\tfrac{1}{2}°$. The angle depends upon the wave length, so that in fact each droplet produces a co-axial family of differently coloured caustic cones. Those drops whose caustics happen to meet our eye produce the rainbow.

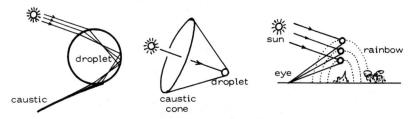

Figure 26. The rainbow is caused by a spectrum of coloured caustic cones produced by each droplet, each cone being a fold-catastrophe.

Catastrophe theory applies to light caustics because light obeys a variational principle : by Fermat's principle the light rays travel along geodesics. Let C be a 3-dimensional neighbourhood of a caustic, and let X be a 2-dimensional surface normal to the incident rays. Each point x in X determines a ray C_x in C, and the union of all these rays, parametrised by X, forms a 3-dimensional manifold.

$$M = \{(c,x);\ c \in C_x\} \subset C \times X.$$

Since all the rays leave X normally, M is given by $\partial f/\partial x = 0$, where $f(c,x)$ is the geodesic distance from x to c. The intensity of light in a volume element dC of C is proportional to dM/dC, where dM is the volume of the inverse image of dC under the projection $M \to C$. The caustic is where this intensity is greatest, namely on the bifurcation set. Therefore stable caustics are stable bifurcation sets, in other words elementary catastrophes. Therefore the classification theorem gives the new result in geometric optics : the only stable singularities a caustic

1. Catastrophe Theory

Figure 27. The caustics obtained by reflecting a point source of light in a parabolic mirror on to a screen. The three pictures correspond to three different positions of the screen, and illustrate three sections of the bifurcation set of a hyperbolic umbilic catastrophe. The catastrophe point is the top-left of the middle section. The top section shows a cusp inside a smooth fold curve; as the sections pass through the catastrophe point, the cusp pierces the curve and is itself transformed into a smooth curve on the outside, while the curve is transformed into a cusp on the inside (see Figure 13(b)). The faint repeated image is an artifact, due to a subsidiary reflection from the front surface of the mirror. Photograph by Warwick University Library Photographic Services.

1. Catastrophe Theory

(a)

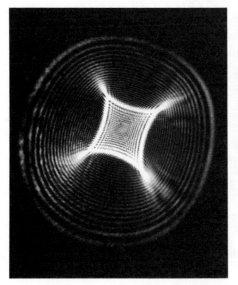

(b)

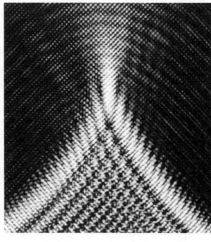

(c)

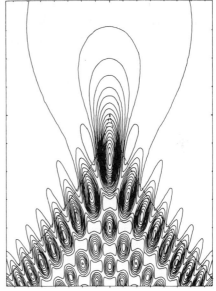

*Figure 28. (a) Caustics obtained by refracting a laser beam in doubly periodic frosted glass. The interference patterns merge into Airy patterns on the fold lines of the caustic, and into Pearcey patterns at the cusp points.
(b) Detail near a cusp, showing quantization of diffraction spots.
(c) Computer calculation of intensity levels of Pearcey pattern.
Photographs by Michael Berry.*

can have, besides cusped edges, are the three types of singular point shown in Figure 13, the swallowtail and the elliptic and hyperbolic umbilics. This discovery of Thom's about caustics was one of the reasons that stimulated him to develop catastrophe theory.

The sections of a hyperbolic umbilic catastrophe illustrated in Figure 27 were obtained by shining a torch in an old searchlight mirror. The other two types of singular point can be obtained using cylindrical and spherical lenses, such as beakers and electric light bulbs filled with water. Different sections near a singular point can be explored by moving the screen or the light.

The photographs of Michael Berry in Figure 28 were obtained by refracting a laser beam in frosted glass. Since the frosting was periodic in two directions the global picture is the projection of a torus in the plane (each little square of frosting is projected onto the whole picture). Each corner is a section of a hyperbolic umbilic. To analyse the fine structure of interference patterns merging in Airy and Pearcey patterns on the caustic, it is necessary to pass from geometric optics the more delicate wave optics. Berry uses the frosted glass as an optical analogue to study the scattering of beams of particles from a solid surface.

Example 12. Euler buckling.

Figure 29 shows a horizontal elastic strut subjected to a vertical load a and a horizontal compression b. (For an easy experiment hold a 1" × 4" strip of thin cardboard between the thumb and forefinger.)

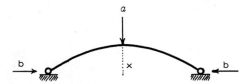

Figure 29. A free elastic strut under load a, and compression b. The resulting vertical displacement is x.

We measure the vertical displacement of the strut by the first harmonic x (i.e. the first Fourier coefficient). Then x satisfies a cusp catastrophe with normal factor $(-a)$ and splitting factor b. Buckling occurs at the

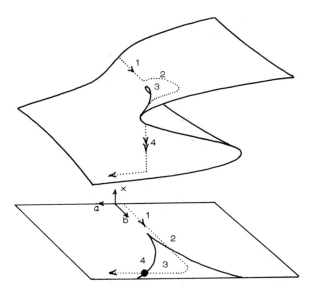

Figure 30. In the free strut the load a is the normal factor, and the compression b the splitting factor, controlling the first harmonic, x. The dotted path shows the strut (1) compressed (2) buckling (3) loaded and (4) snapping.

cusp point, which Euler showed was $a = 0$, $b = \pi^2 \lambda / \ell^2$, where ℓ = length and λ = modulus of elasticity. The dotted path in Figure 30 shows the strut (1) remaining straight under increasing compression (2) buckling upwards (3) supporting an increasing load until (4) it snaps downwards.

Now suppose that the ends of the strut are fixed, so that in an unloaded state it is buckled at height x. Put on the load a, offset from the centre by a distance c (to simulate a manufacturing imperfection) and measure the displacement now by the <u>second</u> harmonic y (see Figure 31). The second harmonic can be seen easily with the strip of cardboard, or more dramatically in the western arch of Clare College bridge, Cambridge.

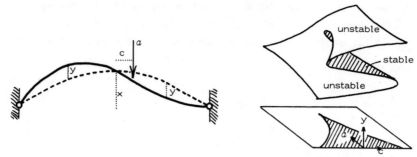

Figure 31. For the pinned strut the imperfection c is the normal factor, and the load a the splitting factor, controlling the second harmonic, y. Therefore the load carrying capacity is imperfection sensitive.

Then y satisfies a dual cusp catastrophe with c as normal factor and $(-a)$ as splitting factor (dual means only the middle sheet represents stable equilibria, and the rest unstable). Therefore if the load a is gradually increased the strut will snap downwards at the cusp line. If c = 0 the critical load is $a = 3\lambda\pi^4 x/2\ell^3$, whereas if $c \neq 0$ the critical load drops sharply by the $\frac{3}{2}$-power-law of the cusp. This phenomenon is called imperfection sensitivity to c.

Several authors including W.T. Koiter, M.J. Sewell, J.M.T. Thompson and G.W. Hunt, have extended these ideas to general theories of elasticity and buckling, in which the classification theorem can be used. For example Figure 32 shows that the failure locus of a stiffened elastic panel is a hyperbolic-umbilic-catastrophe : the diagram was computer-drawn by Hunt from equations of V. Tvergaard. Here the three controls are

a = Euler imperfection

b = compression

c = local panel imperfection.

Figure 32 shows that if $a > 0$ the panel not only fails, but is imperfection sensitive to c; whereas if $a < 0$ and c small, then it can sustain a much higher compression b without failing. Consequently it is sometimes useful to incorporate an imperfection $a < 0$ into the design.

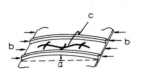
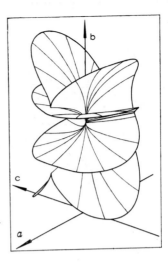

Figure 32. The hyperbolic-umbilic-catastrophe determined by Giles Hunt in the theoretical response of a stiffened elastic panel. The lower region of the surface gives the panel strength as a function of the Euler and local manufacturing imperfections. This picture is diffeomorphic to Figure 13(b).

Example 13. Shock waves.

In fields like gas dynamics, traffic flow etc. the conservation laws for mass, momentum etc. give rise to quasi-linear partial differential equations for density, velocity etc. The solutions of these equations propagate along characteristics in space-time. If the characteristics converge then they form a folded surface with a cusp-catastrophe, whose upper and lower sheets represent compression waves and whose middle sheet represents an (unrealisable) rarefaction wave. Since the solution must be single-valued, a shock wave has to develop inside the cusp, whose speed (and hence position) is determined by the Rankine-Hugoniot condition, which can be deduced from the conservation law (but not from the partial differential equation).

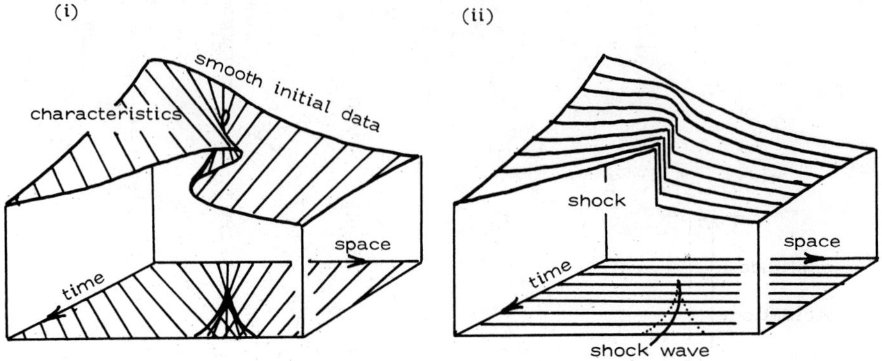

Figure 33. (i) *Propagation of the initial data along the characteristics of the partial differential equation gives a cusp catastrophe.* (ii) *Physically only one solution is possible, and so a shock wave has to develop, whose speed is determined by the conservation law.*

Example 14. Forced non-linear oscillations.

Forced oscillations can be modelled by Duffing's equation

$$\ddot{x} + k\dot{x} + x + \alpha x^3 = F \cos \Omega t,$$

where $k > 0$ is a small damping term, α a small non-linear term ($\alpha = -\frac{1}{6}$ for a simple pendulum), and $F \cos \Omega t$ is a small periodic forcing term with frequency Ω close to 1, the frequency of the linear oscillator. The amplitude A of the resulting oscillation depends upon the parameters, and Figure 34 shows the graph of A as a function of α and Ω (keeping k and F fixed). There are two cusp-catastrophes with α, Ω as conflicting factors. At each cusp the upper and lower sheets represent attractors (stable periodic solutions) while the middle sheet represents saddles (unstable periodic solutions). If the frequency of the forcing term is gradually changed so as to cross one of the cusp lines, going from the inside to the outside of the cusp, then the amplitude A will exhibit a catastrophic jump. There will also be a sudden phase-shift at the same time.

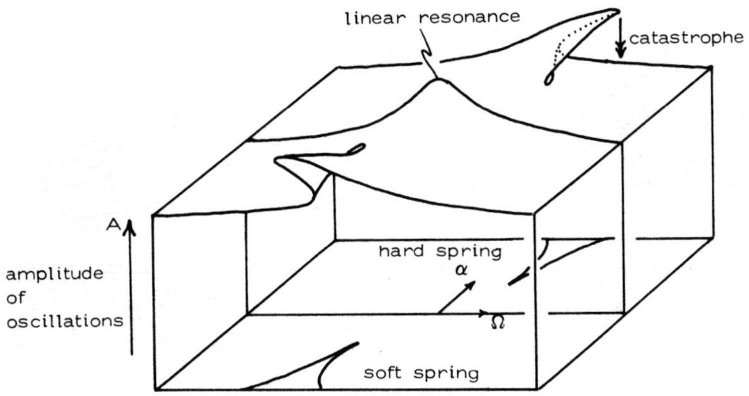

Figure 34. The oscillation of a forced non-linear oscillator bifurcates according to the cusp-catastrophe.

SUGGESTED FURTHER READING.

Books of Thom.

R.Thom, *Structural stability and morphogenesis*, Benjamin, New York, French 1972; English trans. by D.H. Fowler, 1975.

R.Thom, *Modèles mathématiques de la morphogénèse*, Union Générale d'Editions, Paris, 1974.

Introductory mathematics.

V.I.Arnol'd, Critical point of smooth functions, *Proc. Int. Cong. Math., Vancouver*, 1, (1974), 19-39.
T.Bröcker, *Differentiable germs and catastrophes*, Camb. Univ. Press, 1975.
J.Callahan, Singularities and plane maps, *Amer. Math. Monthly* 81 (1974) 211-240.
D.R.J.Chillingworth, *Differential topology with a view to applications*, Pitman, London, 1976.
P.J.Holmes & D.A.Rand, The bifurcations of Duffing's equation : an application of catastrophe theory, *J. Sound & Vib.* 44, 2 (1976) 237-253.
Y.C.Lu, *Singularity theory and an introduction to catastrophe theory*, Springer-Verlag, New York, 1976.
T.Poston & I.N.Stewart, *Taylor expansions and catastrophes*, Pitman, London, 1976.
A.E.R.Woodcock & T.Poston, *A geometrical study of the elementary catastrophes*, Springer Lecture Notes in Maths, 373 (1974).
This book, papers 18, 19.

Anorexia, brain and behaviour.

H.Bruch, Anorexia nervosa, *American Handbook of Psychiatry;* Basic Books, New York, 4 (1975) 787-809.

A.H.Crisp, Deception, obstinacy and terror in anorexia nervosa, *Mims Magazine* 15 (March 1975) 42-53.

K.Lorenz, *On aggression*, Methuen, London, 1966.

P.D.MacLean, *A triune concept of the brain and behaviour*, Hinks Memorial Lectures, Toronto Univ. Press, 1973.

This book, papers 8,9,12.

Some applications in the physical sciences.

M.V.Berry, Cusped rainbows and incoherence effects in the rippling-mirror model for particle scattering from surfaces, *J. Phys. A* 8 (1975) 566-584.

M.V.Berry, Waves and Thom's theorem, *Advances in Physics* 25 (1976) 1-26.

M.V.Berry & M.R.Mackley, The six roll mill : unfolding an unstable persistently extensional flow, *Phil. Trans. Roy. Soc.* (1977).

M.V.Berry & J.F.Nye, Fine structure in caustic juncations, *Nature* 267 (1977) 34-36.

M.V.Berry, Focusing and twinkling : critical exponents from catastrophes in non-Gaussian random short waves, *Advances in Physics* (1977).

M.V.Berry &J.H.Hannay, Umbilic points on Gaussian random surfaces, *J. Phys. A* (1977).

T.B.Benjamin, Bifurcation phenomena in steady flows of a viscous fluid, *Proc. Roy. Soc. Lond. A* (1977).

C.DeWitt-Morette, The small disturbance equation, caustics and catastrophes, Univ. of Texas, Austin, Texas (1975).

J.Guckenheimer, Caustics and non-degenerate Hamiltonians, *Topology* 13 (1974) 127-133.

K.Jänich, Caustics and catastrophes, *Math. Ann.* 209 (1974) 161-180.

J.Komorowski, On Thom's idea concerning Guggenheim's one-third law in phase transition, Inst. Math. Methods in Phys., Warsaw University (1977).

P.D.Lax, The formation and decay of shock waves, *Amer. Math. Monthly* 79 (1972) 227-241.

J.C.Maxwell, Last essays, in L.Campbell & W.Garnett, *The life of James Clerk Maxwell*, MacMillan, London (1882) 434-444.

J.F.Nye, Disclinations and catastrophes in the vector and tensor fields of sea ice, Univ. Bristol (1976).

L.S.Schulman & M.Revzen, Phase transitions and catastrophes, *Collective phenomena*, 1 (1972) 43-47.

M.J.Sewell, Some mechanical examples of catastrophe theory, *Bull. Inst. Math. & Appl.* 12 (1976) 163-172.

R.Thom, Phase transitions as catastrophes, in *Statistical mechanics : new concepts, new problems, new applications,* Univ. Chicago Press (1977) 93-105.

J.M.T.Thompson & G.W.Hunt, *A general theory of elasticity*, Wiley, 1973.

J.M.T.Thompson & G.W.Hunt, Towards a unified bifurcation theory, *J. Appl. Maths. Phys.* 26 (1975) 581-603.

This book, papers 15 - 17.

Other applications.

At the end of each paper in this book is a bibliography related to that paper, and at the end of paper 21 there is a more general bibliography.

2 Levels of Structure in Catastrophe Theory Illustrated by Applications in the Social and Biological Sciences

Catastrophe theory is a method discovered by Thom [14] of using singularities of smooth maps to model nature.

In such models there are often several levels of structure, just as in a geometry problem there can be several levels of structure, for instance the topological, differential, algebraic, and affine, etc. And, just as in geometry the topological level is generally the deepest and may impose limitations upon the higher levels, so in applied mathematics, if there is a catastrophe level, then it is generally the deepest and likely to impose limitations upon any higher levels, such as the differential equations involved, the asymptotic behaviour, etc. Again, in geometry the complexity of the higher levels may render them inaccessible, so that they can only be handled implicitly rather than explicitly, while at the same time the underlying topological invariants may even be computable. Similarly in applied mathematics the complexity of the differential equations may sometimes render them inaccessible (even to computers), so that they can only be handled implicitly rather than explicitly, while the underlying catastrophe can be modeled, possibly even to the extent of providing quantitative prediction.

Therefore catastrophe theory offers two attractions: On the one hand it sometimes provides the deepest level of insight and lends a simplicity of understanding. On the other hand, in very complex systems such as occur in biology and the social sciences, it can sometimes provide a model where none was previously thought possible. In this paper we discuss various levels of structure that can be superimposed upon an underlying catastrophe and illustrate them with an assortment of examples. For convenience we shall mostly use the familiar cusp catastrophe (see [5], [13], [14], [24]).

Published in the Proceedings of the International Congress of Mathematicians, Vancouver, 1974, Volume 2, pages 533-546.

Level 1. Singularities.
Level 2. Fast dynamic (homeostasis).
Level 3. Slow dynamic (development).
Level 4. Feedback.
Level 5. Noise.
Level 6. Diffusion.

Thom's classification of elementary catastrophes belongs to Level 1. Levels 2,3,4 refer to ordinary differential equations, and Level 6 refers to partial differential equations.

Level 1. Singulairties. We begin by recalling the main classification theorem. Let C, X be manifolds with dim $C \leq 5$, and let $f \in C^\infty(C \times X)$. Suppose that f is generic in the sense that the related map $C \to C^\infty(X)$ is transverse to the orbits of the group Diff $(X) \times$ Diff (R) acting on $C^\infty(X)$. (Genericity is open-dense in the Whitney C^∞-topology.) Let $M \subset C \times X$ be given by $\nabla_X f = 0$, and let $\chi: M \to C$ be induced by projection $C \times X \to C$.

THEOREM (THOM). (a) *M is a manifold of the same dimension as C.*
(b) *Any singularity of χ is equivalent to an elementary catastrophe.*
(c) *χ is stable under small perturbations of f.*

The number of elementary catastrophes depends only upon the dimension of C (and not on X):

dim C	1	2	3	4	5	6
elemementary catastrophes	1	2	5	7	11	(∞)

For details of the elementary catastrophes the reader is referred to [10,] [14]. The first complete proof was given by Mather [8], and other references are [1], [17], [18].

REMARK. The classification of singularities goes infinite for dim $C \geq 6$, but the above table can be extended with finite entries, provided the concept of elementary catastrophes is suitably modified, as follows. The singularities correspond to orbits of a group acting on a space of germs (see Level 4 below). In particular the (∞) appears at dim $C = 6$ because there is a stratum of codimension 6 foliated by orbits of codimension 7. Arnold [1] calls the codimension of the foliation, which is 1 in this case, the modality of the stratum. More generally the orbits form a foliated stratification P, which Arnold has shown to be locally finite. The finite numbers of strata of each codimension give the desired extension of the above table.

The reason that catastrophe theory exists is that by a happy accident P is 5-simple, in other words each stratum of P of codim ≤ 5 is simple, that is trivially foliated by a single leaf. These strata correspond to the elementary catastrophes of dim ≤ 5, and hence the latter are finitely classified differential invariants. For most applications it suffices to have dim $C \leq 5$, and so there is no need to worry about the foliation of the higher strata.

Application. Suppose we have some set of objects or events about which we want to test a hypothesis of cause and effect. One of the first things to do is to plot them in cause-effect space, and see if they form a graph. Here C will describe the cause, X the effect, and $f(c, x)$ the probability that cause c will produce effect x. The most likely effects are given by the peaks of probability, where both the gradient vanishes, $\nabla_X f = 0$, and the Hessian is negative definite, $\nabla_X^2 f < 0$. This determines a submanifold G of M (of the same) dimension. Then G will be the desired cause-effect graph in $C \times X$. The events will be represented by a cloud of points clustering near G, with density of clustering depending upon the deviation of the probability distributions.

Consider the first two elementary catastrophes, which occur when dim $C = 2$. The fold-catastrophe occurs at the boundary of G, but since there is no dynamic in Level 1 there may not be any catastrophic jump here—all we can say is that the cloud of points appears to terminate.

The cusp-catastrophe occurs when a probability distribution goes bimodal. In this case observers may implicitly recognise the phenomenon, and capture part of it by either naming the two modes, or alternatively framing some form of words, such as a proverb or a belief. However the cusp-catastrophe can often reveal other facets to the phenomenon, and give a new synthesis of understanding. We illustrate the two alternatives by a couple of examples.

EXAMPLE 1. AGGRESSION [22]. According to Konrad Lorenz [7] fear and rage are conflicting drives influencing aggression. Here the two extreme behaviour modes are attack/flight, and X represents a 1-dimensional spectrum of behaviour varying from neutral to the two extremes at either end. The cause C is 2-dimensional, representing the strengths of the fear and rage drives present in the animal at that moment. Lorenz observes that in the case of dogs the coordinates of fear and rage can be read from the facial expression [7, p. 81]. Rage only causes attack, fear only causes flight, and when both are present the effect is one of two extremes but unpredictable. Therefore the probability goes bimodal, and as a first approximation we might expect our cloud of points to cluster around a cause-effect graph equivalent to the cusp-catastrophe as shown in Figure 1. We shall return to this example, and its uses, in Level 2.

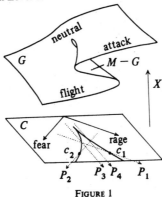

FIGURE 1

Other familiar examples of bimodality which can be modeled by the cusp-catastrophe are (i) liquid/gas [3], [11], [15], (ii) diastole/systole [23], (iii) manic/depressive [25], (iv) dove/hawk [5], or (v) bull/bear ([26], and Example 7 below).

In each case the bimodality is caused either by conflicting factors such as temperature and pressure in (i), or by a splitting factor such as tension in (ii), disease in (iii), cost in (iv) or speculation in (v). Let us now give an example of a proverb.

EXAMPLE 2. MORE HASTE LESS SPEED. This proverb is very familiar in England, although almost unknown in America. Its attraction lies in its brevity and contrariness—it is the opposite of what one would normally expect, especially if the operator is skilled at his task. And this leads to the observation that speed really depends upon two factors, haste and skill, which are conflicting. For, when both factors are present, the probability goes bimodal, because either the operator's skill enables him to increase his speed, or his fumbling haste diminishes it. Therefore again we might expect our cloud of points to cluster around a cusp-catastrophe, as in Figure 2.

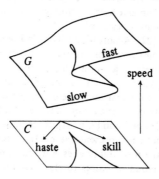

FIGURE 2

We suggest a couple of uses for such a model. Firstly in psychology it might be possible to develop it further into a predictive quantitative model for explicit skills (see Level 2 below). Secondly in sociology it might provide a prototype for reconciling conflicting theories. For, by the theorem, we should expect the cusp-catastrophe to occur in many phenomena, and, although the graph is visually simple, its subtlety is not easy to describe with any brevity in ordinary spoken or written language (see [5, §10]). Therefore although we may often recognise such a phenomenon intuitively, we tend to describe it verbally by an oversimplification, possibly by directing attention only to the unexpected mode. For instance "more haste less speed" directs attention only to the lower front sheet of Figure 2, marked "slow". Similarly two conflicting opinions in a discussion, or two conflicting sociological theories, may in fact each be referring to a single mode of an underlying bimodal phenomenon, and the conflict may sometimes be reconciled by exhibiting the two modes as the two sheets of a cusp-catastrophe growing smoothly out of an area of common agreement.

Level 2. Dynamic. In addition to the generic function $f \in C^\infty(C \times X)$ suppose we are given a dynamic D as follows. Denote the associated function $C \to C^\infty(X)$ by $c \mapsto f_c$. Then $D = \{D_c\}$ is a family of differential equations on X, parametrised by C, such that, for each $c \in C$, f_c is a Lyapunov function for D_c. In other words, f_c increases (or decreases) along the orbits of D_c, and so the maxima (or minima) of f_c are the attractors of D_c. Therefore D_c is gradient-like, and this is the restriction that Level 1 imposes on Level 2.

The graph G now represents the attractors of D. In applications we no longer intuitively imagine a cloud of points clustering statically near G, but points flowing dynamically onto G and then staying there. The model is half dynamic and half static. It is sometimes useful to think of the parameter space C as control, and X as behaviour space. If we slowly move the control c then the behaviour x responds by moving continuously on G for as long as possible; in other words it is a theorem of Level 2 that the system obeys the delay rule of Thom [5], [14].

If c now crosses the bifurcation set, then x may cross the boundary ∂G of G. In this case the dynamic will carry x rapidly onto some other sheet of G. The word "rapidly" assumes that the movement of control is slow compared with the dynamic, and it is the sudden jump that occurs at the fold-catastrophes in Level 2 that is responsible for the name "catastrophe theory".

EXAMPLE 3. The *catastrophe machine* described in [9], [24] is a simple toy made out of a cardboard disk and two elastic bands, which exhibits the catastrophic jump well, and the uninitiated reader is recommended to make one for himself. Here the function f of Level 1 is the potential energy in the elastic given by Hooke's law of elasticity, and the dynamic D is given by Newton's law of motion, suitably damped so as to minimise f.

EXAMPLE 1. Returning to our first example we see that it can be promoted from Level 1 to Level 2. For we may reinterpret X as the space of states of that part of the brain governing mood (perhaps the hypothalamus), and D as the associated dynamic representing neurological activity. Then the attractors of D represent the attacking/retreating frames of mind, providing the background mood against which behavioural decisions are taken. Although X must necessarily be very high dimensional, and D consequently inaccessible in the sense of being only implicit, nevertheless G will still be 2-dimensional. Therefore the cusp-catastrophe can still provide an explicit model, which for individual animals might be made quantitative and predictive. Moreover since it is a Level 2 model, even though D is only implicit, there will be catastrophic jumps of mood, resulting in sudden attacks or disengagements. For example in Figure 1 the path P_1, representing increasing rage at a fixed level of fear, as for instance in a cornered dog, will lead to a sudden attack at c_1 while path P_2 to a sudden disengagement at c_2. Meanwhile paths P_3, P_4 illustrate how nearby paths can lead to divergent behaviour. Similarly humans, when made angry and frightened, are unpredicatable and are denied access to rational behaviour, and may jump from abuse to apology, even from hysteria to tears.

The interest of this example is that it may provide a general model for control of aggression, valid for different species under varying circumstances, and may give insight into how such controls develop and have evolved. More generally it provides a prototype for relating the neurology to the psychology of moods underlying behaviour.

EXAMPLE 2. Our second example may also be promoted from Level 1 to Level 2, because, if we consider the performance of an individual, his tendency to adjust his speed to x, say, within the limitations of his skill and assuming a given amount of haste, is another way of saying there is an implicit dynamic that moves the speed to x.

A path P_1, such as in Figure 1, here represents an increasing skill at a fixed level of haste, as for instance when learning to ride a bicycle, and at the point c_1 a catastrophe occurs when the individual is suddenly able to ride. Moreover the greater the haste—for instance the swifter reactions that are needed to ride a more unstable machine—then the greater the skill needed before the catastrophe occurs. Meanwhile a path P_2 here represents increasing haste at a fixed skill, as for instance a wireless operator trying to read faster and faster Morse code, and at the point c_2 a catastrophe occurs as the performance drops sharply. Moreover the greater the skill, the greater the haste possible before the catastrophe occurs.

In general Level 2 is much easier to test experimentally than Level 1, because the cloud of points more accurately determines G, and the catastrophes determine ∂G. Whenever a phenomenon exhibits any one of the four qualities of bimodality, divergence, catastrophic jumps or hysterisis delays, then it may be possible to model it by the cusp-catastrophe, in which case it may be possible to predict the other three qualities. Sometimes the cusp-catastrophe can also be useful in applications where the control space C is high dimensional, as shown by the following example.

EXAMPLE 4. ECOMOMIC GROWTH. Let X represent the space of states of an economy, and C the external pressures on that economy together with the controls available to the government. Let D represent the implicit response of the economy. We should expect C to be high dimensional, and so at first sight the theorem is of little use. However the evolution of the economy is in fact only a 1-dimensional path in C lifted to a 1-dimensional path in G, and the corresponding 1-dimensional catastrophes are the slumps, inflation explosions, etc.

A typical problem facing the government is the realisation that whereas its present policy is now at control point c_0, it may have to change policy in the next few months to c_1, due to external pressures, balance of payments, etc. The government's freedom of action may be limited merely to choosing the path from c_0 to c_1. However such choice may be critical as we now explain. Suppose there is a choice between two paths P_1 or P_2. For simplicity let us assume that neither path involves a catastrophe. The question that must be asked is: Does the circle $P_1 \cup P_2$ link any codimension-2 stratum Σ of the bifurcation set? For if it does, then a

2-dimensional disk E spanning $P_1 \cup P_2$ will pierce Σ, and the section of G over E will contain a cusp-catastrophe, as shown in Figure 3. The lifts Q_1, Q_2 of P_1, P_2 will exhibit divergence, which could radically affect growth, inflation, unemployment, etc.

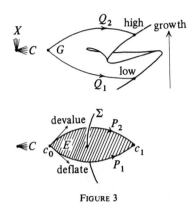

FIGURE 3

For example suppose P_1 represented deflation followed by devaluation (as in the U.K. in 1967), and P_2 the reverse order (as in France in 1968). Then Q_1 could lead to low growth because, with reduced stocks, firms would be unable to exploit the devaluation, whereas Q_2 could lead to high growth, because firms could switch sales of stock from the curtailed home market to the export market, without losing growth momentum. Therefore economists should be concerned not only with the more obvious codimension-1 problems of catastrophe, but also with the more hidden codimension-2 problems of divergence and choice.

Level 3. Development. In addition to $f \in C^\infty(C \times X)$ and the dynamic D suppose that we have time T occurring as one of the axes in the control space C. It is assumed that T is slow compared with the fast time occurring in the dynamic D.

EXAMPLE 5. EMBRYOLOGY. Level 3 occurs in Thom's main application of catastrophe theory to embryology [13], [14], [16], where the control space, $C = S \times T$, represents space-time, and X represents the states of a cell. For instance X may be a bounded open subset of R^n, with several thousand coordinates representing various chemical and physical parameters of the cell. The dynamic D represents the homeostasis of a cell returning it swiftly to equilibrium, and T the slow development of the cells.

An example of a result in this context is the following:

THEOREM [27]. *Whenever a tissue differentiates into two types, the frontier between them first forms to one side and then moves through the tissue before stabilising in its final position.*

The proof uses the cusp catastrophe as illustrated in Figure 4. S is taken to be 1-dimensional perpendicular to the frontier. Development paths of cells are given by lifting time lines to G. The frontier first forms at c_1, then moves as a wave through S along the cusp branch c_1c_2, and then stabilises at c_2, where the cusp touches the time line c_2c_3. Such a wave is often a hidden switching on of genes, and morphogenesis may be caused after some delay by a secondary wave of physical manifestation. For example in [27] detailed models are given for the morphogenesis of gastrulation and neurulation in amphibia, and of culmination in slime-mold.

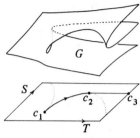

FIGURE 4

Space-catastrophes. The above result depended upon the time-axis not being tangential to the cusp-axis, which can be justified by an appeal to genericity. However to put this type of genericity on a mathematical footing requires a generalisation of the classical theory as follows, which Wassermann [19] calls space-catastrophe theory. (He also studies the dual concept of time-catastrophe theory.)

Let E_n denote the ring of germs at 0 of C^∞-functions $R^n \to R$, m_n the maximal ideal, and G_n the group of germs of C^∞-diffeomorphisms $R^n, 0 \to R^n, 0$. Then G_n acts on m_n, leaving m_n^2 invariant. Classical catastrophe theory [1], [8], [14], [17], [18] consists of analysing the foliated stratification P of m_n^2 by G_n. The elementary catastrophes of dimension s are given by the strata of P of codimension s. Since P is 5-simple, the elementary catastrophes for $s \leq 5$ are finitely classified differential invariants, independent of n (for $n \geq 2$).

For the generalisation we need some more definitions. We say $\alpha \in G_{n+r}$ covers $\beta \in G_r$ if $\pi\alpha = \beta\pi$, where $\pi : R^{n+r} \to R^r$ is the projection. Define

$$G_n^r = \{(\alpha, \beta) \in G_{n+r} \times G_{1+r}; \exists \gamma \in G_r \text{ such that } \alpha, \beta \text{ cover } \gamma\}.$$

Then G_n^r acts on m_{n+r}, leaving $m_n^2 + m_r E_{n+r}$ invariant. For space-catastrophe theory we choose $r = 1$ (representing time), and analyse the foliated stratification Q of $m_n^2 + m_1 E_{n+1}$ by G_n^1. The space-catastrophes of dimension s are given by the strata of Q of codimension $s + 1$.

Wassermann [19] has shown that Q is 2-simple, and hence the 1-space-catastrophes are finitely classified differential invariants, independent of $n \geq 1$. There are exactly four, namely the beginning c_1, the middle, and the end c_2, of the wave in Figure 4, and the "silent" dual of c_2. Therefore the above theorem is valid, and exhibits them all.

However Q is not 3-simple, for Wassermann has shown that the P-strata of swallowtails and umbilics are not only substratified by Q but also foliated. Therefore the number of singularities in 2-space goes infinite, and although the 2-space-catastrophes will still be finitely classifiable, they will no longer be differential invariants. Some will be—for example the cusp-projection of a fold-surface into 2-space (analogous to the fold-projection of a fold-curve into 1-space at c_2 in Figure 4). This example has been used to model the pattern formation of somites in amphibia [27].

However Thom [14], [16] uses the swallowtail, butterfly and umbilics extensively in embryology, and hence it is important to classify the 2- and 3-space-catastrophes. Therefore mathematically we need to analyse the strata of Q up to codimension 4, and to understand the nature of the loss in differentiability implied by their foliation.

Level 4. Feedback. Here we assume that the slow flow is not as simple as merely taking a coordinate in the control space, but may go in different directions on different sheets of G. In fact it may be conceived as a form of feedback:

$$C \xrightleftharpoons[\text{slow feedback } F]{\text{fast dynamic } D} X.$$

More precisely, in addition to f and D, suppose we are given a C^∞-map F: $C \times X \to TC$, where TC denotes the tangent bundle of C, and $F(c, x)$ is a tangent at c, for each $c \in C$, $x \in X$. Therefore D and F together form an ordinary differential equation on $C \times X$ (with the proviso that D is fast and F slow).

EXAMPLE 6. HEARTBEAT AND NERVE IMPULSE [23]. Explicit examples of differential equations in form of feedbacks on the cusp-catastrophe were taken as models. In each case the flow possessed a stable equilibrium, which if suitably disturbed by an "external agent", triggered a catastrophe via D, and a return to equilibrium via F. In the heartbeat the return involved a second catastrophe (relaxation after contraction), whereas in the nerve impulse the return was smooth (repolarisation). These models possess two interesting features. Firstly the feedback does not give a flow precisely on G, but only near G, the order of nearness depending upon the ratio K of fast/slow. If $K \to \infty$ we obtain an idealised flow on G, with instantaneous catastrophes, generalising the relaxation oscillations of electrical engineering. Secondly the words "external agent" above reveal the inadequacies of the models, in being only ordinary differential equations describing the local behaviour of heart muscle and nerve fragment; what is needed is to embed the latter in a larger partial differential equation that describes the global behaviour as waves. We return to this problem in Level 6.

EXAMPLE 7. STOCK EXCHANGES [26]. The cusp-catastrophe is used to model the behaviour of stock exchanges, as follows. The excess demand is the normal factor controlling the rate of change of index, and the speculative content of the market a splitting factor. The dynamic D represents the immediate response of index to investors, and F the somewhat slower feedback. Plausible economic hypotheses lead to a flow that exhibits periodic bull market, recession, bear market and recovery. However, to make this model realistic, we should promote it to Level 5 by including noise.

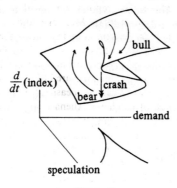

FIGURE 5

EXAMPLE 8. FUNNEL. In classifying the generic low-dimensional feedback-catastrophes, Takens [12] has recently discovered an interesting new type, the simplest of which he calls the funnel. In the associated idealised flow a 2-dimensional piece of G is funneled through a single fold-point P. Figure 6 illustrates the following explicit example:

Fast dynamic $D: \dot{x} = -K(x^2 + 2b)$, K large constant.
Slow feedback $F: \dot{a} = 1$, $\dot{b} = 3a + 4x$.

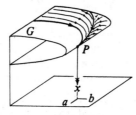

FIGURE 6

Funnels may occur in biological regulation, for instance choosing x, b to model the internal self-regulation of a cell, and a the production by the cell of some hormone for use outside the cell, whose production-rate needs to be funneled precisely.

Level 5. Noise. We may superimpose on $\{f, D, F\}$ stochastic noise in the form of random small displacements of control and behaviour. For most noise the dynamic D carries the state rapidly back onto G, and the slow flow F proceeds as before, and so the noise can be ignored. However in two cases noise can cause catastrophes, firstly if control-noise crosses the bifurcation set, and secondly more interestingly if the behaviour-noise crosses a separatrix.

EXAMPLE 7. In the stock exchange example noise represents external events and consequent jumpiness in the market, and may cause recessions to occur before the bifurcation set is reached.

EXAMPLE 9. RIOTS [4]. This model reports joint work in progress with prison psychologists P. Shapland, C. Hall and H. Marriage, and statistician J. Harrison. We start with a truism: The more tension in an institution the more disorder. This applies not only to institutions such as prisons, universities, firms, or countries but also to individuals. In the case of prisons, an analysis of data suggests that the tension (or distress or frustration) can be measured by the numbers reporting sick, suitably smoothed, and the disorder can be measured by correlating independent assessments of the seriousness of incidents. Alienation (or lack of communication) seems to be a splitting factor, producing the two modes that we have labelled quiet and disturbed in Figure 7, and the data suggest that this may be measured by the numbers of disciplinary reports. The feedback flow represents the increase in tension during quiet (over months) and the release during disturbance (over days). Noise describes incidents, and if the noise level crosses the separatrix AA' at B then the incidents will escalate and spark a riot causing a catastrophe. Some types of prison population (e.g., young long-term) have a higher noise level, and are therefore more susceptible to riots. When the tension has subsided after a few days an incident may cause the reverse catastrophe at B'. The same incident might not have done so earlier, which explains the advantage of playing it cool.

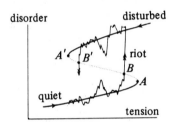

FIGURE 7

EXAMPLE 10. PHASE TRANSITION [3], [11], [15]. If the noise is frequent, and the noise-level high, the state will, averaged over time, seek the absolute maximum (or minimum) of f. This explains why Van der Waals' equation for liquid/gas phase transition has to be supplemented with Maxwell's rule [3], [5], [14], instead of obeying the delay rule. On the other hand, if the noise level is kept low then partial delays can be induced, such as in the supersaturated and superevaporated states of the cloud and bubble chambers. The usual proof of Maxwell's rule in statistical mechanics involves integration by steepest descent, but since this method breaks down near the critical point, it would be interesting if a new abstract proof could be devised, parallel to the proof of Thom's theorem, in order to enhance critical point analysis.

Level 6. Diffusion. The following arises out of joint work in progress with Sharon

Hintze, stimulated by papers of Winfree [20], [21] and Kopell and Howard [6] on the Zhabotinsky reaction. First the mathematics.

Let Y be a manifold and g be a C^∞-vector field on Y. The associated ordinary differential equation is

(1) $$\dot{y} = g(y).$$

In particular we shall be concerned with the type of differential equation given by Level 4, namely $Y = C \times X$ and $g = \{D, F\}$. Suppose now that Y represents the space of local states of some medium in space-time $S \times T$, and that g represents the reaction of that medium. Suppose further that the medium not only reacts but also diffuses. Then, following [6], the global state $y: S \times T \to Y$ of the medium satisfies the reaction-diffusion partial differential equation:

(2) $$\partial y/\partial t = g(y) + k\nabla^2 y$$

where k is a constant (more precisely a vector bundle map $k: TY \to TY$) representing the different rates of diffusion of the various components of Y. We are particularly interested in whether or not the medium can sustain stable periodic wave trains, or stable pulses (isolated waves). If it can, and θ is the speed, then the global state y can be factored $S \times T \to R \to Y$ such that $\partial y/\partial t = \theta \dot{y}$ and $\nabla^2 y = \ddot{y}$, where the dot denotes differentiation with respect to R. Therefore the partial differential equation (2) reduces to the ordinary differential equation

(3) $$\theta \dot{y} = g(y) + k\ddot{y}.$$

This equation (3) is the central interest of Level 6; compare it with equation (1) above. If k is small then (3) can be regarded as a singular perturbation of (1), but in important applications k is large, and so new methods are needed.

For instance (3) can be regarded as a flow on TY, with the same fixed points as (1), on the zero section Y. An attractor of (1) may be a saddle point of (3), and a homoclinic orbit of this saddle will represent a pulse solution of (2). Meanwhile a closed orbit of (3) represents a wave train solution of (2). Therefore we seek homoclinic and closed orbits of (3), that are stable with respect to (2). As yet relatively little is known, even when g represents a canonical elementary catastrophe with the simplest form of feedback.

EXAMPLE 6. In the heartbeat and nerve impulse dynamics [23] Conley and Carpenter [2] have shown the existence of homoclinic and closed orbits, and the next problems are to prove stability and fit data.

EXAMPLE 11. ZHABOTINSKY REAGENTS. Belousov discovered a mixture of chemicals that oscillates in colour at about twice a minute, and later Zhabotinsky and Zaiken observed that circular wave trains would propagate through this reagent, entraining the oscillation. Winfree [21] then modified Belousov's reagent by adding a little more bromide and a little less acid, so as to stop the oscillation. He called his mixture the Z-reagent, after Zhabotinsky and Zaiken, and showed that it could sustain both pulses and rotating scroll-shaped wave trains. In [21] Winfree offers equations which beautifully explain the geometry of the patterns, but which can be mildly criticised on four counts. Firstly his dynamic is discontinuous, and the obvious way to make the model differentiable is to approximate it by a catastrophe model. Indeed as Kopell and Howard [6] point out there are both fast (fractions of a second) and slow (minutes) reactions, as well as a very slow (hours) loss of energy. Therefore one would normally expect the reaction dynamic to belong to Level 4. Secondly Winfree's equations do not illustrate the modification Belousov → Z. However this can be illustrated naturally in catastrophe theory, by modifying one constant, causing a Hopf bifurcation, as we show below. Thirdly his equations exhibit a jump return, like the heartbeat, whereas his photographs illustrate a smooth return, blue → red, as opposed to the catastrophic hard edge, red → blue, more like the repolarisation of the nerve impulse. This feature can be accommodated by using the cusp-catastrophe [20], [23]. Fourthly he does not offer mathematical proof of existence and stability.

As Winfree has pointed out [20], the first two criticisms are answered by a 2-dimensional fold-catastrophe model as follows (cf. [23, Figures 7, 9]). Let $Y = R^2$, and let g be given by

$$D, \text{ fast dynamic:} \quad \dot{x} = -(x^3 - 3x + a),$$
$$F, \text{ slow feedback:} \quad \dot{a} = \varepsilon(x - \lambda),$$

where ε, λ are constants and ε is small. For the Belousov reagent choose $\lambda < 1$, and for the Z-reagent $\lambda > 1$. Then by [23] the decrease of the parameter λ past the value $\lambda = 1$ gives the Hopf bifurcation. The resulting flows are illustrated in Figure 8, with the catastrophe slow manifold shown dotted. For the Belousov reagent a theorem of Kopell and Howard [6] ensures the existence and stability of closed orbits for equation (3) near the Van der Pol attractor, but only provided diffusion is sufficiently small. For the Z-reagent Conley and Carpenter [2] have proved the existence, but not yet the stability, of homoclinic and closed orbits, provided ε is sufficiently small, for the case of a large diffusion of x. What is needed is to handle both cases together and prove stability for large diffusion, giving an estimate on ε, the ratio of slow/fast. Then extend the results to the cusp-catastrophe [23, Example 8]. Finally identity the equations with the explicit chemical reactions, make a quantitative model, and predict the speeds of the various waves.

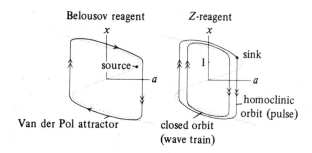

FIGURE 8

References

1. V. I. Arnold, *Singularities of differentiable functions*, these PROCEEDINGS.
2. G. A. Carpenter, *Travelling wave solutions of nerve impulse equations*, Thesis, University of Wisconsin, Madison, Wis., 1974.
3. D. H. Fowler, *The Riemann-Hugoniot catastrophe and Van der Waal's equation*, Towards a Theoretical Biology 4, edited by C. H. Waddington, Edinburgh Univ. Press, 1972, pp. 1–7.
4. C. Hall, P. J. Harrison, H. Marriage, P. Shapland and E. C. Zeeman, *A model for prison disturbances*, Jour. Math. and Stat. Psychology (to appear).
5. C. A. Isnard and E. C. Zeeman, *Some models for catastrophe theory in the social sciences*, Use of Models in the Social Sciences, edited by L. Collins, Tavistock, London, 1974.
6. N. Kopell and L. N. Howard, *Pattern formation in the Belousov reaction* (A.A.A.S., 1974, Some Math. Questions in Biology, VIII), Lectures on Math. in the Life Sci., vol. 7, Amer. Math. Soc., Providence, R. I., 1974, pp. 201–216.
7. K. Lorenz, *On aggression*, 1963; English transl., Methuen, London, 1967.
8. J. N. Mather, *Right equivalence*, Warwick University, 1969 (preprint).
9. T. Poston and A. E. R. Woodcock, *Zeeman's catastrophe machine*, Proc. Cambridge Philos. Soc. **74** (1973), 217–226.
10. ———, *A geometrical study of the elementary catastrophes*, Lecture Notes in Math., vol. 373, Springer-Verlag, Berlin, 1974.
11. L. S. Shulman and M. Revzen, *Phase transitions as catastrophes*, Collective Phenomena **1** (1972), 43–47.
12. F. Takens, *Constrained differential equations*, Dynamical Systems, Warwick, 1974, edited by A. K. Manning, Lecture Notes in Math., vol. 468, Springer-Verlag, Berlin, pp. 80–82.
13. R. Thom, *Topological models in biology*, Topology **8** (1969), 313–335. MR **39** #6629.
14. ———, *Stabilité structurelle et morphogénèse*, Benjamin, New York, 1972.
15. ———, *Phase-transitions as catastrophes*, Conf. on Stat. Mechanics, Chicago, Ill., 1971.
16. ———, *A global dynamical scheme for vertebrate embryology* (A.A.A.S., 1971, Some Math. Questions in Biology, IV), Lectures on Math. in the Life Sci., vol. 5, Amer. Math. Soc., Providence, R. I., 1973, pp. 1–45.
17. D. J. A. Trotman and E. C. Zeeman, *The classification of elementary catastrophes of codimension* ≤ 5, Lecture Notes, Warwick University, 1974.
18. G. Wassermann, *Stability of unfoldings*, Lecture Notes in Math., vol. 393, Springer-Verlag, Berlin, 1974.
19. ———, *(r, s)-stability of unfoldings*, Regensburg Universität, 1974 (preprint).
20. A. T. Winfree, *Spacial and temporal organisation in the Zhabotinsky reaction*, Aakron Katchalsky Memorial Sympos., Berkeley, Calif., 1973.
21. ———, *Rotating chemical reactions*, Scientific American **230** (1974), 82–95.
22. E. C. Zeeman, *Geometry of catastrophes*, Times Literary Supplement, 1971, pp. 1556–1557.
23. ———, *Differential equations for heartbeat and nerve impulse*, Dynamical Systems, edited by M. M. Peixoto, Academic Press, New York, 1973, pp. 683–741.
24. ———, *A catastrophe machine*, Towards a Theoretical Biology 4, edited by C. H. Waddington, Edinburgh Univ. Press, 1972, pp. 276–282.
25. ———, *Applications of catastrophe theory*, Manifolds, Tokyo, 1973, Univ. Tokyo Press, 1975, pp. 11–23.
26. ———, *On the unstable behaviour of stock exchanges*, J. Math. Economics **1** (1974), 39–49.
27. ———, *Primary and secondary waves in developmental biology* (A.A.A.S., 1974, Some Math. Questions in Biology, VIII), Lectures on Math. in the Life Sci., vol. 7, Amer. Math. Soc., Providence, R. I., 1974, pp. 69–161.

BIOLOGICAL SCIENCES

BIOLOGICAL SCIENCES

Paper		Page
3.	Differential equations for the heartbeat and nerve impulse.	81
4.	Primary and secondary waves in developmental biology.	141
5.	A clock and wavefront model for the control of repeated structures during animal morphogenesis (with J. Cooke).	235
6.	Gastrulation and formation of somites in amphibia and birds. (Addendum by R. Bellairs.)	257
7.	Dialogue between a Biologist and a Mathematician.	267
8.	Brain modelling.	287
9.	Duffing's equation in brain modelling.	293 - 300.

Paper 3 concerns dynamical systems in which there are intrinsically two time scales, a fast time and a slow time; such systems fit naturally within the framework of catastrophe theory. In particular systems that display a stable equilibrium, a threshold for triggering an action, and a return to equilibrium, are modelled by flows on the cusp catastrophe. The models are applied to the heartbeat and the nerve impulse.

Papers 4 - 7 are a series of four papers on embryology. The central idea, suggested by a cusp catastrophe model, is that when a developing piece of tissue differentiates into two types of cell, then the frontier formed between those two types moves before stabilising. This movement is called a primary wave, and may in fact be an invisible wave of cell determination; however it may manifest itself several hours later through local secondary effects, causing a visible secondary wave of morphogenesis or organogenesis. In paper 4 the model is introduced, and then applied to amphibian gastrulation and the formation of somites, and to the culmination of slime mould fruiting bodies. Paper 5, written in collaboration with Jonathan Cooke, compares this model for somite formation with other models, evaluates it in the light of experimental evidence, and describes some of his own experiments testing the model. Paper 6 extends the model from amphibia to birds, and is followed by an Addendum by Ruth Bellairs, describing supporting experimental evidence. Paper 7 discusses the philosophy of catastrophe modelling, contrasts the languages of biology and mathematics, explains the choice of hypotheses for this particular model, and describes some of the successful predictions resulting from it. This paper is dedicated to C.H. Waddington.

Paper 8 outlines a general approach to brain modelling, and paper 9 elaborates a simple example of this approach. Assuming the brain to be a collection of coupled oscillators, and knowing that non-linear oscillators like Duffing's equation can bifurcate according to elementary catastrophes, then the oscillators of the brain can be expected to exhibit sudden amplitude jumps and sudden phase-shifts. Various psychological phenomena are discussed in the light of this expectation, including manic-depression.

3 Differential equations for the heartbeat and nerve impulse

We abstract the main dynamical qualities of the heartbeat and nerve impulse, and then build the simplest mathematical models with these qualities. The results are flows on R^2 and R^3, which are related to Thom's cusp catastrophe [1–3]. In the case of the heart we explain how many of the complexities of the beat can be deduced from a simple behaviour of the muscle fibre. In the case of the nerve impulse we fit the model to the experimental data of Hodgkin and Huxley [4, 5]. The model provides an alternative to the latter's own equations, and suggests alternative underlying chemistry.

In these two examples catastrophe theory provides not only a better conceptual understanding, by giving a single global picture that enables an overall grasp of the phenomenon, but also provides explicit equations for testing experimentally. The novelty of the approach lies in modelling the *dynamics* (which is relatively simple) rather than the *biochemistry* (which is relatively complicated). This approach might be useful for a large variety of phenomena in biology, whenever there is a trigger mechanism leading to some specific action. In an appendix we suggest how it might be applied to evolution.

Mathematically the equations that we derive are interesting, because they are generalizations of the Van der Pol and Liénard equations.

Part One

1.1. THREE QUALITIES

The three dynamic qualities displayed by heart muscle fibres and nerve axons are:
(1) stable equilibrium;
(2) threshold, for triggering an action;
(3) return to equilibrium.
The third quality can be divided into two cases according as to whether or not the return is smooth:
(3a) jump return (heart);
(3b) smooth return (nerve).
In the first half of the paper our objective will be to take these three qualities as 'axioms', and derive mathematical models by means of qualitative mathematical

Published in Towards a Theoretical Biology (Edited C.H. Waddington) Edinburgh University Press, Volume 4, 1972 pages 8-67.

argument. We state the models in Theorems 1, 2 in sections 1.7 and 1.10 respectively, and apply them *qualitatively* to heart and nerve in sections 1.8 and 1.11. In order to keep the mathematics as translucent as possible, we only touch upon the physiology where necessary.

In the second half of the paper the models are developed *quantitatively* into a form that could be used for prediction, and for this we need to go more deeply into the physiology and numerical data. In sections 2.1 to 2.3 we discuss the heart, and in sections 2.4 to 2.10 the nerve, concluding both cases by suggesting how the models might be used in experiments. But first we must explain why we chose these three qualities.

There are two states to the heart, the *diastole* which is the relaxed state, and the *systole* which is the contracted state. If the heart stops beating it stays relaxed in *diastole*, which is therefore the stable equilibrium of Quality 1. What makes the heart contract into systole is a global electrochemical wave emanating from a pacemaker. As the wave reaches each individual muscle fibre, it triggers off the action of Quality 2, that makes the fibre rapidly contract. Each fibre remains contracted in the systole state for about 1/5 second, and then rapidly relaxes again—in other words the fibre obeys the jump return to equilibrium of Quality 3a. Therefore the three qualities describe the *local* behaviour for an individual muscle fibre, and our first objective is to describe this by an ordinary differential equation, which we achieve in Theorem 1. In section 2.1 we deal with the *global* structure of the heart and its four chambers, and in section 2.2 describe the pacemaker wave by a separate geodesic flow.

Now for the nerve impulse. A *neuron*, or nerve cell is in some ways like a tree (*see* Figure 1).

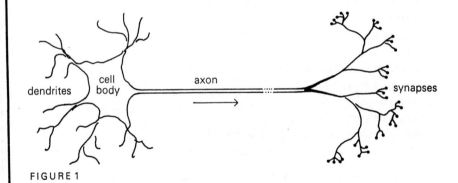

FIGURE 1

Corresponding to the roots of the tree are the *dendrites*, through which the neuron receives messages. Out of the cell body grows the *axon*, which is like the trunk of the tree. Just as the trunk of a tree divides into many branches, ending in thousands of leaves, so the axon divides into many branches, ending in thousands of *synapses*, which touch the dendrites of other neurons. It is the axons that connect up the whole nervous system, and comprise the white matter of our brains, while the dendrites and cell bodies comprise the grey matter. Messages travel along the axon to the synapses, and are thus transmitted to other neurons. This is the basic mechanism underlying the working of the brain, and so it is interesting to examine the nature of the message. A message consists of a series of *spikes* (*see* Figure 2).

Each spike is an electrochemical phenomenon that lasts about 1 msec ($=10^{-3}$ sec), and travels very fast along the axon. The velocity is between 10 and 100 metres per second, depending upon the type of nerve. The simplest way to observe a spike is to put electrodes inside and outside the axon in order to measure the potential difference v of the inside relative to the outside, and then plot v against time (*see* Figure 3). While no message is passing the membrane surrounding the axon is polarized, and v remains constant about -65 mV (1 mV=10^{-3} volts). This constant is called the *resting potential*, and represents the stable equilibrium of Quality 1. As the spike moves along the axon it triggers off a rapid depolarization of the membrane, called the *action potential*, which causes v to swing to $+40$ mV, representing the action of Quality 2. When the spike has passed, the membrane repolarizes relatively slowly, representing the smooth return to equilibrium of Quality 3b. These features can be seen in Figure 3, which is taken from Hodgkin [5, p. 64]. Notice that the potential V shown in Figure 3 is relative to the resting potential,

$$V = v + 65$$

Therefore the resting potential is $V=0$, while the maximum action potential is $V=115$. This notation will be convenient later. The lower graph in Figure 3 is experiment, whilst the upper graph is given by the Hodgkin–Huxley model. The analogous graph given by our model is shown in Figure 26 below (see p. 60), with a calculated velocity of 21·9 m/sec.

As in the case of the heart muscle fibre, the three qualities represent the *local* behaviour at a single point of the axon, and our first objective will be to represent this by an ordinary differential equation in section 1.11. Later we shall incorporate into the model the *global* propagation wave by a partial differential equation ; we shall find in section 2.8 Theorem 3 an explicit solution for the wave, and show

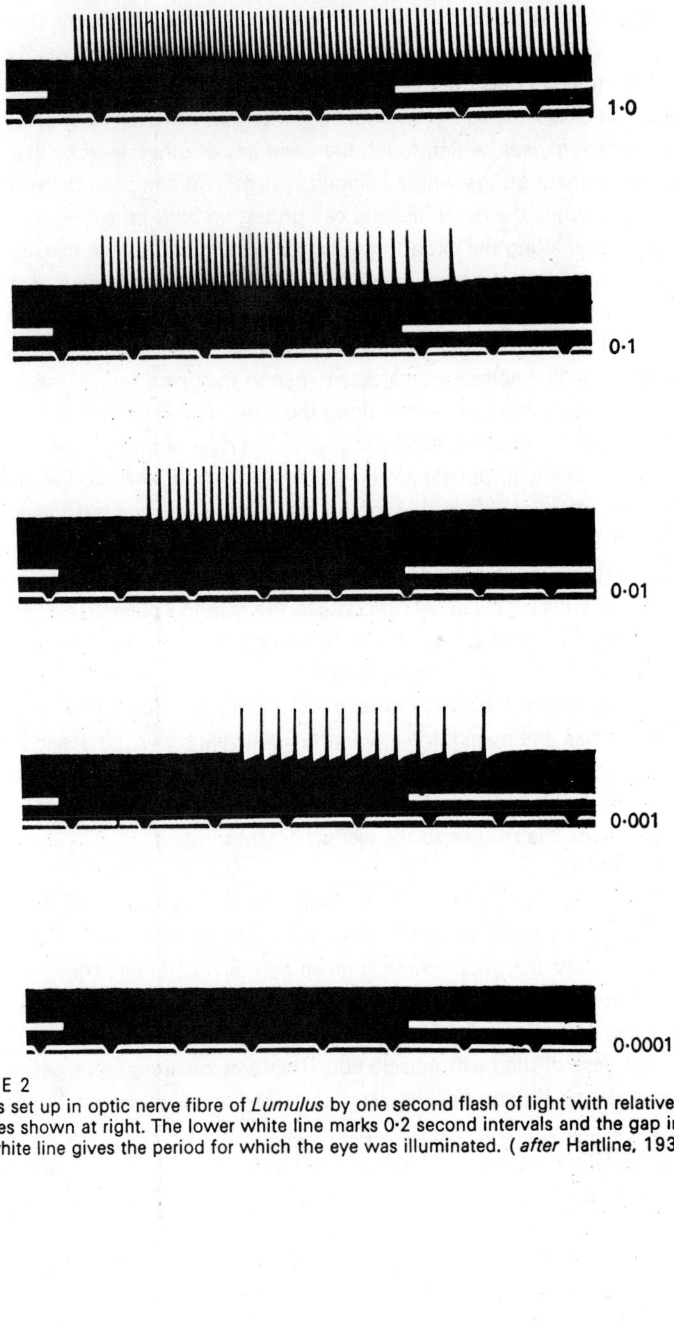

FIGURE 2
Impulses set up in optic nerve fibre of *Lumulus* by one second flash of light with relative intensities shown at right. The lower white line marks 0·2 second intervals and the gap in the upper white line gives the period for which the eye was illuminated. (*after* Hartline, 1934)

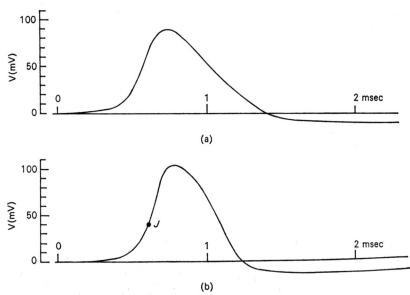

FIGURE 3
Propagated action potentials in (a) theoretical model, and (b) squid axon at 18·5°C. The calculated velocity was 18·8 m/sec and the experimental velocity 21·2 m/sec.

how it triggers the local action. In section 2.4 we emphasize the importance of keeping the local and global ideas separate.

One word of warning about nomenclature : in the literature the phrase *action potential* is used ambiguously in three different senses ; first as the value $V = 115$, secondly as the graph pictured in Figure 3, and thirdly as incorporating all the accompanying biochemical events as well. We shall tend to use it in the second sense, and use the phrase *nerve impulse* for the third sense. We shall also use the word *action* in an abstract mathematical sense (*see* section 1.6 and Figures 8 and 13). In Figure 3 the action takes place during the steep ascent.

1.2. DYNAMICAL SYSTEMS ON THE PLANE R^2

In this section we shall start from the three qualities, and derive mathematical models. We shall proceed by giving a sequence of examples, starting with trivial examples that do not possess the desired qualities, and gradually making them more complicated step by step, until they do. We shall show that each step is necessary in order to achieve the qualities, and in this way we shall arrive at the simplest model.

First we try and represent the qualities by a differential equation in the Euclidean plane, R^2. (The words *dynamical system, differential equation, flow, vector field* all mean the same thing.) We shall show eventually that using R^2 implies 3a, and for 3b it will be necessary to use R^3. However in this section we confine ourselves to R^2. Quality 1 requires a fixed point, which we can take to be the origin $0 \in R^2$. Let

$$\dot{v} = Av$$

be the linear approximation to the differential equation at 0, where A is a real 2×2 matrix. Since the equilibrium is stable, the two eigenvalues of A each have negative real part. We shall now give an argument to show that one has large negative real part, and the other small negative real part, implying that they must both be real.

Consider Quality 2 : what is an *action* ? It is something that is noticeable, and therefore must involve a reasonably large change in one of the coordinates, and take a reasonably short time. Therefore an action must be represented by large vectors of the vector field, or in other words by part of an orbit along which the flow is fast. More briefly we shall call this a *fast orbit*. Quality 2 implies there is a fast orbit near O, and the presence of large vectors near O means that one of the eigenvalues must have large modulus. We call this the *fast eigenvalue*. Meanwhile it must be large in comparison with something else, and so the other eigenvalue must have relatively small modulus ; we call this the *slow eigenvalue*.

▶ *Example 1. Trivial linear case.* Let x, b be coordinates in R^2. Admittedly it is unusual to use letters at opposite ends of the alphabet, but this will emphasize the difference between fast and slow, and will be later useful notation for the cusp catastrophe (section 1.10). Choose a small positive constant ε.

The equations are

$$\varepsilon \dot{x} = -x$$
$$\dot{b} = -b \ .$$

Therefore the matrix

$$A = \begin{pmatrix} -\dfrac{1}{\varepsilon} & 0 \\ 0 & -1 \end{pmatrix}$$

The fast eigenvalue is $-\dfrac{1}{\varepsilon}$, and we call $\varepsilon \dot{x} = -x$ the *fast equation*, because x decays exponentially fast by $e^{-t/\varepsilon}$. The slow eigenvalue is -1, and we call $\dot{b} = -b$ the *slow equation*, because b decays more slowly by e^{-t}. The orbits are shown

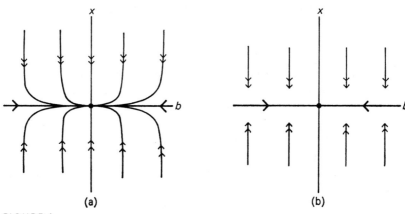

FIGURE 4

in Figure 4a, but we shall find it more convenient to represent the flow symbolically by Figure 4b, for reasons which we shall now explain.

1.3. THE SLOW MANIFOLD AND THE FAST FOLIATION
In all the examples we shall give, the fast equation will always be of the form $\varepsilon \dot{x} = -f(x, b)$, where $f(x, b)$ is a function of x, b that vanishes at the origin. We define the *slow manifold* by putting $\varepsilon = 0$; in other words the slow manifold is the curve through the origin given by $f(x, b) = 0$. In Example 1 the slow manifold is given by $x = 0$, namely the b-axis. In our diagrams we shall always indicate the slow manifold by a thick line with single arrows. Meanwhile define the *fast foliation* to be the family of lines parallel to the x-axis. Except in the neighbourhood of the slow manifold the vector field is approximately parallel to the fast foliation, because the x-component dominates the b-component. Therefore in our diagram we shall always indicate the fast foliation by double arrows parallel to the x-axis, oriented towards x positive where $f < 0$, and oriented in the opposite direction where $f > 0$. If equilibrium is disturbed, we can approximately describe the way the system returns to equilibrium as follows: *the system first homes rapidly down the fast foliation towards the slow manifold, and then homes relatively slowly along the slow manifold back to equilibrium.*

As $\varepsilon \to 0$ the fast foliation speeds up, and the system homes more rapidly towards the slow manifold. In the limit $\varepsilon = 0$, the system jumps instantaneously onto the slow manifold. Therefore, in effect, the system is confined to the slow

manifold, where its behaviour is governed by the slow equation. Summarizing we can say:
(1) the fast foliation is parallel to the x-axis;
(2) the fast equation determines the slow manifold;
(3) the slow equation determines behaviour on the slow manifold.

We emphasize that these statements are precise when $\varepsilon=0$, and first order approximations in ε when $\varepsilon \neq 0$. They are justified by the following lemma.

▶ *Lemma 1*

The tangents at the origin to the fast foliation and the slow manifold are, to first order in ε, the eigenspaces of the fast and slow eigenvalues, respectively.

Corollary 1. The slow manifold is transversal to the fast foliation in the neighbourhood of the origin.

Corollary 2. The slow manifold is an approximation of the invariant submanifold determined by the slow eigenvalue.

Proof. The matrix for the linear approximation of the differential equation at the origin is

$$A = \begin{pmatrix} -\dfrac{a_{11}}{\varepsilon} & -\dfrac{a_{12}}{\varepsilon} \\ a_{21} & a_{22} \end{pmatrix}$$

where $f(x, b) = a_{11}x + a_{12}b +$ higher terms,

$$b = a_{21}x + a_{22}b + \text{higher terms},$$

and each $|a_{ij}|$ is small compared with $\dfrac{1}{\varepsilon}$. Since both eigenvalues are negative, one large and one small, the trace of A must be large negative, and so $a_{11} > 0$. Hence to first order in ε, the fast and slow eigenvalues are $\dfrac{a_{11}}{\varepsilon}$ and $\dfrac{a_{11}a_{22} - a_{12}a_{21}}{a_{11}}$ respectively, with corresponding eigenvectors $(1, 0)$ and $(a_{12}, -a_{11})$ which are tangents to the fast foliation and slow manifold, as required. Corollary 1 follows from the fact that $a_{11} \neq 0$, and Corollary 2 from the lemma.

We now illustrate these ideas with some more examples.

▶ *Example 2*

$$\varepsilon \dot{x} = -(x+b)$$
$$\dot{b} = -b$$

Here the slow manifold is $x+b=0$. The reader can verify that the precise invariant submanifold determined by the eigenvalue -1 is in fact the line $(1-\varepsilon)x + b = 0$, to which the slow manifold is an approximation.

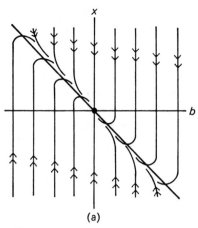 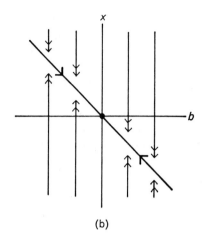

(a)   (b)

FIGURE 5

▶ *Example 3*
$$\varepsilon \dot{x} = -(x+b)$$
$$\dot{b} = x$$

The change from Example 2 to Example 3 is slight but subtle. The slow manifold is the same because the fast equation is unchanged. The behaviour on the slow maniiold is the same, because on the slow manifold $-b=x$, and so the slow equation, which determines the behaviour, is in effect the same. Thus they have the same picture, Figure 5. When we reach Example 6 we shall explain why the change was mathematically necessary at this stage in order to later achieve Quality 3. (*See* Remark after Theorem 1 in section 1.7.) Biologically that change has subtle implications, and it is worthwhile digressing for a moment to explain why.

1.4. BIOLOGICAL DIGRESSION

When a cell develops in the embryo its initial task is to achieve an identity, as a nerve cell, or liver cell, or whatever. Unless it does so, the species will become unstable and die out. This implies a very strong homeostasis. Therefore if we model the underlying chemistry by a dynamical system on R^n (with n perhaps very large), the equilibrium point must be very stable, and so all the eigenvalues have large negative real part. Once a cell has achieved its identity, its next task is to develop its dynamic, specific to its job: a nerve cell must conduct impulses, a liver cell must manufacture glucose, and so on. To do this the cell must weaken its equilibrium, so that some of the internal chemistry can vary according to a very specific sequence of reactions, while the rest of the metabolism is directed at holding the identity in a tight homeostasis. Mathematically this is represented in the dynamical system by a slowing down of one eigenvalue, allowing the system to vary along a specific 1-dimensional slow manifold M. Meanwhile the homeostasis is represented by a fast foliation of $(n-1)$-dimensional disks transversal to M, which rapidly damps down any perturbations off M. For instance we could represent this situation by adjoining to Examples 1, 2, or 3 another $n-2$ fast equations of the form $\varepsilon\dot{y}=-y$.

Suppose now that each coordinate represents the concentration of some enzyme, or the productivity of some enzyme-system (relative to equilibrium). Then Example 1 describes a metabolism in which each enzyme-system is an independent self-correcting mechanism for stabilizing its concentration. In Example 2 the enzyme-systems are no longer independent, because, although b is still slowly self-correcting, any perturbation of b causes a rapid change of x, and x only returns slowly to normal with b. In Example 3 the subtle difference is that the two enzyme-systems become linked : in other words b is no longer self-correcting, but a perturbation of b induces a rapid response in x, which is then the mechanism for restabilizing b. The mathematics will show that it is necessary to develop a linkage between 2 or 3 enzyme-systems before a cell can develop the three desired qualities. The same remarks apply not only to development, but also to evolution (*see* section 2.11). This ends the digression.

1.5. NON-LINEAR EXAMPLES ON R^2

▶ *Example 4*

$$\varepsilon\dot{x}=-(x^3+x+b)$$
$$\dot{b}=x$$

The interest of this example is the non-linearity. At the origin Example 4 has Example 3 as its linear approximation, and so the local behaviour is as before. Globally the slow manifold is curved, and away from the origin the situation is not unlike Example 1. From the point of view of the orbit structure, all the first four examples are diffeomorphic. Figure 6a shows the orbit structure, and Figure 6b the symbolic diagram.

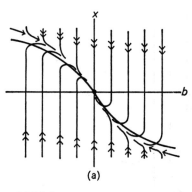 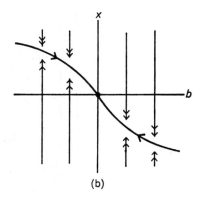

(a)   (b)

FIGURE 6

▶ *Example 5.* Liénard and Van der Pol.
$$\varepsilon\dot{x} = -(x^3 - x + b)$$
$$\dot{b} = x$$

Example 5 is obtained from Example 4 by a change of sign in one of the terms of the fast equation. This has the effect of bending the slow manifold into the S-shaped cubic curve, as shown in Figure 7. In so doing we have lost Quality 1, because the origin has switched from being an attractor to a repellor (from being a sink to a source). However, mathematically the example has several points of interest. Not only has the origin switched, but, from the point of view of the foliation, the entire middle piece of the slow manifold has switched from being an attractor to a repellor—that is why we have indicated it by a dashed line in Figure 7b. At the same time an attracting closed orbit (or limit cycle) has appeared, as shown in Figure 7a. The flow is diffeomorphic to that of the Van der Pol equation. In fact if we eliminate b by differentiating the fast equation and substituting the slow equation, we obtain a single second-order equation in x

$$\varepsilon\ddot{x} + (3x^2 + 1)\dot{x} + x = 0$$

which is an example of the Liénard equation and differs only from the Van der Pol equation by a change of coefficients.

The points T, T' where the fast foliation is tangent to the slow manifold are given by $3x^2 = 1$. Let M' denote the attracting part of the slow manifold, which is given by the equation and inequality

$$x^3 - x + b = 0, \quad 3x^2 \geq 1 \quad .$$

3. Differential Equations for the Heartbeat and Nerve Impulse

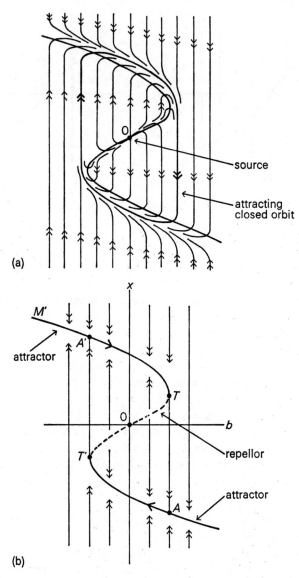

FIGURE 7

If we speed up the fast equation by letting $\varepsilon \to 0$, then in the limit $\varepsilon = 0$ the system is confined to M', and is given by the single equation

$$(3x^2 - 1)\dot{x} + x = 0$$

which is none other than the equation of relaxation oscillations of electrical engineering. This approach explains why the domain M' of relaxation oscillations is disconnected, and Figure 7 illustrates the instantaneous jumps $T \to A$, $T' \to A'$ that occur during the oscillation cycle $TAT'A'$ each time the boundary of M' is reached.

The change from Example 4 to Example 5 can be made continuous, for write the fast equation as $\varepsilon \dot{x} = -(x^3 + ax + b)$, and allow the parameter a to run from $+1$ to -1. The resulting change in the flow is none other than Hopf bifurcation [6]. When applied to evolution such types of bifurcation can look remarkably Lamarckian (*see* section 2.11).

Although Example 5 seems to have diverted us from our main theme, because we have lost Quality 1, nevertheless it will transpire that Example 5 is useful in several other respects. First, it gives insight into the points of tangency of fast foliation and slow manifold, where the latter changes quality from attractor to repellor; we shall need to study these next in order to capture Quality 2. Secondly it explains, as we have seen, the close relationship to the classical equations. Finally the closeness of Example 5 and Example 6 has interesting biological implications, as follows. A small chemical change inside a heart cell (Example 6) or nerve cell could convert the cell into an oscillator (Example 5). In particular such a change might occur during normal metabolism if the cell is not used. This would furnish a simple explanation of ectopic pacemakers in the heart (secondary pacemakers that arise spontaneously if the main one fails) and self-firing neurons (such neurons occur in the visual cortex – *see* section 2.10, paragraph 7).

In skeletal muscle cells such changes might be caused by fatigue and stress, and cause oscillation such as the sewing-machine legs of a climber, or the muscular tremors of a religious dancer.

1.6. THRESHOLD

We tackle the problem of incorporating Quality 2 into the differential equation. This means giving a mathematical interpretation of the words 'threshold, for triggering an action'. We have already interpreted *action* to mean a piece of fast orbit, sufficiently long to be noticeable, and this has given rise to the discussion of fast and slow. What is a *trigger*? We suggest that a trigger is a specific perturbation away from the equilibrium position, imposed by an external agent (external to the differential equation under discussion). We have already considered the local behaviour near the origin: given a small perturbation the system returns by fast orbit to the slow manifold, and then returns slowly to equilibrium. Such local behaviour can hardly be called an action, because the length of fast orbit involved is less than the size of the trigger, whereas the words 'trigger' and 'action' usually carry the overtone that the action is noticeably larger than the trigger. Certainly this is the case for both heart and nerve.

Now the slow manifold is transversal to the fast foliation near the origin, by Corollary 1 to Lemma 1, and this local behaviour persists as long as it remains transversal. Therefore the non-local behaviour required by Quality 2 implies the existence of a point T of non-transversility, where the fast foliation is tangent to the slow manifold. We define this point to be the *threshold*. We can assume the tangency at T is generic, that is to say quadratic (as in Figures 7a and 8), because otherwise, if it were cubic or higher, this would be a non-generic situation, unstable, and removable by an arbitrarily small perturbation of the differential equation. Since the tangency at T is generic, then by continuity of the fast foliation, the slow manifold must change from attractor to repellor at T. Summarizing, we have shown:

▶ *Lemma 2*

A dynamical system in R^2 possessing Qualities 1 and 2 must have the qualitative picture shown in Figure 8 near the equilibrium position.

▶ *Remark 1*. Notice that Figure 8 differs from Figure 7b in the orientation of the flow on the slow manifold near T.

▶ *Remark 2*. Why not bend the fast foliation rather than the slow manifold? Differentiably it would be equivalent, but there is a biological reason for not doing so. To bend the fast foliation would need large first partial derivatives of the vector field, and ultimately these are bounded by the energy available in the underlying chemistry. On the other hand to bend the slow manifold requires comparatively little energy. It is the energy bound on the partial derivatives that makes the fast orbits approximately parallel, and justifies extending the fast foliation outside the immediate neighbourhood of 0.

1.7. SOLUTION IN THE CASE OF THE JUMP RETURN

Finally we must elaborate Figure 8 so as to incorporate Quality 3, the return to equilibrium. The only way to stop a fast orbit is to catch it at the bottom point A with an attracting piece of slow manifold. Then the only way to achieve a smooth

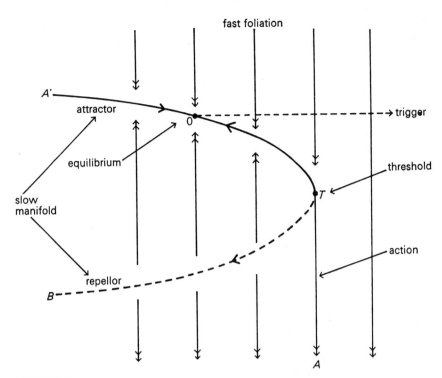

FIGURE 8

return would be to glue this piece of slow manifold back onto the loose end at A'. However this is impossible because in so doing we should have to bend the slow manifold round and unwittingly create another threshold point T', where, not only would the slow manifold switch from attractor to repellor and be impossible to glue onto A', but also we should shoot off onto another piece of fast orbit T'A', disrupting the smooth return. Therefore we have to make a jump return, and :

▶ *Lemma 3*

In R^2 a smooth return is impossible.

Having admitted that another threshold T' is inevitable, the simplest way of constructing a slow manifold with all the desired properties is to glue the resulting new piece of repellor onto the loose end B. The simplest algebraic equation of such a curve is the cubic that we used in Example 5 and Figure 7, namely

$x^3 - x + b = 0$. The only difference from Example 5 is the behaviour on the slow manifold, which is determined by the slow equation. All we need to do is to move the equilibrium point from the repellor part to the attracting part of the slow manifold, that is to say from $x=0$ to $x=x_0$, say, where $x_0 > \frac{1}{\sqrt{3}}$.

Summarizing, we have shown:

▶ **Theorem 1**

There exists a dynamical system on R^2 possessing Quality 1, 2, and 3a. The simplest example is the following:

▶ **Example 6.** (Heartbeat.)
$$\varepsilon \dot{x} = -(x^3 - x + b)$$
$$\dot{b} = x - x_0$$

where x_0 is a constant greater than $\frac{1}{\sqrt{3}}$.

The orbit structure and symbolic diagram are shown in Figure 9, which can be compared with Figure 7.

▶ **Remark.** Let (x_0, b_0) denote the equilibrium point. Consider the effect of replacing the slow equation by $\dot{b} = -(b - b_0)$. If $|b_0| < \frac{2}{3\sqrt{3}}$ then the line $b = b_0$ meets the slow manifold in three points, two sinks and one source. After the action, the system would be caught by the other sink, violating Quality 3. Therefore this would not be a solution to the problem. This explains why it was necessary to pass from Example 2 to Example 3 in our development.

1.8. APPLICATION TO THE HEARTBEAT

Having obtained the solution in the case of the jump return, we must now test Example 6 as a model for the heartbeat. That is to say, we must identify the variables x, b with measurable qualities and use the differential equation for prediction. Obviously x is going to be the length of the muscle fibre (+ a constant), so that the action represents the contraction, and the jump return represents the relaxation. Meanwhile b is going to be some form of electro-chemical control, and there may be several ways of measuring b. The pacemaker wave will change the control from b_0 to b_1, say, thus triggering off the heartbeat cycle as shown by the dotted line in Figure 10.

We shall go into this in more detail in section 2.1, because the tension in the fibre caused by the blood pressure turns out to be a second important control variable. Meanwhile we observe that one of the simplest ways of measuring b may be to measure the potential across the membrane of the muscle fibre (see

3. Differential Equations for the Heartbeat and Nerve Impulse

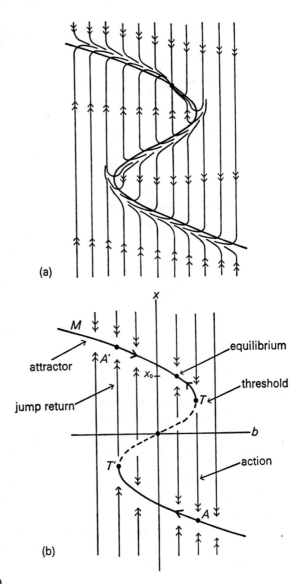

FIGURE 9

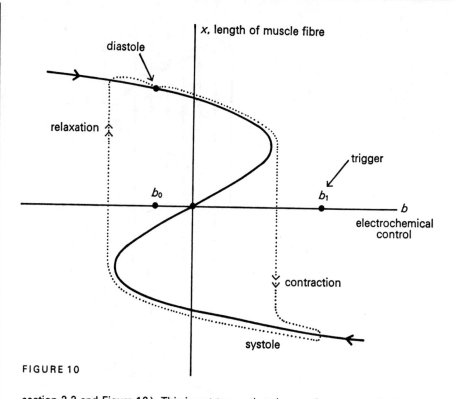

FIGURE 10

section 2.2 and Figure 18). This is not to say that the membrane potential is necessarily a fundamentally important part of the muscle contraction, but only that it may be a convenient artifact for measuring the chemical control, in the same way that colour is a convenient artifact used by a blacksmith to measure the temperature of an iron bar. Alternatively it may be better to measure b by a concentration or a flow of some particular chemical. Indeed we know that muscle chemistry is very complicated, involving not only membrane depolarization, but also changes in permeability, transport of ions, synthesis of actomyosin, breakdown of ATP (adenosine triphosphate) to release the energy for the contraction, amongst other things. Even if we knew the whole story, and were able to represent it by a vast dynamical system in R^n, yet the heartbeat cycle would still be represented by a 1-dimensional path in R^n, consisting of 2 pieces of slow orbit and 2 pieces of fast orbit. And if we chose convenient coordinates x, b such that the projection of the path into the (x, b)-plane was an embedding, then these two coordinates would be sufficient for experimental predictions about the cycle. In fact if our objective is to study the dynamics of the heartbeat rather than the underlying chemistry (which nature may have made unnecessarily complicated), then it may be more efficient to use the R^2-system based on the dynamics, rather than some R^n-system based on approximations to the chemistry.

This completes the discussion on the jump return 3a, and we now turn to the problem of the smooth return 3b.

1.9. DYNAMICAL SYSTEMS ON R^3

We have seen in Lemma 3 that a smooth return is impossible in R^2, and so we try R^3. First observe there can be only one fast eigenvalue. For otherwise, if there were two fast eigenvalues, then the fast foliation would be a family of planes, while the slow manifold would be a curve, as before, and when we consider the points of tangency we should run up against the same argument as before, implying a jump return.

Therefore, for a smooth return there must be only one fast eigenvalue which we choose to be $-1/\varepsilon$, as before, and two slow eigenvalues, which may be real or complex. In the simplest solution it will turn out that they are complex. Choose coordinates x, a, b for R^3. The fast foliation will be the family of lines parallel to the x-axis, while the slow manifold will be a surface cutting the fast foliation transversally near the equilibrium point. First a trivial linear example, analogous to Example 1.

▶ Example 7

$$\varepsilon \dot{x} = -x$$
$$\dot{a} = -a + b$$
$$\dot{b} = -a - b$$

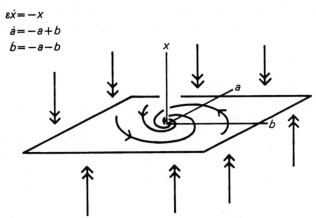

FIGURE 11

The single fast equation determines the slow manifold, which is the plane $x=0$. The pair of slow equations determines that flow on the slow manifold, which is spiralling in towards the origin, because the eigenvalues are $-1 \pm i$. Putting $z = a + ib$, the equations can be written $\dot z = -(1+i)z$, with solution $z = z_0 e^{-(1+i)t}$. It is the spiral quality associated with complex eigenvalues that we shall find useful in negotiating the smooth return.

To find the solution in the case of the jump return we laboriously argued our way from Example 1 to Example 6. Now in the case of the smooth return we can telescope these arguments and proceed straight from Example 7 to finding the solution in Example 8.

Let M denote the slow manifold, which will be a smooth surface in R^3. Choose a constant $k<0$, and let C be the plane $x=k$. Then C is perpendicular to the fast foliation. We can use (a, b) as coordinates for C. The letter C stands for *control space*, because sometimes it is convenient to think of a, b as parameters or controls. Let $\chi : M \to C$ denote the orthogonal projection of M onto C. The only reasons for choosing $k<0$ is that in diagrams it is convenient to draw C as a horizontal plane below M, and not meeting M.

As before, a *threshold* T is a point of tangency of M and the fast foliation, or in other words a point on a fold curve of χ. Therefore Quality 2 implies the existence of a fold curve separating a piece of slow attractor A' from a piece of slow repellor B, as shown in Figure 12. Meanwhile Quality 3 implies the existence of another piece of slow attractor A to catch the action.

Quality 3b implies that A, A' can be connected by an arc in M not crossing any fold curves, in order to achieve the smooth return. The problem is to find the simplest M with these properties. Our criterion of simplicity is to find a fast equation of the form

$$\varepsilon \dot x = -f(x, a, b)$$

where f is a polynomial of lowest degree. Therefore M will be given by the equation $f=0$. Now f must be of degree at least 3 in x, because in Figure 12 there are lines parallel to the x-axis meeting M in 3 points. Further f must be of degree at least 1 in a, b because otherwise, if f did not contain either a or b, then we should be back to the 2-dimensional problem, which is impossible by Lemma 3. These lower bounds for the degree of f turn out to be not only necessary, but sufficient, for consider the surface given by the equation

$$f \equiv x^3 + ax + b = 0 \ .$$

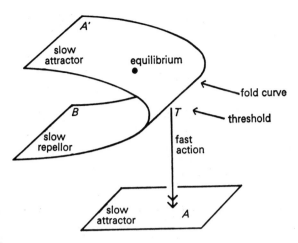

FIGURE 12

This has the required properties, as can be seen from Figure 13. The fold curves are given by:
$$\frac{\partial f}{\partial x} = 3x^2 + a = 0 \quad .$$
When the fold curves are projected onto C they form a cusp, which is obtained by eliminating x from the two equations above, giving
$$4a^3 + 27b^2 = 0 \quad .$$
Outside the cusp M is single sheeted, and inside the cusp M is 3-sheeted. The singularity of the map $\chi : M \rightarrow C$ at the origin 0 is the well-known Whitney *cusp-singularity*. Sometimes we abuse language and call the singularity itself a *cusp*, although in fact M is smooth at 0, and the cusp really exists only in C. In addition to having the Whitney singularity the surface M has the property that the upper and lower sheets are attractors, while the middle sheet is a repellor.

Let us pause for a moment to consider what we are doing. We have found a surface of lowest degree possessing all the required properties. But is this really the 'simplest' example, and is there any virtue in having found the simplest? The topologist regards polynomials as rather special, and tends to turn his nose up at so crude a criterion of simplicity as choosing a polynomial of lowest degree. Moreover in biology of all subjects we should least expect Nature to be so obliging as to use polynomial equations. So perhaps we ought to consider *all possible* surfaces. Now comes the truly astonishing fact: when we do consider

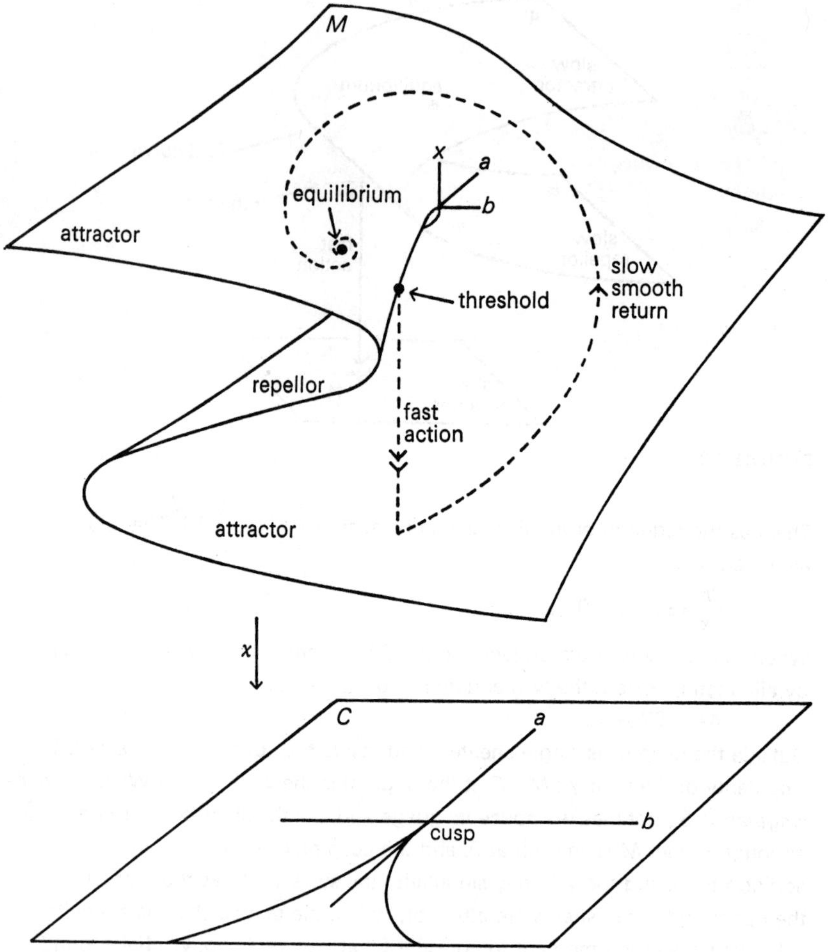

FIGURE 13

all surfaces, not only is this particular surface the *simplest* example, but in a certain sense it is the *most complicated* example, in other words it is the *unique* example. Herein lies the punch of the deep and beautiful catastrophe theory created by René Thom. We must digress to explain precisely the sense in which this surface is unique.

1.10. THE CUSP CATASTROPHE

The reader familiar with Thom's work will recognize the formulae above as canonical formulae of the cusp catastrophe. See, for instance, David Fowler's paper in this volume [7]. The cusp catastrophe is defined as follows.

Consider the potential function in the real variable x,

$$\varphi_0 = \tfrac{1}{4}x^4 \ .$$

This potential has a unique minimum at $x=0$ but is not generic, because nearby potentials can be of a different qualitative type. For instance, there are nearby potentials with two minima. Let φ denote the 2-parameter family of potentials

$$\varphi = \tfrac{1}{4}x^4 + \tfrac{1}{2}ax^2 + bx \ ,$$

where a, b are real parameters. It transpires that this family contains all the different possible types near φ_0, and therefore φ is called an *unfolding* of φ_0. Although φ_0 was not generic, it will turn out that φ is generic, in the sense defined below, and this is what justifies the name unfolding. Meanwhile consider the gradient dynamical system associated with φ,

$$\begin{aligned}\dot{x} &= -\operatorname{grad} \varphi \\ &= -\frac{\partial \varphi}{\partial x} \\ &= -(x^3 + ax + b) \ .\end{aligned}$$

This of course is none other than the fast equation considered above in Figure 13, putting the constant $\varepsilon = 1$. Our slow manifold is therefore none other than the set of stationary values of φ. In more detail

attractor surface of M = minima of φ
repellor surface of M = maxima of φ
fold curve of M-origin = cubic stationary points of φ
origin = the quartic stationary point of φ.

We define the *cusp catastrophe* to comprise the set of five things : the potential φ ; the fast equation $\varepsilon \dot{x} = -\operatorname{grad} \varphi$; the surface M ; the map $\chi : M \to C$; and the cusp-singularity of χ. We use the term ambiguously to refer to any one of these things, provided the context is clear. In a moment we shall define equivalence between potentials, and in some contexts it is convenient to use the term cusp catastrophe ambiguously to refer not only to φ, but also to the equivalence class containing φ, or any member of that class ; in such contexts we refer to φ as the *canonical* cusp catastrophe.

The reason that Thom chooses to use the word *catastrophe* is that in many applications the fast jump between the two attractor surfaces represents a discontinuity in space−time, frequently as a shock-wave (in space) or as a catastrophic change of behaviour (in time).

Consider now the space Ψ of all smooth 2-parameter families of potentials. Since we have fixed x, a, b as coordinates for R^3, any function on R^3 determines an element of Ψ. Therefore Ψ is the same as the space of smooth functions on R^3 (here smooth means that the second partial derivatives are continuous). Given $\psi \in \Psi$, the set M_ψ of stationary values of ψ is defined by $\partial \psi / \partial x = 0$. Given ψ, $\psi' \in \Psi$ define them to be *equivalent* if there exist diffeomorphisms $f: R^3 \to R^3$ and $g: C \to C$ commuting with the projection $R^3 \to C$, and such that $fM_\psi = M_{\psi'}$. Define ψ to be *generic* (or stable) if it has a neighbourhood of equivalents in Ψ. It is a theorem that the generic elements form an open-dense subspace of Ψ, and so from the point of view of applied mathematics the non-generic elements can be ignored. It is a theorem that if ψ is generic, then M_ψ is a smooth surface, and any nearby element will have a qualitatively similar surface. For example, the cusp catastrophe φ is generic and therefore any nearby ψ will have a surface M_ψ with the same qualities as M; namely two-fold curves over a cusp, single sheeted outside the cusp and 3-sheeted inside with two attractors and a repellor in the middle. Intuitively any such M_ψ can be obtained by bending M, so as not to disturb these qualities. We can now state the classification theorem.

▶ *Theorem (Thom)*

If ψ is generic, then the only singularities of the projection $M_\psi \to C$ are folds and cusps. Locally near any cusp, ψ is equivalent to the cusp catastrophe φ.

In other words φ represents the most complicated thing that can happen locally. Of course globally M_ψ may have many cusps and be many sheeted. Also if we enlarge to 3 or more parameters the situation can become more complicated [see 1, 2], but for this paper it is sufficient to stick to 2-parameters. Therefore the theorem is the key mathematical fact behind our whole approach. This ends the digression on the cusp catastrophe.

We now go back to our main problem of finding the simplest dynamical system for the smooth return. If we consider *all possible* generic fast equations, it follows from Thom's theorem that the slow manifold must be a smooth surface with only folds and cusps. It would be possible to embed Figure 12 in a surface without cusps (for consider a looped hosepipe) but this would be both topologically more complicated, and biologically more complicated, in the sense of section 1.4 as a development from Example 7. Therefore the criteria of topological and biological simplicity imply the existence of a cusp. We have seen in Figure 13 that one cusp is sufficient, and by Thom's theorem any cusp is equivalent to the

canonical cusp catastrophe. Therefore we are justified both mathematically and biologically in choosing the canonical fast equation
$$\varepsilon \dot{x} = -(x^3 + ax + b) \quad .$$

Finally we come to the slow equations. Normally the cusp catastrophe has no slow equations, because a, b are regarded merely as parameters or controls. However when we add slow equations this changes the nature of the variables a, b by giving them a dynamic role, and if x appears in the slow equations this can be regarded as a form of feedback on the cusp catastrophe.

There is nothing unique about the slow equations. In section 2.7 we shall take pains to fit the Hodgkin–Huxley data, and shall derive relatively complicated slow equations, one being an electrical capacitance equation, and the other an *ad hoc* piece of data fitting. But our present objective is to make them as simple as possible in order to provide conceptual insight. Therefore we chose the slow equations to be linear with complex eigenvalues, and we fix the coefficients so as to spiral the slow return smoothly round the cusp.

▶ *Theorem 2*

There exists a dynamical system on R^3 possessing Qualities 1, 2, and 3b. The simplest example is given in Example 8.

▶ *Example* 8
$$\varepsilon \dot{x} = -(x^3 + ax + b)$$
$$\dot{a} = -2a - 2x$$
$$\dot{b} = -a - 1$$

The equilibrium is given by putting $\dot{x} = \dot{a} = \dot{b} = 0$, and is the point $x = 1$, $a = -1$, $b = 0$. To first order in ε, the eigenvalues at the equilibrium point are $-2/\varepsilon$ and the complex cube roots of 1.

Assuming that the trigger increases b, then the threshold is reached when $b = 2/3\sqrt{3}$. The smooth return is shown in Figure 14. The figure shows the flow on the slow manifold as seen from above. The words 'as seen from above' mean that inside the cusp the top attractor surface hides both the middle repellor and the bottom attractor surface. This explains why only the right side of the cusp is visible, and why the flow appears to be discontinuous across this fold line because the left side refers to the upper surface and the right side to the lower. In fact the flow is smooth everywhere.

1.11. APPLICATION TO THE NERVE IMPULSE

If we are to apply Example 8, or a refinement of it, to the nerve impulse, we must identify the dynamical variables x, a, b with measurable quantities. We have

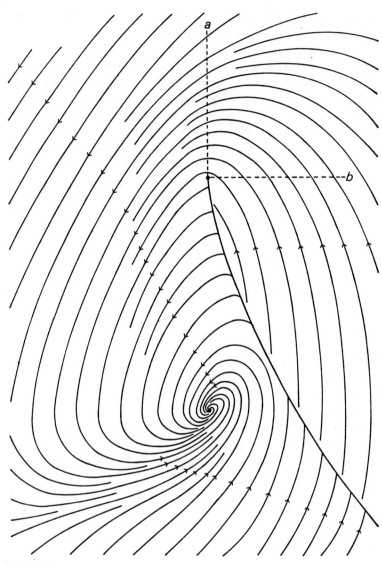

FIGURE 14

the same proviso as in section 1.8, that the experimentally most convenient qualities to measure may in fact be only artifacts, mirroring the dynamics, and we return to this point more deeply in section 2.10.

The natural candidate for b is the membrane potential, because we have assumed that the trigger is given by changing b, and what triggers off the nerve impulse is a depolarization of the membrane. The two other most important variables have been pinpointed by Hodgkin–Huxley [4] as the permeabilities of the membrane to sodium and potassium ions. Of all the changes that occur during the nerve impulse, the most dramatic is the sharp rise in sodium conductance at the beginning of the impulse (*see* Figure 27), and so it is natural to identify this with the action, and to correlate sodium conductance with $-x$. This leaves a to be identified with potassium conductance, and sure enough the latter does not begin to change until after the action has finished, and then it rises and falls smoothly during the smooth return (*see* Figure 27). Therefore Figures 12 and 13 present all the correct qualitative features observed during experiments. However Example 8 is too elementary to predict the correct quantitative features, and in order to build a better quantitative model (diffeomorphic to Example 8), we need to go into the physiology in more detail in section 2.4. This completes the first half of the paper and our objective of deriving qualitative mathematical models of the three dynamic qualities. For the second half of the paper we return to the physiology of the heart.

Part Two

2.1. TENSION AND BLOOD PRESSURE

The purpose of this section is to derive the *local heartbeat equations* from Example 6, allowing for the tension in the muscle fibres due to blood pressure, and to explain a variety of associated phenomena including Starling's Law.

In 1957 Boris Rybak [8] originated the following experiment. I am indebted to him for demonstrating it to me. If the heart (of a frog say) is taken out then it stops beating. However, if it is then cut open into a flat membrane and stretched under slight tension, it then starts to beat again and will continue for some hours. If the tension is relaxed the beating stops. Alternatively if the pacemaker is removed then the beating stops. (This is not quite true because there is a weak secondary pacemaker, and if both pacemakers fail then after a while spontaneous new ones may arise, which are called ectopic pacemakers.) Therefore both factors are necessary for the heartbeat, tension and the pacemaker wave. The tension *in situ* is of course provided by the blood pressure. However the pressure is different in different parts of the heart, and varies during the beat cycle, and so we must examine the global structure.

The heart has four chambers and four valves. At first sight this seems unnecessarily complicated, because a pump needs only one chamber and one valve, but if we examine the pressures involved the reasons become clear. There are two circuits for the blood, one round the lungs to pick up oxygen, and the other round the body to deliver it. The first is a low pressure circuit, so as not to damage the delicate membrane in the lungs, while the second is a high pressure circuit, in order to get down to the feet, through the capillaries, and up again. The right side of the heart is the low-pressure pump for the lungs, and the left side is the high-pressure pump for the body. The maximum/minimum pressures that occur in each part of the heart during normal heartbeat are shown in Figure 15 (in millimetres of mercury). We follow the usual medical convention of interchanging left and right in the picture, because we imagine we are looking at someone else's heart.

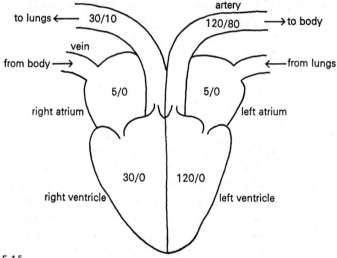

FIGURE 15

Each pump has a main pumping chamber called the ventricle, with an inlet and an outlet valve. It has to have an inlet valve to prevent flowback up the veins while pumping, and an outlet valve to prevent flowback from the arteries while filling. Now comes what would be a problem if there were only one chamber: since the heart is made of non-rigid material it only has the power to push out, and no power to suck in (unlike the lungs which have the ribs). Therefore it would have to depend for filling upon the very feeble venous return pressure. To get a good pump it is necessary to completely fill the ventricle (*see* Starling's Law *below*), and so Nature has installed a small prechamber called the atrium, whose job it is to pump gently beforehand, just enough to fill the ventricle, but not too much to cause a flowback up the veins. In fact 60 per cent of the filling of the ventricle is done by the feeble venous pressure before the atrium begins to beat (*see* [9, p. 107] and the last diagram in Figure 20), and so the atrium does only the final topping up, and consequently does not need to be as big as the ventricle.

The pacemaker sits at the top of the atria, and the pacemaker wave first spreads slowly over the atria, causing them to contract and push the blood into the ventricles. As the inlet valve closes, the wave focuses at a node in between the atria and ventricles, and then spreads rapidly over each ventricle carried by fast fibres (the bundle of His) causing the whole ventricle to contract simultaneously and deliver a big pump of blood down the arteries. Then the muscles relax in the same order, and the cycle repeats itself. The way the left ventricle is made into a high-pressure pump is to have a thicker criss-cross structure of muscle fibres.

Another interesting feature, known as *Starling's Law of the Heart* [10, p. 122], is that the more the muscle fibres are stretched before beating, the more forcible is the beat. It is an excellently designed system for meeting emergencies: suppose fear or rage causes adrenalin to be injected into the blood stream; then the adrenalin causes the arteries to contract and the pulse rate to increase, which in turn causes the blood pressure to rise, and the atria to push more blood into the ventricles; finally Starling's Law describes how the stretched ventricles then give a bigger beat, overcoming the increased arterial back-pressure and circulating the blood faster.

On the other hand if the ventricles are overstretched beyond a certain point, as can happen for instance when a person with high blood pressure receives a sudden shock, then the heart may fail to beat, or only beat feebly, and cardiac failure may ensue.

At the other end of the scale, if the bloodstream is made to bypass the heart during an operation, so that there is no longer any blood pressure in the heart, then the beat becomes sluggish, and the heart tends more to heave without altering much in size. This phenomenon is similar to the Rybak experiment, in which the beat ceases when the tension drops.

Summarizing, the individual muscle fibres can behave very differently, depending upon circumstances, and where they are situated. We propose that most of this difference in behaviour can be attributed to difference in tension, caused by varying blood pressure, and we shall show that by a slight addition to the equations of Example 6 we can explain a large variety of phenomena, including :
(1) Rybak's experiment, and the sluggish beat of the bypassed heart ;
(2) the small atrial beat, followed by the squeezing action of blood through the inlet valves ;
(3) the large ventricular beat, obeying Starling's Law ;
(4) overstretching causing cardiac failure.

We argue as follows. Rybak's experiment shows that the heart can stop beating even though the pacemaker is still going. Therefore if a fibre is no longer under tension the pacemaker wave no longer triggers a sudden contradiction. In other words if the tension drops, the threshold must disappear. Conversely as the tension rises the threshold appears. Now we originally constructed the threshold in Example 5 by changing the sign of x in the fast equation. Therefore the natural place to put the tension is as the coefficient of x in the fast equation. Consequently we are led, once again, to the cusp catastrophe :
$$\dot{x} = -(x^3 + ax + b)$$
where this time (up to multiplicative and additive constants) x=length of fibre, b=chemical control, $-a$=tension.

From the point of view of catastrophe theory a, b are the two parameters of control in the control space C. In Figure 16 the slow manifold M illustrates how these two controls determine the length of the muscle fibre. Without such a picture it is difficult to visualize how the two controls interrelate (as students of anatomy will confirm). The feedback on M is determined by the slow equation, to which we now turn our attention.

Suppose that the chemical control b takes values b_0 in diastole and b_1 in systole. More precisely $b=b_0$ is the equilibrium position, and the trigger moves b from b_0 to b_1, after which the slow equation returns b to b_0. Therefore during diastole b is indeed fixed at b_0, but during systole b is changing, with a maximum at b_1, as illustrated in Figure 10.

We shall assume that $b_0 < 0 < b_1$. The justification for this assumption is that it leads to the simplest explanation for the variety of phenomena, but of course it is a hypothesis that needs to be tested experimentally. It may be that b_0, b_1 depend upon other factors (for example, oxygen lack may increase b_0), and an investigation into the consequences of changing these constants might give insight into

3. Differential Equations for the Heartbeat and Nerve Impulse

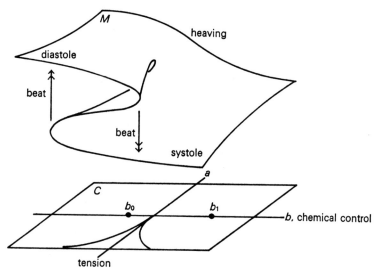

FIGURE 16

pathological behaviour of the heart. But for now let us assume $b_0<0<b_1$. Under given tension a, the diastole length x_a of the fibre will be given by the equilibrium condition $\dot{x}=0$. Therefore x_a is the solution of

$$x_a^3+ax_a+b_0=0 \ ,$$

lying on the upper diastole surface, in other words satisfying the inequality

$$x_a > \sqrt{-\frac{a}{3}} \ .$$

We can now write down the slow equation

$$\dot{b}=x-x_a \ .$$

We do not postulate a slow equation for a, because the main reasons for tension in a muscle fibre are not the internal chemistry but external forces caused by the blood pressure and the pushing and pulling of the different organs of the heart against each other.

Summarizing: *The local heartbeat equations*

$$\varepsilon\dot{x}=-(x^3+ax+b)$$
$$\dot{b}=x-x_a$$

where (up to multiplicative and additive constants) $x=$ length of muscle fibre, $b=$ chemical control (possibly membrane potential), $-a=$ tension.

The behaviour of a fibre under a particular tension a_1 is given by taking the

3. Differential Equations for the Heartbeat and Nerve Impulse

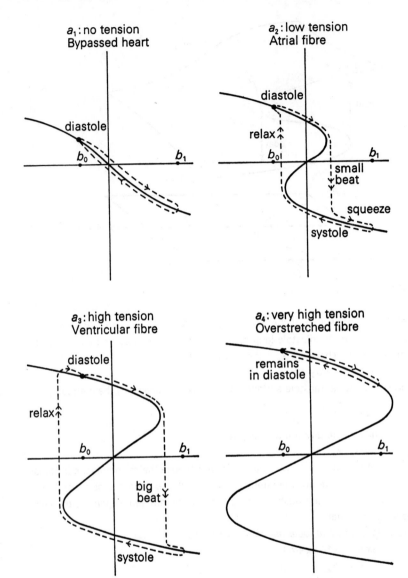

FIGURE 17

section $a=a_1$. To explain the variety of phenomena listed above, we draw four sections in Figure 17 for four values $a_1 > a_2 > a_3 > a_4$ corresponding to the four situations :

a_1 : no tension, bypassed heart ;
a_2 : low tension, beat of an atrial fibre ;
a_3 : high tension, beat of a ventricular fibre ;
a_4 : very high tension, cardiac failure.

In each case the slow manifold is indicated by a thick line and the beat cycle by a dotted line.

Notice the following features :

Case (1). Each fibre does contract a little, but slowly rather than sharply. Therefore the pacemaker wave induces a muscular reaction that can be observed as a wave of sluggish contraction spreading over the heart, rather than a sharp contraction.

Case (2). After the small beat the atria continue to contract slowly for a while, thereby squeezing the blood through the inlet valve.

Case (3). After relaxation the ventricles expand beyond equilibrium, helping them to do most of the filling before the next beat begins. If the tension is increased, then the S-shape of the slow manifold becomes more pronounced, so that the fibre is stretched longer in diastole and contracted shorter in systole. Hence the action is bigger and faster, making the beat more forcible, and thereby explaining Starling's Law.

Case (4). This case occurs if the tension is sufficiently large. Mathematically the condition on a_4 is :

$$a_4 < -(27b_1^2/4)^{\frac{1}{3}}$$

The condition moves the threshold beyond b_1, so that the trigger no longer reaches the threshold ; consequently the fibre remains in diastole. If a few of the ventricular fibres manage to contract first, then they help to overstretch the rest and prevent the rest from contracting, resulting in a feeble ventricular beat. Meanwhile the atria increase the strength of their beat by Starling's Law, further overfilling the ventricles, and possibly causing permanent damage to the latter by overstretching.

2.2. THE PACEMAKER WAVE

The mechanism underlying the pacemaker wave P is not fully understood, even though it is the basis of the extensive field of electrocardiography. We propose a tentative model for P, and argue the importance of separating the mathematics of P from the local heartbeat equations.

P is primarily an electrical phenomenon, probably because electrons are the only physical objects that can traverse tissue at sufficient speed. When P reaches an individual muscle fibre it causes a depolarization of the cell membrane potential, which triggers off the sequence of chemical events leading to the contraction. Figure 18 is taken from Sampson Wright's *Applied Physiology* [10, p. 96]. The upper curve is the time-graph of the membrane potential of a frog ventricular fibre, beating at the rate of 1 per second. If we identify the membrane potential with our chemical control, b, then the thin steep rising line of depolarization corresponds to the trigger, changing b from b_0 to b_1 and causing contraction, while the thick descending curve of repolarization corresponds to our slow equation, ending in relaxation.

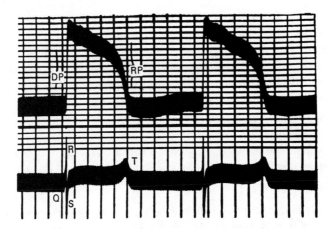

FIGURE 18
A normal cardiac transmembrane potential and its relation to the surface electrocardiogram. DP=depolarization. RP=repolarization. Time lines, 0·1 sec. (*after* Hecht, 1957)

Meanwhile the lower curve is the skin ECG (electrocardiograph), which is the summation over all fibres in the heart. An ECG can be recorded between any pair of points on heart or skin, because the electrical changes are conducted throughout the body fluids, although of course different pairs of points show slightly different readings. The physician uses the ECG to check whether, and when, the atria and ventricles are beating. However we are interested not in the global summation, but in the local membrane potential b, which we can regard as a function of time and position on the heart H. We shall represent the pacemaker wave P as a wave in b.

Now the speed of P depends upon the conductivity of the heart tissue, which is not the same in all directions. Perhaps the best way to model this is to represent H by a surface (with singularities) and represent conductivity by a Riemannian structure on H. We can then define P to be the solitary wave, of amplitude $b_1 - b_0$, emanating from the pacemaker, and propagating over H according to Huyghens' principle, at speed determined by the Riemannian structure. Consequently *the wave front is given by the geodesic flow on H*, emanating at a given moment from the pacemaker.

We summarize the difference between the pacemaker wave P and the muscle fibre contraction. The former is global and the latter local. The former triggers the latter. Both depend upon electrochemical mechanisms, but the former is primarily electrical, while the latter is primarily chemical. Consequently the membrane potential b is a fundamental variable for the former, but may only be an artifact in the latter. However it is convenient to use the same variable for both, because this provides the link between the two mathematical models :

(1) *the global geodesic flow for P* ;
(2) *the local heartbeat equations*.

Mathematically it would be possible to combine these two into a single partial differential equation, but this might lead to fundamental errors, as is shown by the following experiment.

From such a *combined* equation we should be able to extract two mathematically similar waves :

P : the pacemaker wave and wave of muscular contraction ;
Q : the wave of muscular relaxation.

In the normal heartbeat of man Q follows P after about 1/5 second, and in the frog example in Figure 18 the interval is about 1/2 second. If we make a cut across the heart tissue just before the arrival of P, then this would stop P, because P propagates by Huygens' principle, and so could not jump across the cut. On the other hand if we were to make the cut just before the arrival of Q, then this would *not* stop Q, because Q propagates not by Huyghen's principle, but as a wave of local events occurring 1/5 second after P. The mathematical similarity of the two waves would deceptively obscure their physical difference.

2.3. CONCLUSION AND EXPERIMENTS

There are three types of complexity in the heart :
(1) the global complexity of the anatomy ;
(2) the local complexity of the underlying biochemistry ;
(3) the dynamic complexity of the beat.

Most research has been directed at unravelling (1) and (2), with the ultimate objective of understanding (3). What we have tried to do is bypass (1) and (2) and isolate a central simplicity in the mechanical behaviour of muscle fibres. We have expressed this simplicity mathematically in the local heartbeat equations, and have shown that a large part of the dynamic complexity can be deduced fairly simply from these equations, and does not depend upon the complexity of (1) and (2). This does not mean we escape from (1) and (2) ; on the contrary the next thing to do is to relate (1) and (2) to the central simplicity.

First we must test the equation against the underlying chemistry. For example, testing fibre length against membrane potential and tension should verify the qualitative features of the cusp catastrophe, and provide the real refined version of the fast equation, or a functional relationship between the membrane potential and chemical control. Testing the fibres dynamically would give a refined version of the slow equation. It may transpire that the slow equation is determined by ionic flow, as in the nerve impulse, in which case it could be misleading to have fibre length in the slow equation. But the point is that the equations provide a nucleus of theory around which to design a series of experiments.

Secondly we must relate the local simplicity of fibre behaviour to the global geometry of the anatomy. We do this by visualizing how the heart moves over the surface of the cusp catastrophe during the beat cycle, and in this way we shall derive the beat from a smooth motion in the control space (*see* Figure 20). The objective here is to gain a better conceptual understanding of the heartbeat as a whole, which should lead not only to design of experiments but also to a better understanding of heart failure, various forms of heart blocks, and perhaps improved diagnosis.

In Figure 17 we have illustrated the difference between atrial and ventricular fibres. The picture of the atrial fibre is reasonably accurate, because the blood pressure in the atria does not vary greatly, and therefore neither does the tension in atrial fibres. However the picture of the ventricular fibre is an oversimplification because the ventricular pressure varies considerably during the beat cycle [10, p. 107].

At a given time t, each fibre is subject to some tension and chemical control,

3. Differential Equations for the Heartbeat and Nerve Impulse   117

and so there is a map $f_t : H \to C$ from the heart H to the control space C. The muscular state of H is given by lifting f_t to a map $g_t : H \to M$, where M is the cusp catastrophe surface, or slow manifold. Lifting means that Figure 19 is commutative. Although χ^{-1} is 3-valued over the interior of the cusp, the lifting is uniquely defined by past history and the delay convention. The *delay convention* (which is an immediate consequence of the fast equation) says to each fibre 'keep on the upper or lower surface as long as possible, and only jump onto the other one if necessary, because of crossing a threshold'. In Figure 20 we illustrate a lifting, representing symbolically one of the ventricles by a triangle, the corresponding atrium by a circle, and the node between them by a line.

The diagram shows that even though f_t is continuous, g_t may not be. The pacemaker wave has already carried the top half of the atrium past the threshold,

FIGURE 19

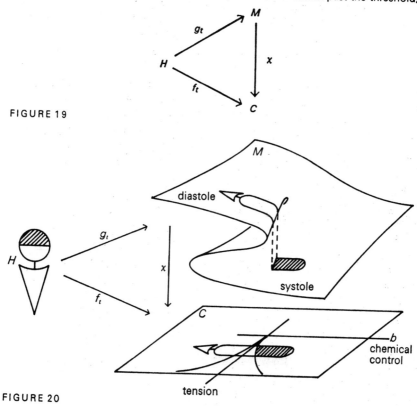

FIGURE 20

118 3. Differential Equations for the Heartbeat and Nerve Impulse

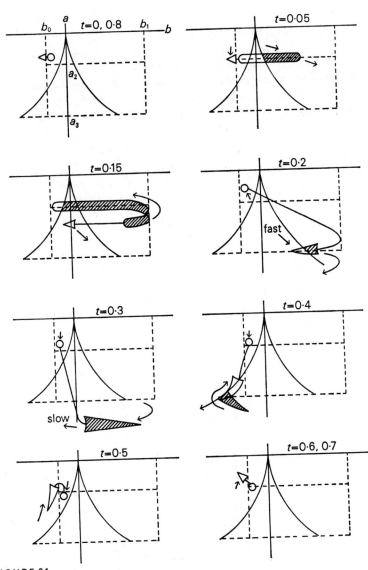

FIGURE 21

and so the image under g_t has dropped off the diastole surface onto the systole surface. In more conventional language, the wave of contraction is already half way down the atrium, so that the top half of the atrium (shown shaded) is contracted in systole, while the bottom half and the ventricle are still relaxed in diastole. Figure 20 represents the situation about 1/20 second after the pacemaker has initiated the pacemaker wave.

In Figure 21 we illustrate a series of eight pictures of C, representing f_t at various times during the heartbeat cycle in man. The pictures are drawn for a pulse-rate of 75, and so the cycle lasts 0·8 seconds. The pacemaker initiates the beat at $t=0$. Recall that the variable a measures tension in the muscle fibres, or what is roughly the same thing, the blood pressure. The constants b_0, b_1 and a_2, a_3 are the same as used previously in Figure 17, and are indicated by dotted lines. The region in systole is shown shaded in each case.

During $0 < t < 0 \cdot 2$ the atrium is slowly beating, with both contraction and relaxation occurring at roughly the same pressure, a_2. During the last quarter of this period the pressure rises sharply in the ventricle to a_3. Then at $t=0 \cdot 2$ the pacemaker wave spreads rapidly over the ventricle, causing the whole ventricle to contract almost simultaneously. During $0 \cdot 2 < t < 0 \cdot 4$ the ventricle remains in systole, while the pressure rises even higher forcing the blood out of the outlet valve. Meanwhile b decreases by the slow equation. Relaxation of the ventricle occurs at $t=0 \cdot 4$, by which time the pressure has dropped to a_3 again. During $0 \cdot 5 < t < 0 \cdot 8$ both chambers are back in diastole with the outlet valve closed. During $0 \cdot 6 < t < 0 \cdot 8$ the pressure in the ventricle drops dramatically to slightly lower than that in the atrium, causing the inlet valve to open and enabling 60 per cent of the filling of the ventricle for the next beat to take place before the beat starts.

In addition to the oversimplified picture we have given, there are pressure and tension changes caused by the geometry of each organ pushing and pulling on its neighbours, as it contracts and expands.

2.4. PROPAGATION AND TIMING OF THE NERVE IMPULSE

We now turn attention back to the nerve axon. By comparison with the heart (*see* section 2.3) the problem is simpler because there are essentially only two types of complexity:

(1) the local complexity of the underlying biochemistry;

(2) the dynamic complexity of the action potential.

Anatomically the axon can be thought of as a long thin cylindrical tube filled with a fluid called axoplasm. To get an idea of size, a typical axon in man might be 1 μ in diameter (1 $\mu = 10^{-3}$ millimetres) and a few millimetres long. In different organs and in different species axons can vary in diameter from 0·1 μ to 500 μ, and in length from a fraction of a millimetre to several metres [see 5, p. 15]. For example the data in Figures 3, 22, and 24 were recorded from a giant squid axon of diameter 476 μ.

The axoplasm can conduct electric current, but the wall membrane acts as an insulator. Most axons in man are surrounded by myelin sheathing, which reinforces the insulation, and increases the velocity of the action potential. Every millimetre or so, there are gaps in the myelin sheathing called the nodes of Ranvier, which have different electrochemical properties [see 5, chapter 4], but as a first approximation we shall ignore them, and assume that the walls are homogeneous.

As explained in section 1.11 the important variables are the membrane potential V, and the membrane permeabilities to sodium and potassium ions. Changes in V cause, and are caused by,
(1) flow of electrons along the axon;
(2) flow of ions through the membrane.
As in the case of the heart (see section 2.2) we shall model these two processes mathematically by
(1) *a global propagation wave;*
(2) *the local nerve impulse equations.*
We have not discussed the propagation wave yet and shall do so in detail in section 2.8 below, and compute the velocity. The local equations we have already discussed qualitatively in section 1.11 and Example 8, but we now get down to a detailed quantitative version in section 2.7 below.

Meanwhile we emphasize the importance of keeping the two equations separate, and not combining them into a single partial differential equation. Otherwise we should be able to extract from the combined equation two mathematically similar waves, about 1 millisecond apart:

P : propagation wave, depolarizing the membrane,

Q : repolarization wave.

The mathematical similarity would obscure the physical difference between P and Q, which is demonstrated by cutting the axon just before the arrival of P, which would stop P, and cutting it just before Q, which would *not* stop Q. The reason is that Q is a wave of local events, originally triggered by P, but thereafter locally determined by the local equations.

This is verified experimentally by an elegant result of Hodgkin and Huxley [4],

3. Differential Equations for the Heartbeat and Nerve Impulse   121

as follows. Trigger an impulse in an isolated segment of axon, by means of a short shock administered simultaneously to the entire segment; then the segment displays the normal action potential without any propagation effects along the axon. In other words it obeys the local equations only. The result is shown in Figure 22, which is taken from Hodgkin [5, p. 65]. The numbers attached to the curves denote the shock strength in $m\mu$ coulomb/cm^2. If the shock is small there is a delay before the action potential is triggered, which we shall explain by means of a saddle point (*see* section 2.7 below, and section 2.10, Remarks 6 and 7).

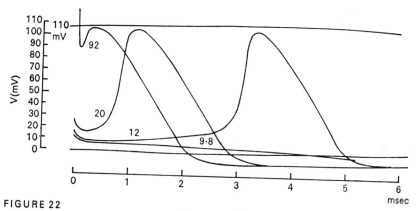

FIGURE 22

Hodgkin [5, p. 14] suggests a very good metaphor in order to stress the fact that the energy for the nerve impulse is released locally rather than propagated globally: he points out that nerve conduction resembles the burning of a fuse of gunpowder rather than the propagation of an electric signal along a cable. In our model, the propagation wave is analogous to the flame running along the fuse setting it alight, while the local equations are analogous to the way it burns locally. There is only one aspect in which this metaphor can be misleading, and that is in the matter of timing. It is worthwhile estimating roughly the time intervals involved.

Suppose we have a segment of axon 1000 times as long as it is wide, for example 1 μ in diameter and 1 mm in length. Suppose that the velocity of propagation is 100 metres/sec. Then the propagation wave traverses this segment in 10^{-2} msec. Meanwhile the complete action potential takes about 1 msec, which is 100 times as long. To put it another way, the wave length is 10 cm, which is much longer than most axons. Of course in different organs and different species these times vary—for example, Figures 3 and 22 refer to a giant squid axon, in which the velocity is only about 20 metres/sec. But the point to be taken is that propagation along the axon is much swifter than the local potential and permeability changes, whereas in the gunpowder fuse it is the exact opposite, because propagation along the fuse is much slower than local burning. Mathematically the timing is very important, because it implies that the propagation wave dominates only at the beginning, after which the local equations dominate. This enables us to keep the two pieces of mathematics fairly separate, and justifies the conceptual device of regarding the end of the propagation wave as the trigger, or initial conditions, for the local slow equations (*see* section 2.9 below).

2.5. VOLTAGE CLAMP DATA

The next three sections are devoted to deriving the *local nerve impulse equations*, which are stated in section 2.7. These are a quantitative refinement of the qualitative model Example 8 developed in sections 1.10 and 1.11. We use the electrical theory and voltage clamp data of Hodgkin and Huxley [4, 5].

The voltage clamp technique evolved by Cole consists of isolating a segment of axon, clamping the voltage difference across the membrane to various fixed values by means of electrodes inside and outside, and measuring the currents carried by the resulting flows of potassium and sodium ions across the membrane. From this can be calculated the large changes that occur in the potassium conductance g_K and sodium conductance g_{Na} of the membrane. Meanwhile the chlorine conductance g_{Cl}, and that of other ions, remain relatively small and constant. What happens is illustrated on the cusp catastrophe in Figure 23. We have identified coordinates

$b \sim V$, voltage

$a \sim g_K$, potassium conductance

$-x \sim g_{Na}$, sodium conductance,

where $\sim$ indicates linear or functional relation between.

When the clamp is switched on, the potential V jumps from resting potential 0 to clamp potential V_c, and so the state is displaced from equilibrium E parallel to the V-axis to the position F in the clamp plane $V=V_c$. Then the fast equation carries the state rapidly to the point G on the slow manifold. Finally the slow equations (or more precisely the $\dot{a}$-component of the slow equations) carry the state from G to H, where H is the point where $\dot{a}=0$. In other words H is the point whose slow vector is perpendicular to the clamp plane. Therefore H is the equilibrium position under the clamp, but if the clamp is released then the state returns to equilibrium E along the dotted flow lines of the slow equations. The effect of the voltage clamp on the sodium conductance g_{Na} is a fast increase along the path FG, followed by a slow decrease GH, while the effect on the potassium conductance g_K is a delay FG, followed by a slow increase GH.

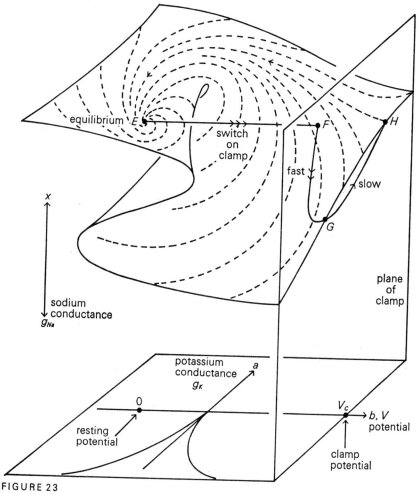

FIGURE 23

This agrees qualitatively with the Hodgkin–Huxley data, shown in Figure 24, which is taken from Hodgkin [5, p. 61], and given in more detail in [4].

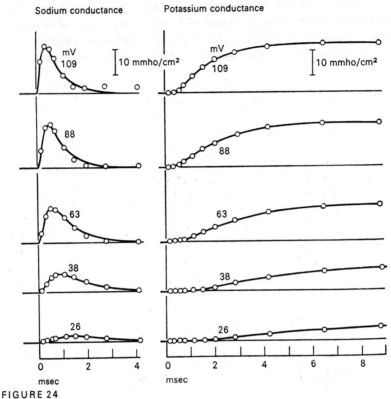

FIGURE 24

In Figure 24 the numbers indicate the voltage clamp, the circles are the experimental data, and the smooth curves are solutions of Hodgkin and Huxley's own theoretical equations [4, p. 518] designed to fit the data. Our aim is to find alternative equations by fitting the cusp catastrophe to the data, and there are two possible procedures.

Procedure (1). Use the data from the time of maximum sodium onwards to plot g_{Na} as a function of V and g_K. Then see if the resulting fragment of surface can be extended smoothly to a surface differentiably equivalent to the cusp catastrophe. The data gives the $\dot{a}$-component of the slow flow on the fragment of surface, while the $\dot{b}$-component is given by the electrical capacitance equation (*see* section 2.6 *below*). The advantage of this procedure is that the construction of the fragment is direct, but the disadvantages are the non-uniqueness of the extension, and the unlikelihood of being able to express the results algebraically. Therefore we could not write down equations. However for the purposes of prediction the results could probably be encoded in a computer programme.

Procedure (2). Retain the canonical fast equation, and hence the canonical cusp catastrophe surface. Then fit the data to this canonical surface by juggling the position of the equilibrium point and the functional relations between b, a, x and V, g_K, g_{Na}. Construct an algebraic equation for $\dot{a}$ to fit the data, and use the electrical equation for $\dot{b}$. The disadvantage of this procedure is that it may not be possible to fit the data precisely, and the juggling does not give a unique answer. However the advantages are that we are working on the canonical surface, which is justified by Thom's theorem (*see* section 1.10), and we obtain usable algebraic equations. In addition the form of these equations may give insight into the underlying chemistry (*see* section 2.10). Therefore we shall attempt the second procedure.

2.6. ELECTRICAL EQUATION FOR IONIC FLOW

We review briefly the electrical theory of Hodgkin and Huxley [4, 5]. Recall the notation of section 1.1.

v = membrane potential
= potential inside axon − potential outside,
$V = v - v_r = v + 65$

where v_r is the resting potential. Thus the resting potential or equilibrium position is given by $V = 0$. In the computation below we use the value $v_r = -65$ mV [*see* 4, p. 520].

It is observed that the concentration of certain ions is very different in axoplasm and in blood [5, p. 28]. Consider first potassium ions, and let K_i, K_o denote concentrations inside and outside the axon. The observed ratio is about $K_i/K_o = 20$. This ratio is maintained by metabolism, and the small flows that occur during the action potential have negligible effect.

Now if the membrane were permeable to potassium, and if the voltage were clamped at $v = 0$, then diffusion would cause an *outward* flow of potassium until the ratio was reduced to 1. However in the resting state the membrane is *not* permeable, and $v_r \neq 0$. Even if the membrane were permeable, the fact that $v_r < 0$ would tend to cause an *inward* flow of potassium ions, since the ions are positively charged. Define v_K to be the theoretical potential at which the outward and inward flows would just balance ; more precisely if the membrane were permeable, and if the voltage were clamped at $v=v_K$, then diffusion would produce an equilibrium at the observed ratio. The value of v_K can be computed by the Nernst formula [5, p. 30]:

$$v_K = \frac{RT}{F} \log \frac{K_o}{K_i} ,$$

where R = gas constant, T = absolute temperature, and F = Faraday. The values we use in the computations below are shown in Table 1 [see 5, p. 28; and 4, p. 520]. The values of b in the last column refer to section 2.7 below.

TABLE 1

Ion		Concentration mmole/kg		v	V	b
		inside	outside			
Potassium	K+	400	20	$v_K = -77$	$V_K = -12$	-1.4
Sodium	Na+	50	440	$v_{Na} = +50$	$V_{Na} = 115$	4.95
Chlorine	Cl−	90	560	$v_{Cl} = -46$	$V_{Cl} = 19$	0.15

The other ions have negligible effect and so we ignore them.

At any given time the flow of potassium ions across the membrane is proportional to

$$v - v_K = V - V_K ,$$

and depends also upon the potassium permeability. The simplest way to measure permeability is electrically, by measuring the *sodium conductance* per unit area, which is defined as follows. If I_K denotes the outward electric current per unit area carried by the outward flow of potassium ions, then define

$$g_K = \frac{I_K}{V - V_K} .$$

Therefore the total outward flow of ionic current per unit area

$$= I_K + I_{Na} + I_{Cl}$$
$$= g_K(V - V_K) + g_{Na}(V - V_{Na}) + g_{Cl}(V - V_{Cl}) .$$

Now the membrane acts as a condenser, of capacity c per unit area, say, and so there is an apparent outward current of $c\dot{V}$, due to storage of charge on the surfaces of the condenser. Therefore if I denotes the total outward flow of current across the membrane per unit area, then

$$I = c\dot{V} + g_K(V - V_K) + g_{Na}(V - V_{Na}) + g_{Cl}(V - V_{Cl}) .$$

To balance this outward flow across the membrane there must be an internal flow of current along the axon (carried by electrons moving through the axoplasm). If we make the *local hypothesis* that the flow along the axon is the same at all points of the axon (although it may be varying in time), then no charge accrues at any point, and so we obtain the *local electrical equation* $I=0$. The local electrical equation provides the slow equation for b in the next section.

Note that $I \neq 0$ during the propagation period, as we shall see in section 2.8, but as soon as the latter is over then $I=0$, and the axon behaves in unison according to the slow equations.

2.7. LOCAL NERVE IMPULSE EQUATIONS

We are now in a position to state the local equations. As notation, let $[y]_\pm$ denote the functions

$$[y]_+ = \begin{cases} y, & y \geq 0 \\ 0, & y \leq 0 \end{cases} \qquad [y]_- = \begin{cases} 0, & y \geq 0 \\ y, & y \leq 0 \end{cases}.$$

The functional relations between physical variables and canonical coordinates are chosen to be:

Potential	$V = (20b + 16)$ mV
Potassium conductivity	$g_K = 2 \cdot 38 [a + 0 \cdot 5]_+$ mmho/cm².
Sodium conductivity	$g_{Na} = (4[x + 0 \cdot 5]_-)^2$ mmho/cm².

We use constants:

Chlorine conductivity	$g_{Cl} = 0 \cdot 15$ mmho/cm² [11, p. 366]
Membrane capacity	$c = 1$ μFarad/cm² [4, p. 520].

Define the *local nerve impulse equations* to be:

$$\dot{x} = -1 \cdot 25(x^3 + ax + b)$$
$$\dot{a} = (x + 0 \cdot 06(a + 0 \cdot 5))(x - 1 \cdot 5a - 1 \cdot 67)(0 \cdot 054(b - 0 \cdot 8)^2 + 0 \cdot 75)$$
$$\dot{b} = -g_K(b + 1 \cdot 4) - g_{Na}(b - 4 \cdot 95) - g_{Cl}(b - 0 \cdot 15).$$

The fast equation for $\dot{x}$ is the canonical cusp catastrophe. The slow equation for b is the electrical equation of section 2.6. The slow equation for $\dot{a}$ is an *ad hoc* equation fitting the voltage clamp data: the first factor vanishes at the clamp equilibrium, the second at the resting potential, and the third adjusts the speed to fit that of the data. The *resting potential* or *equilibrium position* occurs when $\dot{b} = \dot{a} = \dot{x} = 0$, giving

$b = -0 \cdot 8$	and therefore	$V = 0$
$a = -0 \cdot 4$		$g_K = 0 \cdot 24$ mmho/cm² [see 4, p. 509]
$x = 1 \cdot 07$		$g_{Na} = 0$.

Notice that sodium conductance remains zero until $x < -0.5$. (We discuss this sodium cut off in section 2.10, Remark 5.) The propagation wave displaces the state away from equilibrium to the *trigger point* J (*see* section 2.9 *below*) given by:

$b = 1\cdot 2 (V = 40)$
$a = -0\cdot 45$
$x = -1\cdot 2$.

En route the state crosses the *threshold curve* near

$b = 0\cdot 1 (V = 18)$
$a = -0\cdot 4$
$x = 0\cdot 4$

where it jumps by the fast equation from the upper surface to the lower surface of the slow manifold, switching on the sodium flow.

FIGURE 25

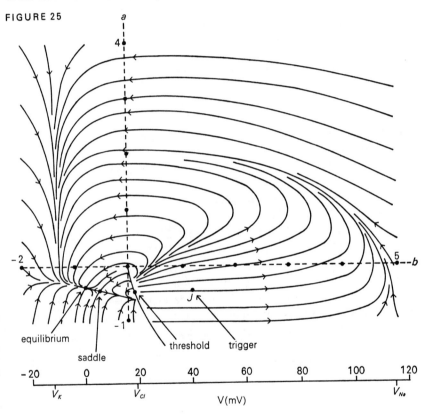

There is also a *saddle point* of unstable equilibrium on the slow manifold at
$$b = -0.55$$
$$a = -0.45$$
$$x = 1$$
which explain the phenomena in Figure 22, that perturbations $V \leqslant 8$ are stable and slowly return to equilibrium, but perturbations $V \geqslant 9$ are unstable and slowly increase, until after about 3 msec the threshold is crossed and the normal action potential ensues. There are more remarks about the saddle point below in section 2.10 (6) and (7).

The calculations in this paper based on the equations have been done by hand, and have been used to draw accurate diagrams of the flow in Figure 25, the action potential in Figure 26 and the conductivity changes in Figure 27. Meanwhile Woodcock [12] has checked the calculations by computer, and suggests that it may be possible to get a better fit of the voltage clamp data by altering some of the coefficients.

Figure 25 shows the flow given by the slow equations on the slow manifold, as seen from above (compare qualitatively with Figure 14).

2.8. PROPAGATION WAVE

Let y denote the coordinate along the axon. The potential V is a function of position y and time t, and let V', $\dot{V}$ denote the partial derivatives

$$V' = \frac{\partial V}{\partial y}, \quad \dot{V} = \frac{\partial V}{\partial t}.$$

Let r = radius of axon, ρ = resistivity of axoplasm.

▶ *Lemma 1*

The axonal current accruing at y is

$$I = \frac{r}{2\rho} V'',$$

per unit area of membrane.

Proof. Let j = current flowing past y, per unit area. The drop in potential from y to $y + \delta$ gives:

$$j\rho\delta = V(y) - V(y+\delta) = -\delta V'.$$

Therefore
$$j = -\frac{V'}{\rho} \ .$$
The amount of current flowing into the segment of axon between y and $y+\delta$
$$= \pi r^2 (j(y) - j(y+\delta))$$
$$= -\pi r^2 \delta j'$$
$$= \frac{\pi r^2 \delta}{\rho} V'' \ .$$
Dividing this by the surface area $2\pi r \delta$ of the membrane of this segment gives the desired result.

▶ **Lemma 2**

If the resting potential is perturbed then the outward ionic current is approximately kV per unit area of membrane, where $k = 0\cdot 388$.

Proof. We use the local nerve impulse equations. At equilibrium $\dot{a}=0$, and if V is perturbed $\dot{a}$ remains small. Therefore we may assume a remains constant at $a=-0\cdot 4$. Therefore near equilibrium
$$g_K = 0\cdot 238$$
$$g_{Na} = 0$$
$$g_{Cl} = 0\cdot 15 \ .$$
Therefore the ionic flow $= 0\cdot 238(V+12) + 0\cdot 15(V-19)$
$$= 0\cdot 388 V, \text{ as required.}$$

We assume that the propagation wave starts at $y=0$ at time $t=0$, and proceeds with constant velocity θ. Therefore before the wave has reached y, the axon is still in equilibrium :
$$V = 0, \quad y \geqslant \theta t.$$
After the wave has reached y the potential is given by the following theorem.

▶ **Theorem 3**

For $\theta t \geqslant y$, the propagation wave is given by
$$V = A \left\{ \exp\left[\frac{2c\rho\theta}{r}(\theta t - y)\right] - \exp\left[-\frac{k}{c\theta}(\theta t - y)\right] \right\}$$
where A is constant.

Proof. Equating the axonal current of Lemma 1 with the membrane current of section 2.6, and using Lemma 2 :
$$I = \frac{r}{2\rho} V'' = c\dot{V} + kV \ .$$

Since the speed of propagation is constant we can substitute $\ddot{V} = \theta^2 V''$ to obtain

$$\frac{r}{2\rho\theta^2}\ddot{V} - c\dot{V} - kV = 0 \ .$$

This linear equation has solutions of the form $V = Ye^{\lambda t}$, where Y is a function of y, and

$$\lambda = \frac{2c\rho\theta^2}{r} \quad \text{or} \quad -\frac{k}{c}$$

to first order in $1/\theta$, assuming that θ is large. Writing down the general solution, using the fact that it is a wave with velocity θ, and imposing the initial condition $y = 0$ when $t = 0$, gives the theorem.

We now go back to the observed action potential in Figure 3 (in section 1.1). There is a point J, prior to which there is exponential growth, and after which the slope $\dot{V}$ is constant for a period, and so there cannot be exponential growth. Therefore we assume that J is the trigger, and prior to J the propagation wave applies. The values of V, $\dot{V}$ at the critical point J are

$V_J = 40$ mV

$\dot{V}_J = 567$ mV/msec.

From these two readings and Theorem 3 we can deduce the propagation velocity θ, for the particular axon to which Figure 3 refers.

▶ Corollary to Theorem 3

$\theta = 21 \cdot 9$ metres/sec.

Notice that this agrees well with the observed velocity of $21 \cdot 2$ m/sec (see Figure 3), and is slightly better than the Hodgkin–Huxley computation of $18 \cdot 8$ m/sec.

Proof. After $0 \cdot 35$ msec the ratio of the growth term to the decay term in Theorem 3 is greater than e^5, and so we can ignore the latter. Therefore

$$\ddot{V} = \frac{2c\rho\theta^2}{r}V \ .$$

Putting in the values at the trigger point J,

$$\theta^2 = \frac{\dot{V}_J}{V_J}\frac{r}{2c\rho} \ .$$

In order to calculate θ in metres/sec we must rewrite the units in terms of metres and seconds, as follows.

$$\frac{\dot{V}_J}{V_J} = 14 \cdot 2/\text{msec} = 14 \cdot 2 \times 10^3/\text{sec}$$

$c = 1$ μFarad/cm^2 = 10^{-2} Farad/m^2.

For the axon under consideration [see 4, p. 528]
$$\rho = 35\cdot 4 \text{ ohm.cm} = 35\cdot 4 \times 10^{-2} \text{ ohm.m}$$
$$r = 238\mu = 238 \times 10^{-6} \text{ m}.$$
Hence
$$\theta^2 = \frac{14\cdot 2 \times 10^3 \times 238 \times 10^{-6}}{2 \times 10^{-2} \times 35\cdot 4 \times 10^{-2}} = 478 \quad .$$

Therefore $\theta = 21\cdot 9$ metres/sec.

2.9. CALCULATION OF ACTION POTENTIAL

To test the model we now calculate the action potential, that is, the graph of V against time. There are two separate phases: We assume that the action potential is determined

(1) before J by the propagation wave

(2) after J by the local nerve impulse equations.

Phase (1) is given by the formula in Theorem 3, putting θ equal to the speed determined by the Corollary. Also put $y=0$, because the action potential is measured at a point, and starts at $t=0$.

The end of phase (1) triggers the beginning of phase (2). More precisely phase (2) is determined by the orbit on the slow manifold (see Figure 25) beginning at the *trigger point* J: $b=1\cdot 2$ ($V=40$); $a=-0\cdot 45$; $x=-1\cdot 2$. The explanation of the trigger point is as follows. During phase (1) the propagation wave overrides the slow equation for b, and finishes at $V=40$ or $b=1\cdot 2$. Meanwhile the slow equation for $\dot{a}$ has marginally reduced a from the equilibrium value $-0\cdot 4$ to $-0\cdot 45$. At the same time the fast equation has been acting to keep the state on the slow manifold, and so the value $x=-1\cdot 2$ is obtained by solving $x^3 + ax + b = 0$.

Notice that during phase (1) four hidden events occur in rapid succession: first the pacemaker wave carries b past the threshold point $b=0\cdot 1$; secondly the fast equation determines the action, which jumps the state from the upper surface to the lower surface of the slow manifold, rapidly reducing x; thirdly as x crosses the value $x=-0\cdot 5$ the membrane becomes permeable to sodium; fourthly the rapid influx of sodium takes over as the main driving force in the slow equation for b during phase (2).

The calculated action potential is shown in Figure 26 and is drawn to the same scale as Figure 3 in section 1.1. Notice that it agrees remarkably well with the observed action potential in Figure 3b, and in fact has a slightly better shape than that given by the Hodgkin–Huxley theory in Figure 3a.

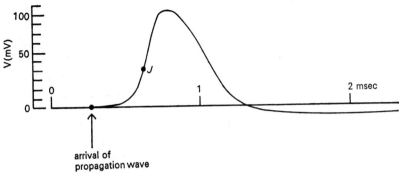

FIGURE 26
Calculated action potential

2.10. CONCLUSIONS AND EXPERIMENTS

We conclude by comparing our equations with the Hodgkin–Huxley equations [4, p. 518 and 5, p. 86]. Both are designed to fit the voltage clamp data, and both have similar time graphs for the potential and conductances, as shown by Figure 27. The graphs are drawn to the same scale, and it can be seen that in our theory the potential rises higher using smaller changes in conductance.

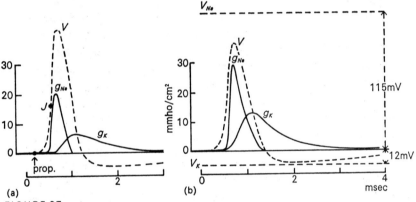

FIGURE 27
(a) Theoretical solution for propagated action potential and conductances.
(b) Analogous results for the Hodgkin–Huxley equations [5, p. 63]

The main points of difference between the two theories are as follows:

(1) Although we do not hold a very strong brief for our explicit equations, nevertheless they do belong to the general mathematical class of flows on the cusp catastrophe, which we deduced from the three dynamic qualities stated at

the beginning of the paper in section 1.1. Thom's deep uniqueness theorem for the cusp catastrophe shows that this class is not arbitrary but natural. We should expect this class to apply not only to the nerve impulse, but to any biological phenomena displaying the three dynamic qualities, such as spreading waves like epilepsy, migraine, or the Leão spreading depression, or discharge phenomena like reflexes, electric eels, or fireflies. Therefore our equations belong to a wider context.

(2) Since it is mathematically natural, and based on the simple dynamical qualities, the concept of representing the nerve impulse by a flow on the cusp catastrophe is well designed to withstand the inevitable buffeting by future discoveries about the underlying chemistry. By contrast the Hodgkin–Huxley equations are specifically based on the known chemistry, and may have to change fundamentally as further discoveries are made.

(3) Besides the electrical equation Hodgkin–Huxley use 9 *ad hoc* equations to fit the voltage clamp data, whereas we use only 1 *ad hoc* equation together with the 1 canonical fast equation.

(4) Hodgkin and Huxley [5, p. 86] assume that the change in potassium conductance is caused by a path for potassium being formed when 4 charged particles move to a certain region of the membrane under the influence of the electric field. As far as I know this 4-particle gate has not been confirmed by observation. If we trace the origin of the number 4 we find it came from data-fitting the delay at the beginning of the potassium conductance curves in Figure 24. Hodgkin and Huxley assume that the two permeabilities are independent of one another, and are governed by linear equations. Their procedure is mathematically equivalent to the following: fit the end of the curve by exponential decay, and then cure the abrupt rise at the origin by squashing it down with a fourth power. Therefore, ignoring constants, they obtain $g_K = n^4$, and $n = 1 - e^{-t}$. The variable n is then given physical interpretation as the probability that one of the gate particles is in the right place. Therefore n^4 is the probability that the 4-particle gate is in the right place.

By contrast we have a completely different explanation of the delay, because we assume the two permeabilities are interrelated. In essence we say that the potassium conductance cannot begin to rise until the sodium conductance has reached its peak. The situation is illustrated by Figure 23: the delay occurs during the initial fast orbit FG, during most of which $\dot{a}$ is small negative, and $\dot{a}$ only becomes positive once the slow manifold is reached at G. Therefore we have no need to assert a 4-particle gate. Similar remarks apply to sodium.

(5) The discussion above prompts the question of what is the physical interpretation of our variables a, x. We have avoided the issue by calling them 'dynamical' variables, and relating them functionally to g_K, g_{Na}. Possibly they may also represent the concentration, or the production-rate, of some chemical that transports the ions. This interpretation would be consistent with the *cut-off* devices that we have built into the functional relations. For instance, the concentration of a has to reach the critical level, $a > 0.5$, before it can begin transporting potassium. Similarly $-x$ has to reach the critical level, $-x > 0.5$, before two particles of x can begin transporting a sodium ion. These are tentative remarks, but perhaps they may suffice to germinate more specific chemical hypotheses that can be tested experimentally. In fact perhaps the most useful purpose our equations can serve is to suggest alternative chemical hypotheses against which to test the Hodgkin–Huxley theory.

(6) An interesting feature of our equations is the very definite cut-off of sodium permeability, well away from the resting potential. For compare

Resting potential : $x = 1.07$

Sodium cut-off : $x = -0.5$.

Therefore in our model there is no inward sodium leak at, or near, the resting potential, nor is there experimental evidence for such. On the other hand there is a great deal of evidence for an outward sodium pump [5, chapter 6] caused by metabolism during the resting potential. It is certainly comforting in our model not to have to have an inward sodium leak and an outward sodium pump going on at the same time. More seriously, *the two surfaces of the cusp catastrophe may in fact reveal the fundamental difference between the two membrane states of sodium pump and sodium leak.* By contrast Hodgkin–Huxley [5, pp. 65, 73] assume a marginal inward sodium flow at the resting potential to explain the data in Figure 22. We explain this by the action of chlorine and other ions, rather than sodium, which mathematically gives rise to the saddle point shown in Figure 25. These features of sodium cut-off and saddle point are consequences of juggling with the position of the equilibrium point and functional relations, rather than being intrinsic to flows on the cusp catastrophe. However they might be a useful area for testing between the two models.

(7) The existence of the saddle point in our model has another interesting consequence, which might provide a fruitful source of experiments. If the constant 1.67 in the slow equation for $\dot{a}$ (*see* section 2.7) is increased to 1.7 then this has the effect of running the saddle and the sink together, and annihilating both. Therefore the stable equilibrium point representing the resting potential

would be replaced by a stable attracting closed orbit, containing the action potential. In other words, the neuron would start firing by itself periodically, without any trigger or propagation wave.

What would cause such a change in constant? There might be a variety of causes because the same effect can be achieved by tinkering with other constants. Possibly use, or non-use, or external control of the neuron could change one or other of these constants. One can speculate on many uses for such a device, for instance:

(a) If the passage of one impulse increased the constant, then a volley of spikes would follow until fatigue reduced it again. In this way the neuron would act as amplifier for feeble messages.

(b) If the neuron had no dendritic input, then normal metabolism might increase the constant. Then the neuron would become a periodic self-firing neuron such as those found in the visual cortex, which possibly underlie the α-rhythm.

(c) The diffuse activating system of the reticular formation in the brain-stem might operate a control over the thalamus of this nature. If so this would help to explain the brain's facility in switching attention to, or from, a particular sense input.

(d) During sleep or dreaming, the brain-stem might operate a crude chemical control over the thalamus of this nature (such as the release of serotonin or noradrenaline by the Raphe system). This could explain the cut-off of sense inputs at thalamic level during sleep. The slowness of chemical control compared with the speed of neurological control in (c) above corresponds to the slowness of falling asleep and arousal compared with the speed of switching attention.

2.11. APPENDIX ON LAMARCKIAN EVOLUTION

We continue the digression of section 1.4. Recall the sequence of Examples 1–5 in sections 1.1 to 1.5 which we illustrate again in Figure 28.

Suppose that this sequence models the evolution of a linkage between two enzyme-systems of a cell. At the beginning (1) the two enzyme-systems are unlinked and are independently in homeostasis. At the end (5) the systems have become linked, and the state varies periodically round the van der Pol limit cycle. In other words the cell has developed a new biological clock.

Now biological clocks are of fundamental importance, not only for governing the timing of cell division, and so on, but also possibly for specifying position in the embryo (*see* recent experiments of Goodwin [13] and Wolpert [9]). In

3. Differential Equations for the Heartbeat and Nerve Impulse

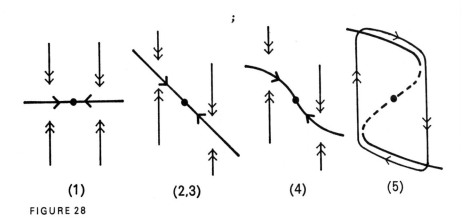

FIGURE 28

fact the appearance of a new biological clock in the cell may lead to a new type of cell, and eventually to a new species. To emphasize the point we shall speak of the formation of the new biological clock as a *change in phenotype*. This is not such an exaggeration as it seems at first sight, for, as we shall see, the formation of the clock is governed by the cusp catastrophe, and generalizes at once to all higher dimensional catastrophes. Now any change in phenotype must depend upon some change in embryo, and Waddington [14] suggests that the mechanisms underlying embryonic developments are chreods, which Thom [1] describes mathematically in terms of catastrophes. Therefore any change in phenotype should have as underlying mathematics a strict generalization of the simple clock we are discussing here.

Summarizing, the passage from (1) to (5) represents both a change in genotype, because the enzyme-systems have to become linked, and a change in phenotype, because of the formation of a new clock. Therefore let us examine where and how the changes occur in the sequence. First we must recap the precise role that mathematics is playing in the discussion. Formally there are three steps involved. The first step from genotype to differential equations is *modelling*. The second step from the equations to the solution and to the attractors is *mathematics*. The third step from the attractors to the phenotype is *interpretation*. Behind the second step lie deep mathematical theorems such as the existence and uniqueness theorems of differential equations, theorems about the attractors of structurally stable systems, and Thom's classification theorem of elementary catastrophes. Therefore if the model is good, allowing identification of equations with genotype, and if the interpretation is good, allowing identification of solution with phenotype, then we should be able to call upon the mathematics to analyze the difference between changes in genotype and phenotype. The purpose of the appendix is to use the simple examples that we happen to have at hand to illustrate this point, as follows.

The identification of equations with genotype gives preferred coordinates, namely the coordinates x, b representing the two enzyme-systems. Therefore the step (1)→(2) represents a change in genotype because in (1) the slow manifold is a preferred coordinate axis, whereas in (2) it is not. As explained in section 1.4, the enzyme-systems in (2) are no longer independent. By (3) the enzyme-systems have become linked, and so the step (2)→(3) is another change in genotype. Both these steps may take a long time in evolutionary terms, and may only be the result of a large number of gene mutations and duplications. The step (3)→(4) is comparatively trivial because (3) is the linear approximation of (4), and therefore we assume that it represents no change in genotype. The step (4)→(5) is of a different nature, because, as pointed out in section 1.5, it can be achieved by a *continuous* change of parameter in the fast equation of the cusp catastrophe. Such a change is generic, and could be caused not by genetical changes but by a continuous change in environment, such as variation in temperature, climate, or food. Moreover such change could be very rapid in evolutionary terms.

Now look at the solutions: There is no change in phenotype from (1) to (4) because the only attractor is the stable equilibrium point. The only change in phenotype is the Hopf bifurcation (4)→(5), creating the clock.

Therefore we have a schema, which, to an observer, would appear remarkably Lamarckian, as shown in Figure 29.

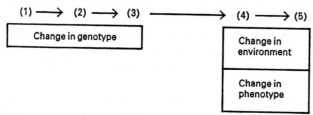

FIGURE 29

Of course it could happen only if the hidden random genetic mutations had stacked the cards ready beforehand, and afterwards the change would be reinforced by Darwinian natural selection on the phenotype. But it might help to explain why fossil records of phenotype can appear static over long periods, and then change relatively fast.

Similar arguments can apply to development as well as evolution. The continuous change (4)→(5) could be caused by diffusion between neighbouring cells coming into contact. For example, there may be a model of this nature associated with the development of mesoderm at the interface of contact between ectoderm and endoderm.

I am grateful to many people, particularly to René Thom and David Fowler for discussions about catastrophe theory, Boris Rybak and Ian Gray for discussions about the heart, Peter Buneman and Ted Woodcock for discussions about the nerve impulse, Francis Crick for suggesting at Serbelloni that catastrophe theory should be applied to the nerve impulse, and to Steve Smale, whose lectures [15] on electrical circuit theory at IMPA suggested how to put feedback on catastrophes into the classical framework of ordinary differential equations. I am indebted to Ted Woodcock [12] for checking the equations against the Hodgkin–Huxley data by computer. I should also like to acknowledge how many of the ideas sprang from A. L. Hodgkin's very stimulating book [5]. Finally may I apologize to my mathematical and biological readers, lest, by attempting to address the one, I inadvertently bore the other.

References

1. R. Thom, Topological models in biology, in (C. H. Waddington, ed.) *Towards a Theoretical Biology 3: Drafts* (Edinburgh University Press 1970).
2. R. Thom, *Stabilité Structurelle et Morphogénèse* (Benjamin, in press).
3. E. C. Zeeman, Applications of catastrophe theory, *Bull. London Math. Soc.* (In press.)
4. A. L. Hodgkin and A. F. Huxley, A quantitative description of membrane current and its application to conduction and excitation in nerve. *J. Physiol. 117* (1952) 500-44.
5. A. L. Hodgkin, *The Conduction of the Nervous Impulse* (Liverpool University Press 1964).
6. E. Hopf, Abzweigung einer periodischen Lösung von einer stationären Lösung eines Differentialsystems. *Ber. Verh. Sächs. Akad. Wiss. Leipzig. Math.-Nat. Kl. 95* (1943) 3-22.
7. D. H. Fowler, The Riemann–Hugoniot catastrophe and van der Waals' equation, in (C. H. Waddington, ed.) *Towards a Theoretical Biology 4: Essays*, pp. 1-7 (Edinburgh University Press 1972).
8. B. Rybak et J. J. Béchet, Recherches sur l'électromécanique cardiaque, *Pathologie-Biologie 9* (1961) 1861–71, 2035-54.
9. L. Wolpert, Positional information and the spatial patterns of cellular differentiation, *J. theoret. Biol. 25* (1969) 1-47.
10. Sampson Wright, *Applied Physiology* (Oxford Medical Publ. 1965, Eleventh edition).

11. A. L. Hodgkin, The ionic basis of electrical activity in nerve and muscle, *Biol. Rev. 26* (1951) 339-409
12. A. E. R. Woodcock, to appear.
13. B. Goodwin, *Temporal Organisations in Cells* (Academic Press : London 1963).
14. C. H. Waddington, *The Strategy of the Genes* (Allen and Unwin : London 1957).
15. S. Smale, On the mathematical foundation of electrical circuit theory (to appear).

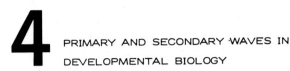

4 PRIMARY AND SECONDARY WAVES IN DEVELOPMENTAL BIOLOGY

ABSTRACT.

Using catastrophe theory, we prove a theorem to the effect that whenever a multicellular mass of tissue differentiates into two types, the frontier between the two types always forms to one side of its final position, and then moves through the tissue before stabilising in its final position. We call this movement a primary wave. Primary waves may sometimes be identified as hidden waves of cell determination, which may not manifest themselves visibly until after a delay of several hours. The visible manifestation will then be a secondary wave of cellular activity, which may cause morphogenesis, for example rolling changes of curvature.

Two applications are worked out in detail, namely models for gastrulation and neurulation of amphibia, and for culmination of cellular slime mold. In the amphibian model the differentiation between ectoderm and mesoderm causes a hidden primary wave, whose visible secondary wave of cells submerging causes not only the morphogenesis of gastrulation but also the formation of notochord and somites. In the slime mold model the differentiation between spore and stalk causes a hidden primary wave, whose visible secondary wave of cells submerging causes culmination and the morphogenesis of the fruiting body. Both models suggest experiments by which they can be tested.

Published in Lectures on Mathematics in the Life Sciences American Mathematical Society, Rhode Island, Volume 7, 1974, pages 69-161.

1. INTRODUCTION.

Our objective is to explain primary waves by catastrophe theory [18], and secondary waves by cell physiology [1,8], and then to use them both together to explain morphogenesis.

By a <u>wave</u> we mean the movement of a frontier separating two regions. We call the wave <u>primary</u> if the mechanism causing the wave depends upon space and time. We call the wave <u>secondary</u> if it depends only upon time, in other words it is series of local events that occur at a fixed time delay after the passage of the primary wave. Therefore, whereas the wave-form of the primary wave is fundamental, the secondary wave only appears to have a wave-form because it follows the primary wave after the fixed time delay. In a sense the wave-form of the secondary wave is accidental because it could be disrupted by mixing up the substrate in between the passage of the two waves. The epidemic example in §2 below illustrates this point. The point is further emphasised by the following difference between the primary and secondary waves : if the substrate is cut before the passage of the primary wave then this stops the primary wave. However if the cut is made between the passage of the two waves then this will not stop the secondary wave, which will appear to jump across the cut.

If the primary wave is invisible, then the secondary wave may appear mysterious. We suggest that this may be a typical situation in developmental biology. For instance a primary wave across a multicellular mass of tissue might consist of the switching on of certain gene systems in each cell, and this may be difficult to detect at the time because biochemical analysis tends to disrupt the delicate dynamics; in fact most experimental evidence that gene systems have been switched on seems to come from observation of some secondary effect after a suitable time delay (see §7 below). The secondary effect in this case is usually some physical manifestation in cell behaviour such as change in chemical composition, change in RNA content, change in oxygen consumption, change in membrane cohesiveness, change in shape, change in amoeboid activity, change in mitosis rhythm, etc. Another common and important secondary effect is for the cell to alter the ratio between the areas of that part of its membrane in contact with other cells and that part comprising free surface of the tissue; for instance the cell can increase contact with other cells by amoeboid action towards them, and decrease its free surface by wrinkling its free membrane (see Figure 21). For convenience we call this process <u>submerging</u>. For example submerging happens during gastrulation (see Figure 20).

It is the secondary wave of physical manifestation that may signal the release of chemical energy to provide the physical energy necessary for morphogenesis. For example submerging cells may push and pull on their neighbours, and thereby alter the overall curvature of the free surface, as described in Gustafson & Wolpert [8]. We suggest that some morphologies that hitherto may have appeared to be explicable may now be explained in terms of secondary waves. If this is the case, this may provide a conceptual framework for the experimental search for hidden primary waves.

The next question is : what causes a primary wave ? The simplest mechanism is diffusion, for instance of chemicals or signals. In order to illustrate the difference between primary and secondary waves, we briefly give elementary examples of epidemics and regulation in §§ 2 and 3, in which the primary wave is caused by diffusion.

However for the rest of the paper we are interested in a more subtle mechanism for producing primary waves. We prove a theorem that the four hypotheses

      I Homeostasis
      II Continuity
      III Differentiation
      IV Repeatability

together imply the existence of a primary wave. Another way of stating the theorem is that whenever a frontier forms between two types of tissue, it first forms off to one side, and then moves as a primary wave through the tissue before stabilising in its final position. The theorem gives no indication of the extent of the movement, although in some applications it appears the movement can be very large, over half the diameter of the embryo. Since the formation of frontiers is common in developmental biology, we should expect such primary waves to be common, and hence morphogenesis to be commonly caused by their secondary waves.

We state the theorem more precisely in §5, and prove it in §8. The proof consists of translating the above four hypotheses into mathematics, thereby making them precise, and then using Thom's classification of elementary catastrophes [18]. Since the latter result is deep, the theorem is a non-trivial description of the mechanism underlying this type of primary wave.

One of the interesting features of the theorem is that diffusion may or may not be present during the passage of the wave : it is irrelevant to the proof. In other words the complexity of biochemical events associated with a particular primary wave, even one that subsequently slows down to halt, may include, for instance, diffusion of chemicals across cell membranes, or diffusion of dynamical signals across the cells entraining some activity in them. On the other hand another primary wave may progress without any diffusion, and without any signals. In this case the wave would be purely kinematic, with each cell behaving according to its own internal clock. The clocks may have been synchronised initially, but, due to some underlying gradient across the tissue, may tick at different speeds. Therefore a switch inside a cell, for instance the hidden switching on of some gene system (analogous to the clock striking), will occur in different cells at different times. The continuity of the underlying gradient will ensure a continuity of these different times across the tissue, and so the switch progresses as a primary wave. The question remains as to what actually causes the switch, and an analysis of the proof of the theorem reveals that the basic cause is, surprisingly, homeostasis. What is homeostasis? - we choose to translate homeostasis into mathematics as a stable equilibrium point of a time-dependent multidimensional dynamical system (see §8).

Returning to diffusion for a moment : of course the continuity of the underlying gradient may itself be due to a much earlier diffusion, for instance in the cytoplasm of the original egg before cleavage. And again if a cell is grafted to alien surroundings, then a new diffusion may occur that upsets its clock. However under normal conditions, it could be possible for both development and primary waves to occur without diffusion. Therefore one implication of the theorem is to alter the possible expectations of the experimentalist concerning "morphogens". A related result from catastrophe theory [26] guarantees the existence of morphogens, but they may not act like classical "organisers". In other words, if we locate the organising centre of some morphology, we may still expect to find a morphogen, that is a chemical, physical or dynamic gradient, whose discontinuities reflect the organisation, but we should not necessarily expect to find an organiser, that is a chemical, physical or dynamic signal emanating from the centre.

The theorem gives both qualitative and quantitative predictions about the shape and speed of primary waves, and consequently also about their secondary waves. We illustrate the theorem by two applications in §§10 - 17, namely the gastrulation of amphibia, and the culmination of slime mold. In both examples we take as hypothesis (for which there is experimental evidence) a secondary wave of cells submerging. In the first example the differentiation between ectoderm and mesoderm causes a primary wave, by the theorem. We take as hypothesis (for which there is experimental evidence) that this hidden primary wave begins at the bottom of the grey crescent. We then deduce that the secondary wave causes, or helps to cause, gastrulation, the dorsal lip, the blastopore, the archenteron roof, the separation of mesoderm from endoderm, the neural folds, and the formation of notochord and somites during neurulation. The fact that nearly all the main morphogenetical movements arise from a single secondary wave, resulting from a single differentiation and the local activity of cells submerging, help to explain why these movements are common to many species.

In the second example of slime mold the differentiation between spore and stalk cells causes a primary wave, by the theorem. We take as hypothesis (for which there is experimental evidence) that the primary wave begins at the tip and proceeds $\tfrac{1}{8}$ of the way along the grex, several hours before culmination. We then deduce the morphogenesis of the fruiting body, giving predictions as to shape and speed.

I am indebted to many people for discussions, particularly mathematicians, René Thom, David Fowler and Klaus Jänich, and biologists C.H. Waddington, Jack Cohen, Lewis Wolpert, Peter Nieuwkoop, Jonathan Cooke and John Ashworth. The main inspiration came from years of conversations with René Thom about applying catastrophe theory to biology. Meanwhile in counterconversations Lewis Wolpert emphasised the inadequacy of using catastrophe theory by itself, because it can only explain the geometry, and not the forces that shape the embryo. On the other hand looking at the local forces by themselves cannot explain the global geometry. Hence the concept or primary and secondary waves grew out of trying to put these two ideas together, the mathematical and the biological.

Discussions with Peter Nieuwkoop about gastrulation were particularly valuable during the germination of the ideas, after an initial presentation of the theorem at a conference in Göttingen in September 1973 organised by Klaus Jänich. Jonathan Cooke stimulated the ideas about pattern-formation and the somites. John Ashworth explained to me the slime mold morphogenesis. I am indebted to the A.A.A.S. and A.M.S. for the opportunity to present the ideas, and to the various authors and journals for permission to reprint their diagrams.

CONTENTS.

1. Introduction.
2. Example : epidemic.
3. Examples : regulation.
4. Example : ecology.
5. Primary waves in embryology.
6. Secondary waves in embryology.
7. Experimental detection of a primary wave.
8. Proof of the theorem.
9. Quantitative aspects of the theorem.
10. Gastrulation of amphibia.
11. Curvature.
12. Neurulation.
13. Pattern formation.
14. Timing of gastrulation and neurulation.
15. Experiments to be done.
16. Slime mold culmination.
17. Experiments to be done.
18. Conclusion.
19. References.

2. EXAMPLE : EPIDEMIC.

This is a simple example to illustrate the difference between primary and secondary waves. The substrate is people. The two regions are those who have been infected by the epidemic and those who have not. The frontier bounds the infected region. This frontier moves forward as a hidden primary wave of infection. This is a simple diffusion wave, that moves steadily forward at the speed of diffusion as each person infects his neighbours (similar to Huyghens' principle). Then after a fixed time delay the visible secondary wave of symptoms follows, assuming that the people have remained stationary. If the people move about in between the waves then the wave-form of the secondary wave will be disrupted. If the substrate is cut before the arrival of the primary wave, in other words the infected population is quarantined, then this stops the primary wave, the spread of infection. However if the substrate is cut between the waves, in other words only the population already showing symptoms is quarantined, then this does not stop the secondary wave, the spread of symptoms.

3. EXAMPLES : REGULATION.

In the heartbeat the pacemaker wave causing muscular contraction is a primary wave, and about half a second later the wave muscular relaxation is a secondary wave [24]. In the nerve impulse the membrane depolarisation along an axon is a primary wave, and about a millisecond later the repolarisation is a secondary wave. On both these cases the primary wave is electro-chemical, and could be described as a diffusion wave (diffusion of electrons) proceeding according to Huyghens' principle.

The main feature of a diffusion wave is that it preceeds at constant speed. By contrast the main feature of the more subtle kind of primary wave that the theorem describes, and which we shall be considering from now on, is that it slows to a halt; in other words the frontier stabilises.

4. EXAMPLE : ECOLOGY.

This is a simple example to illustrate the more subtle kind of primary wave. The wave has complicated causes, but is easy to understand beacuse it is visible rather than hidden. The ecology is greatly oversimplified, but then we are only using it to illustrate the idea.

Consider the ecological development of grass and trees over a continuous environment of soil and climate. Suppose for simplicity that the northern end of the environment is suitable for grass only and the southern end for trees only, so that the former eventually develops into mature grassland, and the latter into mature forest. Suppose that as either vegetation gets established it suppresses the other; trees fail to survive in grassland and grass fails to survive in forest. Suppose at first there is a continuous variation of vegetation, varying from forest in the south, with trees gradually thinning as we proceed north, until grassland is reached. Then at time t_1 the forest will develop a noticeable frontier at latitude s_1, say. This frontier will deepen, in the sense that the difference between the two sides of the frontier will become more marked, due to the suppressive effect of either vegetation upon the other. As the frontier deepens it would be exceptional for it to remain at s_1, the place where it originally formed (exceptional from the point of view of repeatability, as we explain below). Therefore, depending upon the initial conditions, it will either move north as the mass of trees seed themselves into the grassland, or move south as the grassland erodes the forest edge. Suppose that in our case the initial conditions are such that the frontier moves north as in Figure 1.

Figure 1. The frontier of a forest moving as a primary wave.

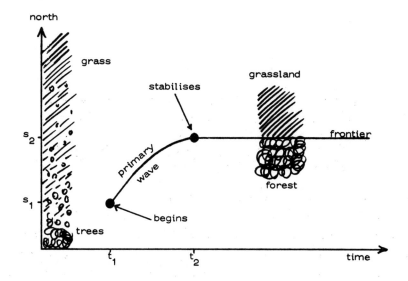

Eventually the northerly expansion of the forest balances out against the unsuitability of the northern climate for trees, and so the northerly movement of the frontier slows down, until it stabilises at s_2 at time t_2. Thereafter the frontier remains in stable equilibrium, and deepens further. The movement of the frontier from s_1 to s_2 during the time interval $t_1 < t < t_2$ is the primary wave.

The primary wave is succeeded by a series of secondary waves representing the spread of various species of flora and fauna that require various time delays of maturity before the forest becomes a suitable habitat for them (woodworm prefer old trees). However in this example we are less interested in the secondary waves.

To show that the primary wave is an illustration of the theorem, we must interpret the four hypotheses in this case.

I : Homeostasis is the tendency of the vegetation in any one place to develop into a stable state, stable with respect to time, that we can name as grassland or forest.

II : Continuity refers to both the continuous environment of soil and climate, and the initial continuous variation of vegetation.

III : Differentiation means that at the end there are two distinct states, mature grassland and mature forest.

IV : Repeatability means that if the initial conditions are varied slightly then the values s_1, t_1, s_2, t_2 may vary slightly but the qualitative behaviour of the frontier remains the same. In other words repeatability means that the whole space-time development, or <u>chreod</u> [22], will be stable under sufficiently small perturbations of the initial conditions. Repeatability is an essential hypothesis for the existence of the wave, because it implies $s_1 \neq s_2$. Otherwise, if $s_1 = s_2$, meaning that the frontier had stabilised where it formed, and so causing no wave, then this would be unrepeatable, in the sense that the initial conditions must have been exceptional, and an arbitrarily small perturbation of them could cause $s_1 \neq s_2$, and hence cause a wave, in other words a qualitatively different development.

Notice that in this example of a primary wave, the wave could be said to be caused by diffusion as the trees seed themselves into the grassland. However the situation is not as simple as in the previous examples, where the speed of the wave was constant and equal to the speed of diffusion, because here the wave slows down and stops as the frontier stabilises. One could make an elementary model of this slowing down by using a linear differential equation with a diffusion term balanced against a survival term, but this would not give insight into the formation and the deepening of the frontier, as does the more sophisticated catastrophe model. Also the two models give different quantitative predictions, which would distinguish between them : for instance in the elementary diffusion model the wave slows down exponentially, but in the catastrophe model it slows down parabolically (see §9). The catastrophe model is more likely to be correct, because of the two processes involved, the initial seeding by diffusion, and the eventual development into mature forest by homeostasis, the latter is the more significant.

To illustrate examples of kinematic waves that depend only upon gradients and internal clocks and not upon diffusion, consider the effects of latitude. For instance the spring blossoming of trees is a wave moving north (in the northern hemisphere) and in autumn the onset of migration by birds is wave moving south.

The reader will easily recognise the existence of many visible primary waves of this nature in ecology, evolution, anthropology and sociology. However in this paper we are more concerned with <u>hidden</u> primary waves in developmental biology, that cause secondary waves of physical manifestations in cells, that in turn may cause morphogenesis. We shall therefore state and prove the theorem in this context.

5. <u>PRIMARY WAVES IN EMBRYOLOGY.</u>

We begin by enlarging a little on the meaning of the four hypotheses in developmental biology. There is no need to give precise definitions at this stage, because the terms are given precision by the way we choose to translate them into mathematics in the proof of the theorem in §8 below. Suppose that E is a multicellular mass of tissue. We are concerned with development of E during a particular time interval T.

I : Homeostasis means that each cell is in stable biochemical equilibrium, an equilibrium that may change with time.

II : Continuity means that at the beginning of T we can represent the chemical, physical and dynamical conditions in different cells by smooth functions on E (the conditions inside a particular cell are represented by the values of the functions at the centre of mass of that cell). In an embryo, where the tissue has developed from an egg by cleavage, the continuity is inherited from the original continuity in the egg, which was due to diffusion in the egg cytoplasm. Any slight discontinuities that arise later tend to be evened out by subsequent diffusion across the cell walls. In aggregates of cells like slime mold, continuity means that the cells have sorted themselves out according to continuous gradients during the aggregation process.

Continuity implies that neighbouring cells will follow nearby paths of development whenever possible. We shall prove that where a frontier stabilises this is not possible, and so across the frontier neighbouring cells will follow divergent paths of development, and large discontinuities will therefore arise.

III : Differentiation means that, whereas at the beginning of T there is only one type of cell (or, more precisely, a continuous variation amongst the cells), at the end of T there are two distinct types, and no continuous variation from one type to the other.

For simplicity we may assume that the tissue E is polarised, that is to say all variation takes place in one direction only (like the north-south line in the previous example in §4). Therefore for mathematical analysis it suffices to consider a 1-dimensional space interval S in that direction. Continuity means that at the beginning of T the cells vary continuously along S. Differentiation means that during T the cells at opposite ends of S develop continuously into different types. At the end of T, since there is no continuous variation between the two types there must be a frontier point in S separating the two types. This implies a frontier surface in E, separating the two types of tissue. If we can show that the frontier point in S moves, then this will imply that the frontier surface in E moves.

IV : Repeatability means that the development is stable, that is to say a qualitatively similar development will take place under sufficiently small perturbations of the initial conditions.

Main Theorem. _Homeostasis, continuity, differentiation and repeatability imply the existence of a primary wave. In other words a frontier forms, moves and deepens, then slows up and stabilises, and finally deepens further._

Therefore whenever a frontier forms, it first forms off to one side and then moves as a primary wave through the tissue before stabilising in its final position. Here by "final position" we mean the position relative to the underlying tissue, which itself may be undergoing morphogenetical movements. The theorem is illustrated in Figure 1, and the proof is given in §8 below.

Remark 1. The theorem is qualitative rather than quantitative; in other words it is a result invariant under diffeomorphisms of space and time. Therefore the theorem cannot predict the extent of travel of the wave, and so in applications the extent must always be taken as an extra hypothesis, and verified experimentally as in §7 below. It appears that some primary waves may travel a large distance, particularly those associated with morphogenesis. For example in gastrulation of some newts the ectoderm/mesoderm frontier travels from the grey crescent at latitude $40°S$ (see Figure 12) to its stabilisation position at $40°N$, which is more then half the diameter of the blastula.

On the other hand the theorem does give quantitative predictions about the initial deepening, and the final stabilisation, of the frontier, because both these obey parabolic laws, and parabolicity is a diffeomorphism-invariant (see §9). These laws should furnish easily testable predictions.

Remark 2. The theorem gives no indication of whether the primary wave is visible or hidden. In embryology primary waves are generally hidden, in the sense of being experimentally undetectable at the time, because they probably consist of the switching on of gene systems, although their passage can sometimes be tracked in retrospect by the grafting experiment described in §7 below.

6. SECONDARY WAVES IN EMBRYOLOGY.

The theorem gives no indication of whether or not a hidden primary wave will result in a visible secondary wave after a time delay; and if it does, the theorem gives no indication of the size of delay, nor of the type of secondary wave. These must depend upon extra detailed biochemical hypotheses about that particular systems that are switched on, what the long term effects these systems have on the cells, how these effects physically manifest themselves, and whether there is a resulting energy release. Broadly speaking these are three possibilities.

In order to describe the three cases it is necessary to be a little more precise about what we mean by the word differentiation. Differentiation can be used in two senses, firstly the hidden determination of the cells into two types that takes place during the passage of the hidden primary wave, and secondly the subsequent development of physical difference between the two types that can be observed. In the hypothesis of the theorem in the last section we used differentiation in the first sense, because this was the best word to capture hypothesis III, the determination of distinct types. In this and the next section we use differentiation in the second sense, because this is the more normal usage. We now describe the three cases.

(i) <u>There is no secondary wave of energy release, and the speed of differentiation is slow compared with the speed of the wave.</u> No release of energy implies no morphogenesis. The slowness of differentiation implies that any secondary wave will probably be unnoticeable. Therefore the only visible effect will be the slow appearance of the frontier between the two types of tissue in its final position s_2. The original primary wave may never be noticed unless looked for.

(ii) <u>There is no secondary wave of energy release, and the speed of the differentiation is fast compared with the speed of the wave.</u> Again no release of energy implies no morphogenesis. However in this case the swiftness of differentiation implies that the frontier between the two tissues will appear before it has stopped moving, and so will present a visible secondary wave. The frontier may not necessarily first appear at s_1, because it may not yet be deep enough to notice, but it will appear at some point s_3, where $s_1 < s_3 < s_2$. It will then move towards s_2, deepening and slowing down according to a parabolic law. This means that near s_2 the speed of the wave s is proportional to $\sqrt{s_2 - s}$ (see §9 Corollary 3). The frontier then stabilises at s_2 and deepens further.

This is a commonly observed phenomenon, and is often described as <u>recruitment</u>. For example* a developing insect eye starts with a few cells, and then enlarges by recruiting neighbouring epidermis cells. The word "recruitment" implicitly suggests that the experimenter should look for an "evocator" or an "organiser" emanating from the existing eye cells causing the recruitment. However if this expanding frontier of the eye were a secondary wave, then perhaps one ought to look for a hidden primary wave passing some hours before, without necessarily any organiser. In mammalian eyes the wave goes the other way: instead of expanding outwards the optic region shrinks in size.

---

* I am indebted to Peter Shelton for this example.

(iii) There is secondary wave of energy release. In this case there will be a dramatically observable secondary wave causing morphogenesis. Geometrically the situation will be complicated by the fact that not only is the wave moving through the tissue, but the tissue itself is also moving and changing shape, with the former causing the latter. It is also likely that the energy release preceeds differentiation. Therefore differentiation may (or may not, as in cases (i) and (ii) above) appear as another secondary wave after the morphogenesis. Therefore two misinterpretations are possible : Firstly, it may be geometrically unobvious to relate the visible secondary differentiation wave to the hidden primary wave, because the morphogenesis will have moved the tissue around in between the two. Secondly it may be tempting to conclude that the differentiation has been evoked by the morphogenesis, or by the new position of the tissue, rather than by the original hidden primary wave. Thus the experimenter's path may be strewn with pitfalls, unless he manages to uncover the primary wave, which possibility we now discuss.

7. EXPERIMENTAL DETECTION OF A PRIMARY WAVE.

Sometimes the hidden **primary** wave may mark a loss in potentiality. In this case, the passage of the wave can be detected by a standard grafting experiment, provided that the tissue is suitable for grafting. For example one can detect the hidden mesoderm wave in amphibian gastrulation by this method[*].

Suppose, as before, that the wave starts at s_1 at time t_1 and stabilises at s_2 at time t_2. At this stage it is hidden and so there is no detectable difference between the cells, but eventually at some later time t_3 differentiation will cause a physically observable difference that we indicate by shading the regions marked A, B in Figure 2. At time t_1, although there is no difference, we can say that a-cells are presumptive A-cells and b-cells are presumptive B-cells. The wave travels from a towards b. The passage of the wave past a cell is indicated by that cell switching from being a b-cell to being an a-cell.

---

[*]I am indebted to P.D. Nieuwkoop [12] for explaining this experiment to me.

Figure 2. Detecting a primary wave by grafting.

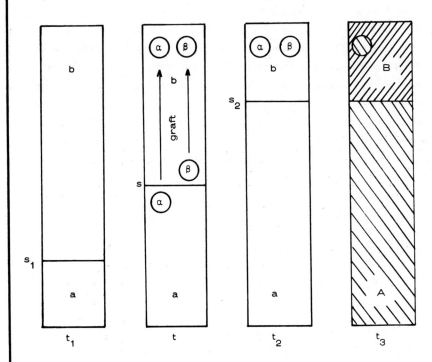

The switch occurs essentially because of homeostasis, as we explain in the proof of the theorem in the next section. Therefore b-cells have the potentiality to develop in either A-cells or B-cells. However a-cells may only have the potentiality to develop into A-cells, and we suppose that this is the case. Therefore the primary wave marks the loss in B-potentiality.

If we want to verify that the hidden primary wave has reached position s at time t, then at time t graft two small pieces α, β from just behind and just ahead of s onto b-tissue, well clear of the wave, as shown in Figure 2. The β-cells are influenced by their new position to remain b-cells, and so by time t_3 develop into B-cells, causing the β-graft to disappear. Meanwhile the α-cells have already switched and lost their B-potentiality and so by time t_3 the α-graft stand out as a patch of A against B.

8. PROOF OF THE THEOREM.

We follow the conceptual ideas of Thom [18, 19]. As explained in §5 above it suffices to use a 1-dimensional interval of space S, transverse to the forming frontier. Let T be a time interval encompassing the development. Let $C = S \times T$ be the rectangle of space-time. Let X denote a manifold representing the states of a cell. One can envisage X as a bounded open subset of n-dimensional euclidean space R^n, where n may be very large (possibly several thousand). The coordinates $\{x_i; i=1,2,\ldots,n\}$ of a point $x \in X$ may represent not only the concentrations of the different proteins in the cell, and the rates of change of those concentrations, but also may include variables representing various physical characteristics of the cell, its membrane, the cell dynamics, etc.

Consider a cell at the point $c \in C$. By Hypothesis I this cell is in homeostasis. We choose to translate homeostasis into mathematics by assuming that the biochemistry of the cell can be modelled by a gradient dynamical system on X

$$\dot{x} = -\operatorname{grad} V_c,$$

where $V_c : X \to R$ is a smooth function and R denotes the real numbers (see Remark 3 below for the meaning of this function). We choose to translate Hypothesis II, continuity, into mathematics by assuming that V_c can be chosen to depend smoothly on c. Therefore we have a function

$$V : C \times X \to R$$

given by $V(c,x) = V_c(x)$. We choose to translate Hypothesis IV, repeatability, into mathematics by assuming that V is generic*. Let $M \subset C \times X$ denote the set of stationary values of V, given by $\nabla V = 0$, where ∇ denotes the gradient with respect to X. Let G denote the closure of the subset of minima, which are given by $\nabla^2 V$ positive definite, where ∇^2 denotes the Hessian with respect to X. Then by smooth genericity, M is a smooth 2-dimensional surface in the (n+2)-dimensional space $C \times X$, and G is a subsurface of M with boundary ∂G. Let $\chi : M \to C$ denote the

---

* Generic means in general position, that is to say the map $c \to V_c$ maps C transverse to the natural stratification of $C^\infty(X)$. Generic V's are open dense in the space of all V's, and therefore both stable, and permissible to use as models.

map induced by the projection $C \times X \to C$. By Thom's classification theorem of elementary catastrophes [18], the only singularities of χ are fold curves and cusp points, since V is smooth and generic. The boundary ∂G consists of fold curves and cusp points.

By homeostasis the state of cell at c is at a minimum of V_c, and therefore represented by a point
$$\sigma(c) \in G \cap \chi^{-1}c.$$
Therefore σ is a section of χ,

$$\chi \downarrow\!\!\uparrow \sigma \quad \begin{matrix} G \\ C \end{matrix}$$

In other words $\chi\sigma = 1$. The interesting point is that, whereas χ is smooth (induced by projection), σ may be forced by χ to be discontinuous. We now use Hypothesis III, differentiation, to analyse this discontinuity and show, using I and IV again, that it implies the primary wave.

Let $S = [s_0, s_3]$, $T = [t_0, t_3]$. Let us analyse the continuity of σ on the boundary ∂C of $C = S \times T$. Firstly σ is continuous along the side $s_0 \times T$ because a-cells are developing smoothly into A-cells. Similarly σ is continuous along $s_3 \times T$ because b-cells are developing smoothly into B-cells. Next σ is continuous along the side $S \times t_0$, because, by Hypothesis II, continuity, we may assume that at the beginning the tissue is continuous.

Finally σ cannot be continuous along the side $S \times t_3$, because, by Hypothesis III, differentiation, at the finish two distinct types of cells A, B have developed, with no continuous variation between them. By Hypothesis II, continuity, A must spread continuously from one end, and B from the other, towards some point of discontinuity, which is, as yet, undetermined.

Figure 3 shows the partial section of G over ∂C that must exist ready to receive the map $\sigma : \partial C \to G$. Now comes the problem of extending σ to the interior of C. First we ask the simpler question: what singularity must the map $\chi : M \to C$ have over the interior of C? Since $M \supset G$, M must extend the section over ∂C shown in Figure 3. Therefore by the classification theorem M must have at least one cusp singularity over the interior of C. A single cusp would be sufficient. Moreover a single cusp is qualitatively the simplest solution of the

Figure 3. Section of G over the boundary of S×T.

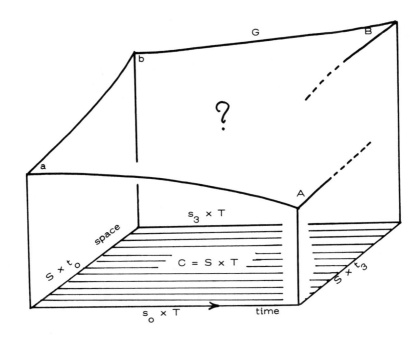

extension problem, and we can justify the simplest solution by again appealing to Hypothesis II, continuity; in other words we assume minimal discontinuity subject to Hypothesis III, differentiation.

In Figure 4 we illustrate two examples of a surface M over C, each with one cusp, and each extending the given section over ∂C. In each case the shaded subsurface indicates M−G (representing saddle-points of V, in other words unstable equilibria of the biochemical dynamic, and so not realisable by homeostasis). If we ignore the product structure of space-time C = S×T, then the two pictures are qualitatively equivalent. However if we take note of the product structure, then Figure 4a is exceptional because the time-axis at the cusp point c_o coincides with the cusp-axis. Hypothesis IV, repeatability forbids this exceptional

160   4. Primary and Secondary Waves in Developmental Biology

Figure 4. The graph G of homeostatic states.

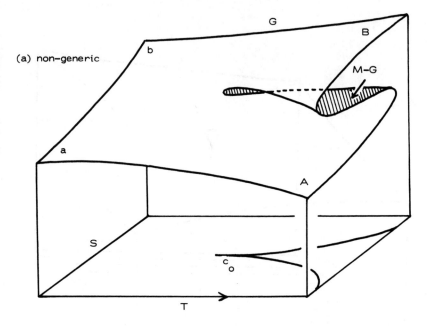

(a) non-generic

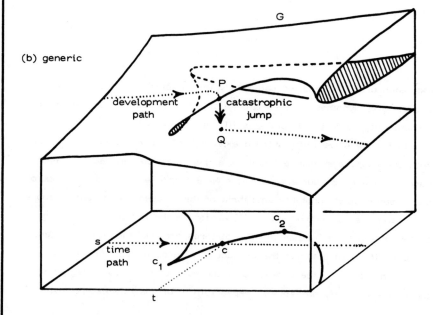

(b) generic

situation and so Figure 4a is ruled out.

The reader may ask why do we choose the rather elaborate Figure 4b, and so we must explain. Firstly the direction of the cusp-axis must have a non-zero S-component (to avoid the fault of Figure 4a) and a positive T-component, because the frontier between the differentiated tissue must occur inside the cusp. Probably in most cases the T-component will be greater, but to emphasise both components in Figure 4b we have drawn them approximately equal, in other words in the perspective drawing of C we have drawn the cusp-axis inclined at $45°$ to the time-axis at c_1. Secondly, as the two branches of the cusp widen out, there will be a unique first point c_2 (on the branch for which t is greater) where the tangent is parallel to the time-axis. These two points c_1, c_2 will mark the beginning and ending of the primary wave, as we shall now prove. We have merely drawn Figure 4b so as to emphasise the qualitative features of these two points.

We now plot the development of each cell by lifting its time-path in C up onto the graph G. Since G is the graph of homeostatic states, the lifted path will represent how the state of the cell changes, in other words will represent its development-path (see Figure 4b).

Since homeostasis is represented by a differential equation $x = -\nabla V$, the development path will be a continuous path on G, held continuously in stable equilibrium by the differential equation, unless the path happens to cross ∂G. Therefore, in the language of Thom [18], the changing state will obey the Delay Rule. Now ∂G is the fold curve of M lying above the two branches of the cusp. Suppose that the time-path of the cell at s crosses the branch $c_1 c_2$ of the cusp at the point c at time t (see Figure 4b). Then the development-path of s will cross ∂G at the point P above c, and at this point the homeostatic stability breaks down, because the corresponding minimum of V_c has coalesced with a saddle (represented by M-G), and disappeared. Consequently the homeostatic differential equation comes into play and carries P rapidly to Q, which is the unique new stable equilibrium on G above c, in whose basin of attraction P lies (see [26]). The rapid change of state from P to Q caused by homeostasis is called a catastrophe, or catastrophic jump; this is the moment when the b-cell at s switches into an a-cell. What we have previously vaguely refered to as "switching on of gene systems" is represented mathematically by the fast flow from P to Q along an orbit of the differential equation in X representing homeostasis. That was why in the introduction and in §7 above we remarked that the switch occurs essentially because of homeostasis. Therefore at time t the cell s marks the frontier between a-cells and b-cells, namely the position of the wave. The depth of the frontier is represented by the length PQ. The catastrophe occurs in all cells lying in the interval $s_1 < s < s_2$ during the period $t_1 < t < t_2$, as their time-paths cross the cusp branch $c_1 c_2$, and this defines the primary wave. Therefore $c_1 c_2$ determines the track of the primary wave in space-time

Meanwhile cells for which $s < s_1$ suffer no catastrophe, but develop from a-cells into A-cells along a smooth development-path. Similarly cells for which $s > s_2$ develop smoothly from b-cells into B-cells. The various development paths are shown in Figure 5.

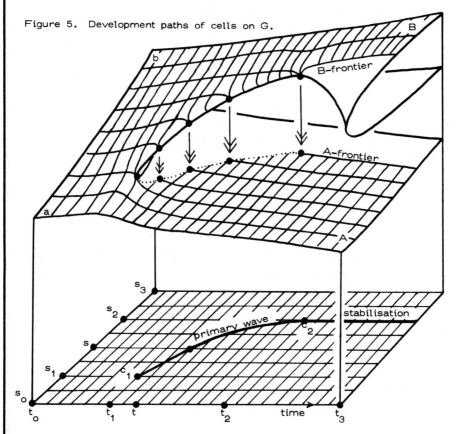

Figure 5. Development paths of cells on G.

From Figure 5 we can read off the qualitative features required in the statement of the theorem, as follows :

At time t_1 the frontier first forms at s_1.

Between t_1 and t_2 the frontier moves from s_1 to s_2 and deepens.

At the time approaches t_2 the frontier slows up, and approaches s_2.

At t_2 the frontier reaches s_2, and stabilises.

After t_2 the frontier deepens further.

This completes the proof of the theorem. Before we proceed to quantitative features of the theorem we make four remarks.

Remark 1. Sometimes the wave does not begin in the middle of the tissue, but on the boundary of the tissue, so that the tissue appears to "grow into" the frontier. This seems to be the case with slime mold (see §15 below), and with the development of chicken wings, for instance. In this case the mathematics is simpler because the space-time track of the tissue does not cross the cusp point, but only the fold curve. In fact there may not necessarily be any cusp point at all. In Figure 5 the slime-mold grex would be represented by $[s,s_3]$ with tip at s and tail at s_3. At time t the wave would begin at the tip, and then proceed along the grex to stabilise at s_2 at time t_2. The front part $[s,s_2]$ eventually develops into stalk-cells A, and the back part $[s_2,s_3]$ into spore-cells B.

Remark 2. Not all frontiers are formed by primary waves of this type, because in some cases our translation of the four hypotheses into mathematics may not be valid. For example in the gastrulation of birds and mammals, or in mixing experiments [1,10], the frontier is caused by migration of different types of cells, sorting themselves out, whereas we have assumed that the cells stay more or less in the same place relative to one another in the tissue. However in some of these cases a primary wave may already have taken place in some underlying gradient, and the migration of cells up or down the gradient may be merely a secondary wave.

Remark 3. There was one drastic simplification that we made in the proof of the theorem, in the way that we chose to translate homeostasis into mathematics. It may well be reasonable to represent homeostasis by a dynamic D on X, but it is not obvious that D should be a <u>gradient</u> dynamic, $\dot{x} = -\nabla V$. In special cases V may represent some potential energy that is minimised, and then it is reasonable. But in general D may be non-gradient, particularly when the cell contains biological clocks. Even then in some cases it is technically possible to reduce D to the gradient case, by chosing V to be a Lyapunov function for D (see [18, 26]). However in other cases this may not be possible; frontiers arising from, or associated with turbulence, for instance, would probably not behave so simply.

Remark 4. We have drawn Figures 3, 4, 5 as if X were 1-dimensional, and as if

$$M \subset C \times R \subset R^3.$$

In fact this is not true because X is an open subset of R^n, where n may be very large, and therefore more precisely

$$M \subset C \times X \subset R^{2+n}.$$

However this does not alter the fact that M is 2-dimensional surface, and therefore our diagrams are indeed rigorous pictures of the map $\chi : M \to C$. Moreover Thom's classification theorem [18] can be modified [26] in an important manner for this context, as follows :

If $\chi : M \to C$ has a cusp catastrophe, then in the neighbourhood of that point we can choose a map $\pi : X \to R$ such that

$$1 \times \pi : C \times X \to C \times R$$

throws M diffeomorphically onto the surface pictured in Figure 4b. Moreover π can be chosen to be the projection of X onto one of the given axes of R^n, that measure concentrations etc. In fact we can choose any axis not perpendicular to the tangent to ∂G at σc_1. Let us call the chemical or physical property that this axis measures, a _morphogen_. Then the vertical axis in Figures 3, 4, 5 measures the morphogen. The morphogen need not be unique, and may only be an artifact. But if the morphogen is easy to measure, then it may be useful for experimental predictions. It is remarkable that Thom's theorem guarantees the existence of a morphogen for each developing frontier.

9. QUANTITATIVE ASPECTS OF THE THEOREM.

From Figure 5 we can deduce some quantitative estimates about primary waves. The estimates are computed to first order in small quantities, and are therefore only accurate near the beginning and the end of the wave.

   Corollary 1.   Initially, when the frontier first forms, it is moving at constant speed.

   Proof.   The path of the wave is the branch $c_1 c_2$ of the cusp, which, near c_1, to first order, can be replaced by the tangent at c_1.

   Corollary 2.   Initially, when the frontier first forms, its depth increases by a square-root law, in other words the depth of the frontier is proportional to $\sqrt{t-t_1}$ (and hence also to $\sqrt{s-s_1}$).

   Proof.   At time t the depth of the frontier is equal to the catastrophic jump PQ in Figure 4. Therefore we must compute PQ. Choose origin O at σc_1, the point of M over the cusp point c_1. Choose two axes ξ, η at O as follows : ξ is measured along the tangent at O to ∂G in X, oriented towards P, and η is measured along the tangent at c_1 to the cusp in C. Let K denote the (ξ, η)-plane. Then K is the osculating plane of ∂G at O. Therefore, by genericity, and ignoring third order terms, ∂G lies in K and has equation $\eta = k\xi^2$, where $k > 0$. Therefore P satisfies $\xi = +\sqrt{\eta/k}$. Meanwhile Q satisfies $\xi = -2\sqrt{\eta/k}$, because M is the diffeomorphic image of a cubic surface, namely the canonical cusp catastrophe. Therefore $PQ = 3\sqrt{\eta/k}$. But η is proportional to $t-t_1$, and hence PQ is proportional to $\sqrt{t-t_1}$, as required.

   Remark.   The initial position and movement of a developmental wave may be difficult to observe, because of the initial shallowness of the frontier. However it might be possible to find the initial position by using Corollary 2 to exterpolate backwards (and hence find the organising centre, if the wave happens to emanate from a point).

   Note that Corollary 1 also remains true for any secondary wave. However we should not necessarily expect Corollary 2 to apply to a secondary wave, because the two waves are of a totally different nature : the primary wave marks the frontier between two diverging types of tissue, whereas the secondary wave makes the onset of a secondary effect within one type of tissue. Therefore the initial movement of the primary wave may be difficult to observe, whereas that of the secondary wave may be easy to observe — for instance the first invagination in gastrulation (see §10 below).

Corollary 3. *Eventually, just before the frontier stabilises, it slows down parabolically. In other words (s_2-s) is proportional to $(t_2-t)^2$, and the speed is proportional to (t_2-t).*

Proof. By genericity the curve c_1c_2 touches the time axis at c_2 with quadratic tangency. Therefore near c_2, ignoring third order terms, the curve has the equation

$$(s_2-s) = h(t_2-t)^2,$$

where $h>0$. The speed is given by differentiating:

$$\dot{s} = 2h(t_2-t).$$

This completes the proof of Corollary 3.

Remark. Corollary 3 should be easy to observe and verify. If there is no morphogenesis, then the same result will hold for any secondary wave. Therefore in a recruitment phenomenon, for instance, the parabolic law might provide a good test to distinguish whether it was a secondary wave or the result of entrainment.

If there is morphogenesis, then the displacement of cells relative to one another may upset the parabolic law for the secondary wave, but Corollary 3 may nevertheless yield other quantitative predictions — see for example the estimate for stalk diameter in the slime mold fruiting body, in §16 below.

Energy release. Some morphogenetical movements begin slowly, and build up to a recognisable climax before finally dying down. This can be seen most clearly in time lapse films, and the reader is especially recommended to see the two Göttingen films of Luther [11] on gastrulation and neurulation, and Gerisch [7] on slime mold. For instance in [11] gastrulation begins by invaginating slowly, then the tissue begins to roll over the dorsal lip, then pours over the entire circle of blastopore lip, until it eventually slows down, and the blastopore gently closes. Similarly neurulation begins with the neural folds appearing slowly, then they rear up and the neural tube snaps shut in the middle, and the closing process runs towards both ends, which eventually seal themselves more gently. In [7] the slime mold fruiting body heaves itself slowly off the ground, then accelerates and rises rapidly up its stalk, then slows down, and eventually the knob at the top gently disappears.

We shall now show, by computing the energy released, that this is the normal pattern for a morphogenetical movement arising from a secondary wave.

Note that in the films it can also be observed that the gastrula gives a final heave before the blastopore closes, and the slime mold fruiting body gives a hiccup halfway up, but these are subsidiary frictional effects that we shall explain later.

Assuming that the primary wave begins at time t_1 and ends at time t_2, let

$V(t)$ = speed of primary wave at time t, $t_1 < t < t_2$;

$A(t)$ = area of wave-front at time t, $t_1 < t < t_2$;

$e(\tau)$ = rate of energy-release by a cell at time-interval τ after the primary wave has passed it. We may suppose that $e(\tau) = 0$ outside an interval $\delta_1 < \tau < \delta_2$, where $0 < \delta_1 < \delta_2$, and where δ_1 is the delay between the primary and secondary waves, and $[\delta_1, \delta_2]$ the period during which energy is released by the secondary effect.

Lemma 1. *The total rate of energy release at time t, where $t_1 + \delta_1 < t < t_2 + \delta_2$, is given by*

$$E(t) = \int_{t-\delta_2}^{t-\delta_1} A(\tau)V(\tau)e(t-\tau)d\tau$$

Proof. The number of cells crossed by the primary wave in the interval $[\tau, \tau + d\tau]$ is $A(\tau)V(\tau)d\tau$, and by the time t each of these is still releasing energy at the rate $e(t-\tau)$. Integrating gives the lemma.

In Figure 6 we sketch the qualitative shape of the graph of E. The assumptions on which the sketch is based are as follows: V is initially constant near t_1, by Corollary 1, and eventually decreases linearly to zero at t_2, by Corollary 3. If we assume that the primary wave emanates from a point, and that the wave-front initially expands linearly, then we deduce that A starts from zero at t_1 and initially increases parabolicially, in other words proportional to a square law. Eventually A becomes constant as the frontier stabilises at t_2. We assume the secondary effect e starts suddenly at δ_1, and then decreases linearly to zero by δ_2. It can be shown, by using the lemma to integrate these assumptions, that E both begins and finishes proportional to cube laws.

Figure 6. Graph of energy released by a secondary wave.

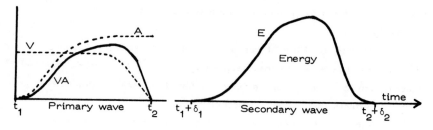

Summarising : this characteristic pattern of energy release during a morphogenetic movement might provide a useful clue that the movement was the secondary effect of an earlier hidden primary wave.

George Oster suggests that the energy released by individual cells might be observed by measuring the heat loss microcalorimetrically.

The ripple ahead of a wave. Consider a fixed time t during the primary wave, $t_1 < t < t_2$. The state of the tissue is obtained by lifting the section Sxt of C up onto G in Figure 5. Consider the variation in the state of the cells as the frontier is approached from either side. On the a-side the state is approximately constant, and so the cells are homogeneous, but on the b-side the variation is parabolic as the frontier is approached.

In some cases this phenomenon might be repeated visibly in a secondary wave. For instance, if we had a situation as in §6(ii) above, where the secondary wave was differentiation, then the phenomenon might be visible as a slight ripple ahead of the wave. For example, suppose the expanding frontier of the insect eye were a secondary wave. Then inside the frontier the already recruited eye cells should appear relatively homogeneous, but outside the frontier, the epidermis cells just about to be recruited might show physiological signs of the impending recruitment. In mammalian eyes the effect might appear on the inside of the frontier, because the wave goes the other way.

APPLICATIONS

10. GASTRULATION OF AMPHIBIA.

For the mathematical reader unfamiliar with gastrulation a recommended introduction is to read Balinsky [1, Chapter 8], and to see the film [11] from the Göttingen film library. The following diagrams are taken primarily from [1,20]. Figure 7 shows photographs of what gastrulation looks like from the outside, while Figure 8 gives diagrams of the main morphogenetical movements going on inside at the same time. Figure 9 shows supporting photographs of sections. Figure 10 gives Vogt's drawings of the 3-dimensional flow of mesoderm cells from the surface into the interior, and Figure 11 the resulting flow of the mesoderm mantle, finishing with its position at the end of gastrulation and beginning of neurulation, with a photograph of the corresponding section. Figures 12 and 13 give Vogt's detailed fate maps. Figures 14 and 15 are diagrams of the subsequent neurulation, leading to the tail bud stage in Figure 16. Figures 17 and 18 show part of the normal tables for newts [9] and frogs [17], giving an idea of the timing involved.

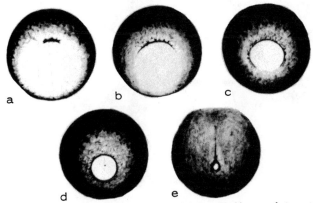

Figure 7 Changes in shape of the blastopore and closure of the blastopore during gastrulation in a frog. From Balinsky [1, p. 187]

170 4. Primary and Secondary Waves in Developmental Biology

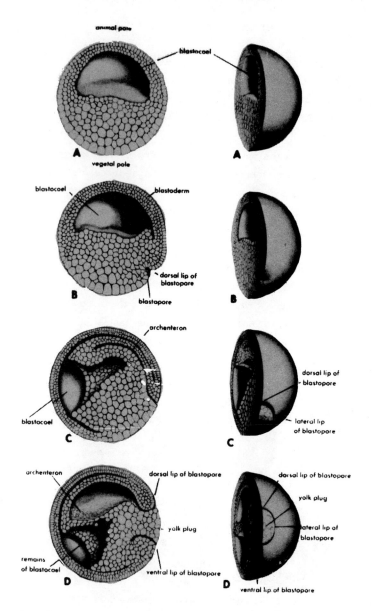

Figure 8 Four stages of development of a frog embryo; A, A,' late blastula stage; B, B,' beginning of gastrulation; C, C,' middle gastrula stage; D, D,' late gastrula stage (semidiagrammatic). Drawings on the left represent the embryos cut in the median plane; drawings on the right represent the same embryos viewed at an angle from the dorsal side (A, B, C) or from posterior end (D). From Balinsky [1, p. 191]

4. Primary and Secondary Waves in Developmental Biology   171

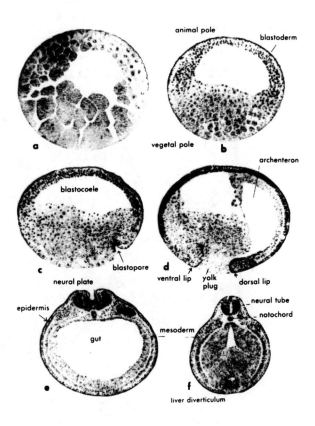

Figure 9 Blastula, gastrulation and formation of primary organ rudiments in the frog. *a*, Late cleavage stage showing difference in the size of the blastomeres; *b*, late blastula; *c*, early gastrula; *d*, middle gastrula; *e*, neurula, transverse section; *f*, transverse section after completion of neurulation. From Balinsky [1, p. 192]

Figure 10 Trajectories of the movements of parts of the marginal zone during gastrulation in amphibians. Thick lines show movements of cells on the surface of the embryo; thin lines show movement of invaginated cells. (From Vogt, 1929.) [20, p. 431] and [1, p. 194]

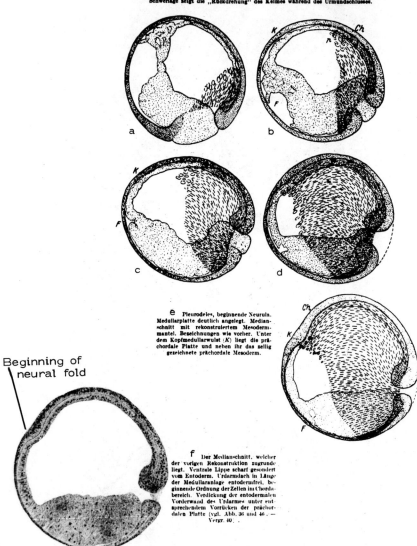

Figure 11 a-e: Flow of the mesoderm mantle over the archenteron roof during gastrulation. f: Photograph of a corresponding median section at the end of gastrulation and beginning of neurulation. From Vogt [20, p. 520].

4. Primary and Secondary Waves in Developmental Biology   173

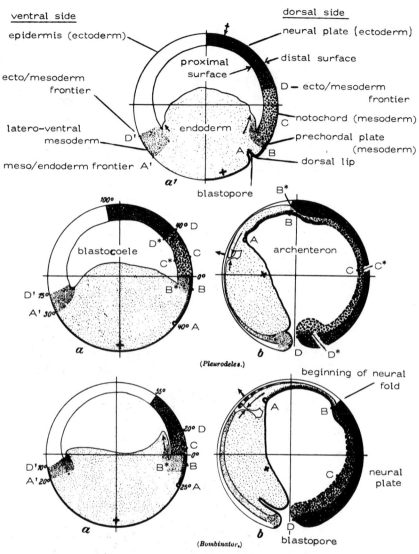

Figure 12. Diagrammatic median sections and fate maps of amphibian embryos (a) before, (a^1) during, and (b) after, gastrulation. The top three diagrams are for the newt, and the bottom two for the fire-bellied toad. The plus sign marks the track of the vegetal pole, A the track of the initial invagination point, B the track of the point opposite the boundary point B* of the proximal surface, C the midpoint of the notochord, and D the presumptive ecto/mesoderm frontier. Notice that the final position of B in Fig. (b) determines the boundary of the neural plate and the beginning of the neural fold. From Vogt [20, p. 657].

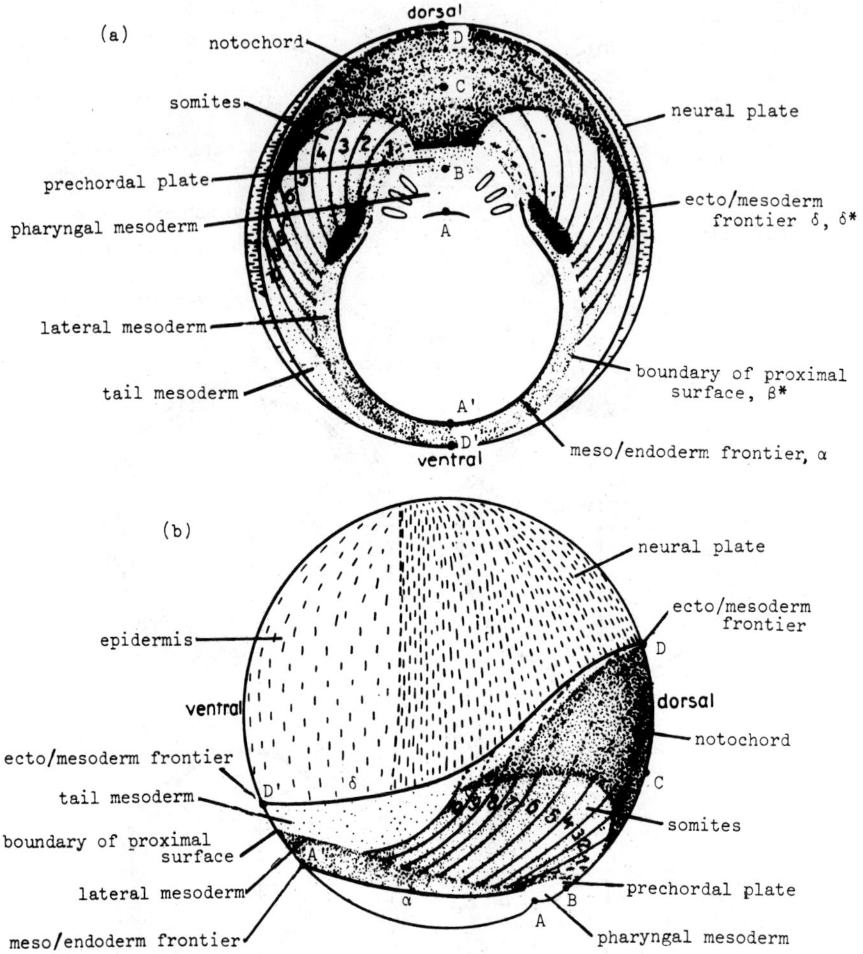

174   4. Primary and Secondary Waves in Developmental Biology

Figure 13. Diagrammatic fàte map for urodeles (newts and salamaders) at the beginning of gastrulation, as seen from the outside (a) from below and (b) from the side. The symbols A,B,.... indicate the same as in Figure 12. Invagination is beginning at A, and extends slit-shaped along the short arc through A shown in (a). Notice that the boundaries between presumptive somites are approximately circular arcs centred at A. We have joined these by dotted arcs running through presumptive notochord to indicate the primary wave-fronts. From Vogt [20, p. 392].

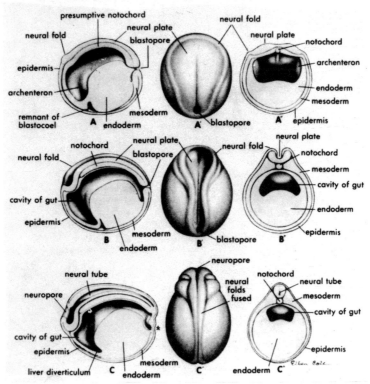

Figure 14 Three stages of neurulation in a frog embryo. The drawings in the middle show whole embryos in dorsal view. The drawings on the left show the right halves of embryos cut in the median plane. The drawings on the right show the anterior halves of embryos cut transversely. A, A', A", Very early neurula; B, B', B", middle neurula; C, C', C", late neurula with neural tube almost completely closed. C shows the blastopore closed; the asterisk indicates the point at which the anal opening will break through. From Balinsky [1, p. 196]

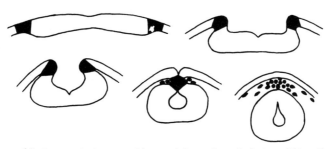

Figure 15 Stages in the formation of the neural plate and neural tube in amphibians. Transverse sections (diagrammatic). Neural crest cells are shown in black. From Balinsky [1, p. 198]

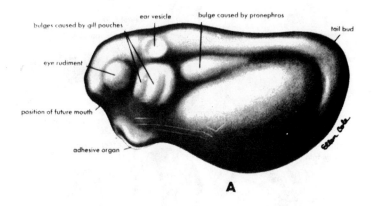
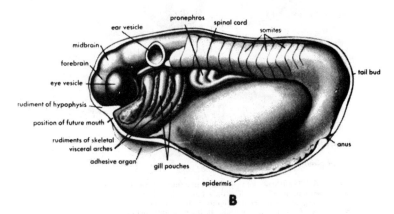
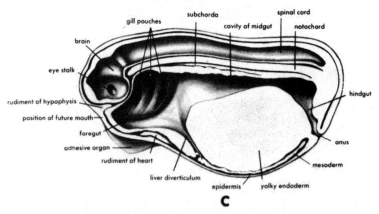

Figure 16. A frog embryo in an early tail bud stage. A, External view; B, same embryo with the skin of the left side removed; C, same embryo cut in the median plane.

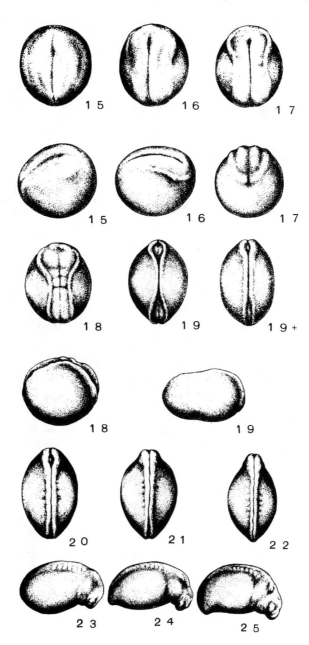

Figure 17. Normal development of the Japanese newt, showing stages of neuralation and formation of somites. From Koyama [9, pp 468-9].

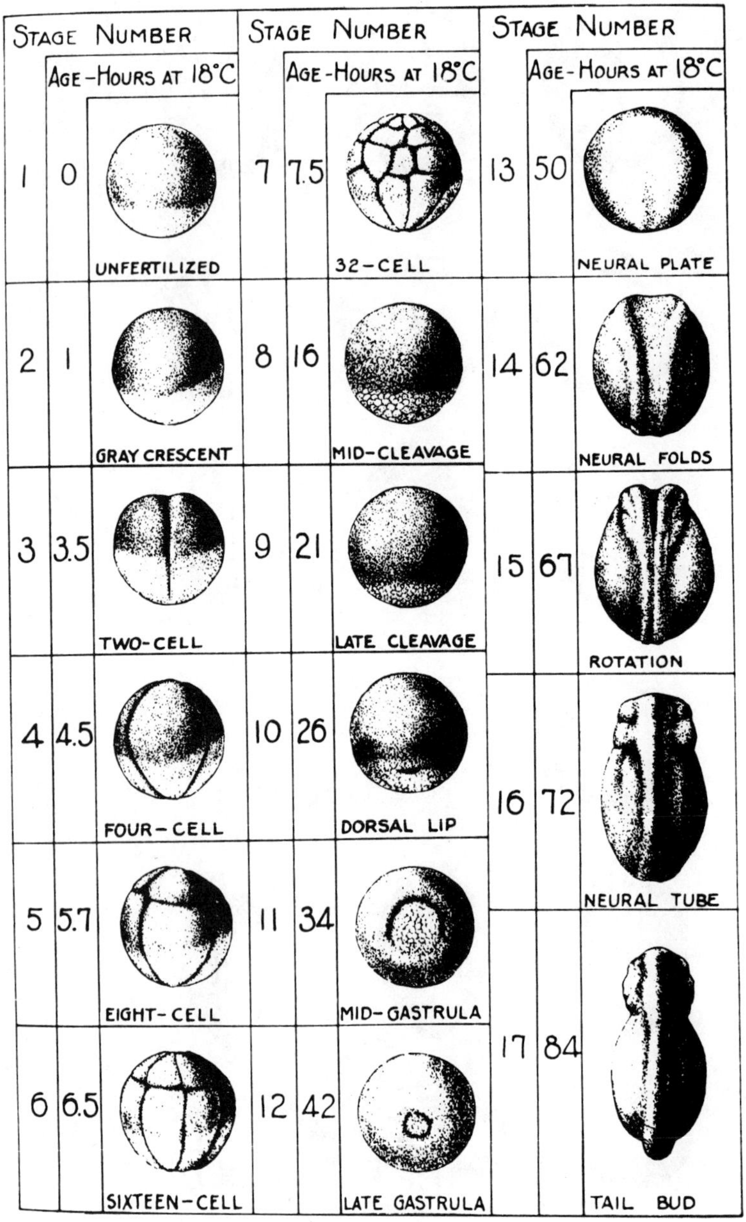

4. Primary and Secondary Waves in Developmental Biology

Stage Number		Stage Number		Stage Number	
\#	Age-Hours at 18°C	\#	Age-Hours at 18°C	\#	Age-Hours at 18°C
1	0 — UNFERTILIZED	7	7.5 — 32-CELL	13	50 — NEURAL PLATE
2	1 — GRAY CRESCENT	8	16 — MID-CLEAVAGE	14	62 — NEURAL FOLDS
3	3.5 — TWO-CELL	9	21 — LATE CLEAVAGE	15	67 — ROTATION
4	4.5 — FOUR-CELL	10	26 — DORSAL LIP	16	72 — NEURAL TUBE
5	5.7 — EIGHT-CELL	11	34 — MID-GASTRULA	17	84 — TAIL BUD
6	6.5 — SIXTEEN-CELL	12	42 — LATE GASTRULA		

Figure 18. Normal development of Rana Pipiens.
From Shumway [17, p 143].

Having described the morphogenesis pictorially, we now get down to the main business of discussing the causes. Gastrulation can last from a few hours to a few days, depending upon the species and the temperature (see Figure 8 and Figure 19 (Stages 10-13), and §14 below). During this period there appear to be two local physiological processes going on at the same time, which at first sight seem to have almost the opposite effect at cell level :

        (1)   cells flattening,

        (2)   cells submerging.

However, whereas process (1) goes on uniformly in all cells of the northern hemispherical shell throughout gastrulation, process (2) does not happen simultaneously everywhere. Our main hypothesis will be that process (2) is secondary wave. Therefore the two processes have radically different global effects upon the shell, as follows :

        (1)   Uniform expansion.

        (2)   Rolling changes of curvature.

    (1) <u>The expansion process.</u> During gastrulation the northern hemispherical shell expands to about twice its area and half its thickness, due to the outer surface cells flattening themselves, and to intercalation amongst the cells of the lower layers, until the shell is only two cells thick [1, Chapter 9]. Moreover this is an active process, because Holtfreter has shown that if the gastrula is prevented from folding inwards then the expansion will force it to fold outwards [21,p.442].

    (2) <u>The wave process.</u> We begin with the fact that the northern hemispherical shell differentiates into ectoderm and mesoderm (we are not concerned with the southern hemisphere because it is already different, being yolky, and destined to become endoderm). Therefore, by the theorem, the differentiation will cause an ecto/mesoderm frontier to form, and move as a primary wave. The final position of the wave on the outer surface is a circle δ, whose position can be read from the fate maps (Figures 12 and 13). On the dorsal side δ goes through the point D, and forms the frontier between presumptive neural plate and presumptive notochord, while on the ventral side δ goes through D' and forms the frontier between presumptive epidermis and presumptive tail mesoderm. For the beginning position of the wave we need a hypothesis :

<u>Hypothesis 1.</u> <u>The primary wave begins at the point A on the dorsal side at the bottom of the grey crescent.</u>

It is no accident that A is same point where several hours later the gastrula first begins to invaginate, because the latter is the beginning of the secondary wave (see Figures 7(a), 8(b), 9(c), 12(a_1), 13 and 20).

Let A' denote the point on the ventral side marking the presumptive meso/endoderm frontier, and let α denote the circle on the outer surface through A and A'. From Figures 12 and 13 it can be seen that α is almost a parallel of latitude.

After beginning at A the primary wave then spreads simultaneously in the following directions : firstly east and west round the circle α towards the ventral side, secondly northwards up the dorsal side, and thirdly inwards from the <u>distal</u> (outer) surface to the <u>proximal</u> (inner) surface of the northern hemispherical shell. In Figure 19 we illustrate this on a simplified version of the fate map if Figure 13(b), showing successive positions of the wave on the distal surface.

Figure 19: Fate map.

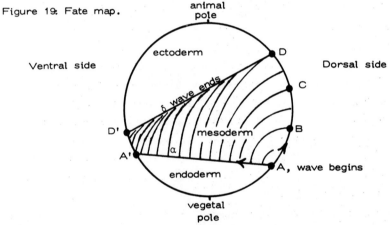

The extent of travel up the dorsal side, AD, is greater than that on the ventral side, A'D', and it differs in different species. Examples of the latitudes of A and D are as follows :

	Newt	Fire-bellied toad	Axolotl
Wave begins at A :	40°S	25°S	30°S
Wave ends at D :	40°N	20°N	20°N.

The figures for newt and toad are from Vogt (Figure 12), and those for axolotl are due to Nieuwkoop [12]. Nieuwkoop confirms that a hidden wave of mesoderm determination can be detected in axolotl by the grafting experiment described in §7 above. We return to the timing of this wave in §14 below.

Before we leave Figure 19 there is a small point to be mentioned. We have labelled the entire region between α and δ as mesoderm, whereas in the literature there is some ambiguity as to whether the immediate neighbourhood of the point A should strictly be called mesoderm. More detailed fate maps tend to leave the meso/endoderm frontier undefined near A, as for example in Vogt's fate maps in Figures 12 and 13. However we suggest that the ambiguity may be resolved by considering the 3-dimensional picture rather than merely the 2-dimensional distal surface, and shall treat this in detail when we come to analyse neurulation in §12 below, and in particular in Figure 23.

We now come to our main hypothesis.

Hypothesis 2. There is a secondary wave of cells submerging.

The time delay between the primary and secondary waves is discussed in §14 below. For example in the case of the frog (Figure 18) we estimate the delay is about 16 hours.

Remarks about submerging. Recall that submerging means that the cell decreases its free surface, and increases the proportion of surface in contact with other cells. From the cellular point of view this is a complicated process, but the advantage of denoting it by a single word is that we can then more easily describe the process proceeding as a wave across the tissue. However, at the same time, we must be cautious about the dangers of over-simplification, because different cells may submerge in different ways. For instance the submerging behaviour may depend not only upon the gene systems that have been switched on by the primary wave, but also upon characteristics of the cell membrane, which may in turn depend upon the cell's position.

For example consider the beginning of the secondary wave : this occurs at the point A because by Hypothesis 1 the primary wave began at A. The submerging cells remain attached to the surface, causing the surface to invaginate, and this is the beginning of gastrulation. The submerging cells are known as flask cells because they become elongated, as can be seen in the diagram in Figure 20(a) and photograph 20(b).

4. Primary and Secondary Waves in Developmental Biology

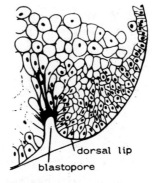

Figure 20(a) Median section through the blastopore region of an early gastrula of a newt, showing the cells at the bottom of the pit streaming into the interior. (After Vogt, 1929.)
From Balinsky [1, p. 187]

Figure 20(b)

Trit. crist. Gastrulationsbeginn, sagittal, etwas schräg geschnitten. Dorsale Lippe paramedian getroffen. In ihr deutet kleinzelliges Material der inneren Randzone, welches nach innen vorzudringen scheint, die erste Mesodermbildung an. Urmundgrübchen ganz von entodermalen, z. T. flaschenförmigen Zellen gebildet. — Vergr. 50×.

From Vogt [20, p. 510]

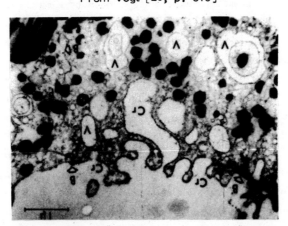

Figure 21 External surface of cells invaginating in the amphibian blastopore (electron-micrograph). Cr, Crypts formed at the surface of the cells; V, vesicles pinched off at the bottom of the crypts; P, pigment granules; B, cell boundaries; Y, yolk platelet.
From Balinsky [1, p. 215]

Possibly the most important part of the underlying biochemical processes in a flask cell may be associated with the contraction near the free surface. By important we mean where the energy is released. The resulting wrinkling of the free surface is shown in the electronmicrograph in Figure 21. This contraction may be responsible for the squeezing of the nucleus and most of the cytoplasm towards the other end of the cell; and this, in turn, may cause the membrane at the other end to bulge out like a balloon, provided it is sufficiently elastic. Hence the characteristic flask-shape. Therefore the amoeboid action of the submerging flask cells streaming into the interior (Figure 20) may not be an active process at all, but merely the passive consequence of the free surface contracting and the fact that the cells have an elastic membrane. All the cells around the circle α (the presumptive meso/endoderm frontier) submerge in the same way.

By contrast, the cells further north on the blastula submerge less dramatically, as can be seen by comparing the development of isolated fractions [1, Chapter 9]. Possibly this is merely because they may have less elastic membranes at the time of submerging. Consequently when the secondary wave hits them they do not balloon out like flask cells, but exhibit more moderate changes of curvature. One might be tempted to say that they submerge more gently, except that the forces involved may be just as powerful.

Again cells on the proximal surface may submerge differently to those on the distal surface, as we shall see when we come to discuss neurulation in §12 below. But the common feature of all submerging cells is that they push and pull on their neighbours. Indeed, as Gustafson and Wolpert [8] have pointed out, the forces that shape the embryo must necessarily originate from cells pushing and pulling on their neighbours, and this is essentially how a secondary wave of submerging cells causes the rolling changes of curvature that produce morphogenesis. Therefore, from the global point of view, the most effective way to describe the results of the secondary wave may be mathematically in terms of changes of curvature. Since curvature will be central to our discussion we digress in the next section to give an elementary mathematical treatment of curvature. But first a remark about the dorsal lip.

The dorsal lip. Mesodermal tissue starts rolling over the dorsal lip about 2 hours after the initial invagination. The main cause appears to be the expansion process (1). Therefore mathematically it should be possible to model the resulting surface flow as the gradient of a potential satisfying Poisson's equation, in other words Laplace's equation with a sink at A and source distributed uniformly over the whole surface north of the circle. This explains why initally there is a flow of tissue from ventral to dorsal side.

The secondary wave of Hypothesis 2 now comes into play, causing the invagination to follow the primary wave both east and west round α, forming the circular lip of the blastopore (shown in Figure 7). The resulting gradient flow on the surface now has a circular sink, as shown in Figure 10. It might be possible to use Poisson's equation to simulate on a computer the time-development of the whole surface flow.

Now consider what happens to the tissue when it rolls over the dorsal lip. First we must explain why the lip itself starts fairly blunt (Figure 20) and then becomes sharper as the flow progresses (Figures 8, 9, 11, 12). We shall show that this is due to the preservation of mean curvature of the shell. We shall then show that the secondary wave, by changing the mean curvature, causes first the concavity of the archenteron roof, and later the formation of notochord and somites during neurulation.

11.   CURVATURE.

Let S be a closed surface, and P a point on S. Let r_1, r_2 be the principal radii of curvature of S at P.

Example 1. If S is a sphere of radius ρ then at each point $r_1 = r_2 = \rho$.

Example 2. Let S be the surface of the mid-gastrula (Figures 7(d), 8(D), 9(d), 10, 11(a), and 18 (Stage 11)), and let P be a point on the dorsal lip. If ρ is the radius of the gastrula, then the average radii of dorsal lip and blastopore are approximately $\frac{\rho}{6}$ and $\frac{\rho}{4}$. Therefore the principal radii of curvature at P are $\frac{\rho}{6}$, $-\frac{\rho}{4}$ (negative because the radius of the blastopore points outwards, in other words out of the tissue).

Define the <u>gaussian curvature</u> of S at P to be $\frac{1}{r_1 r_2}$. In example 1 the sphere has positive gaussian curvature, and in example 2 the dorsal lip has negative gaussian curvature.

Define the <u>mean curvature</u> μ of S at P by
$$\mu = \frac{1}{2}(\frac{1}{r_1} + \frac{1}{r_2}).$$

In example 1 the sphere has mean curvature $\frac{1}{\rho}$ and in example 2 the dorsal lip also has mean curvature
$$\mu = \frac{1}{2}(\frac{6}{\rho} - \frac{4}{\rho}) = \frac{1}{\rho}.$$

It is no accident that these are the same as we shall prove in Corollary 2.1 below.

Suppose now that S is the distal (outer) surface of a shell of thickness ϵ, with proximal (inner) surface S^*. We have in mind the northern hemispherical shell of the blastula (Figures 8, 9, 12). Define the <u>excess</u> e at a point P of S by
$$e = \frac{\text{incremental area of distal surface}}{\text{incremental area of proximal surface}}$$

<u>Lemma 2.</u> $e = 1 + 2\epsilon\mu$.

<u>Proof.</u> Let O_1, O_2 be the two principal centres of curvature of S at P, corresponding to the principal radii of curvature r_1, r_2. Let a_1, a_1^* be two small arcs, centred at O_1 and subtending the same angle at O_1, on the distal and proximal surfaces, respectively, and therefore of radii r_1, $r_1-\epsilon$. Therefore
$$\frac{a_1}{a_1^*} = \frac{r_1}{r_1-\epsilon} = (1-\frac{\epsilon}{r_1})^{-1}.$$

Similarly let a_2, a_2^* be two small arcs perpendicular to a_1, a_1^*, centred at O_2. Then $a_1 a_2$, $a_1^* a_2^*$ are incremental areas on the distal and proximal surfaces, respectively.

Therefore
$$e = \frac{a_1 a_2}{a_1^* a_2^*} = (1-\frac{\epsilon}{r_1})^{-1}(1-\frac{\epsilon}{r_2})^{-1}$$
$$= 1 + \epsilon(\frac{1}{r_1} + \frac{1}{r_2}), \text{neglecting } \epsilon^2,$$
$$= 1 + 2\epsilon\mu.$$

This completes the proof of lemma 2.

Corollary 2.1. If ρ, ρ_1, ρ_2 are the radii of gastrula, dorsal lip, and blastopore at any moment during mid-gastrulation, then

$$\frac{2}{\rho} = \frac{1}{\rho_1} - \frac{1}{\rho_2}.$$

Proof. Consider a piece of mesodermal tissue as it rolls over the dorsal lip. Since the expansion process (1) is uniform and gradual, and since the secondary wave (2) has not yet hit this piece of tissue, we may assume that both the excess, e, and the thickness, ϵ, remain approximately constant* as it rolls over the lip. Therefore by the lemma the mean curvature μ must also be constant. Before the tissue rolls $\mu = \frac{1}{\rho}$, and as it rolls $\mu = \frac{1}{2}(\frac{1}{\rho_1} - \frac{1}{\rho_2})$. Hence the result.

Remark. As the blastopore closes at the end of gastrulation ρ_2 becomes very small, but ρ_1 cannot become too small, because it must at least be greater than the thickness. Therefore μ must become smaller. Therefore ϵ must become greater. This explains the thickening of the lips as the blastopore closes (see Figure 11(f)). Also furrows are often seen on the surface radiating from the closing blastopore, as the distal surface compensates for its increasing negative gaussian curvature. In particular this is probably responsible for the posterior end of the neurulation furrow (see Figure 14 (A') and §12 below).

The archenteron roof. Once the mesoderm has rolled over the lip and got inside it is called the mesoderm mantle. Its mean curvature has changed from being positive to negative, because the distal surface which used to be the convex side of the shell, has now become the concave side, and forms the archenteron roof (see Figures 8,9,10,11,12,14). What causes the change of sign of the mean curvature is the secondary wave of hypothesis 2, progressing along the distal surface shortly after it has rolled over the lip. As the secondary wave hits each cell it will begin to submerge, but it cannot submerge very far because by this time the mesoderm mantle is only two cells thick. Therefore there is no chance of any dramatic behaviour such as displayed by the initial flask cells, which have the luxury of a thick yolky mass of cells in which to submerge. Therefore the main effect of the secondary wave at this stage will be the reduction of free surface membrane (see Figure 21) in other words a reduction in the area of the

---

* We take account of the gradual decrease in the thickness, ϵ, due to the expansion process in §14 Figure 29.

distal surface, and this effect will persist while the mesoderm mantle climbs up over the archenteron roof (Figures 10, 11, 12). If the distal surface area is reduced by a factor λ, then the excess, e, will also be reduced by λ. We can therefore compute λ by Lemma 2.

Corollary 2.1 If ρ is the radius of the gastrula and ε the thickness of the mesoderm mantle, then
$$\lambda = 1 - \frac{4\varepsilon}{\rho}.$$

Proof. By the lemma, before rolling over the lip the excess is
$$e = 1 + \frac{2\varepsilon}{\rho}.$$

After rolling the excess now becomes
$$\lambda e = 1 - \frac{2\varepsilon}{\rho - 2\varepsilon}$$

because the radius of the distal surface is now $\rho - 2\varepsilon$, and the mean curvature negative. Therefore
$$\lambda = (1 - \frac{2\varepsilon}{\rho - 2\varepsilon})(1 + \frac{2\varepsilon}{\rho})^{-1}$$
$$= 1 - \frac{4\varepsilon}{\rho}, \text{ ignoring } (\frac{\varepsilon}{\rho})^2, \text{ as required.}$$

Data on shell thickness. From the photographs of section in Figures 9(d) and 11(f) we estimate $\varepsilon = \frac{\rho}{12}$. From this single piece of datum we shall deduce several corollaries, including estimates of diameter of notochord and somites in the next section. For the moment we deduce
$$\lambda = \frac{2}{3}, \quad e = \frac{7}{6}, \quad \lambda e = \frac{2}{3} \times \frac{7}{6} = \frac{7}{9}.$$

Therefore:

Corollary 2.3 The effect of the secondary wave is to reduce the surface area by one third.

The secondary wave is the onset of the change in mean curvature, and we have assumed that this effect comes on relatively sharply. The effect must then last for several hours in order to enable the mesoderm mantle to slide round the archenteron roof (see Figures 9(d), 10, 11). In the case of the frog at 20°C we estimate that the effect must last about 20 hours (see Figure 18 and §14 below). As the effect dies away the mesoderm mantle gradually loses the energy to hold itself in negative mean curvature, and will begin to try and reassert its original positive mean curvature. This can be seen most clearly in Vogt's careful drawing in Figure 11(d). However the mesoderm mantle is prevented from recovering its original curvature by the enclosing sack of ectoderm. Therefore the entrapped mesoderm will begin to exert counterpressures against the enclosing ectoderm, like a mild spring pressing against the underside of the ectoderm. In fact the pressure is visible from the outside, because the margin of the mesoderm mantle pushes up a ripple in the ectoderm, that can be seen running across the outer surface, and is called the gastrulation wave (see the film [11]).

Another effect of the mild springlike action of the mesoderm mantle is to cause it to tear* away laterally from the endoderm, which by contrast moves passively where it is pushed and pulled by the mesoderm.

A third effect can sometimes be seen towards the end of gastrulation as follows. When the mantle reaches the anterior end (Figure 11(d) and (e)) the increased friction between it and the ectoderm may momentarily interrupt the forward sliding of the mantle, even though at the same time fresh mesoderm will continue to be pushed over the blastopore lip by the expansion process; as a result the blastopore will begin to protrude outwards, pear shaped, until the increased forward pressure on the mantle overcomes the friction, and persuades it to continue sliding forward once more. From the outside, this frictional obstruction and release makes it look as though the gastrula is heaving and gasping for breath (see the film [11]).

A fourth, and probably the most important effect of the pressure of the mesoderm against the ectoderm is that it enables the mesoderm to biochemically induce neural plate in the overlying ectoderm, wherever they touch. Had there been no secondary wave of negative curvature, then the mesoderm mantle, once inside, could not have had the same contact with ectoderm, because, in an attempt to retain its positive mean curvature, it would have had to arch itself away from the ectoderm, leaving cavities, and therefore would not have been able to induce neural plate.

---

* I am indebted to Klaus Jänich for drawing my attention to the topology of this process.

12.   NEURULATION.

Once the neural plate has been induced then neurulation begins. The main feature of neurulation is the neural plate rolling itself up into the neural tube (Figures 14, 15, 17), which eventually becomes the spinal chord and brain. The rolling up is an active process, because isolated pieces of neural tissue tend to curl up by themselves [1, Chapter 10]. It appears that one of the consequences of induction is to make neural cells submerge, thereby causing the mean curvature of the neural plate to change sign and hence causing the rolling up.

However that is not the only event happening during neurulation because at the same time the underlying mesoderm mantle is forming itself into notochord and somites, and this is what we shall be primarily interested in. The notochord is a long thin cylinder running from front to back just beneath the neural tube (see Figures 14, 16, 26, 27). As it forms it elongates itself and goes rigid, stretching the embryo lengthwise. It is only a temporary organ, and later disintegrates and disappears. However while it exists it plays an important role in providing structural support for the formation of spinal chord and skeleton.

The somites are small masses of mesodermal cells, each with a small cavity inside, arranged in two rows on either side of the neural tube and notochord (see Figures 16, 17, 27). They form sequentially from front to back, until there are about 30 or 40 on either side, depending upon the species. Initially they form fairly rapidly and then slow down towards the tail (in frogs initially one pair every 40 minutes, slowing eventually to 2 or 3 a day). Like the notochord, the somites are temporary organs because the cavities soon disappear, and different parts of the somites develop into the vertebrae, muscles and the connective tissue layer of the skin. However, although temporary, the somites do play an important role in laying down the basic pattern, which the subsequent formation of vertebrae, skeleton and spinal chord makes permanent.

Therefore both the notochord and the somites should perhaps be regarded more as templates rather than organs, because as organs they have no function and soon disappear, whereas as templates they create and donate permanent pattern to the animal. The question arises how is this pattern created ? We shall show that it is caused by the same secondary wave that caused gastrulation.

At first sight it might appear a little overambitious to attempt a single explanation for such divers phenomena during both gastrulation and neurulation and so let us put it more tentatively. In attempting to explain gastrulation we were led to postulate Hypothesis 2 in the last section, namely the existence of a secondary wave - and now this same hypothesis leads us to expect further events in the mesoderm mantle that do seem to occur with the correct timing and geometry during neurulation.

For consider what happens when the secondary wave hits the proximal surface of the mesoderm mantle. It may take some time for the secondary wave to cross the mantle, because when the primary wave originally crossed it the shell was considerably thicker (see Figure 9(a)), with thickness comparable to the distance of travel up the dorsal side. Furthermore there is no reason to suppose that the speed of the primary wave through the shell is necessarily exactly the same as its speed across the surface. In the case of the frog we estimate that the time taken to cross the shell is about 16 hours compared with 24 hours to travel up the dorsal side (see §14 below). The secondary wave will follow with the same timing. Therefore by the time the secondary wave hits the proximal surface, its previous effect of reducing the area of the distal surface may have mostly died away. Therefore the mesoderm mantle will suddenly experience a violent reversal of curvature for two reasons : firstly it will be tending to recover its original positive mean curvature as the effect on the distal surface dies away, and secondly this positive curvature will be drastically reinforced as the reduction effect now hits the proximal surface. The mild spring will be suddenly transformed into a violent spring, and the mesoderm will try to coil outwards underneath the neural plate. However we shall show in Corollary 2.5 below that it is impossible for the spring to achieve its desired curvature. As a result the central piece of mesoderm will rip itself off, and roll up separately flexing itself straight and forming the notochord. On either side individual segments will rip off and curl up into somites. Meanwhile the remaining lateral mesoderm remains unaffected by the secondary wave.

To justify these statements we shall first compute the relevant curvature, and then examine the position to which the proximal surface has been carried by gastrulation in order to explain why lateral mesoderm is unaffected. Next we shall analyse the passage of the secondary wave across the proximal surface, in order to explain why the notochord forms and elongates, and why the somites form sequentially from front to back. Then in §13 we explain why the somites are segmented, how their size regulates to the size of the embryo, and why the later somites are smaller and slower to form. Finally in §14 we discuss details of timing. First the curvature; let ρ denote the radius of the embryo.

Corollary 2.4. *When the secondary wave hits the proximal surface the mesoderm mantle will attempt to take up a positive mean curvature equal that of a sphere of radius $2\rho/9$, or a straight cylinder of radius $\rho/9$.*

Proof. By the data on shell thickness in §11, the original excess before gastrulation was $e = \frac{7}{6}$. By Corollary 3 the surface reduction factor imposed by the secondary wave is $\lambda = \frac{2}{3}$. Therefore applying the reduction factor to the proximal surface, the new excess is

$$\frac{e}{\lambda} = \frac{3}{2} \times \frac{7}{6} = \frac{7}{4}.$$

Therefore by the lemma

$$2\epsilon\mu = \frac{e}{\lambda} - 1 = \frac{3}{4}.$$

By the data $\epsilon = \frac{\rho}{12}$, and so

$$\mu = \frac{3}{8\epsilon} = \frac{9}{2\rho}.$$

Therefore if the mesoderm were able to assume spherical form, of radius ρ_1, then

$$\rho_1 = \frac{1}{\mu} = \frac{2\rho}{9}.$$

Alternatively if the mesoderm were able to assume straight cylindrical form, of radius r, then the other principal radius of curvature would be ∞, and so

$$\mu = \frac{1}{2}(\frac{1}{r} + \frac{1}{\infty}) = \frac{1}{2r}.$$

Therefore

$$r = \frac{1}{2\mu} = \frac{\rho}{9}, \text{ as required.}$$

Corollary 2.5. *The mesoderm mantle is unable to achieve the required positive mean curvature without tearing itself.*

Proof. Suppose, on the contrary, that it does achieve required mean curvature without tearing itself; then we shall prove a contradiction. When the neural plate begins to roll up, it has negative gaussian curvature (see Figures 13(B') and 17 (Stage 18)), similar to the dorsal lip (§11 Corollary

2.1). However this time the situation in the underlying mesoderm mantle is aggravated on the one hand by the fact that it is trying to achieve an even greater positive mean curvature, and on the other hand by the neural plate getting in the way preventing it from curving so much. More precisely, let n denote the radius of the transverse section, and r the radius of the median section, at a point on the dorsal side of the distal surface of mesoderm, as illustrated in the sketch of the transverse section in Figure 22.

Figure 22. Diagram of transverse section of neurula.

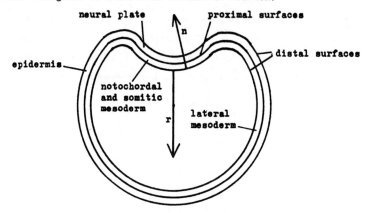

Let ε denote the thickness of mesoderm, as before. Initially the overlying neural plate will have approximately the same thickness (see the photograph in Figure 11f), but as it rolls up it becomes thicker. Therefore $n \geqslant 2\varepsilon$. The principal radii of curvature are n, -r (negative because r points out of the mesoderm), and so the mean curvature is

$$\mu = \frac{1}{2}(\frac{1}{n} - \frac{1}{r}).$$

But the required $\mu = \frac{3}{8\varepsilon}$, by the proof of Corollary 4.
Therefore

$$\frac{1}{n} = \frac{3}{4\varepsilon} + \frac{1}{r} \geqslant \frac{3}{4\varepsilon} ;$$

therefore

$$n \leqslant \frac{4\varepsilon}{3} .$$

But this contradicts $n \geq 2\varepsilon$, and therefore completes the proof of Corollary 2.5.

A consequence of Corollary 2.5 is that when the curvature or sheer forces in the mesoderm become greater than the cohesive forces between neighbouring cells, then the mesoderm mantle will tear itself. In order to explain where the tears will occur, and in what order the torn pieces will curl up, we shall need to examine the way in which the secondary wave travels over the proximal surface. But before we do this, we make a couple of remarks about the validity of Corollary 2.4.

Remark 1. When the notochordal-mesoderm tears free and rolls itself up to form the cylindrical notochord, we might expect its diameter to be $\frac{1}{9}$ that of the embryo, from Corollary 2.4. In fact this is sometimes quite a good estimate (see Figure 26). When the somites form, although they are not exactly spherical in shape, their cross-section is roughly twice that of the notochord, which again is in agreement with Corollary 2.4.

Remark 2. In using the value $\lambda = \frac{2}{3}$ in the proof of Corollaries 2.4 and 2.5 we have implicitly made the rather drastic assumption that cells on the proximal surface submerge in the same way as those on the distal surface, or, more precisely, we have assumed that the effect on the curvature is the same. In fact they may submerge quite differently. However we are justified in using the term submerging, because Mookerjie, Deuchar and Waddington [21,p.451] have shown that the area of contact between cells does increase considerably when the notochord forms. Our implicit assumption is that this local effect imposes the positive mean curvature, which not only forces the cylinder to have small radius, but also to be straight (because if $r < \infty$ in Corollary 2.5, this reduces the mean curvature). Therefore it is the secondary wave of submerging cells that initially forces the notochord to be stiff. Later when the secondary effect wears off the notochord retains its stiffness by the cells vacuolating (that is swelling with fluid) making the whole cylinder turgid, and by the secretion of a thin supporting sheath [21,pp. 262, 450].

The proximal surface, P. Recall that in §10 above we defined the two circles α and δ on the distal surface of the blastula to be the presumptive meso/endoderm and ecto/mesoderm frontiers, respectively. Let X denote the solid piece of blastula bounded by the planes through α and δ. Then X is shaped like a lop-sided barrel, as shown in

Figure 23(X). The surface of X comprises 5 regions as follows:
(i) The annulus at the top bounded by circles δ, δ^*, to which presumptive ectoderm is attached (shown cross-shaded).
(ii) The disk at the bottom bounded by circle α, to which the rest of the presumptive endoderm is attached (shown shaded).
(iii) The inside yolky surface is a disk bounded by the circle β^*, which we have drawn flat for convenience, but which in fact is usually curved (see Figures 9, 12).
(iv) The distal surface is the outer curved annulus bounded by α and δ, and containing the points ABCD on the dorsal side.
(v) The proximal surface P, which we define to be the inner curved annulus bounded by β^* and δ^*, containing the points $B^*C^*D^*$ on the dorsal side. Notice that we do not include the yolky surface (iii) in the definition of proximal surface.

The presumptive meso/endoderm frontier lies in the interior of X, and is approximately the conical annulus bounded by α and β^* (shown dotted); this frontier is already predetermined by the boundary of the yolky region of the egg before cleavage. Therefore the sides of the barrel form the presumptive mesoderm, and the floor of the barrel, which is yolky, forms part of the presumptive endoderm, the rest of which is attached to the bottom disk.

Gastrulation turns the barrel inside out, into the bottle-shape Y shown in Figure 23(Y). For diagrammatic simplicity we have not shown all the details of the endoderm part of Y, for instance where it has been torn away from the mesoderm (along the lateral parts of the dotted annulus), but the shape of the mesoderm in Y is sufficiently accurate for our purpose. Let Q denote the new position of the proximal surface after gastrulation, now lying on the outer surface of Y between the circles β^* and δ^*, and containing the points $B^*C^*D^*$. The main point we wish to emphasise is that whereas the distal surface ABCD, between α and δ, now stretches round nearly the whole embryo, the proximal surface Q only stretches over less than half as much. It is the proximal surface that induces neural plate (and not the yolky surface because that is endoderm), and therefore the position of the former determines the latter (see Figure 12). This explains the pear-like shape of the neural plate (see Figures 14, 17, 18, 23). This also answers a question of Waddington

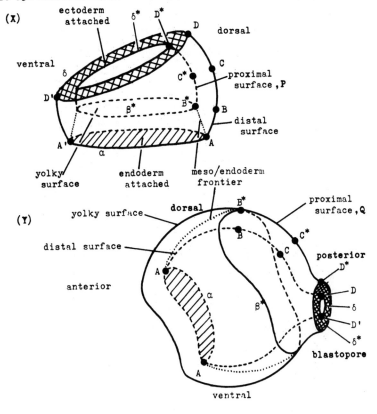

Figure 23. Mesoderm mantle before (X) and after (Y) gastrulation.

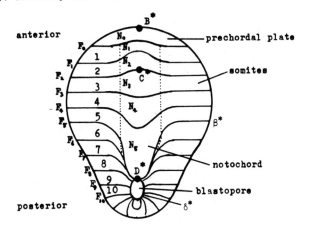

Figure 24. Secondary wave fronts on the proximal surface Q.

[21, p. 467] as to why the neural plate is broad at the front and narrow at the back (which previously seemed to be a paradox on the assumption of an evocator diffusing only from the dorsal mid-line).

The secondary wave-front on Q. Figure 24 sketches the successive wave-fronts as the secondary wave travels across the proximal surface Q, looking down on Q from the dorsal side. Quantitively we should expect this picture to differ slightly for different species, but to justify the picture qualitatively let us work from Vogt's detailed fate maps for newts in Figure 13, as follows.

Consider first the passage of the primary wave, which travels through only the presumptive mesoderm, that is the sides of the barrel X. The primary wave starts at A, by Hypothesis 1, and travels up the distal surface to the circle δ; at the same time it penetrates through the shell, first hitting the proximal surface P at B^*, and then travelling up P to δ^*. Let F be the family of curves on P representing the primary wave-fronts, in other words the successive positions of the primary wave.

Now turn to the fate maps of Figure 13. Admittedly these are drawn on the distal surface showing mesoderm as the annulus between α and δ, but by projecting radially from the centre of the embryo, we can identify P with the subannulus consisting of presumptive notochord, prechordal plate, somites and tail mesoderm. P does not contain the complementary subannulus consisting of lateral and pharyngal mesoderm.

Since the primary wave starts at A, it is reasonable to assume that, near A at any rate, F is approximately the family of semicircles concentric with A. Now the boundaries between presumptive somites are also circular arcs concentric with A, and as we shall explain in the next section this is no accident. But for the moment let us postpone the question of why, when the somites form, they segment themselves along secondary wave-fronts, and merely assume at this stage that the presumptive somite boundaries lie along primary wave-fronts.

Let F_n denote the wave-front separating somites n and $n + 1$. The complete wave-front F_n is a semi-circle, whereas the somite boundaries consist of two arcs, one on either side, and so in order to draw F_n we must join the latter by an arc running through the presumptive notochord, which we have indicated by a dotted line in Figure 13. These dotted lines do not appear in Vogt's original fate map because as opposed to what happens in the somites, the secondary wave-fronts do not manifest themselves physically in the formation of notochord, for reasons which we explain in the next section. Therefore the notochord when it forms appears as homogeneous and unsegmented.

Now consider the effect of gastrulation upon the wave-fronts. Gastrulation turns the barrel X inside out into Y, and induces a diffeomorphism

$$h: P \to Q.$$

The family $\{F_n\}$ of primary wave-fronts in P is mapped by h onto the family of secondary wave-fronts in Q.

Lemma 3. *The secondary wave-fronts are as shown in Figure 24.*

Proof. We examine the diffeomorphism h. Metrically the most striking feature of h is that it shrinks the large circle δ^* into the small blastopore. At the same time h expands areas uniformly by the expansion process (1) of §10 above (mathematically h has constant Jacobean). Therefore since h shrinks the neighbourhood of δ^* longitudinally it must compensate by expanding latitudinally. Therefore the presumptive notochord in P is mapped by h into an elongated roughly rectangular strip on the dorsal side of Q, shown by dotted lines in Figure 24. Meanwhile we know from experimental observation (see Figures 16, 27) that the boundaries between somites are mapped onto lateral lines in Q, which are shown in Figure 24 with the same numbering as in Figure 13. There remains to justify the extension of the wave-fronts into the notochord region in Q.

Since F_o is near to β^* in P, the same is true for F_o in Q. Since F_5 touches δ^* at D^* in P, the same is true for F_5 in Q. The region between F_{n-1} and F_n in P consists of the nth somite and a region of notochord, which let us call N_n. An examination of Figure 13 shows that the areas of $N_1, \ldots, N_5$ are approximately proportional to the numbers $1, \ldots, 5$. Therefore the same is true for their images in Q because h expands uniformly. The only way to extend the curves $F_1, \ldots, F_4$ so as to divide the rectangular strip of notochord in Q in these proportions is as shown in Figure 24. Finally $F_6, \ldots, F_{10}$ all touch δ^* at D, and so the same is true of their images. This completes the proof of Lemma 3.

We can now follow the progress of the secondary wave across the proximal surface and describe its effects.

Lateral mesoderm. First notice that the secondary wave is confined to the proximal surface, because primary wave never crossed the yolky surface, since that was part of the endoderm (see Figure 23). This explains why the thin tongues of lateral and pharyngal mesoderm remain relatively passive during neurulation.

Prechordal plate. The secondary wave first hits the proximal surface at its most anterior point B*. Therefore the first piece of mesoderm to be effected is the prechordal plate, lying between B* and F (see Figure 24). This may be the reason why the anterior neural fold is the first appearance of neurulation (see Figures 11f, 12, 17). Admittedly the neural folds seem to be the initial phase of the neural plate rolling itself up, but Waddington [21,p.476] remarks that it is surprising that the most anterior part of the neural plate should be the last to be induced and yet the first to curl up. We suggest that the prechordal plate may be giving a helping hand underneath. For the mesoderm at this stage certainly maintains close contact with the neural plate (see Figure 11f); and active positive curvature by the underlying mesoderm would have the physical effect of reducing horizontal tension between overlying neural plate cells, providing them with the opportunity of gently increasing mutual contact without the danger of being pulled apart again, and therebye facilitating their columnar formation.

Supporting evidence that the prechordal plate is itself actively curling up is that it begins to tear itself away from the more anterior parts, the pharyngal mesoderm, the tear occuring along β^*. This tear enables it to achieve both positive gaussian and positive mean curvature, and to surround the fore-brain region (see Figures 14, 16). Therefore it is not subjected to the same severe stresses as the more posterior parts of the mantle, which are prevented by negative gaussian curvature from achieving the desired positive mean curvature without further tearing. Consequently the prechordal plate is the only piece of proximal surface able to retain its integrity as a sheath surrounding the neural tube.

Neural folds. We suggest that as the secondary wave proceeds through the successive wave-fronts $F_1, F_2, F_3, \ldots$ it causes the neural folds to spread from the front round to the sides (see Figure 17 Stage 16). It takes much longer for the folds to reach the back; in fact Sedra and Michael [16] remark upon the fading of the neural folds towards the blastopore at late stages of neurulation.

One can ask the question (c.f. [21, p. 417]): if the secondary wave is also spreading over the interior of the proximal surface (Figure 24), why does the curvature only appear at the edges of the neural plate and not in the middle? An answer is suggested by imagining unrolling a roll of stiff paper and trying to hold it flat: the paper refuses to go flat and retains most of its curvature at the ends, where the bending moment is least. So the enclosing sack of ectoderm may be pressing down pssively on the mesoderm mantle at this stage, allowing it only to curl up at the edges (see Figures 14(A"),15, 17 Stage 17). When the neural plate itself becomes active shortly afterwards, its curvature tends to be the same both in the middle and the edges (see Figures 9(e), 14(B"), 15, 17 Stage 18).

As supporting evidence for this point of view notice Shumway's remarks [17,II], that photographs of sections at this stage show the neural folds flatter than they appear in living specimens, and show the neural furrow shallower. This can be explained as follows : the force pushing up the neural folds is due to the energy released by the submerging mesoderm cells, and when these cells are killed during preparation of slides, this force disappears allowing the elastic resistance of the overlying neural plate to push the folds down again,

Neural furrow. As can be seen from Figure 24 the secondary wave travels much faster over the notochord region than along the somite region. By the time the wave-front F_5 is reached nearly the whole of the former has been covered. One of the first results is that the mesoderm tries to form a tight fold along the dorsal mid-line, similar to the dorsal lip (§11 Corollary 2.1), only more so because of the greater positive mean curvature. In so doing it pinches together the overlying neural plate to form the neural furrow (see Figures 14, 15). This is probably the main reason for the neural furrow, because without mesoderm underneath the furrow tends not to appear [23, p.305], and consequently the neural tube tends to develop with a round cross-section rather than elliptical. Supporting evidence that the mesoderm causes the neural furrow comes from time-lapse films*, in some of which the anterior end of the furrow can be seen to disappear with a flip, just at the moment when mesoderm underneath would be freeing itself from the neural plate in order to roll up into notochord.

<u>Notochord.</u> We suggest that by the time the secondary wave has reached the curve-front F_5 the mesoderm has begun to tear itself along the dotted lines in Figure 24. What causes the tearing we shall discuss in a moment. But once the tears have been made the mesoderm between the dotted lines is free to roll itself up into a cylinder to form the notochord (see Figures 14, 16, 26, 27), or, more precisely, the cells in that region are free to submerge further by increasing mutual contact [21, p.451], and thereby form the cylinder. The same increase of cell contact causes the cylinder to flex itself straight, which in turn causes it to also tear itself away from the prechordal plate along F_o, leaving the notochord attached to the rest of the mesoderm only at the blastopore end.

But what causes the tear along the dotted lines? One can avoid the question by merely saying that this region of mesoderm "differentiates into notochord" - but having taken the point of view that notochord and somites are primarily templates rather than organs, we are obliged to look for some more mechanical reason for the tearing and the formation of pattern. We tentatively suggest the the dotted lines may be the locus of maximal bending moment, maximal sheering forces, and maximal weakness of the mesoderm mantle, as follows.

Firstly the bending moment. Figure 25 is a photograph of a transverse section at this stage, showing the notochordal cells submerging away from the proximal surface, inducing positive curvature and bending moment stress in the mesoderm mantle. Meanwhile the increased mutual contact between notochordal cells increases their cohesive strength, and makes the wave-front itself the momentary line of meximal weakness. But from Figure 24 it can be seen that the wave-fronts $F_5, F_6 \ldots$ coincide along the posterior end of the notochord boundary, stabilising the line of weakness, possibly sufficiently long for tearing to occur.

---

* I am indebted to Jack Cohen for showing me his films on axolotl neurulation.

4. Primary and Secondary Waves in Developmental Biology   201

Figure 25. Transverse section of newt neurula showing notochord forming. From Vogt [20, p.544].

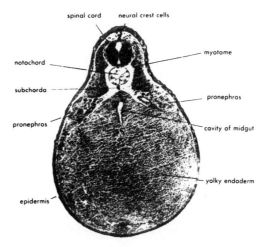

Figure 26. Transverse section of frog embryo showing notochord. By this stage the somites have developed into dermatome, sclerotome and myotome. From Balinsky [1, p.316].

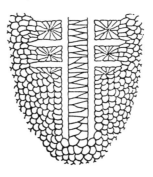

Figure 27. Diagram of cell arrangement in the mesoderm of the newt neurula, showing the developing notochord and somites. From Waddington [21, p.449].

Secondly the sheering forces. The rapid spread of the secondary wave over the posterior end of the notochord region and the consequent submerging will cause the latter to push forward on its anterior part, setting up longitudinal sheering forces. Meanwhile somites $1,\ldots,5$ will already have segmented, as we explain in the next section, in other words will have torn away from each other laterally along the lines $F_0, F_1, \ldots, F_5$ up to the dotted lines in Figure 24. Therefore each of the first five somites on either side will be attached to the notochord region only along a fragment of the dotted line. This fragment will be subjected to both lateral bending moment and longitudinal sheering force, and, if tearing is to occur, will be the most likely place for the somite to tear away from the notochord.

Admittedly we have suggested rather complicated reasons for tearing, but our description of the forces present appears to coincide with experimental observation. The tearing caused by such forces might be somewhat ragged initially, but then the submerging effect would be an automatic self-correcting device for subsequently rounding up the notochord cylinder and each individual somite (see Figures 16, 27).

13. **PATTERN FORMATION.**

Jonathan Cooke [3,4] has shown experimentally, by surgery, that the somites are determined before gastrulation begins, at just about the same time that our hidden primary wave is beginning to travel up the presumptive mesoderm in the blastula shell. Coupling this with the fact that the presumptive somite boundaries coincide geometrically with the semi-circular primary wave-fronts in Figure 13, it is almost irresistible to conclude that the primary wave is connected with the determination of the somite pattern. Stimulated by Jonathan Cooke, we introduce a new theoretical idea in this section to show how such pattern could be formed. The idea is to combine the primary wave with a clock. There are several ways of combining primary waves with clocks to form regular repeating patterns, and we select one method that seems to give the simplest and most appropriate model for this context.

Suppose that during the passage of the primary wave there is a periodic fluctuation in the levels of concentration of certain chemicals in the cell. Normally the homeostasis and the slow development of

Figure 28. (a) Chemical fluctuations impose a ripple on the primary wave.
(b) Fluctuations become dominant.
(c) Continuous deformation from (a) to (b) along parameter θ.
(d) Segmentation of somites.

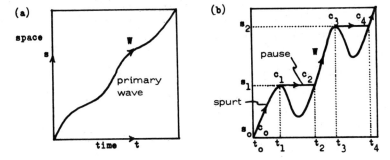

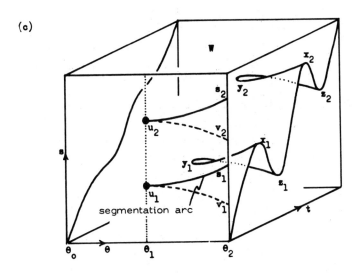

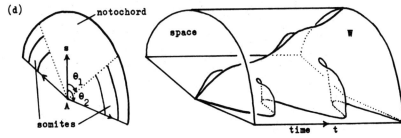

a cell might hardly be affected by such fluctuations, but at the moment that homeostatic stability breaks down and the state of the cell suffers a catastrophic jump (as described in the proof of the theorem in §8 and illustrated in Figures 4(b) and 5), then the cell might be exceptionally susceptible to even small fluctuations. For instance fluctuations could affect theshold, and hence the timing of the jump. More precisely, for any given time t, the state of the fluctuation at that moment will cause the corresponding section of the surface M in Figure 5 to be displaced in the s-direction. As time progresses the section will be displaced to and fro synchronous with the clock. Therefore the clock will impose a ripple on the fold curve of M, and hence upon its projection W in space-time. In other words the clock will impose a ripple upon the track W of the primary wave in space-time, as illustrated in Figure 28.

If the fluctuation becomes dominant, then it will cause the amplitude of the ripple to increase until W is no longer monotonic, as illustrated in Figure 28b. In this case W will no longer represent a continuous wave, because only parts of it will manifest themselves, as is shown by the following lemma.

Lemma 4. If the flutuation is dominant the the frontier alternately moves in spurts and pauses.

Proof. Using the notation of Figure 28b, during the interval $t_o < t < t_1$ the arc $c_o c_1$ represents the frontier moving as a wave from s_o to s_1 and slowing parabolically to a halt at s_1. During $t_1 < t < t_2$ the line $c_1 c_2$ represents the frontier pausing at s_1. During $t_2 < t < t_3$ the arc $c_2 c_3$ represents the frontier moving from s_1 to s_2, and so on. This follows from tracing the development path of each cell on the surface M lying above space-time, as in Figure 4b.

Remark. Notice that the peaks c_1, c_3 of the ripple manifest themselves, whereas the troughs do not. This is because, near a trough, a cell will already have catastrophically jumped into its new equilibrium state (onto the lower surface of Figures 4a, 5), and so is never able to attain the old stable states represented by the interior of the trough. We therefore say the troughs are silent.

Segmentation of somites. Lemma 4 gives the key to understanding why the somites segment. The primary wave alternately moves in spurts and pauses at the wave-fronts $F_o, F_1, F_2, \ldots$ in Figure 13. Therefore the secondary wave will do the same in Figure 24. Therefore during the spurts from F_{n-1} to F_n all the cells in the nth somite on each side will submerge almost simultaneously, and during the subsequent pause the submerging cells will have time to tear themselves away as a mass from the as yet unaffected region, along F_n. The tear is facilitated by the longitudinal tension set up by the elongating notochord. Once the mass has torn free, the continuing submerging of the cells causes the mass to round up into the somite (see Figures 16, 27).

Having presented the main idea of the model of segmentation we now go on to explain the geometry of the segmentation arcs, and why they end where they do. For as we suggested at the end of the last section, this in turn will determine the line of tear of the somites away from the notochord, and hence determine the boundary of the notochord. Next we shall explain what causes the fluctuations, and the timing of the clock, and finally how the size of the somites is regulated to the size of the embryo. But first we must go back to the theory for a moment.

Segmentation of arcs. Let θ be a parameter representing increasing dominance of the fluctuation over the primary wave, in other words increasing amplitude of the ripple. Figure 28c shows θ drawn as a third dimension, varying smoothly from $\theta = \theta_0$, where the fluctuation is subservient as in Figure 28a, to $\theta = \theta_2$, where the fluctuation is dominant as in Figure 28b. The critical value $\theta = \theta_1$, shown by the dotted line is where dominance first appears, in other words where the graph first becomes non-monotonic.

Now suppose that θ is realised as a second space-dimension perpendicular to s. Then in Figure 28c the (s,θ)-plane nearest the eye represents 2-dimensional space, and the cube represents 3-dimensional space-time. The surface W represents the path of the primary wave in space-time. Generically the projection of W onto the space-plane will have fold curves $x_1 y_1 z_1$, $x_2 y_2 z_2$ which project onto the* cusps $s_1 u_1 v_1$, $s_2 u_2 v_2$. The cusp points u_1, u_2 occur at the critical value $\theta = \theta_1$.

---

*Mathematically the cusps here are pheonomenologically different from the cusp catastrophe shown in Figures 4a,5. The latter arose solely from the dynamic, whereas the cusps here arise from interaction between the time-axis and the dynamic. More precisely they are second order tangencies of time-axis with the fold surface, a 1-higher-dimensional analogue of the point c_2 in Figure 5, which was a first order tangency of the time-axis to the fold curve.

Lemma 5. *The primary wave pauses along the arcs s_1u_1, s_2u_2. Therefore these become the segmentation arcs of the secondary wave.*

Proof. From Figure 28b we saw that pauses began at the peaks c_1, c_3 of the ripple. Therefore in Figure 28c pauses begin along the fold curves x_1y_1, x_2y_2. The projections of these curves onto the space-plane are the arcs s_1u_1, s_2u_2, which are therefore the segmentation arcs, as required.

Notice that the fold curves y_1z_1, y_2z_2 represent troughs, which are silent, and therefore their projections u_1v_1, u_2v_2 do not manifest themselves in space. That is why we have drawn them as dashed lines in Figure 28c. Note also that each segmentation does not happen simultaneously, as shown by the following corollary.

Corollary 5.1. *Segmentation starts at s_1 and proceeds towards u_1 along the segmentation arc s_1u_1 in the direction of θ decreasing. Then along s_2u_2, and so on.*

Proof. The projection of the fold curve x_1y_1, into the time axis is in the direction of time increasing.

Geometry of the somites. The above arguments work just as well if (s, θ) are taken as polar coordinates rather than cartesian coordinates, and Figure 28d is the diagram for polar coordinates analogous to Figure 28c for cartesian. The critical value $|\theta| = \theta_1$ is shown by the dotted lines, dividing the region $|\theta| < \theta_1$, where the fluctuation is subservient, from the region $\theta_1 < |\theta| < \theta_2$ where it is dominant. The wave fronts are approximately the circles r-constant, and the segmentation arcs are subarcs given by $\theta_1 < |\theta| < \theta_2$. Comparing with Vogt's fate map in Figure 13 we have a reasonably good qualitative explanation for the geometrical shape of the segmentation arcs between successive somites.

Moreover Corollary 5.1 explains why each segmentation begins at the boundary β^* of the proximal surface and proceeds inwards towards the notochord. Figure 16 illustrates clearly the geometry of the somite mesoderm first torn away along β^* curling up round the neural tube, and then the segmentation proceeding successively from front to back, with each segmentation proceeding inwards towards the notochord. In the case of frogs and toads there is an additional detail that each somite rotates through 90° as it forms [3], and this may be explained by the inward process of Corollary 5.1 coupled with the forces exerted by the submerging cells.

Boundary of the notochord. What determines the limits θ_1, θ_2 of the segmentation arcs in Figures 13 and 28d ? The limit θ_2 is easy to understand because this is determined by the boundary β_* of the proximal surface (see §12). But θ_1 is more subtle, and is interesting because in turn it determines the subsequent boundary between somites and notochord. It is well known that before cleavage the gradients in the egg are dominated by the animal-vegetal axis, for instance the pigment is concentrated towards the animal pole and the yolk towards the vegetal pole. Therefore is not unreasonable to assume that when the primary wave travels in a direction having θ sufficiently small, the animal-vegetal gradient will dominate the fluctuation. Conversely if the egg is relatively homogeneous longitudinally it is not unreasonable to suppose that the fluctuation might dominate when θ is sufficiently near $\frac{\pi}{2}$. These two assumptions imply the existence of a critical angle θ_1 between, which in Vogt's fate map in Figure 13 appears to be initially about 30°. As the primary wave proceeds north and begins to slow down before stabilising the effect of the fluctuation may decrease, and so it is not unreasonable to suppose that θ_1 should be an increasing function of s. Therefore we have an explanation of the halberd-shaped region of presumptive notochord in the blastula.

Fluctuations. What is the most likely cause of fluctuations in the levels of chemical concentrations in a cell ? One obvious answer is cell-division. Indeed Paul Weiss [23, p.77] points out that mitosis is a kind of earthquake for a cell, which monopolises its resources, and during which development is temporarily suspended. He observes [23, p.85] that cellular differentiation and multiplication are two processes which, if not strictly mutually exclusive, are nevertheless markedly antagonistic in their tendencies. Therefore it is not unreasonable to suppose that it is the regular cleavage in the blastula that produces the fluctuations that cause the ripple during the primary wave, and therefore subsequently the segmentation during the secondary wave.

This hypothesis looks plausible from the point of view of timing, for, from Figure 18, we observe that the cleavage in the blastula occurs at intervals of slightly less than an hour in frogs, and according to Cooke [3] this slows down to one every 2 or 3 hours shortly after gastrulation has started. If the time delay to the secondary wave is constant, then we should expect a similar timing in the formation of the somites. And sure enough the early somites form one about every 40 minutes [3], slowing to one every 2 or 3 hours in early tail-bud. In the late tail-bud the final somites slow to 2 or 3 per day, but this effect may be enhanced by an increase in the delay between the primary and secondary waves in the final ventral region.

Note that the above figures may be unreliable in that they refer to different experiments, possibly at different temperatures, but are in sufficient agreement to warrant a careful experimental correlation of the timing of cleavage before and during gastrulation, with the formation of somites during and after neurulation.

Regulation. There are nearly always the same number of somites in individuals of the same species. Therefore the size of each somite must regulate to the size of the embryo. But how does an individual cell know whether it belongs to a big embryo or a little embryo, in order to cooperate with the correct number of neighbours to form a big somite or a little somite? To explain this we need to make an additional assumption, that cells at the beginning and end of the wave have predetermined development. Then :

Lemma 6. The length of each somite is proportional to the length of the embryo.

Proof. Let L be the length of the embryo. The length of travel of the primary wave is proportional to L, but the time taken to travel that length is independent of L, since the development of the end points is predetermined. Therefore the average speed of the wave is proportional to L. But the periodicity of mitosis is independent of L. Therefore the distance travelled by the wave between two mitoses is proportional to L. In other words the length of each somite is proportional to L.

Remark. The above proof appears to be similar to the usual argument for regulation in a standard gradient model for pattern formation, which runs as follows : a longer embryo has a shallower gradient and hence longer somites. However Cooke [3,4] points out that the latter argument gives the wrong answer where comparing somites at the front and back, whereas our model can give the right answer. For near the end of the wave the gradients become shallower, and therefore in the standard gradient model one would expect the somites to become longer, whereas in fact they become shorter. By contrast in our model the shallower gradients are associated with the wave slowing down parabolically (see §9), and so the distance travelled by the primary wave between mitoses decreases causing the somites to become shorter. It would be interesting if it was possible to measure experimentally the parabolic slowing of the primary wave and the decrease in rate of mitosis, for this would then provide a quantitative prediction for the decrease in size of somites. Another way to test the model is to devise a method for altering the speed of either mitosis or the primary wave, without altering the other, because this would then alter the size and number of somites.

14. TIMING OF GASTRULATION AND NEURULATION.

Our whole theory rests upon the existence of a hidden primary wave. As yet there has been little experimental attention directed towards confirming this wave because the very concept of this type of wave first needed the deep mathematics of catastrophe theory. With this mathematics, our theorem predicts the existence of the primary wave, based only upon the fact that mesoderm differentiates from ectoderm.

Our Hypothesis 1 of §10 that the wave begins at A (See Figures 12, 13 and 23) is based on experimental results of Nieuwkoop [12], that in axolotl there is a hidden wave of mesoderm determination travelling up the dorsal side from A to D. This can be detected by the grafting experiment of §7 : if the primary wave has already passed the donor point of the graft, then, by transplanting the graft to the ventral side, the graft will induce a subsidiary invagination after the appropriate delay, as in the classical Spemann-Mangolde experiment [1, Chapter 10]. According to Nieuwkoop the primary wave in axolotl begins at A soon after 7th cleavage, and measured in hours after fertilisation has the following timing.

	Hour	Delay
Primary wave begins at A (7th cleavage)	16	20
Secondary wave begins at A (dorsal lip)	36	
Primary wave reaches D	36	20
Secondary wave reaches D (blastopore closes)	56	

The fact that the delay at A is the same as the delay at D makes Hypothesis 2 plausible, that the secondary wave does indeed follow the primary wave after a 20 hour delay. (The fact that the seondary wave begins at A at the same time as the primary wave reaches D may be accidental because these points are spacially far apart, and may not occur in other species.)

In different species and at different temperature we should expect different timing. For instance from the normal tables for frogs and toads ([13,16,17] and Figure 18) we deduce the following feagures. In each case we have assumed that the hidden primary wave begins at A after 7th cleavage (Stage 7-7½) and reaches D by mid-gastrulation (Stage 11), although of course these assumptions needed to be checked experimentally, as in the axolotl case. Meanwhile the visible secondary wave begins with the dorsal lip (Stage 10) and reaches D with the blastopore closing (Stage 13).

Species	[17] Rana pipiens	[13] Rana sylvatica			[16] Bufo regularis
Temperature	18°C	18°C	15.4°C	10.4°C	25°C
Primary wave begins at A	10	6	*4.7	*11	3.5
Secondary wave begins at A	26	19	24	45	7
Primary wave reaches D	34	24	32	60	10
Secondary wave reaches D	50	36	52	96	14.5
Delay	16	12/13	19.3/20	34/36	3.5/4.5

* Note that the figures for the two lower temperatures for Rana sylvatica are measured in hours after the first cleavage, as opposed to the others which are measured from fertilisation.

In each case the delay at A equals the delay at D, well within the tolerance of experimental measurement. These figures suggest that Hypothesis 2 remains plausible under wide variations, and would appear to justify further experimental work to confirm the existence and timing of the primary wave.

We now turn out attention to neurulation and examine in detail the timing for one species namely Shumway's normal table [17] for the frog Rana pipiens, shown in Figure 18. Assuming that the primary wave takes ⅔ as long to cross the blastula shell as to travel up the dorsal side from A to D, we show that Hypothesis 2 continues to be plausible throughout neurulation as well as gastrulation, from the point of view of timing. In order to be precise let us recap, for the frog, all the assumptions that we have made so far about timing. The letters A, B, etc., refer to the points in Figures 12, 13 and 23.

(i) The primary wave begins at A after 7th cleavage, 10 hours after fertilisation.

(ii) The primary wave takes 8, 16, 24 hours to reach points B,C,D, respectively, on the distal surface.

(iii) The primary wave takes 16 hours to cross the shell to the proximal surface.

(iv) The secondary wave is delayed 16 hours after the primary.

We can now compute the time in hours after fertilisation when the secondary wave hits each of the different points :

Point	A	B	C	D	B*	C*	D*
Hour	26	34	42	50	50	58	66

In order to compute the changes in area of the distal and proximal surfaces we need two more assumptions :

(v) The effect of the secondary wave is to reduce the surface area by one third (by §11 Corollary 3), and this effect dies away linearly over 20 hours.

(vi) The expansion process of the blastula shell (see §10 above) starts at hour 20, and lasts for 30 hours, by which time it has doubled the area.

Note that these six assumptions are guesses based on three peices of data: firstly the timing of the normal table of Figure 18, which Shumway observes can vary up to 10% in individuals; secondly the data on shell thickness at the end of gastrulation, $\varepsilon = \rho/12$, from Figures 9(d) and 11(f); and thirdly the estimate AB = ⅓AD from Figure 12, which refers to newts and toads rather than frogs. Therefore our assumptions must be tentative. However we are not claiming quantitative accuracy at this stage, but only verifying quantitative plausibility of the theory.

Using the six assumptions we can now plot against time, in Figure 29, the expansion and contraction at different points of the distal and proximal surfaces of mesoderm mantle, relative to the initial distal area before gastrulation. The curves are drawn for the point A, and the three pairs of points (B,B^*), (C,C^*) and (D,D^*). The continuous curves represent points A,B,C,D on the distal surface, and the dotted curves points B^*,C^*,D^* on the proximal surface. In the case of A there is no corresponding point A^* because the proximal surface does not extend that far down (see Figures 12 and 23). In the other three cases we can deduce the excess e and hence the mean curvature μ from Lemma 2.

Since the thickness of the shell after gastrulation is $\varepsilon = \frac{\rho}{12}$, and since it was twice as thick before the expansion process, it must initially have been $\varepsilon = \frac{\rho}{6}$ at the beginning of gastrulation. Therefore by the lemma the excess was initially $e = \frac{4}{3}$. Therefore since we are comparing areas with the initial distal area, all the distal curves must start at 1 while the proximal curves start at $\frac{3}{4}$. (In Figure 29 the scale on the vertical axis indicated refers only to the bottom pair D,D^*.)

During the expansion process the distal area doubles, and so the distal curves increase to 2. Meanwhile the thickness halves to $\varepsilon = \frac{\rho}{12}$, reducing the excess to $e = \frac{7}{6}$, and so the proximal curves increase to $2 \times \frac{6}{7} = \frac{12}{7}$. Therefore if there were no secondary wave each pair of curves would be approximately parallel, implying constant positive mean curvature, $\mu = \frac{1}{0}$.

However when the secondary wave hits the distal surface, shown by minus signs in Figure 29, the distal curve suddenly reduces by $\frac{2}{3}$, causing the pair of curves to cross over, and so the mean curvature goes negative, forming (after the initial invagination) the archenteron roof during gastrulation. When the secondary wave hits the proximal surface, shown by plus signs, it is the turn of the proximal curve to suddenly reduce by $\frac{2}{3}$, causing the pair of curves to recross and diverge, and so the mean curvature goes positive, causing the formation of the anterior neural folds, the notochord and early somites. Figure 29 confirms that this takes place mainly during neurulation, between the hours 55 and 70, in agreement with the normal table of Figure 18. Therefore our theory of

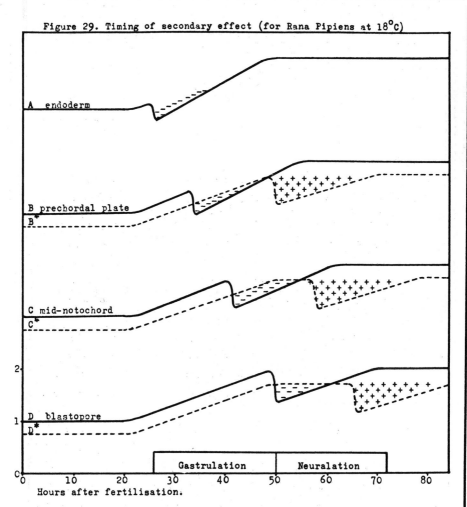

Figure 29. Timing of secondary effect (for Rana Pipiens at 18°C)

neurulation is plausible from the point of view of timing.

After the primary wave has reached D* it continues to spread round to the ventral side. Therefore the secondary wave, after neurulation, continues to spread ventrally over the proximal surface to form the later somites, eventually extending into the tail bud stages.

15.   EXPERIMENTS TO BE DONE.

Summarising, we have given a unified explanation of both the morphogenesis of gastrulation and the morphogenesis of mesoderm during neurulation, assuming only :

        the shape of the blastula,

        the expansion of the shell,

        the differentiation between ectoderm and mesoderm,

        that differentiation begins at the grey crescent and

        that there is a secondary wave of cells submerging.

The theory admits many testable predictions about timing, and some of the experiments that could be done are as follows :

(i) Verify the passage and timing of the primary wave in different species at different temperatures by graft experiments.

(ii) Find the time that the primary wave takes to cross the blastula shell by splitting the shell and grafting slices of distal and proximal surfaces separately.

(iii) Measure the timing of the secondary waves of curvature change more precisely than the existing normal tables.

(iv) Simulate the morphogenetical movements of gastrulation on a computer, by using Poisson's equation for the expansion process, and the secondary wave of curvature change.

(v) Verify that the archenteron roof, notochord and somites are caused by the same type of cellular phenomenon, by grafting distal and proximal slices onto the surface of endoderm and timing the submerging effect.

(vi) Verify the energy release during the secondary effect by microcalorimetric measurements of heat loss, and by oxygen consumption, in distal cells during gastrulation and proximal cells during neurulation.

(vii) Measure how long the secondary effect lasts, and the speeds of onset and dying away. Compare this with the length of time that the archenteron roof retains its negative mean curvature, the time that a somite preserves its shape, and the time before the notochord cells vacuolate.

(viii) Measure and compare the forces exerted by mesoderm (prechordal plate) and ectoderm (neural plate) in pushing up the anterior neural folds.

(ix) Measure the forces involved in the mesoderm mantle during neurulation, comparing the bending moments and sheering forces with the cohesive forces between cells, in order to test the tearing hypothesis.

(x) Confirm experimentally the pattern of secondary wave fronts on the proximal surface, and verify the timing of the formation of notochord, starting slowly at the front, and accelerating towards the back.

(xi) Compare the timing of mitosis during the primary wave with somite formation during the secondary wave.

(xii) Measure the parabolic slowing of the primary wave and the slowing of mitosis, and predict the rate of decrease of size of somites.

(xiii) By some method of interference, alter the speed of either mitosis or the primary wave, without changing the other, and predict the change of size and spacing of somites.

(xiv) By rotating grafts, and staining, alter the primary wave fronts and relative dominance of the animal-vegetal gradients, and verify the resulting change of presumptive boundaries of notochord and segmentation arcs between somites.

16.   SLIME MOLD CULMINATION.

Slime mold is an interesting species because it normally lives as individual cells, but is also capable of multicellular organisation. When the food runs out the cells stop dividing, and aggregate into a slug-like object called the <u>grex</u>, which first migrates and then culminates into a fruiting body of spores (see Figure 30). When the fruiting body bursts the spores land, and begin life again as individuals.

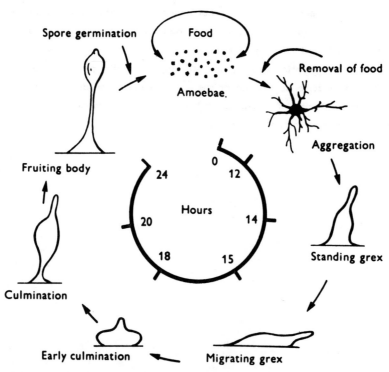

Fig.30 Life cycle of the cellular slime mould *Dictyostelium discoideum*. The times refer to development on Millipore filters (Sussman, 1966).

[6, p. 408]

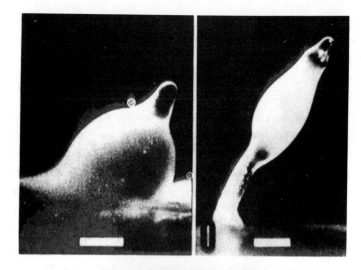

FIGURE 31 N, early culmination about one hour from the end of migration; O, fruiting body erecting about 3 hours after N; P, eight hours after O, fruiting body with stalk, spherical spore mass and tip differentiating into spores. Length marks: N and O, 1/10mm; P, 1/2mm.
From Robertson [15, p. 58]

For background reading the reader is recommended to read [2,5,6,10, 14,15], and also to see the Göttingen film [7] of G. Gerrisch. Figure 30 is from [6] after M. Sussman, and the photographs in Figure 31 are from [15]. Most mathematical modelling has been concerned with the aggregation process, but here we shall be concerned with culmination. The morphogenesis of the fruiting body is surprisingly complicated (see [2,5,21]) considering that for most of the time the species lives as single cells. We shall explain the morphogenesis by means of a hidden primary wave, followed several hours later by a secondary wave.

The fruiting body of spores stands on a long stalk made of vacuolated cells, which die after they have performed their function of supporting the fruiting body. Basically there is just one differentiation, which divides the grex into spore and stalk-cells. There may be a slight variation between cells at the foot of the stalk and those higher up the stalk, but since this variation is probably continuous we are justified in saying that all the stalk cells form one type, and all the spore cells another type. Therefore, since there are only two types, this is one of the simplest species to which we can apply the main theorem. The theorem states that the frontier between spore and stalk must move before stabilising, and therefore gives a primary wave. The final position of the wave is the presumptive stalk/spore frontier, which Raper [2,6,14] found experimentally to be about ⅓ of the way along the grex, the front ⅓ becoming stalk, and the back ⅔ becoming spore. As in the case of gastrulation we need two hypotheses :

Hypothesis 1. The hidden primary wave begins at the tip of the grex several hours before culmination.

Hypothesis 2. There is a secondary wave of cells submerging, and then exuding a coating of slime. Note that in this example the submerging cells do not retain any portion of their membrane on the surface of the tissue as in the previous example (Figure 21), but, by amoeboid action towards their neighbours, submerge themselves completely into the interior of the tissue. Therefore the cells do not keep the same topological positions relative to one another.

We shall first discuss the experimental evidence for the primary wave (Hypothesis 1). We shall then explain how the secondary wave (Hypothesis 2) causes culmination and the formation of the fruiting body.

The primary wave. Initially all the cells may have double potentiality, but as the primary wave passes each cell, that cell switches from pre-spore into pre-stalk, and loses spore-potentiality. This is confirmed by the classical experimental results of Raper [14], and others [2,5,6,10,15] in which pieces were cut off the front and back of the migrating grex, as follows.

Firstly the front pieces culminated into stalks only, with no fruit. We explain this by the loss of spore-potentiality of the entire front third. In the geometrical language of catastrophe theory, the piece is caught on the lower surface of Figure 5, and no part of it can get back up onto the upper surface.

Secondly the back pieces culminated normally, that is to say formed normally proportioned fruiting bodies but took about twice as long to do it. We explain this by assuming all the cells in the back piece to be pre-spore. Consequently the back piece behaves just like the initial grex, only smaller in proportion, so that a new primary wave starts at the front of the piece, and finished $\frac{1}{3}$ the way down the piece. This takes some time, which explains the delay in culmination.

Thirdly, if the front pieces were cut off 24 hours before culmination, then the pieces this time culminated normally. We explain this by assuming that the primary wave had not yet started, and so when it did eventually start if stopped $\frac{1}{3}$ along the piece. This puts an upper bound of 24 hours on the delay between the primary and secondary waves. The experimental data was not oriented towards looking for hidden primary waves, and so is not yet sufficient to determine the delay accurately.

Remark about regulation. These results raise the question of regulation : why does the primary wave always stop $\frac{1}{3}$ along the grex ? We have chosen to take this as an experimental fact but if we wanted to explain this fact, we should have to make an additional hypothesis concerning the nature of the gradient underlying the primary wave. In this example there is strong experimental evidence that this gradient is caused by diffusion, as opposed to the previous example where the gradients in the amphibian blastula were largely inherited from the oöplasmic organisation of the egg before cleavage, rather than caused by subsequent diffusion between the blastomeres. But in the slime mold we start with a homogeneous collection of cells, and the simplest way to create a gradient is by diffusion. And indeed there is evidence of both chemical diffusion, for instance of cyclic-AMP [2,5,6,15], and dynamic diffusion, that is to say periodic activity of cells entrained by signals emanating from a pacemaker [5,15]. In particular the aggregation process seems to be initiated by chemical signals emanating from one particular cell, towards which the others aggregate, and which then acts as pacemaker. Subsequent periodic surges of movement, entrained by the pacemaker, can be seen during aggregation in the film [7]. Mathematical models of aggregation suggest that the chemical diffusion and the dynamic diffusion are interrelated [15]. When the grex begins to migrate, the pacemaker and source of diffusion is situated at the tip [5], while the sink of diffusion is situated at the tail. Moreover each cell appears to seek its "own" position relative to the diffusion gradient [10], possibly by means of a relative movement induced by comparison between the entrainment clock and some internal clock.

An additional hypothesis sufficient to explain regulation would be that cells at the diffusion source and sink should develop into specific types of stalk and spore cells, independent of the size of the grex. Then diffusion would induce continuity, and impose a different development upon each of the cells in between, according to its relative position. In other words the two ends are regulated by the additional hypothesis, and the final position of the frontier between them is regulated by diffusion.

Culmination.

(a) Culmination begins when the grex stops moving and settles down into an onion shape, as shown in Figures 30 and 32(a). The tip of the grex becomes the tip of the onion, and the tail of the grex the base of the onion, and so what were the successive wave-fronts of the primary wave settle into horizontal shallow saucer-shaped layers, roughly concentric with the tip. The wave fronts are sketched (qualitatively rather than quantitatively) in Figure 32(a).

We shall now apply Hypothesis 2. We shall show that when the secondary wave hits this sequence of wave-fronts, from the tip downwards, causing them successively to submerge and exude slime, then this will cause the erection of the fruiting body. Moreover we shall show that this causes several of the various qualitative features displayed in Figure 32. The following sequence of paragraphs (b), (c),...,(g) refer to the sequence of stages pictured in Figure 32.

(b) The first few wave-fronts have submerged, and form a roughly egg-shaped mass M, that floats below the surface. As each layer submerges it pulls the boundary of the next layer together at the top,

4. Primary and Secondary Waves in Developmental Biology 221

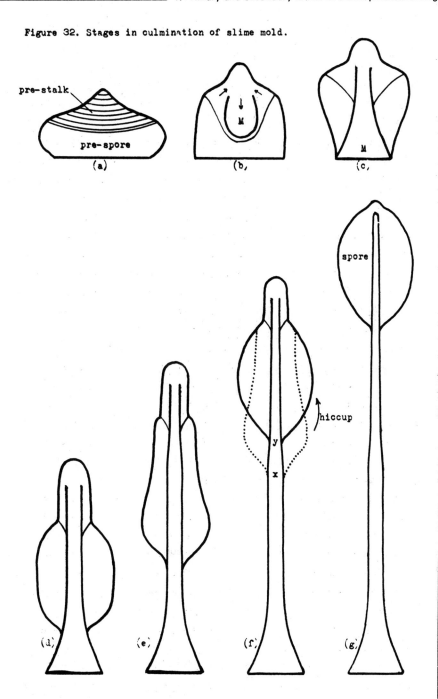

Figure 32. Stages in culmination of slime mold.

like pulling closed the neck of a sack, enclosing M. The layer that is pulled together at the top, in turn pulls inwards and upwards on its successors, and that causes an equal and opposite downwards push on M. Therefore M begins to move downwards, pushing the centres of the subsequent layers downwards ahead of it, thereby changing them from saucer-shape into cup-shape.

Soon after the cells of M have submerged they begin to exude slime, and so M develops a cup-shaped coating of slime around it, shown by the thick curve in Figure 32(b). This coating of slime facilitates the downwards movement of M, under the downward pressure, and enables M to rupture the subsequent layers, so that they become shaped like the sides of a cup without the bottom. Therefore topologically each layer is first punctured at the bottom, into an annulus, and then pulled together at the top, back into a disk again. This topological puncturing and glueing does not in any way interrupt the passage and timing of the secondary wave, because the latter is merely a set of local events in cells occuring at a fixed time delay after the primary wave.

(c) By this time the downward pressure on M has pushed it clean through the onion to the ground, where it bulges out to become the foot of the stalk. The slime coating eventually dries to form a **cellulose** tube round the stalk, giving it structural strength. But as yet, especially near the top of the stalk, the slime is still slippery, and so as the secondary wave hits each successive wave-front, causing it to submerge, it cannot adhere to the slimy walls of the stalk, and so is forced to adhere to the top of the stalk. Therefore the stalk starts growing upwards, pulling the rest of the onion with it. Once the latter pulls clear of the ground, it slides easily over the foot because the latter is conical in shape and offers no frictional resistance. Therefore the onion slides into shape (d).

(d) By this stage the characteristic knob has appeared at the top of the growing fruiting body (see the photograph Figure 31), and persists through stages (e) and (f). The structure of the knob is sketched in detail in Figure 33. In order to describe the geometry of what is going on in the knob, we shall assume discrete intervals of time, and have marked in successive wave-fronts accordingly. Assuming the wave is travelling at constant speed at this stage, then the layers B,C,D, ... etc

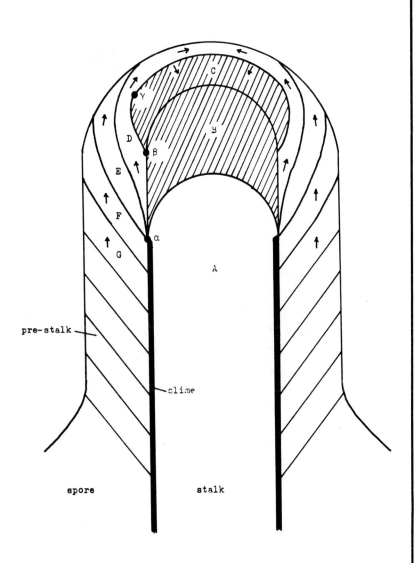

Figure 33. Structure of the knob.

between successive wave-fronts have equal volume. This procedure is not quite as artificial as might at first be supposed, because the estimates of size below show that layers B,C may be only cell thick. Nevertheless during the following description, let us assume that each layer behaves as a qualitative whole.

When the secondary wave hits a cell it becomes active for a period during which it actively submerges by amoeboid action into the interior of the tissue, by Hypothesis 2; it then ceases to be active and exudes slime. The active cells are in the shaded layers B and C in Figure 33. Therefore, going through the layers in detail we have:

A is static stalk, no longer active, with a cylindrical cellulose wall of exuded slime, shown by the thick line in Figure 33.

B is active, and the submerging cells of B cause B to minimise its surface area, at the same time as adhering to the top of A. Therefore if B has the volume of a sphere of diameter δ, then B will adopt a hemispherical top of radius δ, and cylindrical walls.

Lemma 6. *The height of B, measured along this axis, is $\frac{2}{3}\delta$.*

Proof. The volume of a sphere is $\frac{2}{3}$ that of an enclosing cylinder of equal height. Therefore a cylinder of equal volume has $\frac{2}{3}$ the height.

Estimates of size. The typical grex might contain 15,000 cells (it can vary from 500 to 50,000). About $\frac{2}{3}$ of these become spore cells, and half the rest comprise the foot, leaving 2,500 in the stalk itself. The stalk is about 1mm long (see the photograph in Figure 31(P), and a cell is about 0.01mm in diameter [15], and so the stalk is about 100 cells long and 25 cells in cross-section. Therefore the stalk is about 5 cells wide on the average, and so δ is about 0.05mm, which agrees with the photographs in Figure 31. Therefore the volume of B is 5^3 times the volume of a cell, and consequently the number of active cells in B and C together is about 250.

Meanwhile the grex has ratio of length to width about 5:1, and allowing for the fact that cells in the grex are slightly elongated, the grex must be about 60 cells long and 16 cells wide. Therefore a cross-section of the grex contains about 256 cells. Therefore the number of cells active during the secondary wave is of the same order as the number of cells across a wave front of the primary wave. These estimates help to show that our proposed knob structure is at least plausible. We now return to the business of analysing the layers of this structure.

C is also active, and the cells have submerged by pulling the D layer over the top. C adheres to the surface of B. Since C has the same volume as B, and is at least one cell thick, and assuming B is 5 cells across, then C can only cover the hemispherical top of B (because a sphere of diameter 7 has roughly 3 times the volume of a sphere of diameter 5). Therefore the lower rim of C is at the point β in Figure 33.

D is inactive because the secondary wave has not reached it yet. The cells of D are pulled radially inwards over the top of C, by the submerging cells of C, and are stretched taut in the process. The cells of D cannot slide over the cells of B and C, otherwise they would elastically slip back, exposing C again. Therefore the bottom rim of B cannot slide any higher than the point α in Figure 33, because this is the top-most point of the wall of slime. Therefore the layer D is stretched over the whole of the top of B and C, and, being the same volume, must be only half as thick as C. Therefore the cells of D must be stretched and elongated until they are only about half their normal width, as is confirmed histologically.

E, F, ... are as yet inactive, and are pulled up by D. Therefore the outer cylinder of the knob consists of all the rest of the layers of stalk cells, which explains why the knob is cylindrical. Meanwhile the spore cells are less elastic than the stalk cells, and so at the presumptive stalk/spore frontier, which is the final wave-front, the knob broadens out into the shoulders of the fruiting body.

The above structure that we have proposed for the knob is somewhat complicated, and although we have indicated how the structure might have developed initially, the question remains whether the knob can preserve this structure. The answer is given by :

Lemma 7. <u>The structure of the knob is time invariant.</u>

Proof. We must verify that in one unit of time, each layer adopts the shape of its predecessor, translated a distance ⅔δ upwards. We examine each layer separately :

A remains inactive and fixed.

B goes inactive and exudes slime. This enables the bottom rims of layers D,E,F to slide up to β, which is now the top of the newly made slime.

C remains active, and the effect of the submerging cells of C causes C to minimise its surface area, and round up to the same shape as B.

D becomes active as the secondary wave hits it, and submerges by pulling the cells of E radially inwards over the top, to form a closed layer over the top of D. Meanwhile the cells at the bottom rim of D submerge, by amoeboid action towards their neighbours in D, and therefore reduce the surface area of D by migrating from β up to γ.

E is not yet active, and is stretched over D with bottom rim held at β.

F, G, ... are pulled up.

Therefore the entire structure is reproduced isomorphically ⅔δ higher.

Lemma 8. *The diameter of stalk and rate of growth are proportional to the size of the fruiting body. Hence the time taken to culminate is independent of size.*

Proof. Let T be the unit of time. Let A be the area of cross-section of the grex. Let V be the speed of the primary wave along the grex — we may assume for the moment that V is constant, since we are dealing with mid-growth stage. Then in time T the primary wave crosses volume VAT of the grex. This must equal the volume of one layer. But B has the volume of a sphere of radius δ. Therefore

$$\frac{4}{3}\pi\delta^3 = VAT.$$

$$\delta = \sqrt[3]{\frac{3VAT}{4\pi}}$$

Therefore the rate of growth of the stalk is

$$v = \frac{\frac{2}{3}\delta}{T} = \sqrt[3]{\frac{2VA}{9\pi T^2}}$$

Now T is a constant, because 2T is the time interval of the secondary effect, in other words the period that a cell spends actively submerging.

Suppose that we increase A by a factor λ, and the length of the grex by a factor μ. (Normally the ratio of the length to width of grex is 5:1, and so $\lambda = \mu^2$, but it is not necessary to assume this.) Then the length of travel of the primary wave is increased by μ, because of regulation, while the timing of beginning and end of the primary wave is unchanged, because it is determined by the development of tip and tail cells, independent of the size of grex. Therefore the speed V is increased by μ. Therefore VAT is increased by $\lambda\mu^2$, which the increase in volume of the grex. Therefore δ, ν are both increased by $\sqrt[3]{\lambda\mu^2}$, which is the increase in linear size of the fruiting body. This proves the lemma.

Remark. It might be possible to measure T microcalorimetrically. It is easy to measure A, and it might be possible to estimate V from detection of the primary wave. Hence one could predict and verify δ and ν.

Lemma 9. The length of the knob shrinks at ⅓ the speed that it grows upwards.

Proof. This lemma depends upon the experimental observation that, initially at any rate, the diameter of knob is twice the diameter of stalk [7]. Therefore area of cross-section of knob is 4 times that of stalk. Therefore the area of cross-section of the cylinder surrounding the stalk is 3 times that of stalk. But the upward growth is merely transference of pre-stalk material from cylinder to stalk. Hence the rate of growth of stalk is 3 times the rate of shrinkage of knob.

This concludes our discussion of structure of the knob and stage (d) of Figure 32; we now move on to the next stage.

(e) This stage occurs slightly after that shown in the photograph Figure 31(O), and the qualitative shape of the spore-mass is slightly more pronounced. We must explain this shape. The spore cells form and elastic mass, that, if it were allowed to float freely in a liquid of the same density, would minimise its surface area by adopting spherical shape. If this elastic mass is suspended in air from the top, then gravity would cause it to adopt a lemon-shape, as in Figures 31(P) and 32(g). At the stage we are considering in Figure 32(e) the spore mass is being pulled up from the top, but there is also a considerable frictional drag against the cylindrical stalk, and it is the combination of the top suspension together with downward drag, acting on the elastic mass, that produces the characteristic shape.

The stalk is cylindrical, because this stage corresponds to a period during which the primary wave had constant velocity V, and so by Lemma 8 the stalk has constant diameter δ. The friction is increased by the fact that the slipperiness of the slime tends to get used up leaving a dry cellulose tube, so that the drag is more pronounced at the bottom. It should be possible to simulate these simple forces of elasticity, suspension, gravity and drag on a computer, and produce quantitative predictions about the qualitative shape of Figure 32(e).

(f) At this stage there occurs the characteristic hiccup*, as the spore mass suddenly flips from the dotted profile, similar to the previous stage (e), into the lemon-shape, similar to the secondary stage (g). The hiccup can be seen very clearly in the time-lapse film [7]. When the hiccup occurs the knob and shoulders do not move. Therefore the hiccup cannot be caused by a change in the upward supporting force, and must be caused by a sudden reduction of the downward frictional drag. And it is easy to see why, because the point x in Figure 32(f) is the point at which the stalk changes from being cylindrical to conical. Measurement of the stalk diameter at x and y, before and after the hiccup, [7], reveals that the diameter drops from about ½ to about ⅕ of the knob diameter. The change to conical shape at x causes the sudden disappearance of the maximal frictional drag of the cylinder on the bottom of the spore mass, and so friction suddenly becomes negligible compared with gravity in determining the shape of the spore mass. Hence the hiccup. Moreover it should be possible to simulate the hiccup by a minor adjustment of the previously suggested computer programme.

There remains the question of why the stalk diameter begins to decrease. In the proof of Lemma 3 we showed that the diameter depended upon V, A, and T. Now T is constant, being a property of the individual cell, and A is constant at this stage, since it is the cross-sectional area ⅕ along the grex. Therefore our attention is drawn

---

* I am indebted to John Ashworth for pointing out the hiccup to me, after I had suggested that the morphogenesis might be caused by a secondary wave.

to V. And, sure enough, the primary wave slows down parabolically just before it stabilises by §9 Corollary 3.

This gives the correct qualitative explanation of why the diameter decreases. However we must be cautious about applying §9 Corollary 3 quantitatively, because in this application the cells do not stay topologically in the same place relative to one another during the morphogenesis. Therefore although the quantitative result is true for the primary wave, it may no longer apply to the secondary wave. Indeed, as the secondary wave comes to a halt, the continuity on which the estimate was based, becomes less important than the timing inside individual cells in determining their final position. This is an interesting and delicate point, and so let us enlarge upon it.

Suppose, on the contrary, that §9 Corollary 3 was applicable to the end of the secondary wave. Let V be the speed of the primary wave at time τ before its end, and let δ be the diameter, v the speed, and η the distance of the stalk below its eventual top, at time τ before the end of the secondary wave. Then $V \sim \tau$, by §9 Corollary 3, and so

$$\delta \sim v \sim V^{1/3} \sim \tau^{1/3}$$

by Lemma 8 above, implying

$$\eta \sim \int v d\tau \sim \tau^{\frac{4}{3}} \sim \delta^4 .$$

Therefore this would predict a very blunt 4th power top to the stalk. Now the initial narrowing of the stalk, when it changes from cylindrical to conical shape at the point x in Figure 32(f), may indeed obey a 4th power law, and it would be interesting to measure this experimentally. However when the stalk has reduced to a width of 1 or 2 cells then this result no longer applies because of the individuality of cells, as follows.

The final wave front of the primary wave contains about 250 cells. Therefore we may expect that the last few of these cells will experience their catastrophic switch from pre-spore to pre-stalk not all together, but spaced out, at individual times. How few, and how spaced-out, will depend upon the noise level (for instance the amount of irregularity in cell size and original environment). Therefore as the secondary wave hits each of these last few cells in succession, it will induce it to slide up and adhere to the top of its predecessor. Therefore the final piece of stalk will be only one cell thick, and possibly of length comparable to the length of the final spore mass (which is about 30 cells). This can be seen in the final enlarged shot of the film [7], for instance, where the stalk is only 1-cell wide at the base of the spore mass.

(g) As the last stalk cells submerge, they pull spore cells over the top of the knob. Therefore in its final stages the knob consists of spore cells, rather than stalk cells. Therefore it is rounded rather than cylindrical. The final disappearance of the knob is due to the elasticity of the spore mass trying to reduce its surface area, which is a weaker force than that created by cells submerging. Added to this is a renewed frictional drag, because the top part of the stalk is a cylinder again, one cell wide. Therefore the final disappearance of knob takes a long time, as in the film [7], and it may not entirely disappear. This effect will be exaggerated in fruiting bodies orginating from an irregular environment. This completes our description of the seven stages in Figure 32.

17. <u>EXPERIMENTS TO BE DONE.</u>

Summarising, we have explained the morphogenesis of the culmination of cellular slime mold, assuming only :

    the shape before culmination,

    the differentiation between spore and stalk,

    that differentiation begins at the tip of the grex, and

    that there is a secondary wave of cells submerging and

    exuding slime.

Our theory agrees with the classical experimental observations of Bonner [2, and 21 p. 4363] and with many qualitative details found by Gerisch [7] and Farnsworth [5]. The new contribution of our theory, compared with previous theories, is that it offers an explanation of <u>why</u> the local cellular forces occur in the order that they do, and <u>how</u> they create the surprisingly complicated sequence of global shapes of the culminating fruiting body. In other words it provides a link between the local and the global in space-time.

The theory admits several testable predictions about timing and the structure of the knob, and some of the experiments that could be done

(i) Establish the timing and speed of the primary wave by cutting different proportions off the front of the migrating grex at different times. It might be possible to first slice the grex lengthways, in order to be able to use half as control.

(ii) Verify and time the secondary wave, by grafting pieces of pre-stalk onto pre-spore and observing when the cells submerge.

(iii) Measure the duration of the secondary effect microcalormetrically.

(iv) Verify the diameter and growth speed given by Lemma 3.

(v) Verify the knob shrinkage speed given by Lemma 4.

(vi) Simulate the spore mass shape and hiccup on a computer.

(vii) Turn the culminating fruiting body upside down. Then after the hiccup, when gravity becomes dominant, the fruiting body should be fatter because it is suspended from below.

(viii) Cut off half the knob during culmination. Then the spore mass and the remaining cut cylinder of knob stump should slide down a little, leaving an exposed stump of stalk. Then when the secondary wave hits the knob stump it should submerge and form a submerged solid torus of stalk cells. Similarly one should be able to design several tricks to test the knob structure, by making the secondary wave submerge in various different directions, (as for example in [5]).

18.   CONCLUSION

The point of view in this paper, and the explanations, designs of experiments and predictions in the two examples, arise from the use of catastrophe theory. If primary and secondary waves are confirmed to be as widespread as it seems they must be, then this by itself would be a useful contribution of catastrophe theory to biology. But we have only used the two simplest elementary catastrophes, the fold and the cusp. To glimpse the potential riches offered by the use of the higher dimensional catastrophes, one only has to glance the astonishing writings [18,19] of the creator of this theory, René Thom.

REFERENCES

1. B.I. Balinsky, An introduction to embryology, (Saunders, Philadelphia & London, 1965) 2nd edition.

2. J.T. Bonner, The cellular slime moulds, (Princeton Univ. Press, 1967).

3. J. Cooke, conversations.

4. J. Cooke, Some current theories of the emergence and regulation of spatial organisation in early animal development, Annual Rev. of biophys. and bioeng. (1975), to appear.

5. P. Farnsworth, Morphogenesis in the cellular slime mould Dictyostelium discoideum; the formation and regulation of aggregate tips and the specification of developmental axes, J. Embryol. exp. Morph. 29 (1973), 253-266.

6. D. Garrod & J.M. Ashworth, Development of cellular slime mould, Dictyostelium discoideum, Symp. Soc. Gen. Microbiology XXIII, Microbial differentiation (1973), 407-435.

7. G. Gerisch, Dictyostelium discoideum (Acrasina) Aggregation und Bildung des Sporophors (Institut für den Wissenschaftlichen Film, Göttingen), Film E 631 (1963).

8. T. Gustafson & L. Wolpert, The forces that shape the embryo, Discovery 22 (1961), 470-477.

9. J. Koyama, Normal table of the development of the Japanese newt Diemictylus (Triturus), Zoo. Mag. (Tokyo) 42 (1930), 465-473.

10. C.K. Leach, J.M. Ashworth & D.R. Garrod, Cell soring out during the differentiation of mixtures of metabolically distinct populations of Dictystelium discoideum, J. Embryol. exp. Morph. 29, 3 (1973), 647-661.

11. W. Luther, Entwicklung des Molcheies (Institut für den Wissenshaftlichen Film, Göttingen) Film C. 939 (1967).

12. P.D. Nieuwkoop, conversations.

13. A.W. Pollister & J.A. Moore, Tables for the normal development of Rana sylvatica, Anat. Rec. 68 (1937) 489-496.

14. K.B. Raper, Pseudoplasmodium formation and organisation in Dictyostelium discoideum, <u>Jour. Elisha Mitchell Scientific Society,</u> 56, (1940), 241-282.

15. A. Robertson, Quantitative analysis of the development of cellular slime mold, (A.A.A.S., 1970, Some Mathematical Questions in Biology, V,), <u>Lectures on Maths. in the Life Sciences,</u> Vol. 4 (Amer. Math. Soc., Providence, U.S.A., 1972), 47-73.

16. S.N. Sedra & M.I. Michael, Normal table of the Egyptian Toad, Bufo regularis Reuss, <u>Československá Morfologie,</u> 9 (1961), 333-351.

17. W. Shumway, Stages in the normal development of Rana pipiens I, <u>Anat. Rec.</u> 78 (1940), 137-147; II, <u>Anat. Rec.</u> 83 (1942), 309-315.

18. R. Thom, <u>Stabilité structurelle et morphogénèse,</u> (Benjamin, New York, 1972).

19. R. Thom, A global dynamical scheme for vertebrate embryology, (A.A.A.S. 1971, Some Mathematical Questions in Biology, VI), <u>Lectures on Maths. in the Life Sciences,</u> Vol. 5 (Amer. Math. Soc., Providence, U.S.A., 1973) 3-45.

20. W. Vogt, Gestaltungsanalyse am Amphibienkeim mit örtlicher Vitalfarbung, II, Gastrulation und Mesodermbildung bei Urodelen und Anuren, <u>Roux Arch.</u> 120 (1929), 385-706.

21. C.H. Waddington, <u>Principles of embryology,</u> (Allen & Unwin, London, 1956).

22. C.H. Waddington, <u>The stratergy of the Genes,</u> (Allen & Unwin, New York, 1957).

23. P. Weiss, <u>Principles of development,</u> (1939; reprinted Hafner, New York, 1969).

24. E.C. Zeeman, Differential equations for the heartbeat and nerve impulse, <u>Towards a theoretical biology</u> 4, (Ed. C.H. Waddington, Edinburgh University Press, 1972), 8-67.

25. E.C. Zeeman, Differentiation and pattern-formation, (Appendix to Reference 4).

26. E.C. Zeeman, Gradients and catastrophes in developmental biology, (in preparation).

3 ↑ *Numbering in this book.*

5 A Clock and Wavefront Model for Control of the Number of Repeated Structures during Animal Morphogenesis

J. COOKE†

*National Institute for Medical Research,
The Ridgeway, Mill Hill, London NW7 1AA, England*

AND

E. C. ZEEMAN

*Institute of Mathematics, University of Warwick,
Coventry, Warwick, England*

(*Received* 17 *June* 1975, *and in revised form* 1 *October* 1975)

Most current models for morphogenesis of repeated patterns, such as vertebrate somites, cannot explain the observed degree of constancy for the number of somites in individuals of a given species. This precision requires a mechanism whereby the lengths of somites (i.e. number of cells per somite) must adjust to the overall size of individual embryos, and one which coordinates numbers of somites with position in the whole pattern of body parts.

A qualitative model is presented that does admit the observed precision. It is also compatible with experimental observations such as the sequential formation of somites from anterior to posterior in a regular time sequence, the timing of cellular change during development generally, and the increasing evidence for widespread existence of cellular biorhythms. The model involves an interacting "clock" and "wavefront". The clock is is a smooth cellular oscillator, for which cells throughout the embryo are assumed to be phase-linked. The wavefront is a front of rapid cell change moving slowly down the long axis of the embryo; cells enter a phase of rapid alteration in locomotory and/or adhesive properties at successively later times according to anterior–posterior body position. In the model, the smooth intracellular oscillator itself interacts with the possibility of the rapid primary change or its transmission within cells, thereby gating rhythmically the slow progress of the wavefront. Cells thus enter their rapid change of properties in a succession of separate populations, creating the pattern.

It is argued that the elements, a smooth oscillator, a slow wavefront and a rapid cellular change, have biological plausibility. The consequences of combining them were suggested by catastrophe theory. We stress the necessary relation between the present model and the more general concept of positional information (Wolpert, 1969, 1971). Prospective and ongoing experiments stimulated by the model are discussed, and emphasis is placed on how such conceptions of morphogenesis can help reveal homology between organisms having developments that are very different to a surface inspection.

† All correspondence to J. Cooke.

1. Introduction

(A) THE PROBLEM OF NUMBER CONTROL IN MORPHOGENESIS

Among the spatial patterns of cell activity occurring in early development is one type that poses particular problems for theories of morphogenesis. This is the occurrence, within the body patterns of some animals, of series of equivalent or homologous structures, their number (on the order of 10–50) being relatively invariant amongst individuals of a given species. The two most important cases are the segmentation of the arthropod embryo and vertebrate somite formation. The former segmentation is fundamental to the arthropod body pattern, since this pattern actually consists of the subsequent unique specialization of each of the originally demarcated segments, and the ordinal number of the segment undergoing each specialization is quite reliably the same in all genetically normal individuals. Development of other than the normal number of segments is rare (i.e. number is modally canalized—see Maynard Smith, 1960). There are senses, both morphogenetic (Lawrence, 1970) and genetic (Garcia-Bellido & Santamaria, 1972), in which the segments are repetitions of a developmental unit. Somite formation, by contrast, involves just two longitudinal tracts within one of the three basic cell layers of the embryo, and the coefficient of variation for number of somite blocks developed anterior to the tail region (up to some 30, dependent on species) is some 4% (Maynard Smith, 1960, and our own data on *Xenopus* larvae). The repeating pattern is so developed as to be in relatively or absolutely close register with other aspects of the body pattern, such as limb rudiments or internal organs. We shall refer to this as the co-ordination between, say, the somite pattern and the "whole body pattern". Thus we normally find a constant number of the repeated structures (spinal nerve roots, vertebrae), associated with somite formation, lying between pairs of distant markers in the anatomy such as the base of the skull and the rear of the pelvic girdle. Derivatives of somites having particular ordinal positions in the series therefore quite reliably pursue similar specializations in each individual (Straznicky & Szekely, 1967; Mark, 1975), though number of units formed in the remaining tail material is often more variable, and abnormal temperature regimes during development can bias the overall number (e.g. Lindsey, 1966).

In both the animal groups mentioned, development is known to be regulative in the classical embryologist's sense, although in arthropods this condition may only obtain at earliest stages, before cellularization of the embryo through cell boundary formation (Herth & Sander, 1973; Sander, 1975). In such embryos, if material is removed during early phases of development before cellular determination, a normally ordered and proportioned whole-body pattern of differentiation is nevertheless achieved within the remaining material, This is accomplished without special cell migration or re-assortment, but rather by cells re-adjusting their trajectories of development according to their new relative positions in the whole material. Compensatory cell division does not at first restore the normal cell population, so fewer cells are available to form each part of the body pattern in the remaining embryonic tissue. In amphibian (vertebrate) development, following removal of nearly half the cells of the early embryo, such regulation causes development of normally formed bodies out of an initially smaller-than-normal cell population (Spemann, 1903; Schmidt, 1933). The question can now be posed as to whether, in this regulated development, the somites follow the strict rules of classical regulation, or some other rule. Is the number of somites in the pattern kept similar to normal in small embryos (necessarily by reducing cell number in each somite at its formation), or is somite size kept at a species-typical value in morphogenesis so that (again necessarily) only a number of somites proportional to the reduced body dimension can be formed?

In fact, numbers of somites are essentially normal in these circumstances, the distance in terms of cells between the separating fissures that define them being reduced accordingly (Cooke, 1975a,b). The comparable aspect of arthropod (insect) development has not to our knowledge been tested as explicitly. It now seems very likely, however (Sander, 1975, including symposium discussion), that the answer to the equivalent experimental question is the same; that number of segment units is kept constant by similar, or the same factors as regulate the whole body pattern, rather than that size of individual units is kept "species-normal". The implications of this phenomenon for theories of spatial organization in biology can now be discussed.

Each unit (somite or arthropod embryonic segment) of repeating patterns is qualitatively similar, at least at the cellular level, at its initial formation. There is apparently a regular spatial alternation between a few modes of cell behaviour, such that successive populations of cells become separated from similar populations forming immediately before and after them in spatial and temporal sequence down the body axis. The process is essentially the regular segmentation of material already there, rather than one of serial addition at one end by true growth (though in regeneration, in some organisms, the identical result can be produced by just such an alternative process). Turing and various later authors have developed theories whereby spatially periodic "prepatterns" might arise within embryonic tissue and then control such patterns (Turing, 1952; Gierer & Meinhardt, 1972; Wilby & Ede, in press). The concept of prepattern refers to a hypothetical spatial distribution of some quite simple cellular variables, such as concentrations of morphogen substances, the distribution being isomorphic with the final pattern of cell activities to be seen. Cells would then respond by behaving according to local values of prepattern variables, in making simple "decisions" of a threshold character between alternative behaviours. So where patterns consist of controlled numbers of similar structures, the prepattern must consist of the same controlled number of spatial repetitions (e.g. peaks of morphogen concentration). Cellular response is conceived as having a rather all-or-nothing character, the problem of prepattern control thus being one of positioning periodically, in real space, threshold values for the variables that control the switch in cellular response. The fact that somites (and usually insect segments) form in regular series in time as well as space, presents no problem in itself, as some of the more recent prepattern models produce a series of peaks distributed along another, overall morphogen gradient (Gierer & Meinhardt, 1972).

Prepattern theories can now be modelled in terms of principles known to regulate biosynthesis in cells, and their performance simulated by computation. On the basis of this, when biologically plausible degrees of control over the various interacting parameters are considered, cases such as somites seem very demanding because of the great regularity in size, or "wavelength" in the units, and the relative accuracy in control of a large number (Maynard Smith, 1960; Bard & Lauder, 1974). Furthermore, all current models for periodic prepatterns seem to us to share a particular rather deep property, whereby they can be challenged experimentally. They all postulate allosteric feedback interactions (positive or negative) between particular large and small molecules underlying morphogen synthesis, and then diffusion of such molecules for spatial interactions in time. For any actual such system therefore, with its particular molecular hardware, the "wavelength" generated between successive repetitions should be constant on a space or cell-number basis, and so the number of final pattern units should be closely dependent on the extent of the tissue in which morphogenesis is occurring. Variation in overall dimension at early stages, much greater than one normal pattern unit in extent, is nevertheless followed by development of normal numbers of somites. To achieve this, the actual parameters describing allosteric interaction and substance diffusion, and thus wavelength, would need to be modifiable according to some local correlate of overall body dimensions.

In an abnormally small whole amphibian embryo the width and depth of the somite-forming columns of cells has also been regulatively reduced, since these are features of the whole body pattern. At an imaginative stretch one could conceive of the local "diffusion" kinetics within such columns being appropriately altered, to give adjustment of the unit "wavelength" of a Turing type prepatterning. Experiments in which the width and depth of the predifferentiated column is experimentally reduced, on one side only of a normal sized embryo (Cooke, 1975b), probably dispose of such an idea. The results of these early cell removal operations show that number (thus lengths) of somites formed is controlled only with reference to the embryo's long dimension, and not by the width or depth of tissue in which segmentation is occurring. This is also counterevidence to the suggestion (e.g. Waddington & Deuchar, 1953) that the width/depth/length proportions of somites in a given species are due to local control mechanisms of some sort, acting as each block of cells is individualized from the remaining material.

The positional information gradient concept developed by Wolpert (1969, 1971) provides a satisfying and plausible formal explanation as to how pattern elements might each be adjusted in extent to changes in size of the whole (the process referred to above as regulation). In its present form it does this, however, only for patterns such as that of the basic body plan (mentioned earlier), where during one period of development cells are assigned a small number of unique determinations, e.g. as heart, fore and hind-limb rudiments, kidney, etc., according to relative position within the whole. Extending earlier gradient theories (e.g. Child, 1946), this model supposes that some quite simple early variable is spatially distributed in monotonically graded fashion across the entire developing system. Its local value can be perceived by cells, and determines the choice between the small available number of developmental pathways. This positional information (hence, p.i.) variable is preserved at particular "boundary" values in special organizing regions of the embryo, and its profile of change in space between such regions is mediated by cellular interactions.

Diffusion of a morphogen substance between local source and sink cells provides just one particular realization of this concept. Many others are possible (Wolpert, 1971; Goodwin & Cohen, 1969; Cohen, 1971). Their chief feature is that removal or internal transposition of cells is followed by restoration of all normal gradient values in their normal order, though the gradient in cell state, per cell may be more or less steep (after size alteration). This in turn mediates pattern regulation in later morphogenesis, since a normal range of values for the p.i. variable across the system results in a normal spatial distribution of the developmental pathways followed by the embryonic cells. A variety of mechanisms [such as that of a few, bistable, genetic switching systems (Kauffman, 1973, 1975)] could plausibly control cellular entry into one or other developmental pathway according to thresholds in values of such a variable, "read" directly by the cells and interpreted accordingly. It is when we consider the co-ordinated development of the many-times-repeated aspects of body patterns that such direct interpretation becomes implausible.

We know that numbers of somites are (a) usually in close register with the whole body pattern and (b) approximately normal in abnormally small bodies. Naively, a mechanism for achieving this would be to have the pre-somitic cells, which are two longitudinal tracts down the body axis, responsive to a particular regular succession of values of the p.i. variable or body gradient, to which alone they react by performing the activity that defines either the centres of somites or the fissures between them. But this requires an implausible performance for any interpretative mechanisms working within cells, all of which are in the same state of non-determination at the time of laying down the pattern. Cells would be required to respond with a particular programme of activity (say, de-adhesion from their neighbours, or extra close adhesion) to each of some 15 to 30 discrete values of a variable, while ignoring or reacting differently to all others in between. Furthermore, somites are always regular in size within each embryo, and yet their number is not absolutely invariant, relative to whole body pattern, within each species. We observe complete body-patterns within which there happens to have developed one more or one fewer somite than usual. This observation is incompatible with a strict p.i./interpretation mechanism, since it implies that spatial regularity is a deeper property of somites than their precise relation to other body parts.

There is striking evidence from haploid amphibian embryos (Hamilton, 1969), that the "wavelength" of the somitogenic process, which ultimately determines somite number in the body, is controlled as a spatial extent of cellular material rather than as a particular cell number. The more recent results just discussed show that this "wavelength" in the repeating programme of cell processes is itself adjusted, according to the overall dimensions within which morphogenesis is occurring. The profile (i.e. steepness) of a gradient in a variable, registering relative position in the body, is strongly implicated in such control.

We next describe briefly the cellular anatomy of somitogenesis as it appears in the amphibian *Xenopus*, to show how it lends itself to quantitative study and to the pursuit of a theory of control. We then present the outline of such a theory. This involves postulating a wavefront of sudden cell change, passing along the pre-somite material and controlled in its time course by the whole body p.i. gradient. It is then proposed that all the cells are also coupled oscillators with respect to an unknown "clock" or limit cycle in the embryo, periodically modulating the effect of the wavefront as the latter progresses. The additional problem of co-ordination of somitogenesis in the tail rudiment (Cooke, 1975a) where the process progresses for the first time into truly growing tissue, will not be dealt with in detail.

(B) THE ANATOMY OF SOMITOGENESIS

Somitogenesis in the amphibian *Xenopus* is described at the light microscope level from wax embedded material, by Hamilton (1969). Her findings have been confirmed from glutaraldehyde-fixed and araldite-embedded embryos (Cooke, unpublished) and also extended to *Bombina*, another anuran amphibian. Figure 1 shows, schematically, how presomite mesoderm all along the body first migrates towards the midline to form the two longitudinal columns of spindle-shaped cells, one cell wide transversely and many cells deep dorso-ventrally. The actual width (i.e. degree of stretching of the long axes of the cells) and depth (number of ranks of cells in transverse section) both decrease smoothly in posterior regions, and the cells are also smaller there when the later somites are formed, presumably because more cell divisions have occurred since onset of egg cleavage. The process of somitogenesis itself then occurs in the successive isolation, by transverse fissures where cell de-adhesion occurs, of blocks of spindle-shaped cells. Each block is initially a regular number of cell-widths in extent, as counted along the animal's longitudinal axis, but then rotates through 90°, the medial edge moving forward. It thus becomes a single bundle of spindle-shaped cells, now a defined number of cells wide at any particular horizontal level in the embryo, and of course as many cells deep as the original presomite column. The process from visible onset of de-adhesion to completion of rotation does not normally occupy more than two successive blocks or presumptive blocks in the series at once. Each block is therefore half-rotated by the time the subsequent block is first defined by the next de-adhesion. In principle the positions of many subsequent fissures might already be determined in a "prepattern" sense, behind the latest visible morphogenesis at any one time, but there is no evidence that this is so.

The cells almost never undergo division (mitosis) once "blocked" and rotated. We know moreover, from histological sections, that each somite just after its formation shows a number of cells, in its width, comparable to that seen there after many subsequent somites have been formed. This number therefore remains a record of the local "wavelength" (in spindle-cell widths) of the somitogenic process as it occurred in each part of the body, even though stretching of the larva finally lengthens the spindle cells considerably (see Fig. 1). Somites actually form, within each embryo, with a clock-like regularity in time. This knowledge comes from experiments in which a large population of sibling *Xenopus* embryos, set to develop under constant conditions and synchronized at onset of somitogenesis, was sampled at regular intervals of laboratory time by fixing and dissecting batches of five embryos. Within no batch did number of somites formed vary by more

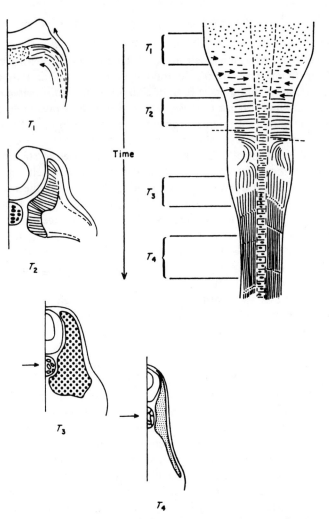

FIG. 1. Representation of the cellular anatomy of somite formation in *Xenopus*. Somite formation is represented in horizontal, longitudinal section through the notochord (mid-dorsal axial skeleton). Gradation from top to bottom represents advance in development with time at any one level of the head–tail axis. Each level passes stages T_1, T_2, T_3 and T_4, which are each shown in unilateral transverse section. Note initial demarkation of the mid-dorsal notochord cells from the mesoderm (see section 3) and then progressive migration towards the midline of the elongating pre-somite mesoderm cells to form into columns of spindle-shaped cells as at T_2, while the notochord starts differentiation. Formation into somite blocks, by de-adhesion then rotation of cell groups, is shown, completed by T_3. Stretching during continued notochord differentiation (vacuolation) increases greatly the lengths of formed somites, making their cells more slender as they differentiate (T_4). Numbers of cell-widths visible, at the notochord level marked in T.S., remains unchanged between T_3 and T_4, however.

than one, over formation of the first 30 somites! At any one temperature, this temporal regularity is accompanied by great repeatability in size of successive somites (say 9·5 cells' "wavelength" in anterior body regions, derived from averaging horizontal section cell counts for each somite within single embryos). Both features are reflected in the smoothness of an experimental curve such as that of Fig. 2, of cumulative cells in somites against time at 21°C. One of us (J.C.) found somites to form at one per 40 min anteriorly, slowing to about one per hr posteriorly, but this may have been due to a cooling laboratory. The real plot for somites formed against developmental time may be linear (Murray Pearson—personal communication).

Despite the above-mentioned regularity of size, somites do become smoothly and systematically smaller (fewer-celled) in posterior body regions after about somite 15. The decrease in slope, or parabolic shape of the curve of cells per time, as in Fig. 2, thus has one certain cause, and may have a second (slower formation of each somite) acting synergistically. Because

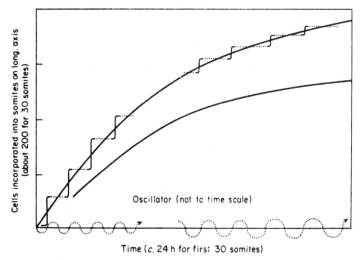

FIG. 2. The curves for rate of cell involvement in somite formation, experimentally determined and for a hypothetical small embryo. Cumulative numbers of spindle-cells, in horizontal sections, incorporated into somites (ordinate), are plotted against laboratory time (abscissa). Upper curve is experimentally produced from a control *Xenopus* population (see section 2), while the lower one is that expected for a regulated, experimentally size-reduced embryo. Hypothetical oscillator advances and retards recruitment of cells into new behaviour (oscillations not to time scale), so that the curve is actually manifested as a step function whose vertical components are the synchronous rotation of blocks of cells adhering together. The curve slope (cells per time) definitely decreases with time, because later somites are smaller and form either slightly less often (our data) or at the same regular rate as earlier ones. Result: the plot of cells per somite, per somite in the series is a near straight line.

cells in later formed somites are themselves smaller, histology shows that in terms of material between fissures, the wavelength decrease is even more accentuated than it is in terms of cell number. Examination of newly forming somites at widely differing body levels supports the following; that whereas width and depth of the somite column, and "wavelength", each decrease smoothly along the body axis, the proportions in these dimensions are by no means kept constant. In other words the small blocks of posterior regions are not simply scaled-down isomorphic versions of anterior ones. In fact the very first somites, where the wavelength is maximal, nevertheless have abnormally reduced widths and depths when formed, because the pre-somite column here fits in between head structures and is reduced. This is further presumptive evidence (see also section 1), that the deployment of fissures to establish the total number of somites is controlled with respect to the length of the whole body, and not by a local proportioning process.

We are ignorant of the biochemistry or even cell biology of the organized changes in adhesion and locomotory behaviour whereby this morphogenesis is brought about. Thus it is only appropriate at present for models of the pattern formation to be formal ones, that can be mapped onto whatever machinery is found to mediate cell behaviour. Anatomy of somite formation in the tailless amphibia is clearly secondarily simplified, in the evolutionary sense, but the more basic vertebrate version can be related to it easily enough. The essential problem of creating by cells' behaviour a set of discrete, self-adhesive populations separated by boundaries, is the same in the formation of more radially symmetrical "rosettes" as in salamanders, birds and mammals; the process is just harder to study quantitatively.

(C) THE MODEL

By the close of gastrulation, some hours before the earliest anterior somites are formed, the embryo is already a mosaic of regions, each determined as capable only of differentiating into particular parts of the whole body pattern (Holtfreter & Hamburger, 1955, for review). Gastrulae chopped in two in the future transverse plane at this stage, for instance, develop into quite well-formed anterior and posterior half-larvae, with no regulation and with numbers of somites that together add up to about the normal for whole bodies. Let us suppose that a longitudinal gradient of the whole body p.i., whatever its nature, is used at this early stage to determine two aspects of the future development. It sets cells to differentiate in qualitatively unique ways in different places, and also sets rates for the process of intracellular development, according to its local value, along the columns of cells that are determined as pre-somite material. There will thus be a rate gradient, or timing gradient along these columns, and we shall assume a fixed monotonic (not necessarily linear) relation between rate of an intracellular evolution or development process, and local p.i. value experienced by a cell at the time of setting that rate.

Sufficiently complex molecular descriptions are not currently available for us to know the nature of the control of timing, direction and execution of the developmental programmes undergone by early embryonic cells. How many variables are involved in such control? Are many of the determining variables in the "metabolic" state-space of the cell (see, e.g., Goodwin, 1963), or are nearly all of them of macromolecular "switch-like" specificity? In view of this ignorance an adequately general description of the developing cell is as a multi-dimensional state space, or manifold. In this state space, the vectors must be such that a relatively small number of attractor domains exists, each corresponding to determination and then biochemical and hence cell-behavioural execution of one of the differentiation pathways open to cells of the species concerned. At the phenomenological level, cells in early embryos characteristically behave "smoothly" and stably with little overt change for considerable periods, interspersed with relatively rapid overt changes of behaviour and state of determination.

Catastrophe theory has been used (Thom, 1973; Zeeman, 1974) in describing the possible intracellular and intercellular processes during development. The trajectory of development within a cell is modelled by the movement of a point within the manifold that represents the intracellular state space. Smooth motion along homeostatic surfaces within such manifolds may be interspersed with relatively sudden, unstable jumps to new such surfaces (i.e. catastrophes). Such a choice of description is by no means essential for understanding the present model but does lend itself well to thinking about ways in which developmental switches and programmes of activity may be ordered in space and time during morphogenesis. We are suggesting that the p.i. cellular variable itself sets the time, within individual somite cells, at which an instability will be reached followed by the catastrophe or by immediate competence to undergo the catastrophe if triggered, say by a catastrophe-associated signal from an anterior neighbour cell. According to preference, the catastrophe may be imagined as the switch-on or switch-off of a single, "executive" unit of genetic activity, or just the forceful entry of the cell state into a new attractor domain or configuration of metabolic-genetic activity (e.g. Kauffman, 1969). But let us assume that the sudden change in locomotory, and adhesive behaviour that occurs as cells participate in formation of a somite, is the expression of such a programmed discontinuity in their development.

Now a fixed relationship of p.i. (gradient) variable to cell development rate is proposed, and we know (at least on the positional information theoretical paradigm) that after size reduction of an early embryo the gradient will be restored to a normal range of values. Thus, a wavefront in space of sudden cell behaviour change will pass down the body pattern in the somite column, and the rate of passage of this will be such as to traverse the entire column (relative to other markers in the head–tail axis) in unit time, whatever the size of the whole body pattern. This time period will be a characteristic of the species at any one temperature, because of (a) the fixed relation of intracellular development rate to p.i. value and (b) the normal gradient profile between boundaries of the regulated body pattern. Speed of the wavefront, in terms of cells per time, will be proportional to the length (cell number) of the embryo, for each part of the whole body pattern, provided that morphogenesis has regulated to wholeness after any early disturbance.

Such a wavefront might arise from any of the variety of underlying mechanisms. For instance, it might partake of the character both of a true wave, involving propagatory interactions between cells, and of a purely kinematic "wave" controlled, without ongoing cellular interaction, by a much earlier established timing gradient. The kinematic wave might be set up with respect only to onset of intrinsic competence for catastrophe, while a catastrophe-associated intercellular signal might mediate propagation of the actual rapid cell change down the body. But a gradient in the cells' intrinsic rates of development must limit the speed of the wavefront to account for regulation of the somite pattern. The smoothness of a purely kinematic wave might seem to depend upon extreme accuracy of control, over many hours of intracellular development rates in response to the early gradient. However, we might expect "thermal" fluctuations to be smoothed out, meanwhile, by intercellular diffusional communication for variables (substance concentrations) that decisively affect timing of the catastrophe. Indeed, the local reversal of developmental gradients of behaviour following grafting experiments (Waddington & Schmidt, 1933; Abercrombie, 1950) is evidence for such local interactions.

A smooth antero-posterior timing gradient (thus, wavefront) with respect to morphogenetic movements and differentiation in general, is a widespread empirical feature of early vertebrate and arthropod development. Moreover, in the case of experimental small *Xenopus* embryos, the process of formation of somites progresses from the head to the tail over the same period as in synchronously developing unoperated controls. So in a sense this element of the model is simply a surface description of a reality that itself requires an adequate explanation.

The morphogenesis we observe involves recruitment of cells, in discrete successive populations of regular size, into an activity which otherwise they might enter in smooth succession as a result of the wavefront. In fact, the control problem is that of the repeatability of a number, so that the large unit time we propose as that occupied by transit of the wavefront must be partitioned by a regular series of much shorter time intervals, also characteristic of the species at a given temperature. We propose that the latter are provided by an oscillator, shared by all the pre-somite cells, with respect to which they are an entrained and closely phase-organized population, because of intercellular communication. Probably, this time period is that actually observed to count out the very regular morphogenesis of somites down the axis. But note that in principle both wavefront and "clock" might be earlier, very much faster processes than those observed in the final cell behaviour. By programming the latter behaviour, these hidden processes would then in effect be setting up a periodic pre-pattern, and visible somite formation would be a "secondary process" in the terms used by Zeeman (1974).

We conceive the oscillator as interacting with the wavefront by alternately promoting and then inhibiting its otherwise smooth passage down the body pattern. It could do this by affecting periodically either the onset of catastrophe in cells, or else those particular expressions of the rapid change which cause the new locomotory-adhesive behaviour. This can be visualized (Fig. 2) as converting the course of the wavefront into a step function in time, in terms of the spread of recruitment of cells into post-catastrophe behaviour. The observed distribution of somite centres or of the fissures of de-adhesion between them, would result because each population of cells had become self-adhesive and discrete, due to almost synchronous behavioural change, at a time significantly after the preceding such population, but significantly before the subsequent cells were competent to participate. Although the biochemistry of the control of mutual recognition, motility and adhesiveness in cells is little understood, it seems reasonable from the look of the process in various animals that this is how each somite is individualized.

The essential property of number preservation during morphogenesis follows from the species-typical, invariate values for both the transit time of the wavefront along the body pattern, and the period of the "clock" during this transit. The lengths of the cell populations recruited by clock and wavefront interaction, like the rate in cells per time of the wavefront itself, will be proportional to embryonic length.

The postulation of an endogenous oscillator of cell behaviour as controlling animal development is not as *ad hoc* as might seem. Such oscillations, with a period of minutes, are now known to be fundamental to slime mould morphogenesis (e.g. Robertson, 1972), where they were first discovered because the initially single amoebae aggregate across distances involving vigorous motile behaviour. But the cells among which most early animal morphogenesis occurs are quite closely stacked in sheets, or else held in presumably contact-inhibited mesenchymal layers. Unpublished observations of time-lapse cinemicrography (J.C.) suggest in a preliminary way that when moving extra vigorously as in neurulation or bird gastrulation, such cells may propagate short-term periodicities of behaviour. Although most overt biological oscillators studied hitherto have had cycle times on the order either of fractions of a second (neuronal) or of many hours (circadian systems) we need not assume there is any *a priori* problem in postulating the intermediate, an hour long oscillator. Relaxation times of cyclical processes involving known levels of biosynthetic control machinery can accommodate intracellular oscillations of such periods (Goodwin, 1963). But known oscillators show independence from inhibition of macromolecular synthesis and from many metabolic inhibitors (Enright, 1971), so that universal models depending upon ion/membrane transport phenomena are becoming increasingly plausible (Njus, Sulzman & Hastings, 1974).

For readers who enjoy conceptualizing visually, though maybe not for others, the development of the embryo can be graphed in three axes as in Fig. 3. In Fig. 3(a), real space and real (developmental) time are horizontal, and the state-space or manifold representing intercellular states is collapsed onto one, vertical axis. The progress of the whole embryo is then represented by a folded surface (see also Thom, 1973; Zeeman, 1974), in a space where the analogue of the overall vector driving development in both its stable and catastrophic phases would be the force due to gravity. Only a sector of the space axis some three somites long is shown, so that the fold edge marking the catastrophic instability in intracellular development recedes in time through, say, 3 hr within this sector of tissue. Considered as a wavefront in real space, the sudden change or competence to undergo it spreads through the tissue in 3 hr. We then express the limit cycle of the oscillator as shown, in real time and in some dimensions of the intracellular state space, assuming that cells are so closely phase linked as to be effectively locally synchronous on a developmental time scale. The oscillator interacts with the homeostatic surface that includes the catastrophe fold in each cell, so that as the cells move in smooth succession towards this fold (according to spatial gradient of developmental rates organized by p.i.), their entry into instability and fast change is rhythmically gated in time. Alternatively the catastrophe fold is visualized as parallel to the real space axis as in Fig. 3(b). Then a head–tail row of cells has to be represented as points moving through time on the upper and lower surfaces, obliquely to the fold edge. On the catastrophe fold, the undulation in time ensures that cells arriving at the unstable fold edge undergo rapid change in regular groups.

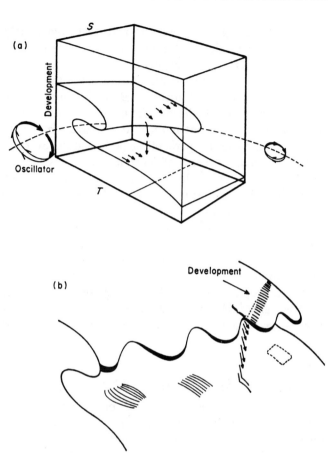

Fig. 3. Topological representations of the model for control of somite number. (a) A section of the embryonic axis a few somites long graphed in real space (S) in head–tail axis, real developmental time (T) (i.e. onset of somite formation at each level) and a dimension representing intracellular development (vertical, with gravity as analogue of the vectorial nature of development). The fold in the descending surface, representing onset of fast unstable cell change involved in somitogenesis, is oblique to the time and space axes. Thus of any longitudinal string of somite-forming cells, some will not yet have changed, a group will be changing and the rest will be in a new era of slow development (differentiation) following change. The hypothetical oscillator that controls the grouping of these cells is represented as a point describing a limit cycle, in real time and in some of the intracellular biochemical dimensions (vertical). The dashed line shows that it involves oscillation in the position of the instability or fold-edge. (b) The same surface is shown, but with real space and time represented by the rippled shape of the fold-edge and by drawing a string of cells travelling through their development (i.e. time) at an angle to the axis of the fold-edge, so that all must meet it. Cells thus undergo change in a succession of discrete, synchronized groups in time and space. A formed and differentiating, a just-formed, and a forming block are shown, while still on the upper surface are the presumptive cell-group of the subsequent somite (dotted lines and bracket).

At this point it is tempting to digress, and note that immediately before the first somites are formed, the pre-somite columns themselves have been demarcated in the mesoderm by a longitudinal pair of fissures of cell de-adhesion, separating them off from the mid-dorsal strip of the cells that then differentiate as the unsegmented notochord. Both the formation of these fissures and the first stages of notochord differentiation progress very quickly indeed down the long axis, so that notochord demarcation is nearly synchronous over at least the anterior parts of the body pattern. So mesodermal cells are capable of creating fissures of de-adhesion at this time, and if a clock is utilized in somite formation we should expect it already to be operating in cells at slightly earlier stages. Why does the notochord column not segment? We can say that the wavefront of cell change is many times faster-moving in this case, as compared with somitogenesis, either because the p.i. gradient is of very shallow profile in the midline or because the relation between p.i. and the timing of cell change is different during notochord formation. One can imagine on many grounds, especially if genuine propagation of catastrophe events were involved, that a fast progressing wave might not be susceptible to periodic interruption, by a given cycle of events in the cells, that would convert a more slowly travelling one into a step-function.

(D) TESTS OF THE MODEL

Just as the model has two elements, two types of experiment are suggested by it. They attempt to alter the number and size distribution of somites in ways consistent with having disturbed a gradient profile and a clock, respectively. After early operations transposing sectors of presomite material between disparate levels of the body axis, we might expect regions containing abnormal numbers and sizes of somites. These would be understandable as regions where abnormal slope or steepness of the p.i. gradient, and thus rate of passage of the wavefront of cell change, had obtained at the time of somite formation. They would be the equivalent in somitogenesis of the miniature patterns of cuticle types seen in the regions of grafts in insect epidermis (Stumpf, 1968), where graft/host interaction mediating gradient regulation is incomplete at the time that the gradient sets the pattern (Lawrence, 1970). Such experiments on amphibian gastrulae have been started and will be reported elsewhere, but the work presents difficulty, histologically and in interpretation of results. Abnormally long and short somites are seen, however, in regions of experimental larvae which show by their anatomy that host–graft interaction, rather than self-differentiation of graft and host material, have occurred.

Most classical grafting experiments studying amphibian morphogenesis have been performed either too early or too late to be of interest in the present context. In the first case, the whole-body p.i. gradient regulates its profile completely, as judged by normal morphogenesis. After late transpositions, pre-somitic tissue presumably merely self-differentiates with regard to somite number and to time, and the sequence of morphogenesis can also "cross" gaps made late in the pre-somite column, to proceed with normal timing posterior to them (Deuchar & Burgess, 1967). Such results suggest that the timing gradient is determined early on in cells by p.i. and furthermore that cells do not require immediate signalling from neighbours for passage of the wavefront of cell change, even though signalling might normally be involved in ensuring continuity and smoothness of the process. We stress that any empirical demonstration, that somite size can be varied in a way related to steepness of a hypothetical gradient after grafting operations, is not distinctive of the model presented here as opposed to a direct positional information/interpretation model (see section 3). Such demonstration would merely reinforce the belief that somites cannot be controlled by a repeating prepattern of the Turing class.

The much more distinctive, second experimental approach is that of attempting to unlink an oscillator from its normal tight co-ordination with the timing gradient, so producing abnormal somite numbers and sizes, possibly in otherwise normal animals. Somite number is highly constant across a wide range of developmental temperatures in *Xenopus*, even though development of the pattern may take $2\frac{1}{2}$ times as long at 17°C as at 26°C, and similarly, concentrations of respiratory inhibitors that slow development somewhat, leave the somite number normal if overall development is normal. Thus we must assume that the oscillator is really quite deeply embedded in the intracellular developmental process, maybe itself measuring out or setting the longer term rates of other aspects of development, including the catastrophe underlying the wavefront. But we cannot cling to such a saving clause through too many empirical failures to unlink normal somite number from normal development, lest we be accused of ascribing the harmony of morphogenesis to the harmonious revolutions of the heavenly spheres.

A small list of substances is becoming known to affect the free-running frequencies of a wide variety of circadian and other biological oscillators (Enright, 1971; Pittendrigh, Caldarola & Cosbey, 1973). Chief among them is heavy water (D_2O) which slows the frequencies of many systems at concentrations tolerable for development.

Development of embryos in heavy water media until they have formed about their first 20 somites does indeed alter the number of somites found between the base of the skull and the hind limb-bud, to a small but significant degree. The situation is complicated, however, by the effects of the D_2O upon the stretching processes in the embryos and also, apparently, on the cell-division schedule in the pre-somite cells (Cooke, unpublished data).

A further approach to be pursued consists of attempts to perturb the hypothetical "clock" by very short stimuli of a shock nature. The disturbances in final somite pattern that follow such perturbations, at the population level, might be found to correspond with the phenomenology ("temporal topology") that has been established and reviewed by Winfree (1975) as distinctive of perturbed biological oscillators in a variety of systems. These oscillators appear to be describable as limit cycles in a space that defines phase and amplitude around a singular state.

A critical-sized perturbation delivered at critical phase, rather than phase-resetting the oscillator as other similar perturbations do, brings the system into the neighbourhood of the singular state, following which amplitude is reduced or obliterated over many cycle-times, or else the system recovers with random phase. These types of disturbance should each be distinctively noticeable on appropriate local examination of the somites, if the appropriate mode of short-term perturbation can be found.

(E) THE DISTRIBUTION OF SOMITE SIZES IN THE NORMAL EMBRYO

Characteristically, in the early embryos or larvae of a variety of vertebrate types, somites are equally sized in the anterior and middle parts of the body-pattern, and then form a smooth gradation of decreasing sized populations of cells posteriorly into the tail. Because of the simple anatomy and lack of mitosis in somite muscle cells described in *Xenopus*, the numbers of cells cut off in the longitudinal dimension, between all the intersomitic fissures for about the first 30 somites, can be recorded in horizontal histological sections. Using these in conjunction with the curve for rate of somite formation per somite (see section 2), derived from the same population of embryos, two curves can be plotted. These are for total cells in the long axis incorporated into somites, against time (see experimental curve of Fig. 2), and against number of somites formed. The decreasing rate of wave front progression posteriorly is expressed as fewer-celled somites there; especially so if possibly slower tempo of the clock (see earlier) is taken into account. Ignoring cell boundaries and considering only space or distance across cell surface, as mentioned before (section 2) we have another component of slowing of the wavefront, not incorporated in such curves, due to the smaller cells when somites are formed posteriorly. None of the curves reaches zero slope, but somites were still being made in the tail at the time of preparation for histology in these embryos.

What factors could underlie such a trajectory of the wavefront? We suggest two alternative sorts of explanation. Firstly, we deduced from the coherent head to tail sequence of somite formation that the relation between the initial p.i. value and the local timing of the catastrophic change in cells was monotonic, but it need not be linear. Figure 4, in which the wavefront of the catastrophe in real space is drawn passing into a graded surface

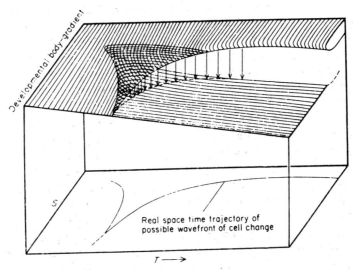

FIG. 4. A cusp-catastrophe unfolding into a graded surface. This shows a possible way in which the dynamics or shape of the surface in the state space within cells may condition the trajectory in real space (S) and time (T) for morphogenetic changes occurring as a wavefront in a linked sheet of cells. In the present instance, such a concept may help in understanding the size distribution of original somite masses in the embryonic body, though it has more general significance (see text section 5, and Zeeman, 1974).

(i.e. the length of the embryo as a p.i. organized gradient of decreasing development rate), itself suggests how the wavefront trajectory might describe a branch of a cusp. Projected down onto the space/time plane only, this cusp branch eventually comes parallel to the time axis; that is, a stable frontier forms in space between changed and never-to-be-changed cells (see Zeeman, 1974). Now since somites form to the tail-tip, our frontier here is virtual, as it were beyond the growing tail-tip. While the range of gradient values actually found within the embryo never prohibits the spread of somite morphogenesis, there could exist in posterior regions an increasingly non-linear relation between p.i. value and the successively later onset of competence for the rapid cell change, reflected in slower progress of the wavefront and smaller somites. This might occur even if the p.i. gradient itself were better expressed as a linear change in real space, the non-linearity coming in the relationship of catastrophe timing within the cell to p.i. value, whereby the wavefront in real time could describe a section of one branch of a cusp as shown in Fig. 4.

Alternatively, we can have no meaningful *a priori* expectation that the profile of p.i. in the body axis is in any sense linear, rather than convexly non-linear which would also produce the observed distribution of somite sizes. A "steeper" progression of gradient values in tail regions would lead to a slower wavefront progression, thus shorter somite populations. The mode of maintenance of positional cues in growing tissue is controversial (Summerbell, Lewis & Wolpert, 1973; Wolpert, 1975, see symposium discussion) but certain experiments on amphibian tail rudiments suggest that after about the level of somite 15, there is no longer any longitudinal cellular interaction in maintaining the position gradient, but rather a special terminal posterior zone of tissue where cells of new values are generated (Cooke, 1975a). In this case any "shape" of gradient is plausible as there are no longer any stability (diffusion) problems.

The locus of such non-linearity (i.e. is it in real space, the gradient, or in the intracellular machinery of response to the gradient?), may never become meaningfully definable, biochemically, but possibilities like those outlined above enable us to conceive how varied types of patterns of cell determination may be controlled synchronously within the embryo, so as to remain spatially co-ordinated as is observed. The replicability of the timing of somite formation, in embryos beginning development synchronously, is evidence with which the overall timing of cells' activities can be controlled. The known "accuracy" of biological clocks in other systems is very great (Winfree, 1975; Enright, 1971), such that interaction of wavefront and clock could quite plausibly produce a co-efficient of variation in somite number, within species, as low as that observed.

2. Conclusion

Although abstract, we do not feel that these hypothetical ways of thinking about development are vacuous. They promote searches for the entities (in this case "clock" and "wavefront") that they postulate, at least at the phenomenological level. If validated at this level, then even without a molecular analysis, they allow us to proceed with considering how form might be re-created reliably in each ontogeny, as well as changed during evolution. For instance, although our model (Zeeman) and experiments (Cooke) refer in the first instance to Amphibia, a similar model would appear equally valid for vertebrates as different, superficially, as birds or mammals in their early development. In such ways we might see more clearly the homology between the superficially very different early morphogenesis of distantly related organisms, which is only hinted at on comparative anatomical criteria.

APPENDIX A

A Threshold Model

We have seen that it is generally difficult for models based on gradients, thresholds, etc., to allow the degree of number control observed for somites and arthropod segments. However, one type of threshold model that would behave appropriately has been suggested by Graeme Mitchison, of the MRC Laboratory of Molecular Biology, Hills Road, Cambridge. By analogy with particular positional information models for *Hydra* morphogenesis (Wolpert, 1971, for review), imagine two gradients, in substances P and I, running in homopolar manner from head to tail of the embryo. Assume linearity and assign an arbitrary normal range of values (1 to 0) to each gradient. Suppose that P (non-diffusible) is fixed within cells throughout somite formation, while I can diffuse rapidly to equilibrate in concentration across local groups of cells when they are coupled (but separated from the source/sink boundary regions). At first all cells become uncoupled, before the "wave" of cell change is initiated in its passage down the embryo. Passage of the wavefront then involves successive recoupling of cells. As a group of coupled cells grows in size by recruitment of those posterior to it, the joint "coupled" value of I will fall, so that eventually the difference $(I-P)$ will rise to some threshold value within the tailward cells of a coupled group. If the response to this threshold is failure to couple, of the cells where it obtains, with those behind them then a population of cells will become functionally isolated as a unit. If boundary values of the body gradient P and I and for the $(I-P)$ threshold are fixed, unit number will regulate over different lengths. This model still postulates a wavefront, but not an endogenous oscillator whose temperature coefficients must co-ordinate with that of the wavefront transit in poikilothermic animals.

REFERENCES

ABERCROMBIE, M. (1950). *Phil. Trans. R. Soc. Ser. B.* **234,** 317.
BARD, J. & LAUDER, I. (1974). *J. theor. Biol.* **45,** 501.
CHILD, C. M. (1946). *Physiol. Zoöl.* **19,** 89.
COHEN, M. H. (1971). *Symp. Soc. exp. Biol.* **25,** 455.

COOKE, J. (1975a). *Nature, Lond.* **254,** 196.
COOKE, J. (1975b). in press, U.C.L.A. Squaw Valley Winter Conference; Developmental Biology (Walter Benjamin, Menlo Park, California).
DEUCHAR, E. M. & BURGESS, A. M. C. (1967). *J. Embryol. exp. Morph.* **17,** 349.
ENRIGHT, J. T. (1971). *Z. vergl. Physiol.* **72,** 1.
GARCIA-BELLIDO, A. & SANTAMARIA, G. (1972). *Genetics,* **72,** 87.
GIERER, A. & MEINHARDT, H. (1972). *Kybernetic,* **12,** 30.
GOODWIN, B. C. (1963). *Temporal Organisation in Cells.* London: Academic Press.
GOODWIN, B. C. & COHEN, M. H. (1969). *J. theor. Biol.* **25,** 49.
HAMILTON, L. (1969). *J. Embryol. exp. Morphol.* **22,** 253.
HERTH, W. & SANDER, K. (1973). *Roux'Arch. Entwicklunpmech.* **172,** 1
HOLTFRETER, J. & HAMBURGER, V. (1955). In *Analysis of Development* (Willier, Weiss & Hamburger, ed.) 230. Philadelphia: Saunders.
KAUFFMAN, S. A. (1969). *J. theor. Biol.* **22,** 437.
KAUFFMAN, S. A. (1973). *Science N.Y.* **181,** 310.
KAUFFMAN, S. A. (1975). In CIBA Symposium no. 29, *Cell Patterning.* Amsterdam: Elsevier.
LAWRENCE, P. A. (1970). *Adv. Insect. Physiol.* **7,** 197.
LINDSEY, C. C. (1966). *Nature, Lond.* **209,** 1152.
MARK, R. (1975). CIBA Symposium no. 29. *Cell Patterning.* Amsterdam: Elsevier.
MAYNARD SMITH, J. (1960). *Proc. Roy. Soc.* **152,** 397.
NJUS, D., SULZMAN, F. M. & HASTINGS, J. W. (1974). *Nature, Lond.* **248,** 116.
PITTENDRIGH, C. S., CALDAROLA, P. C. & COSBEY, E. S. (1973). *P.N.A.S.* **70,** 2037.
ROBERTSON, A. D. J. (1972). *Am. Math. Soc.* **3,** 47.
SANDER, K. (1975). CIBA Symposium, no. 29. *Cell Patterning.* Amsterdam: Elsevier.
SCHMIDT, G. A. (1933). *Wilhelm Roux Arch. Entw Mech. Org.* **129,** 1.
SPEMANN, H. (1903). *Wilhelm Roux Arch. Entw Mech. Org.* **16,** 552.
STRAZNICKY, K. & SZÉKELY, G. Y. (1967) *Acta biol. hung.* **18,** 449.
STUMPF, H. (1968). *J. exp. Biol.* **49,** 49.
SUMMERBELL, D. LEWIS, J. H. & WOLPERT, L. (1973). *Nature, Lond.* **244,** 492.
THOM, R. (1973). *Am. Math. Soc.* **5.**
TURING, A. M. (1952). *Phil. Trans. B.* **237,** 37.
WADDINGTON, C. H. & DEUCHAR, E. M. (1953). *J. Embryol exp. Morph.* **1,** 349.
WADDINGTON, C. H. & SCHMIDT, G. A. (1933). *Wilhelm Roux Arch. Entw Mech. Org.* **128,** 521.
WILBY, O. K. & EDE, D. A. *J. theor. Biol.* (1975). **52,** 199.
WINFREE, A. T. (1975). *Nature Lond.* **253,** 315.
WOLPERT, L. (1969). *J. theor. Biol.* **25,** 1.
WOLPERT, L. (1971). In *Current Topics in Development.* pp. 183. N.Y. and London: Academic Press.
WOLPERT, L. (1975). CIBA Symposium no. 29. *Cell Patterning* Amsterdam: Elsevier.
4 ZEEMAN, E. C. (1974). *Am. Math. Soc.* **7,** in press.
ZEEMAN, E. C. (1975). *Ann. Rev. Biophys. Bioeng.* **4,** 210.

Numbering in this book.

6. GASTRULATION AND FORMATION OF SOMITES IN AMPHIBIA AND BIRDS

The talk first gave an exposition of primary and secondary waves [5], and then extended the ideas from amphibia to birds. We give here a brief summary.

Differentiation is an embryo refers to regions that are initially alike and subsequently become different; in other words cells in those regions follow divergent paths of development in the space of states of a cell. Divergence implies a cusp catastrophe over space-time, and genericity requires that the cusp axis be inclined to the time axis. This implies a wave of cell determination which we call a primary wave, and which eventually slows to rest, forming the frontier between the two regions. The primary wave may be hidden, in the sense that the changes in each cell may be too delicate to detect at the time, but the cell may manifest some visible secondary effect after a delay of several hours. Although this secondary effect may be purely intracellular, nevertheless since it has been programmed by the primary wave, it will appear as a visible secondary wave. Furthermore, if the secondary wave causes morphogenesis, then the wavefronts may themselves be geometrically distorted by the morphogenesis. Consequently any tertiary effect will produce a tertiary wave that may appear as somewhat mysterious because it will have to follow the distorted pre-programmed wavefronts.

Published in Structural Stability, the theory of catastrophes and applications in the sciences, Springer Lecture Notes in Mathematics, Volume 525, 1976, pages 396-401.

Amphibia. In [5] these ideas were applied to the differentiation between ectoderm and mesoderm in amphibia. The hidden primary wave of mesoderm determination results in a secondary wave of invagination, causing gastrulation, and a tertiary wave of adhesiveness, causing the formation of notochord. Furthermore, in a model worked out jointly with Jonathan Cooke [2,3,5], the formation of somites can also be explained by superimposing a smooth clock during the passage of the primary wave. The clock allows the primary wave to move smoothly along the dorsal axis, but laterally causes it to move periodically in jerks. Therefore the resulting tertiary wave of adhesiveness causes the notochord to grow continuously along the dorsal axis, but laterally causes periodic groups of cells to adhere to one another, thus forming the somites. One advantage that this model has over previous models is that it permits regulation of somites, up to the observed precision.

Experiments. Subsequent experiments with frogs stimulated by the wave + clock model have produced results compatible with the model. Cooke has grown frogs in heavy water, which apparently slows the clock but not the wave, thus producing slightly fewer larger somites. Elsdale, Pearson and Whitehead [4] have momentarily interrupted the clock, but not the wave, by raising the temperature (from $25°$ to $37°$ for 5 minutes) during the primary wave determining somites; then when the somites form 6 hours later two or three of them are meshed together, while earlier and later somites are unaffected in position or timing.

Birds. I am indebted to Ruth Bellairs for explaining the development of birds to me (see [1], Chapters 4,5,6). The morphogenesis of birds is not so easy to understand as that of amphibia, because the bird embryo, instead of floating freely in 3-dimensions as the amphibian embryo, is crushed down onto the 2-dimensional surface of the yolk. Nevertheless a similar wave + clock model would seem to fit the observed facts, and it would be interesting to test the model by repeating and developing the above experiments with chicks.

In Figures 1-6 we sketch the model applied to the gastrulation of chicks. Figure 1 shows the hidden primary wave of mesoderm determination beginning at the embryonic shield and travelling across the area pellucida, through the presumptive mesoderm, which consists of presumptive notochord and somites. We have symbolically drawn eight wave-fronts to indicate successive positions of the primary wave; the even spacing indicates the regularity of the wave at this stage, before any morphogenesis has taken place (regulated by diffusion prior to cleavage). When the primary wave approaches wavefront number 8 it slows to rest, and forms the frontier between mesoderm and ectoderm.

We assume that the secondary effect causes cells to invaginate, thus producing the primitive streak and causing gastrulation. We also assume that the mesodermal sheet retains it topological integrity (apart from the temporary contact of the two sides of each wavefront as it rolls through the streak). Figure 2 shows wavefront 1 rolling through the streak, with the embryonic shield already folded underneath. Figure 3 shows* wavefront 5 rolling through the streak, with wavefronts 1-4 already folded underneath, and shown dotted. Figure 4 shows the end of gastrulation, with the secondary wave coming to a halt, and the rolling through the streak halted at wavefront 8.

---

* We have not shown what is happening to the surrounding area opaca, because it is not clear whether the mesoderm tears away or remains attached. If it remains attached along the boundary 0 then a tongue of endoderm will be pulled out of the area opaca underneath the mesoderm, and move in an anterior direction parallel to the streak.

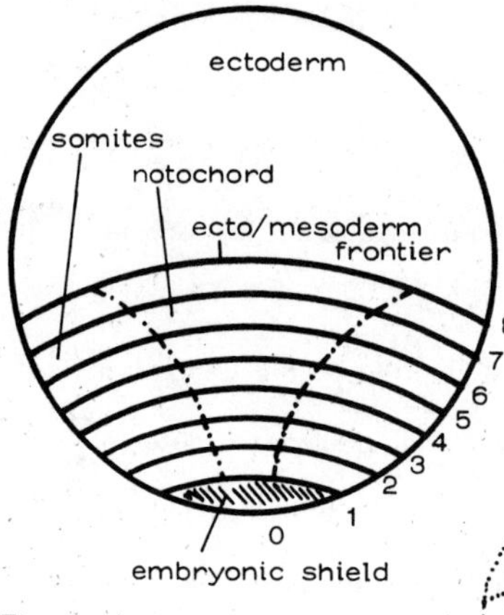
ectoderm
somites
notochord
ecto/mesoderm frontier
embryonic shield
0
1
2
3
4
5
6
7
8

Figure 1.
AREA PELLUCIDA

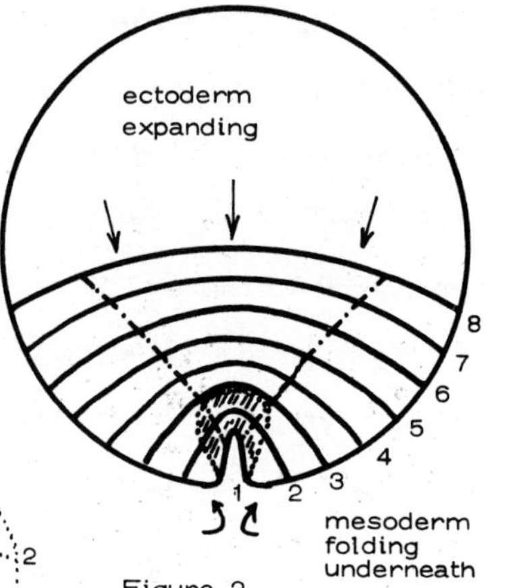
ectoderm expanding
mesoderm folding underneath
1
2
3
4
5
6
7
8

Figure 2
SHORT PRIMITIVE STREAK

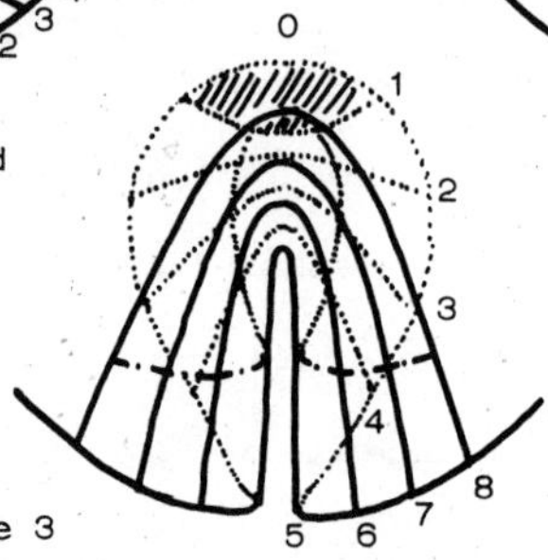
0
1
2
3
4
5
6
7
8

Figure 3
MEDIUM PRIMITIVE STREAK

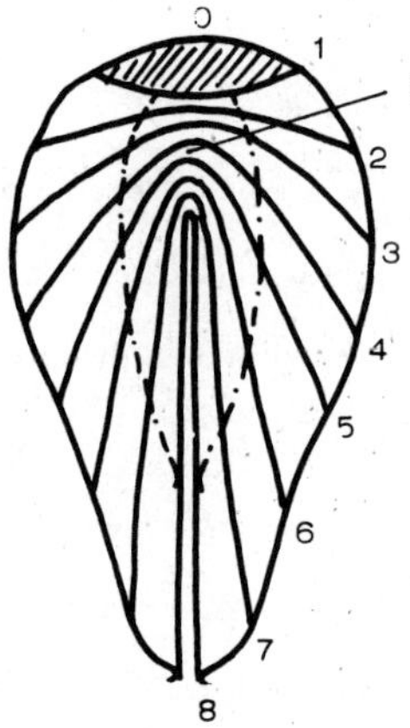
0
1
2
3
4
5
6
7
8

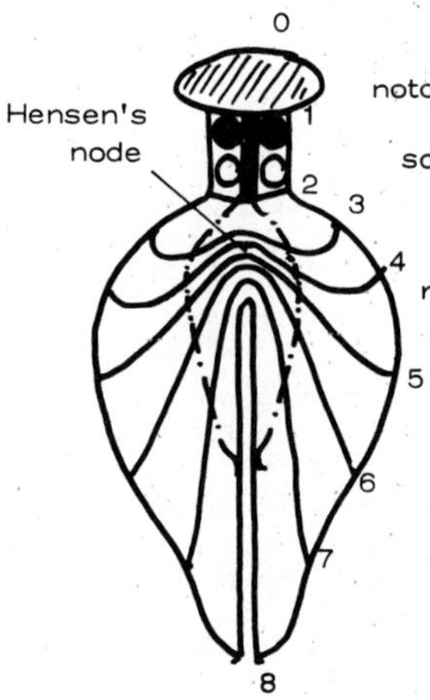
Hensen's node
0
1
2
3
4
5
6
7
8

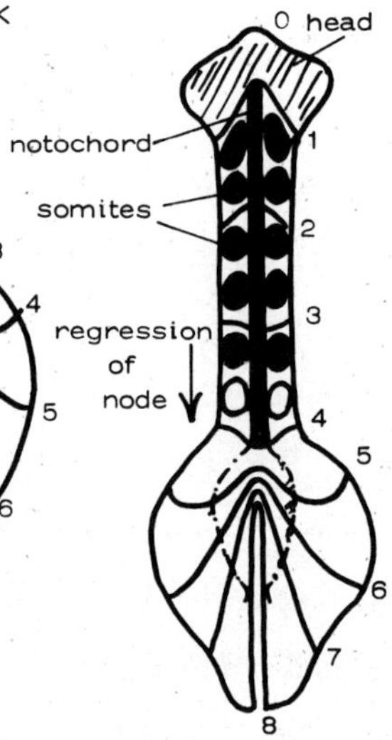
0 head
notochord
somites
regression of node
1
2
3
4
5
6
7
8

Figure 4
END PRIMITIVE STREAK

Figure 5
HENSEN'S NODE

Figure 6
FORMATION OF NOTOCHORD AND SOMITES

ISBN 0-201-09014-7, 0-201-09015-5 (pbk.)

By now the whole mesoderm has folded underneath, and so in Figure 4 we have drawn it in black lines rather than dotted. We can gain some idea of the geometric distortion of both mesodermal and ectodermal sheets at this stage by imagining the original disk of area pellucida folded along wavefront 8 into a semi-disk, and then pulled into a disk by bringing the two ends of the fold together. Such a process would stretch the perimeter and compress the centre. The resulting bulge at the centre is called Hensen's node. Meanwhile all the earlier wavefronts have been distorted and bunched together at the node, as shown in Figure 4.

We now assume that the tertiary effect causes each cell to become adhesive and increase its contact with its neighbours. Thus we have a tertiary wave of adhesiveness following the distorted wavefronts. Figure 5 shows the cells of the central region between wavefronts 1 and 2 adhering together to form the beginning of the notochord. At the same time the tertiary wave moves in jerks across the lateral regions to form the first few somites (we have symbolically drawn 2 pairs). The resulting contraction of wavefront 2 causes further bulging of the material between wavefronts 2 and 4, thus enhancing the bulge of Hensen's node. Figure 6 shows the tertiary wave reaching wavefront 4, having caused the notochord to grow continuously, and the next few somites to form. The contraction of wavefront 4 now causes bulging of the material between wavefronts 4 and 6, and so Hensen's node begins to regress. Meanwhile the sideways contraction of material into the notochord causes the latter to elongate, pushing the head process forward, and Hensen's node backwards, accelerating its regression.

Thus our assumption of a primary wave together with secondary and tertiary effects suffice to explain the whole geometry of embryonic axis. In particular Hensen's node is merely the bulge where the tissue has been forced to thicken by the way that the secondary wave caused it to infold through the streak, and this bulge contains the distorted pre-programmed wavefronts all bunched together. Naturally if Hensen's node is transplanted, then the tertiary wave will duly follow its programme, and form a complete embryonic axis in the host. Usually Hensen's node is called an "organiser", which can be a dangerous euphemism, because it may lead to a fruitless search for non-existent morphogens that diffuse and "organise". For example [1,p.112] reports the conclusions of Grabowski and Tsung et.al. that "Hensen's node is initially a head organiser, but that as it retreats it becomes successively a trunk and then a tail organiser". We suggest that the concept of a bunch of distorted pre-programmed wavefronts may provide a better conceptual framework for testing experimentally.

REFERENCES

1. R. Bellairs, Developmental processes in higher vertebrates, (Logos Press, London, 1971).

3. J. Cooke, Some current theories of the emergence and regulation of spatial organisation in early animal development (Appendix : E.C. Zeeman, Differentiation and pattern formation), Annual Rev. of Biophys and Bioengineering, 4 (1975) 185-217.

3. J. Cooke & E.C. Zeeman, A clock and wavefront model for control of the number of repeated structures during animal morphogenesis, J. Theoretical bio. (to appear). [5]

4. T. Elsdale, M. Pearson & M. Whitehead, Abnormalities in somite segmentation induced by heat shocks to xenopus embryos, (to appear).

5. E.C. Zeeman, Primary and secondary waves in developmental biology, Lectures on Mathematics in the Life Sciences, (Amer. Math. Soc., Providence, USA), 7 (1974), 69-161. [4]

↑ — *Numbering in this book.*

ADDENDUM by R. Bellairs.

In chick embryology there is now evidence to support the following assumptions:-

1. That the folding along wavefront 8 stretches the perimeter and compresses the centre of the area pellucida. This is supported by the morphological investigation of Bancroft and Bellairs [Anat. Embryol. 147 (1975) 309-335].

2. That there is a wave of changing adhesiveness and that this leads to somite segmentation. This is supported by the findings that mesoderm cells which have become segmented into somites are significantly more adhesive than those which have not yet become segmented. [Bellairs, Curtis, Portch and Sanders, in preparation]. The change occurs as the cells become spindle-shaped and arranged in a radial manner within each somite. It seems unlikely that the increase in adhesiveness is spread evenly over the surface of each somite cell; it is more probable that each cell becomes slightly polarised with regions of high adhesiveness and regions of low adhesiveness. Probably the cells then aggregate at their most adhesive regions, thus forming the centre of each somite. If they simultaneously draw apart from neighbouring cells at their least adhesive regions, the developing somite will become separated from the adjacent tissues. It may be significant that desmosomes form at the centre of each somite soon after segmentation; desmosomes are specialised regions on the cell surface whose main function is to help cells to remain closely attached to one another.

The above are scanning electron micrographs taken from two regions of a stage 12 chick embryo (45 hours incubation). The ectoderm has been removed so that the dorsal surface of the mesoderm is visible. The fibres are extracellular material.

A. Mesoderm which has not yet segmented into somites. The cells are flat and contact one another by small projections. × 3000.

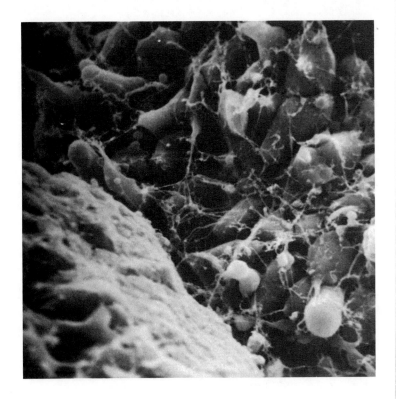

B. Mesoderm from a somite which has just segmented. The cells appear to be smaller and more rounded. The white mass in the bottom left corner is the side of the neural tube. × 3000.

C. Scanning electron micrograph of a single somite in longitudinal section, taken from a chick embryo of the same age. The cells are spindle-shaped and arranged radially, probably with regions of high adhesiveness at the centre, and regions of lower adhesiveness facing outwards. ×1000.

7 A Dialogue between a Biologist and a Mathematician

Dedicated to C.H. Waddington

The use of mathematical models in biology can pose a problem of communication between the two fields, particularly if the mathematical proofs behind the models are sophisticated, as in catastrophe theory [18,20]. On the one hand the mathematical arguments may appear as unnecessary black magic to the biologist, while on the other hand it may take a long time for the mathematician to gain sufficient experience in biology before he is able to judge the soundness of his biological hypotheses. In order to try and minimise these dangers I sometimes use the following format when writing down a catastrophe model, although the order of exposition may be different from the order of discovery. I try to separate the mathematics from the biology as much as possible.

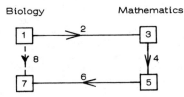

(1) Choose biological hypotheses that are acceptable to experimentalists.
(2) Translate these into mathematical hypotheses (3).
(4) Prove mathematical theses (5).
(6) Retranslate these back into biological conclusions (7).

The composition of the three steps (2,4,6) gives the deduction (8), going from biological hypotheses to biological conclusions. Thus the difficult mathematics is isolated in step (4). Meanwhile any doubt about the biological conclusions (7), or any experimental contradiction of them, can be thrust back either upon the biological hypotheses (1) or upon one of the two translation processes (2) and (6).

This method of exposition is generally the reverse of the order of discovery. Usually the model begins by <u>observing</u> the biological phenomenon (7), then <u>describing</u> it by the mathematical model (5,6), and finally <u>explaining</u> it by the argument (1,2,3,4). Sometimes these three processes may develop alongside one another over a long period, during which the experimental testing of the model may begin.

By Occam's razor, it is good mathematical practice to minimise the mathematical hypotheses (3) necessary to make the mathematical proof (4). It is then that the difference between mathematical modelling in biology and physics becomes apparent, because in physics the mathematician may usually select his physical hypotheses (1)

Published in Biosciences Communications, 1977.

from amongst a few fundamental principles such as the conservation of energy, the maximisation of entropy, or a variational principle, and from amongst a few fundamental theories such as gravity or electromagnetism. Biology, on the other hand, has as yet fewer well-established fundamental principles, and so the mathematician may have to consult the biologists at some length before selecting his hypotheses.

I will illustrate the process by describing the evolution of one particular model in catastrophe theory, concerning primary and secondary waves. The model began in 1967 with vague ideas at the first of Waddington's four conferences on theoretical biology at Serbelloni [22], and then appeared seven years later with the publication of my first paper on embryology [26]. It has led to the present ongoing successful prediction of a variety of experiments [2,3,5,6,7,8,9,10,11,12,16,28]. I will tell the story by means of a dialogue between a Biologist and a Mathematician, in order to show how the two languages, with their contrasting philosophies, may at first present a barrier to communication.

Of my two characters the Mathematician is supposed to be me (influenced by Thom [18,19] and other mathematicians), while the Biologist is an amalgam of several biological friends and coworkers, together with some books and papers read, lectures heard, and a few arguments of my own. If this amalgamated character says anything sensible, my sources may claim the credit, but when he speaks with foolish tongue the blame is mine. The dialogue is supposed to portray how the model gradually evolved as the two languages slowly meshed together, leading, on the one hand, to the formulation of hypotheses, and on the other to prediction.

I am particularly indebted to the biologists Ruth Bellairs, Jonathan Cooke, Tom Elsdale, Boris Rybak, C.H. Waddington, Paul Weiss, and Lewis Wolpert, and to the mathematicians David Fowler, Jacek Komorowski, Larry Markus, David Rand and René Thom. I would like to dedicate this paper to the memory of Wad, not only because he was paramount in initiating dialogues between biologists and mathematicians, but also because he himself was a man of many parts.

BIOLOGIST. We would like to know how the information in the genetic code determines development.

MATHEMATICIAN. That sounds a good question.

B. I mean we would like to have some models.

M. Ah well then, what precisely do you mean by development?

B. For example there are various different processes going on in the embryo such as differentiation and morphogenesis.

M. What do those words mean?

B. By differentiation we mean the process of change in cells, tissue and organs as they become different from one another, and by morphogenesis we mean the development of form and structure [1].

M. Do these two processes go on in the same place at the same time?

B. Well yes and no; I mean sometimes one precedes the other, and at other times they overlap. It all depends. But I would think that some aspect of differentiation always precedes morphogentic changes.

M. What makes you say they are different processes?

B. There are some things that we would say were definitely one rather than the other. For instance when a cell changes from being a mesoderm cell into being a notochord cell we would call that differentiation, whereas the movement of a mass of cells, as when the whole embryo rolls inside itself during gastrulation, we would definitely call a morphogentic movement [1]. However they are not precise terms, and you should not read any more into them than their usage; it might be a mistake to get too involved in trying to define them, or to use them as a basis for deduction. For instance some biologists tend to say that differentiation includes morphogenesis, while others tend to put it the other way round.

M. On the other hand it might be quite interesting to get involved in trying to define them, because that way we might be able to establish a more formal connection between them.

B. Could mathematics show that?

M. I don't know. Anyway not without some hypotheses.

B. What sort of hypotheses do you envisage?

M. Well let's try and define differentiation.

B. As I said, the cells in a piece of tissue start more or less alike and then become different.

M. Do you mean that in a particular place, and at a particular time, a frontier forms, such that all the cells on one side of that frontier become different from all those on the other side?

B. Well its generally more gradual than that.

M. How do you know the cells are different?

B. Sometimes they look different; one lot may look rounded and the other lot may look like cornflakes [6]. Another way to distinguish them is to take them out and grow them in tissue culture for some time, and then observe some difference in their behaviour that can be measured, and might indicate, for example, different adhesive properties between the cells [5]. Sometimes we know that cells are different but we can't tell they're different.

M. Then how do you know they're different?

B. I mean that they are already determined - we could grow them in culture until they manifest their difference. For example we know that at a certain stage some mesoderm cells are pre-notochord while others are pre-somite, although at that stage the precise nature of the difference is apparently too delicate, or as yet too difficult, to measure.

M. Well then, in order to get started, could we define differentiation to be a frontier gradually forming?

B. Not even that I'm afraid, because sometimes the cells first become different, and then migrate to where they know they've got to go.

M. That's ridiculous - how can they "know" anything?

B. They may be programmed to move in response to cues from the environment [25].

M. You mean they move.

B. Yes.

M. But I thought you said that movement was part of morphogenesis rather than differentiation.

B. Ah no, I said a morphogenetic movement was when a mass of cells move around, as in gastrulation [1]. The migration of individual cells takes place for example if you mix up epidermal and mesenchyme cells : they will sort themselves out by the former migrating to the outside and the latter to the inside of the aggregation of cells [4].

M. Does this also happen in the real embryo?

B. Yes and no. For example neural crest cells migrate all over the place. But you do have a point, because I know of no histological patterns that arise in vivo by the sorting-out of random mixtures of cells in different states [25, p.15]. On the contrary, pattern formation in one of its fundamental aspects always precedes morphogenesis. Cells become different (including their locomotory and adhesive properties) according to their position within a cell sheet. Some of them may thus be turned into migratory individualist cell types, which only then carry out those migrations. Such movements, as well as the folding and bending of cell sheets, are all really part of morphogenesis. When the cells first differentiate it is probably true to say that there is a frontier, at least in the majority of cases, and since it is difficult to pinpoint the exact time of the first appearance of that frontier it seems reasonable to call it gradual.

M. Good. Let us define differentiation to be the gradual formation of a frontier, which is at first fixed.

B. Oh, but it doesn't have to be fixed, because the tissue in which it is forming might itself be moving around due to some other morphogenetic movement.

M. I meant fixed relative to the cells.

B. Well there again, that's not always the case because the cells may not all differentiate at the same time; some may be delayed. For instance the notochord grows from front to back [4], and so if you took your frontier to be that between the cells that have already become notochord and those that are about to become notochord, then this frontier would actually move backwards down the embryo, relative to the cells themselves.

M. That's interesting. Is there any connection between the movement of this frontier, and for instance the morphogenetic movement of gastrulation?

B. No, gastrulation takes place earlier.

M. I mean do the two movements occur in the same place?

B. No, the movement of the notochord frontier is in the interior, whereas gastrulation, in amphibians for instance, is essentially a flow of cells over the dorsal lip, which is on the exterior.

M. I meant relative to the cells of course. Regarding the embryo as a 3-dimensional manifold, and the morphogenetic movement as an isotopy of this manifold in the ambient space, can the pull back in the manifold of both the movement of the frontier, and the flow over the dorsal lip, be represented by the same isotopy of a surface in the manifold?

B. Can you put that question more simply?

M. Well, first, can we regard the embryo as a manifold? What did you mean by a "mass" of cells when you were talking about morphogenetic movements [1]? Do the cells maintain their spacial relationship next to one another? I mean, if two cells touch, do they continue to touch?

B. Yes and no. There are several types of junction, some of which persist, while others may be repeatedly made and broken. Some cells walk hand-in-hand, as it were, while others may move like a crowd going through a door, where two neighbours probably remain close but may not necessarily remain in contact. It is not easy to tell.

M. Perhaps it might be better to concentrate on a single cell. So let me start again, and put a slightly different, but related question. Is the time interval between going over the dorsal lip, and becoming a notochord cell, the same for every notochord cell?

B. That's a funny question.

M. Why?

B. Because I've never thought of it before. As a matter of fact it is quite an interesting question.

M. Why?

B. Because I can do experiments to check it. Why did you ask it?

M. Well I thought perhaps both events might be secondary effects of some hidden primary wave of determination, that took place much earlier, while the embryo was still round, and while everything was much more stationary, and still retained some of the smoothness that must have been originally set up in the egg by diffusion. You see at this stage the geometry might be still quite simple, and such a wave merely a kinematic effect, due, as you said, to delays. The delays might be caused by a single gradient going from northpole to southpole, and the determination might be an event connected with the DNA. Then after a fixed time-delay each cell might do something observable, like contract some of its microfilaments or become sticky or something. If it contracted some microfilaments, then what you'd actually observe might be a wave of change of curvature, causing the embryo to roll in on itself, as in gastrulation, and you'd call it morphogenesis.. On the other hand if it became sticky, then what you'd actually observe might be a moving frontier, where fresh cells were sticking onto a growing column of already sticky cells, as in the notochord, and you'd call it differentiation.

B. This gets better and better, because you have now postulated three waves. I think I might be able to design some experiments to test their connection, and which might disprove the existence of such a hypothetical primary wave as you propose. For instance, I would interrupt it, or chop bits out, and see if the secondary waves were affected.

M. Meanwhile let me see if I can show there is a wave associated with any differentiation.

B. How could you show that?

M. I would like to develop a model, based on some hypotheses, on which I hope we could agree.

B. OK, what sort of hypotheses would you like?

M. Lets try once more to define differentiation. Suppose that differentiation takes place in a given piece of tissue, during a given
(Hyp.I) time interval. Our first biological hypothesis is that the tissue is continuous at the beginning of the time interval, that the two ends of the tissue develop continuously during the time interval, and that there is a fixed frontier between two types of cells at the end of the time interval. Then I can translate this into a mathematical hypothesis about the boundary values of the morphology in our little piece of space-time [26,29].

B. I agree to the fixed frontier, and I agree to the continuous development of the two ends of the tissue, but I'm not sure what you mean by the continuity at the beginning. After all, tissue is made up of cells, and there may also be a lot of intercellular matter.

M. I meant the scale of the tissue is much larger than the scale of a cell, and if we want to model global events at tissue level, then its sometimes useful to assume continuity at the local level. For instance in fluid dynamics precise predictions can be made using differential equations, in spite of the fact that at local level fluids are made of molecules. I know that in our case the contrast between the two levels is not as great, but you said yourself you couldn't detect any difference between cells at the beginning, and so any gradient on the tissue would presumably be represented by a smooth graph.

B. In that sense of course I agree.

M. The second biological hypothesis is the homeostasis of cells.

B. I agree to that, too.

M. Wait a minute, because I want to translate homeostasis into mathematics in a special way. Consider everything that could possibly be measured in a cell by continuous variables, such as the concentrations of all the proteins, their rates of change, the amplitude and periodicity of internal clocks, the temperature, pressure, density, viscosity, length, breadth, weight, stickiness, etc., - lets say a million variables. Geometrically we can

(Hyp.II) represent these variables by coordinates in a million-dimensional space, X, and then the cell will be represented by a point of X. Assume that the interaction between all these variables can be represented by a differential equation on X (which depends smoothly on space-time). We translate the biological hypothesis of homeostasis into the mathematical hypothesis that the cell is at a stable equilibrium point of this differential equation [26].

B. But I couldn't measure a million variables!

M. I'm not asking you to measure them.

B. Then why ask me to swallow such a crazy hypothesis?

M. Well suppose you modelled a cell with the two differential equations $\dot{x} = A-x$, $\dot{y} = B-y$. Here x,y represent the concentrations of two important proteins, and A,B are their homeostatic levels. Would you think that was a good model?

B. Maybe. It depends what for.

M. Well it strikes me as a lousy model, because it looks far too simple. After all, the actual mechanism of homeostasis must surely involve a complicated chain of of events, and so why not include all those events in the model?

B. Because I can't measure them.

M. For measurement purposes we could always project my million dimensions onto your two dimensions (provided we forget my dynamic). What I am saying is that the mathematics allows us the luxury of n-dimensions if we want it. If it worries you, why don't you privately think of n as 2, while I'll privately think of n as a million.

B. I have another objection. The most important thing in the cell is the DNA, and this is a combinatorial structure, not measureable by a continuous variable. So your model is leaving out the most important thing.

M. Of course I agree wholeheartedly to the importance of the DNA and the genetic structure. I even admit to the possibility of the managerial gene (that fond brainchild of the managerial molecular biologist).

B. Then you agree to the inadequacy of continuous variables?

M. On the contrary, it reinforces my belief in the use of continuous variables. The genes manifest themselves through the production of proteins, and so, from the dynamic point of view, the genetic activity may well be best measured by the varying concentrations of those proteins in the cell.

I am really following the philosophy of Paul Weiss here [24, 1969 Foreword]. He suggests that the transfer of combinatorial information along the chain DNA → RNA → protein → ... stops at macromolecular structures. Thereafter, at cell level and beyond, new types of system come into play, whose structural stability may depend essentially upon the local randomness of the substrate, for instance the freedom of position and velocity at molecular level. Examples of measurement of such systems might be not only the protein levels and their rates of change, but also the periodicity of clocks, the temperature and the pressure, etc., and those are the types of variable that I am using in my model. The justification for the use of continuous variables is the local randomness (which is essential for the global stability). The eventual answer to your original question, on how the information in the genetic code determines development, may lie in the necessity to change the nature of the modelling at this juncture, due to the change in the nature of the processes.

The cell resembles a chemical factory, and if you want to study the internal statics of the factory then by all means use combinatorial methods. However if you want to study the dynamic of the whole animal, then it may be more fruitful to place more emphasis on, and perhaps even to restrict attention to, the outputs of the factory, which in some aspects are a great deal simpler than the factory itself, and the simplicity of which may be best captured by the use of continuous variables.

B. Alright, I'll accept your second hypotheses about homeostasis, but with philosophical reservations.

M. I have some reservations too, but of a more mathematical nature. An equilibrium point may be too simple. For instance the dynamic might possess a cyclic attractor, representing a biological clock in the cell. I have mathematical ideas on how to deal with that case, but in more complicated cases there might be strange attractors [14,17], which we don't know enough about yet.

And sometimes the number of cell divisions appears to be significant, for instance in the organogenesis of the limb-bud [25]. Paul Weiss says that mitosis is a kind of earthquake for the cell [24, p.76] and so each cell division may cause a redistribution of the protein populations. To handle this case the homeostasis might be made more sophisticated by introducing population dynamics and the theory of chaos [15].

B. Stop! Lets do the simplest case first.

M. The third biological hypothesis is repeatability : if we repeat the experiment it will be qualitatively the same.

B. I accept that.

M.   We translate this into the mathematical hypothesis that the model
(Hyp.III) is time-stable. That is to say sufficiently small perturbations of
the model are equivalent under diffeomorphisms of space-time
that preserve the foliation of space-time by time-paths [23,27].
In other words this formalises the concept of the stability of the
developmental paths of cells. This is the crucial hypothesis
that puts us into the framework of catastrophe theory.

B.   So far your hypotheses seem to be so general that I can't
imagine how you are going to get anything out of them.

M.   We will only need one more hypothesis, and I agree it will have
to be of a rather different character. So far Hypotheses II and III
have determined the nature of the model, that it is a catastrophe
model, while Hypothesis I describes the morphology that we can
observe on the boundary of our little piece of space-time. What I
want to do is to try and deduce what the morphology looks like
inside our little piece of space-time, to that you can go and do
some experiments to look for it. So far there are many
possibilities that would satisfy our first three hypotheses,
but most of them involve irrelevant things like other frontiers
appearing and disappearing again in the middle of the time interval.
What I need is some sort of hypothesis that would guide us towards
the simplest case. That's obviously the thing you ought to go and
look for first. And if that doesn't work then we can go back
and investigate other cases.

B.   Why couldn't you just say that the morphology is to be as
continuous as possible subject to the one constraint of differentiation?

M.   That's an excellent way of putting it. We will translate your
(Hyp.IV) biological hypothesis of "continuous as possible" into my mathematical
hypothesis of simplicity, that the morphology should have the fewest
possible catastrophes [29]. That will be our fourth and last
hypothesis.

B.   Good; then I'll accept all four.

M.   Good; then I'll describe the conclusion by drawing you a picture on
a cusp catastrophe [26]. There's no need to explain the proof
unless you want me to [26,29]. The main deduction is that the
frontier moves before stabilising, and its movement is the primary
wave associated with the differentiation. If you want to use the
model for prediction, then depending upon exactly how you want to
use it, we may have to extract a few details from the proof, in
order to explain where it's tough and where it's flexible, where it
might be used quantitatively and where it can only be used
qualitatively.

B. Could you use the model to explain the formation of somites? They are the templates for the vertebrae [4]. They form in a sequence running from front to back, and they seem to be determined at about the same time as your hypothetical primary wave. Could you explain how it is possible to control the number of somites? The exact number depends of course upon the species, but is the same for all the individuals in that species - lets call it 40 for argument's sake. How can the number be so precise?

M. We could add one more hypothesis of a smooth clock in each cell, such that nearby cells were approximately synchronised. Then because of the catastrophic nature of the events taking place at the wavefront, this would cause the primary wave to move in jerks [26]. Therefore the secondary effect would be to jerk free one somite at a time.

B. What do you mean by the events being catastrophic?

M. I meant merely that the equilibrium of one of the homeostatic states of a cell breaks down, causing the cell to switch relatively quickly into its other homeostatic state. This was implied by our earlier hypotheses.

B. But how would such a model regulate to temperature and size, so that you always got exactly 40 somites?

M. Because the wave is really caused by delays, and so both wave and clock are regulated by time. Raise the temperature and they both speed up.

Perhaps a good way to think of it is to imagine a row of several hundred alarm clocks, all ticking at the same speed, but with the last one set to go-off 40 minutes after the first, and all the others set to go-off at some time in between, according to their position. Then, as you watch them, the ringing of the alarm clocks going-off will proceed as a kinematic wave going down the whole row. You now add an extra refinement by attaching a second-hand to each alarm clock in such a way as to stop the bell from ringing immediately the alarm clock goes-off, and to delay it until the next time that the second-hand reaches the top. What you will then observe is that each minute, on the minute, a whole bunch of alarm clocks will start ringing together, like a somite forming. Moreover if you made the row twice as long, analogous to having an ambryo of twice the size, and still had the last alarm-clock set to go-off 40 minutes after the first, meaning that the head and tail-bud develop at prescribed stages, then the wave would have to go twice as fast in order to travel twice the distance in the same time. Each minute twice as many alarm clocks would start ringing, making each somite twice as big. So there would still be exactly 40 of them.

B. That's fantastic, because this is the only model of somite formation that can give the observed precision [7,10]. I can now go and do some experiments.

(later) ....

B. I have measured the beginning and the end of the wave of somite formation. Your prediction that the velocity is constant at the beginning was correct [8,9,10]. The prediction of the parabolic slowing down at the end looks right, but is more difficult to say for certain in this case, because the tail-bud is still growing, and the clock is slowing, and the point of stabilisation is difficult to locate in space-time.

....

B. I think I have managed to slow the clock but not the wave by growing the frogs in heavy water, because they have three fewer somites [8,9,10]. However the experiment is not conclusive because it might be open to other interpretations.

M. Why not grow them in a gently oscillating temperature bath? The oscillations might entrain the clock, but leave the average speed of the wave unaltered. Therefore if you ran the oscillations slightly faster than the normal clock for the first 20, and slightly slower for the last 20, then the poor frog ought to have 20 slightly smaller somites up front, and 20 slightly larger ones at the back. And if the experiment worked, it might be susceptible to some quite precise quantitative testing.

B. OK, next spring is when the frogs spawn in the ponds.

M. I bet the clock is connected with mitosis. In which case the rate of cell division at a particular place at the time of the primary wave should be equal to the rate of somite formation at the same place, several hours later. That's a real tough prediction.

B. Yes, but that might be a rather difficult experiment to perform. Mitotic rate is difficult to measure without interfering with later processes. For instance I have to measure the rate of somite formation rather indirectly by taking a synchronous population of embryos, fixing a statistically significant number of individuals at various stages, and later counting the number of somites formed in each individual under a microscope. Measurements of mitotic rate are usually done by a similar process of fixing.

M. Why not just take a time-lapse film of one individual?

B. You'd have to remove the neural plate in between the primary and secondary waves, otherwise you couldn't see the somites.

M. Why not use a chick embryo because that's transparent?

B. You'd better find a chick expert who likes making films. Anyway chick embryology is quite different from that of amphibians. Instead of a dorsal lip they have a primitive streak and Hensen's node.

M. What's Hensen's node?

B. Hensen's node is initially a head orgaiser, but as it regresses it becomes successively a trunk and then a tail organiser [13,21].

M. Another movement, eh?

....

B. Couldn't we just call it a clock + wavefront model [10], and drop all this catastrophe business, because it makes it so much more difficult to sell to other biologists? After all I can make perfectly good predictions, and verify them experimentally, without reference to any complicated mathematics.

M. I can think of at least five reasons. Firstly we might never have thought of the model in the first place.

B. I know, but once thought of, lets discard the redundant debris.

M. Secondly the catastrophic nature of the wave front is necessary to explain the jerks.

B. Just call it a wavefront of relatively fast determination.

M. Thirdly, I find I have to continually go back to the proof whenever you keep coming and asking me questions about the model, to find out where it's rigid and where it's flexible.

B. That's what you're for.

M. Yes, but you must agree that it's the "complicated mathematics" that guides me. Fourthly it applies to all differentiations satisfying our hypotheses.

B. Surely it can't apply to all differentiations. I'm sure there are some cases in which the frontier does not move.

M. That may be because it doesn't satisfy one of our hypotheses, and the model may be useful in directing our attention to find out which one; for instance our interpretation of homeostasis may not be sophisticated enough. And anyway the model doesn't say how far the frontier will move. If the movement was less then one cell width, you'd never see it, and you might think it hadn't moved at all.

B. That's really cheating. With that sort of argument you can get away with anything.

M. On the other hand it may open your eyes to look for movement where previously you might never have thought of looking for it. The model may sometimes offer a testable alternative to euphemisms like information, message, programme, morphogen, to organise, evoke, induce, or recruit. For example insect eyes are said to grow by recruiting neighbouring skin cells. That dangerous word "recruit" may send you off on a wild goose chase, looking for a morphogen diffusing from the eye cells into the skin cells in order to induce their recruitment, whereas in fact the growth may be just a secondary wave, following the path of the original hidden kinematic primary wave associated with the differentiation of eye cells.

B. But I prefer "recruitment" to a hidden kinematic primary wave because its easier to visualise.

M. So you like going off on a wild goose chase do you?

B. Point taken. And the fifth reason?

M. Suppose enough evidence accumulated over the years to establish several cases of frontiers moving before stabilising. Sooner or later someone would ask if there was a general principle of this nature in developmental biology. And someone else would want to speculate on how such a mechanism could have evolved. It is then useful to thrust such principles back upon more fundamental hypotheses, as we have tried to do. We have done it using catastrophe theory and Thom's theorem [18,20,26].

....

B. I have established the existence of your hidden primary wave by giving the embryo a heat shock, in other words raising the temperature from $25°$ to $37°$ by putting it in warm water for 15 minutes. Then after a specific time-delay of about 6 hours, at the place where the primary wave must have been passing at the time of the shock, a somite forms which is garbled up with the next two. Meanwhile the later ones form perfectly as if nothing had happened [11,16].

M. That result also supports my suggestion that the wave might be kinematic.

B. I agree. And we are now doing some more follow up experiments. The reason why we find your ideas interesting is that they take out the vitalistic element, and give a feel for how the whole process could develop under its own steam. The notion of a dynamic unfolding is very appealing. Also I like the idea of temporal-information as well as positional-information, the cells setting their count-down clocks by position, and then acting autonomously for a while, independent of their environment. The dorsal lip is a case in point, because it seems to roll up when the time comes wherever you put it. The trouble with positional-information alone is that it makes the cells into automata depending only on the environment, whereas the interplay with temporal-information gives richer possibilities. What you are doing is introducing space-time into biology.

M. Any credit in that direction should go to Thom.

....

B. I tried to test your predictions [26] about what would happen if you snip off the knobs of slime mould fruiting bodies, to see if the culmination was due to a secondary wave, but its very difficult to do by hand. I couldn't find a pair of scissors small enough. We will have to try some other method. Meanwhile we measured 100 migrating grexes, and its seems you were right about contractions of the slime sheath, although I still find it difficult to see how this could be a major propulsive force for the movement [12].

....

B. Have you thought about birds yet?

M. Yes. In terms of the model, the primitive streak appears to be just a folded wavefront, like the dorsal lip in amphibians, except that it's squashed down flat onto the surface of the huge yolk. The squashing down must stretch the periphery, and compress the centre, distorting the trace of the hidden primary wave into a bulge. And probably that's all that Hensen's node really is, no mysterious organiser, but just a bulging bunch of distorted wavefronts, waiting to manifest themselves by becoming sticky in the right order. And when they do, that creates the geometry of the embryonic axis [28].

....

B. I checked your suggestion about compression and stretching, and you were right. We found morphological evidence in support [3]. However we have done some preliminary experiments that throw doubt upon your other suggestion that the cells become more adhesive when the somites form. I dissected somite and pre-somite mesoderm from the embryo and grew it in tissue culture; and from the outgrowth of cells it looks as if the somite cells don't stick to one another as much as the others do.

M. I can see also that your photographs seem to be against me [6]. Before segmentation all the cells look like cornflakes, and surely they couldn't look like that unless the dominant forces on their membranes were the tensions at the junctions with other cells. Meanwhile after segmentation they look so round that surely the dominant force must be the internal hydrostatic pressure. That's a great pity, because had they become more sticky when the secondary wave jerks across, this would have explained why the somites form. Could the photographs possibly be misleading?

B. No I don't think so, the difference is too striking to be the result of any artefact of preparation. However there may be interpretations other than those that you have just given. Also there might possibly be reinterpretations of the experiment that I just mentioned. But if you still don't believe me, and feel that strongly about it, then I'll go and design some other experiment and prove you are wrong.

....

B. You were right after all. We measured the adhesiveness in a viscometer, and the results supported your predictions [5]. Probably the cells not only become more adhesive, but also slightly polarised, with regions of high adhesiveness towards one end, and regions of lower adhesiveness towards the other end. Then probably the regions of high adhesiveness aggregate at the centre of the somite, as the cells become spindle shaped, arranged in a radial manner within each somite. The photograph that you saw showed the outside of a somite; the visible parts of the cells that you were looking at were the outward facing
regions of lower adhesiveness, and they probably looked round to you because those regions had separated away from other cells. The effect can be seen in the third photograph of a somite cut in half [6]. When we extended our earlier experiments, we found that the results were due to the motility of the cells rather than their adhesiveness [5], which explains why at first we had thought you were wrong. I'm glad you stuck to your guns.

....

B. In your paper on primary and secondary waves [26] I didn't believe that first example about the frontier of a forest moving as a wave.

M. But I only meant it to be a simple intuitive example in order to explain the idea to embryologists.

B. No wait. Your model gave me the idea to go and measure some data in the glacial valleys of New Zealand where natural regeneration of vegetation and re-afforestation is taking place. I was intending to prove you wrong, but to my surprise I discovered that you appear to be right after all. I measured both the density of certain species, and the movements of their frontiers, by counting the numbers of trees of different ages in various transects. The frontiers slowed down parabolically before stabilizing as you had predicted, and they also exhibited your predicted ripple ahead of the wave [2].

M. That's very interesting. Since you have such good data, you might be able to observe another little subtlety, namely a sudden increase in the rate of deepening of the frontier at the moment of stabilisation. This is associated with the ripple ahead of the wave, which disappears after the frontier has changed its character from being moving to being fixed. Mathematically it is the angle at the stabilisation point between the fold curve and the intersection of the equilibrium surface with its tangent place [26, Figure 5].

. . . .

B. I think Wad would have enjoyed listening in on our dialogue.

M. Actually he even joined in on some of it [22]. But that's one of the nicest compliments you could have paid me.

REFERENCES

1. M. Abercrombie, C.J. Hickman & M.L. Johnson, <u>A dictionary of biology</u>, (Sixth edition), Penguin, 1977.

2. K. Ashton, Forest community boundaries - an application of catastrophe theory, Report Series 106, Univ. Auckland, New Zealand.

3. M. Bancroft & R. Bellairs, Differentiation of the neural plate and neural tube in the young chick embryo, <u>Anat. Embryol.</u> 147 (1975) 309-335.

4. R. Bellairs, <u>Developmental processes in higher vertebrates</u>, Logos, London, 1971.

5. R. Bellairs, A.S.G. Curtis, P.A. Portch, & E.J. Sanders, Cell adhesiveness and somite segmentation, (to appear), University College, London.

6. R. Bellairs, Electron scanning micrographs of chick somite cells, in <u>Catastrophe theory</u> (by E.C. Zeeman), Benjamin, to appear.

7. J. Cooke, The emergence and regulation of spatial organization in early animal development (Appendix by E.C. Zeeman, Catastrophe theory and biological patterns), <u>Annual Rev. Biophys & Bioeng.</u> 4 (1975), 185-217.

8. J. Cooke, Control of somite number during morphogenesis of a vertebrate, <u>Xenopus laevis,</u> <u>Nature</u> 254 (1975) 196-199.

9. J. Cooke, The control of somite number during amphibian development; experimental analysis and a model, <u>Proc. UCLA Winter Conference in Developmental Biology</u> 1975, (Walter Benjamin, Menlo Park, Cal.).

10. J. Cooke & E.C. Zeeman, A clock and wavefront model for control of the number of repeated structures during animal morphogenesis, <u>J. Theoret. Biology,</u> 58(1976) 455-476.

11. T. Elsdale, M. Pearson & M. Whitehead, Abnormalities in somite segmentation induced by heat shocks to <u>Xenopus</u> embryo, <u>J. Embryol. exp. Morph.</u> 35 (1976) 625-635.

12. D. Garrod, (letter 1976). University of Southampton.

13. C.T. Grabowski, The induction of secondary embryos in the early chick blastoderm by grafts of Hensen's node, <u>Am. J. Anat.</u> 101 (1957), 101-134.

14. J. Guckenheimer, A strange attractor, The Hopf bifurcation (Edited by J. Marsden & M. McCracken), Springer Grad. Texts in Appl. Math. (to appear).

15. J. Guckenheimer, Comments on catastrophe theory and chaos, Preprint, University of California, Santa Cruz.

16. M. Pearson & D. McLaren, Primary and secondary waves and catastrophe modelling of the differentiative process in amphibian development, (to appear), M.R.C. Cyto-genetics Unit, Edinburgh.

17. D. Ruelle, Turbulence and the Lorenz attractor, Turbulence and the Navier-Stokes equations, Springer Lecture Notes, 565 (1976) 146-156.

18. R. Thom, Structural stability and morphogenesis, Benjamin, New York, French edition, 1972; Eng. transl. by D.H. Fowler, 1975.

19. R. Thom, A global dynamical scheme for vertebrate embryology, Lectures on Maths. in the Life Sciences, 5 (1973) Amer. Math. Soc., Rhode Island, USA, 3-45.

20. D.J.A. Trotman & E.C. Zeeman, The classification of elementary catastrophes of codimension ≤ 5, Structural stability, the Theory of Catastrophes and Applications in the Sciences, Springer Lecture Notes in Maths, 525 (1976), 263-327.

21. S.D. Tsung, I.L. Ning & S.P. Shieh, Studies on the inductive action of the Hensen's node following its transplantation in ovo to the early chick blastoderm. II. Regionally specific induction of the node region of different ages. Acta Biol. Exp. Sinica 10 (1965), 69-83. English abstract in Excerpta Medic. section XXI 6 (1966) 2856.

22. C.H. Waddington, Towards a Theoretical Biology, Edinburgh Univ. Press, 4 volumes, 1967-1971.

23. G. Wasserman, Stability of unfoldings in space and time, Acta. Math. 135 (1975) 57-128.

24. P. Weiss, Principles of development, Hagner, New York, 1939; Reprinted with new Foreword, 1969.

25. L. Wolpert & J.H. Lewis, Towards a theory of development, Fed. Proc. 34 (1975), 14-20.

26. E.C. Zeeman, Primary and secondary waves in developmental biology, Lectures on Maths. in the Life Sciences, 7 (1974) Amer. Math. Soc., Rhode Island, USA, 69-161.

2   27. E.C. Zeeman, Levels of structure in catastrophe theory, <u>Proc. Int. Cong. Maths. Vancouver (1974)</u> 2, 533-546.

6   28. E.C. Zeeman, Gastrulation and formation of somites in amphibia and birds, <u>Structural stability, the theory of Catastrophes, and Applications in the Sciences</u>, Springer Lecture Notes in Maths. 525 (1976), 396-401.

      29. E.C. Zeeman, An extension problem involving cusps, (to appear), University of Warwick, Coventry.

Numbering in this book.

8 BRAIN MODELLING

What is needed for the brain is a medium-scale theory. To explain what I mean by "medium-scale", consider an anology with the knee. A small-scale theory of the knee would include a description of all the biochemical events involved in muscular contraction, whilst a large-scale theory would include a description of all the movements of the knee in conjunction with other joints during complicated manoeuvres like running and leaping. Neither theory is particularly helpful in capturing the basic essentials of the knee because they are both too complicated; fortunately in this case we have a simple medium-scale model of the knee as a few bones and muscles acting as levers, pivots and elastic threads. Not only does the medium-scale model give the simplest picture, but it also provides a link between large and small.

Now let us return to the brain. The small-scale theory is neurology : the static structure is described by the histology of neurons and synapses, etc., and the dynamic behaviour is concerned with the electrochemical activity of the nerve impulse, etc. Meanwhile the large-scale theory is psychology : the static structure is described by instinct and memory, and the dynamic behaviour is concerned with thinking, feeling, observing, experiencing, responding, remembering, deciding, acting, etc. As with the knee, it is difficult to bridge the gap between large and small without some medium-scale link. Of course the static structure of the medium-scale is fairly well understood, and is described by the anatomy of the main organs and main pathways in the brain - for instance an example of an organ is the

---

Published in Structural Stability, the theory of catastrophes, and applications in the sciences, Springer Lecture Notes in Mathematics, Volume 525, 1976, pages 367-372.

hypothalamus, and an example of a pathway is that running from the amygdala through the hippocampus to the hypothalamus. But what is strikingly absent is any well developed theory of the dynamic behaviour of the medium-scale. True, there have been several models concerned with groups of cells and computer simulations, but none have really matched up to the two main requirements of providing a framework for prediction and experiment, and providing a link between the large and small. On the one hand the network theories of neurons and the combinatorial theories of synapses seem unable to escape from the small-scale, and therefore appear to be unrelated to psychology. On the other hand the computer simulations of perception and problem-solving seem unable to escape from the large-scale, and therefore appear to be unrelated to neurology.

Question : what type of mathematics therefore should we use to describe the medium-scale dynamic? Answer : the most obvious feature of the brain is its oscillatory nature, and so the most obvious tool to use is differential dynamical systems. In other words for each organ O in the brain we model the states of O by some very high dimensional manifold M and model the activity of O by a dynamic on M (that is a vector field or flow on M). Moreover since the brain contains several hierarchies of strongly connected organs, we should expect to have to use several hierarchies of strongly coupled dynamics. Such a model must necessarily remain implicit because it is much too large to measure, compute, or even describe quantitatively. Nevertheless such models are amenable in one important aspect, namely their discontinuities.

To explain what I mean, let me digress for a moment to look at the simplest possible situation of one oscillator driving another. The classical example of the forced damped non-linear oscillator is Duffing's equation (see* [1,6]), and one of the most striking qualities of Duffing's equation is its cusp catastrophe; for instance take the non-linearity and the forcing frequency as 2 controls, and then almost any measurement of the resulting oscillation (for instance its amplitude or its phase) will exhibit the cusp catastrophe.

Returning to the brain we would expect some parts of the dynamics to be structurally stable, and to exhibit catastrophes, both elementary and generalised. Therefore, in principle, we should expect at least some changes in brain activity to be modellable by elementary catastrophes.

Let $\eta: Y \to C$ be such a model, where η is an elementary catastrophe germ between two k-dimensional manifolds, and for simplicity assume $k \leq 5$ [see 3]. Here Y is implicitly sitting inside $M \times C$, where M is some very high dimensional manifold modelling the states of some organ O in the brain, and Y is describing the bifurcation of some attractor of some dynamic on M parametrised by C, modelling the activity of O. Here C resembles the experimental hypothesis, because C measures the k factors that we hypothesise are the important ones causing the changes in the behaviour of O that we are interested in. For each $c \in C$ we have a flow F_c on M. Then Y is constructed as follows : in M choose a disk D transverse to the attractor that is bifurcating, let Y_c be the (finite) set of points where D is pierced by the nonwandering set of F_c, and define $Y = \cup Y_c$. Finally $\eta: Y \to C$ is induced by the projection $M \times C \to C$.

So what? If everything is so implicit and unmeasurable, can the model be any use at all? I claim that the answer is yes. By the classification theorem [2,3] η is equivalent to some canonical model $\zeta: Z \to C$ of the elementary catastrophe, where $Z \subset \mathbb{R}^2 \times C$, and $\mathbb{R}^2$ is the euclidean plane.

---

* Do <u>not</u> see [5], because this contains a mathematical mistake which is corrected in [1].

(If it is a cuspoid we can further reduce $\mathbb{R}^2$ to $\mathbb{R}$). We take Z as an explicit model for the large-scale psychology. We label the various sheets of Z with psychological words, describing the behaviours that are observed when the states of O lie on the corresponding sheets of Y. For coordinates in $\mathbb{R}^2$ we seek two psychological indices that correlate with the labelled sheets. Then Z is an explicit quantitative psychological model for testing experimentally.

Meanwhile we use the classification theorem in another way to relate Y to the small-scale neurology as follows.

<u>Lemma.</u> For an open dense set of maps $f:M \to \mathbb{R}^2$, the product $f \times 1 : M \times C \to \mathbb{R}^2 \times C$ maps Y diffeomorphically into $\mathbb{R}^2 \times C$ (onto X say).

<u>Proof.</u> Let Q^q be the kernel of the derivative $D\eta$ of η. Then q is the corank of the elementary catastrophe, and so $q \leq 2$ by the classification theorem, since $k \leq 5$. Therefore $f \times 1$ maps the germ diffeomorphically if and only if the kernel of Df is transverse to Q^q, which is an open-dense condition.

Consequently almost any pair of measurements of M will reproduce a faithful copy X of Y. Therefore the EEG recordings from almost any pair of probes in or near O will suffice, in spite of the fact that such recordings may be mere peripheral artifacts compared with the multitudinous important events contributing to the total activity of O. In spite of being based on an artifact X is an explicit quantitative neurological model for testing experimentally.

Summarising we have the commutative diagram

| Small-scale neurology explicit quantitative | Medium-scale dynamic implicit qualitative | Large-scale psychology explicit quantitative |

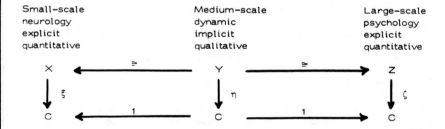

The composition $X \cong Z$ gives the required diffeomorphism between the neurology and the psychology.

There are two important warnings. Firstly a diffeomorphism is in general non-linear, and so there is no quantitative connection between the neurological measurements and the psychological measurements. That is why the neurologist says to the psychologist "your measurements must be unscientific because they are based upon such vague indices", and the psychologist replies "your measurements must be irrelevant because you are measuring such a tiny part of the brain". The mathematician intervenes with the observation "both your models are diffeomorphic, and therefore share the same diffeomorphism invariants, or qualitative properties". What are these? They are the singularities, catastrophes, bimodality, multimodality, hysterisis, divergence, inaccessibility, etc. For example this explains why the changes in α-rhythm occur at the same time as changes in behaviour, however casually one measures the α-rhythm or however crudely one measures the behaviour. Admittedly that is not very surprising, but what will be more impressive will be the future experiments confirming diffeomorphisms between neurological and psychological cusps, butterflies and double-cusps.

The second warning concerns the dynamic. Only the medium-scale has the dynamic, that makes some sheets of Y stable and others unstable. Implicitly the dynamic arises from the neuronal network, the synaptic connections and the metabolism of O, imparting a homeostatic tendency to certain oscillations, and thereby creating the attractors of the flows on M, which we have modelled by points of Y. It is true that there is also a gradient dynamic lurking in the background of the canonical model Z, but the classification theorem makes no connection between the two. Indeed the classification theorem makes no reference to any dynamic in either its thesis or its hypothesis, because it is implicitly utilising the existence of some Lyapunov function. Therefore it is a mistake to look for an explicit dynamic in either X or Z. The reason that the neurological state obediently follows the stable sheets of X and obediently jumps catastrophically at fold points onto other stable sheets, and the reason that the psychological state does the same on Z, must remain locked implicitly in Y.

The point of view expressed in this paper has gradually evolved from ideas in [2,4]; the new emphasis here is on the possibility of experimental prediction. The sort of applications that I have in mind are to test neurologically various psychological models, for example the cusp catastrophe models of aggression in animals and man, and the butterfly catastrophe models of psychotherapeutic treatment of nervous disorders, such as anorexia nervosa [7].

REFERENCES

1. P.J. Holmes & D.A. Rand, The bifurcation's of Duffing's equation: an application of catastrophe theory, J. Sound & Vibration (to appear).

2. R. Thom, Stabilité Structurelle et morphogénèse, Benjamin 1972.

3. D.J.A. Trotman & E.C. Zeeman, The classification of elementary catastrophes of codimension ≤ 5, this volume.

4. E.C. Zeeman, Topology of the brain, Mathematics and Computer Science in Biology and Medicine, Medical Research Council, 1965.

5. E.C. Zeeman, Catastrophe theory in brain modelling, Intern. J. Neuroscience, 6 (1973), 39–41.

6. E.C. Zeeman, Duffing's equation in brain modelling, Symp. for J.E. Littlewood's 90th birthday, 1975, Bull. Inst. Math. and Appl. (to appear).

7. E.C. Zeeman, Catastrophe theory, Scientific American (to appear).

Numbering in this book.

9. Duffing's Equation in Brain Modelling

WHICH branches of mathematics should be used to model the brain? I do not mean local models of the individual neuron, such as the Hodgkin–Huxley equations, but global models concerned with thinking and behaviour, that attempt to relate brain to mind. A brain model ought to be implicitly based on the underlying neurology, and explicitly capable of predicting some psychological behaviour. Models that only describe without predicting lay themselves open to the charge of being useless. And indeed under this criterion most published brain models to date are in fact useless, including most of my own.[1] Nevertheless, in this paper I shall argue the case in favour of nonlinear ordinary differential equations, and although there is still a considerable gap between the tentative predictions described here and those that would be required by the experimental scientist, at least the type of model proposed has qualities that would appear to be potentially useful in the design of experiments.

I am indebted to Larry Markus for many interesting discussions about differential equations. He once observed to me that the two prototype equations whose qualities characterise the nonlinear theory are the classical oscillators of Van der Pol[2] and Duffing.[3] These particular equations are so rich that it is not surprising that they caught the attention of Littlewood in the early 1940's, when he was already in his late fifties. They occupied much of his energy during the subsequent 25 years, and began his famous partnership with Mary Cartwright.[4] We shall not need to use the detailed analytic estimates found by Cartwright and Littlewood, but only some of the more elementary qualitative properties.

1. The classical oscillators

We briefly recall their elementary properties.

(i) The harmonic oscillator

$$\ddot{x} + x = 0.$$

This induces a flow on the phase-plane $\mathbf{R}^2$, whose coordinates are $x, \dot{x}$. The orbits of this flow are concentric circles (Fig. 1). The flow is structurally unstable because arbitrarily small damping changes the types of orbit.

(ii) The Van der Pol oscillator

$$\ddot{x} + \varepsilon(x^2 - 1)\dot{x} + x = 0, \quad \varepsilon \text{ small constant} > 0.$$

This is a structurally stable perturbation of the harmonic oscillator. The non-wandering set of the resulting flow on $\mathbf{R}^2$ consists of a repeller point at the origin, and an attractor of amplitude approximately 2, in other words an attracting closed orbit, or limit cycle, lying near the circle of radius 2 (see Fig. 2). Here "near" means of order ε. This can be estimated by assuming $x = A \cos t$, computing A, ignoring ε^2 and putting $\dot{A} = 0$. The rigorous proof follows by transversality (see references 5 and 6).

Fig. 1. Harmonic oscillator

Fig. 2. Van der Pol oscillator

(iii) The Duffing oscillator

$$\ddot{x} + \varepsilon k \dot{x} + x + \varepsilon \alpha x^3 = \varepsilon F \cos \Omega t,$$

where ε is a small constant > 0; k, F are constants > 0; $\Omega = 1 + \varepsilon \omega$ and α, ω are parameters. This is another structurally stable perturbation of the harmonic oscillator. It is the simplest nonlinear forced damped oscillator. Being non-autonomous, it induces a flow not on the phase-plane $\mathbf{R}^2$, but on the solid torus $\mathbf{R}^2 \times T$, where T is a circle representing periodic time, with period $2\pi/\Omega$, where Ω is the frequency of the forcing term. For sufficiently small values of the parameters α, ω the non-wandering set of the flow consists of either one attracting limit cycle, or else two attractors and one saddle-type limit cycle. The amplitude A and phase lag ϕ of these limit cycles can be estimated (to order ε) by substituting

$$x = A \cos(\Omega t - \phi)$$

into the equation, ignoring ε^2, equating the coefficients of $\cos \Omega t$ and $\sin \Omega t$ and solving for A and ϕ, giving

$$A^2(\tfrac{3}{4}\alpha A^2 - 2\omega)^2 = F^2 - k^2 A^2, \tag{1}$$

$$\tan \phi = \frac{4k}{3\alpha A^2 - 8\omega}.$$

Equation (1) is called the *Duffing amplitude relation*. For rigorous treatments see references 5, 6 and 7. (Do *not* see reference 8 because it contains a mistake.) Equation (1) gives the graph of A as a function of the parameters α, ω as shown in Fig. 3. The graph has two cusp catastrophes.[7,9,10] The cusp points are found by differentiating equation (1) twice with respect to A^2, and eliminating A, giving

$$(\alpha, \omega) = \pm \left(\frac{\sqrt{(3)}k}{2}, \frac{32k^3}{9\sqrt{(3)}F^2} \right).$$

At each cusp the upper and lower sheets represent attractors, and middle sheet saddles. When $\alpha = 0$ the equation is linear and there is always a unique attractor, whose amplitude reaches a maximum $A = F/k$ when $\omega = 0$, i.e., when the frequency of the forcing term equals that of the original oscillator, $\Omega = 1$, causing resonance.

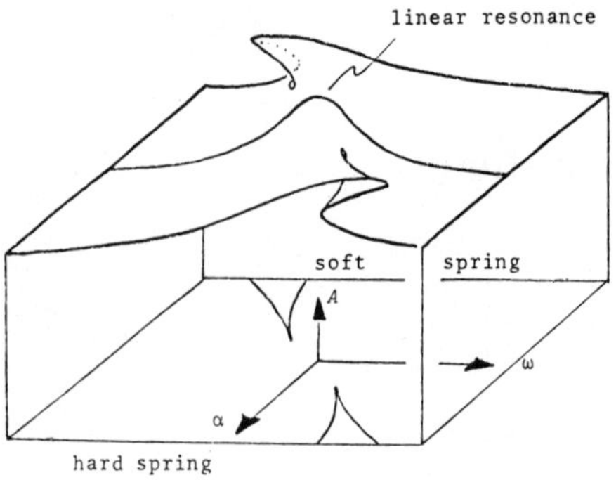

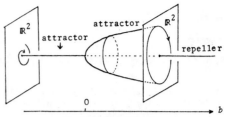

Fig. 3. The Duffing amplitude relation has two cusp catastrophes

Fig. 4. The Hopf bifurcation

The case $\alpha > 0$ is called a *hard spring*. If, further, $\alpha > \sqrt{(3)}k/2$ then the graph becomes folded over because of the cusp catastrophe. If ω is slowly increased from negative to positive values, then A smoothly increases to the maximum $A = F/k$ at a point inside the cusp given by $\omega = 3\alpha F^2/8k^2$ (which is given by the vanishing of the left hand side of equation (1)). If ω is increased further then at the right hand side of the cusp the larger attractor will coalesce with the saddle and disappear, causing a catastrophic jump into the smaller attractor, a catastrophic drop in amplitude and a catastrophic shift in phase. Imagine shaking a small tree in resonance and then increasing the frequency until the tree suddenly "turns against" the shaker. Conversely a decrease in ω will cause a catastrophic increase in amplitude and phase-shift, at the left hand side of the cusp. For the *soft spring*, $\alpha < 0$, events happen in the symmetrically opposite way.

If we add further nonlinear terms, for example, replace αx^3 by $\alpha_1 x^3 + \alpha_2 x^5 + \ldots$, then the graph of A over the enlarged parameter space exhibits higher dimensional catastrophes such as the butterfly. The important conclusions are:

(1) *Nonlinear oscillators typically bifurcate according to the elementary catastrophes.*

(2) *Smooth changes in frequency* (of the forcing term) *can cause both smooth and catastrophic changes in amplitude and phase* (of the oscillator).

(iv) *The Hopf bifurcation*

An immediate word of warning is necessary, because not all stable bifurcations of oscillators are elementary catastrophes; it depends upon whether or not there exists a stably bifurcating Lyapunov function. As yet the non-elementary bifurcations are unclassified. The most famous counterexample is the Hopf bifurcation,[11] which is the one-dimensional bifurcation exhibited by the parametrised Van der Pol oscillator, with parameter b:

$$\ddot{x} + \varepsilon(x^2 - b)\dot{x} + x = 0.$$

When $b < 0$ the flow in the phase-plane $\mathbf{R}^2$ has only an attractor point at the origin. When $b > 0$ the origin turns into a repeller and an attracting limit cycle appears of radius approximately $2\sqrt{b}$. Thus the non-wandering set, as b varies, consists of (or more precisely is differentially equivalent to and within ε of) a paraboloid and its axis (Fig. 4).

(v) *Van der Pol with large damping*

Now consider what happens when the damping ε becomes large. Let us replace ε by K to indicate its largeness. The phase-plane with coordinates $x, \dot{x}$ is no longer a good geometrical way to represent the oscillator because even although x may remain bounded the velocity $\dot{x}$ becomes very large. Therefore it is better to use the "dual" phase-plane with coordinates $x, \int x$, as follows. We begin with the Van der Pol oscillator

$$\ddot{x} + K(3x^2 - b)\dot{x} + x = 0$$

where K is a large constant, b a parameter and the factor 3 is put in for convenience. Suppose that $x, \dot{x}$ take initial values $x_0, \dot{x}_0$. Let

$$a_0 = x_0^3 - bx_0 - \frac{1}{K}\dot{x}_0,$$

$$a(t) = a_0 - \frac{1}{K}\int_0^t x(\tau)\,d\tau.$$

Then

$$\dot{a} = -\frac{1}{K}x.$$

Substituting in the oscillator

$$\ddot{x} + K[3x^2\dot{x} - b\dot{x} - \dot{a}] = 0.$$

Integrating

$$\dot{x} + K[x^3 - bx - a] = \text{constant}$$

$$= 0, \quad \text{initially, by choice of } a_0.$$

Hence in the dual phase-plane, with coordinates a, x, the oscillator is represented by the flow given by the first order equations:

$$\begin{cases} \text{Fast equation} & \dot{x} = -K[x^3 - bx - a], \\ \text{Slow equation} & \dot{a} = -\frac{1}{K}x. \end{cases}$$

The qualitative difference between "fast" and "slow" is determined by the size of the damping K.

We call the curve in the (a, x)-plane given by

$$x^3 - bx - a = 0$$

the *slow manifold*. If we now allow b to vary then the same equation gives a surface in (a, b, x)-space which is none other than the canonical cusp catastrophe surface, with normal factor a and splitting factor b (see Fig. 5). The slow manifolds are the sections of this surface given by $b = $ constant. Off this surface the fast equation ensures that orbits are nearly parallel to the x-axis (where "near" means of order $1/K$). Thus the fast equation acts as a catastrophe dynamic for the variable x, making the upper and lower sheets (given by $3x^2 > b$) into attractors, and the middle sheet (given by $3x^2 < b$) into a repeller. This fast dynamic does three things: first, it rapidly carries any point on to the attracting surface (or more precisely near the surface); second, it holds the point on (or near) the attracting surface for as long as possible; third, when this becomes no longer possible, for example, when the point crosses one of the fold curves (given by $3x^2 = b$) bounding the attracting surface, then the dynamic causes a catastrophic jump on to the other attracting sheet.

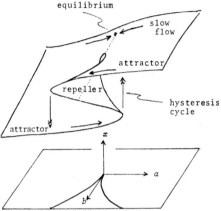

Fig. 5. Van der Pol with large damping is a cusp catastrophe with feedback flow

Once x lies on (or near) the surface, then $\dot{x} = 0$ (or is of order $1/K$), and so the slow equation comes into its own, causing the point to flow slowly along (or near) the slow manifold. In the language of catastrophe theory, the slow equation is a feedback flow of the fast "behaviour" variable x upon the slow "control" parameters a, b. When $b < 0$ the slow equation makes the origin $a = x = 0$ into a unique attractor point. When $b > 0$ the origin turns into a repeller point and the slow equation creates a new hysteresis cycle, consisting of two portions of slow flow along the upper and lower attracting sheets of the slow manifold, alternating with two fast catastrophic jumps between the two sheets. (See reference 12 for more details and pictures.)

Thus the Van der Pol oscillator with large damping, and its accompanying Hopf bifurcation, are both represented as additional structure superimposed upon the elementary cusp catastrophe, even although the Hopf bifurcation itself is non-elementary. The important conclusion is:

(3) *Nonlinear oscillators with large damping can sometimes be interpreted as slow feedback flows on elementary catastrophes.*

Having extracted some general conclusions about the way that nonlinear oscillators can behave and bifurcate and, having expressed them in the language of catastrophe theory, we now return to the brain. I like to approach the brain as a collection of strongly coupled oscillators, driving one another. The stability of our instincts, habits and memories indicate strong stability of some of the oscillators and the swiftness of our reactions indicates a coexisting instability due to strong coupling. But before we proceed further, it is a good idea to mention briefly a few neuro-psychological experiments to support this point of view.

2. Neuro-psychological experiments

(*i*) Shutting the eyes and relaxing induces the α-rhythm. The α-rhythm is an observed frequency of about 10 cycles per second in the EEG (electroencephalograph) pattern. It probably indicates that large parts of the brain are oscillating in resonance.

(*ii*) I once asked a mathematician under EEG to calculate a complicated homotopy group. The recording needles stopped dead for a minute while he thought, and then started again once he had given me the answer. This probably indicated a lack of resonance during the specialised cortical activity.

(*iii*) During epilepsy operations Penfield[13] touched the surface of the brain with a small electric oscillator and patients reported induced memories and a double reality. Moreover the same memories recurred when the experiment was repeated.

(*iv*) Adey[14] recorded oscillations from a number of electrodes implanted in a cat's brain, in the limbic system near the hippocampus. When the cat was relaxing the frequencies varied between 4 and 7 cycles per second, but when a trap door opened leading to food, then all the oscillations locked on to one frequency (of about 6 cycles per second), with a specific phase ordering. If the cat made a wrong turning at a T-junction, then the phases hunted, until the cat turned back to the correct turning when they locked on to the correct ordering again.

3. What is an oscillator?

What is the connection between the classical equations and the behaviour of the complex biological systems that we have so glibly called oscillators? We cannot possibly measure all the important events taking place in the brain, or even those in one organ of the brain. Nevertheless, it is possible to imagine a mathematical dynamical system of sufficient complexity to model those events. From general theory we know that such dynamical systems can have attractors, which can bifurcate. Moreover, although we cannot measure those attractors, nevertheless we can sometimes catch their bifurcations by means of artifacts. For example, the EEG pattern obtained from electrodes placed on the outside of the skull would appear to be an artifact, whose behaviour could not possibly give any significant information about the important dynamics within. Yet when the attractor inside makes a catastrophic jump, the artifact outside may well display a sudden qualitative change in behaviour. Such qualitative changes are easy to recognise in the resulting EEG pattern. Moreover, when they occur at the same time as psychological behavioural changes, such as in the experiments described above, then it strongly suggests that the artifacts are echoing the important bifurcations and catastrophic jumps within. Extending this argument to more than one dimension, if an attractor is bifurcating according to a higher dimensional catastrophe, then the measurements of the artifact may reveal a diffeomorphic copy of the same catastrophe.[15]

Hence, although the artifact may be but a pale shadow of the internal dynamics, yet its catastrophes may furnish a brilliant reflection of significant events. In this sense the artifact may provide a non-trivial qualitative model for the underlying neurology. Therefore the procedure of measuring and using neurological artifacts to predict psychological behavioural changes is scientifically tenable.

Let us now put the above discussion on a more concrete mathematical footing. Suppose that we model a biological system B by a mathematical dynamical system, that is a multidimensional manifold M, together with a vector field X on M. In the case of the brain we are quite prepared for the dimension of M to be as high as 10^{10}, the number of neurons, or 10^{14}, the number of synapses. The attractors of X represent the homeostatic state of B. The C^0-density theorem of Smale,[16] Shub[17] and de Oliveira[18] says that, by making an arbitrarily small C^0-perturbation if necessary, we can assume that X is structurally stable and that the only attractors of X are points (= stable equilibria) or closed orbits (= limit cycles).

Point attractors are easy to understand, and if B is parametrised, or driven, by another system then those points will bifurcate according to only the elementary catastrophes.

However, there are two reasons why the closed orbit attractors are more important than the point attractors. First, the EEG evidence suggests that periodicity is the rule, and static equilibrium the exception. Second, from the evolutionary point of view, the brain that can respond more swiftly than its neighbours to the environment has an evolutionary advantage. And if the dynamical system of the brain only had point attractors, it would remain stable when weakly coupled to any other stable system (representing some part of the environment or some sensory input) and hence the brain could not respond. On the other hand, a system with closed orbits can resonate with, or lock-on to, the attractors of the other system, however weakly coupled, thus enabling the brain to respond swiftly. (See references 19 and 20 for the underlying theorems.) Hence we would expect the brain to evolve non-gradient dynamics and limit cycles. By contrast the developing embryo does *not* want to be too perturbed by the environment during the crucial stages of development; hence we would expect it to evolve gradient dynamics and equilibrium states, as indeed it has.

We now bring this abstract multidimensional theory down to earth by relating it to ordinary differential equations, such as the classical oscillators that we started with.

Lemma. If y is a measurement of a closed orbit of an arbitrary dynamical system, then there exists a second order differential equation having y as its unique attractor.

Proof. Let C be the given closed orbit in M. By suitable choice of time unit, we can assume that C has period 2π. Let T denote $\mathbf{R}$ modulo 2π, representing periodic time, and let $T \to C$ be the diffeomorphism giving the timing round the orbit. Let $M \to \mathbf{R}$ be the given measurement. For instance a point in M might represent the brain state and its image in $\mathbf{R}$ the resulting potential difference across two electrodes on the skull, measured by the EEG.

Let y denote the composition
$$T \to C \hookrightarrow M \to \mathbf{R}.$$
Then $y(t)$, $t \in T$, will be the periodic function of time recorded by the EEG. Let $\dot{y}(t)$, $\ddot{y}(t)$ denote the derivatives with respect to t. Let $x, \dot{x}$ be coordinates in $\mathbf{R}^2$. Define
$$\psi: \mathbf{R}^2 \times T \to \mathbf{R}$$
by $\psi(x, \dot{x}, t) = \ddot{y}(t) + 2(\dot{y}(t) - \dot{x}) + 2(y(t) - x)$. Then
$$\ddot{x} = \psi(x, \dot{x}, t)$$
is the required differential equation. It can be verified by substitution that the general solution of the equation is
$$x(t) = y(t) + A e^{-t} \cos(t - \phi), \quad A, \phi \text{ constants}.$$
Hence all the solutions decay to $y(t)$, which is therefore the unique attractor, as required.

4. Summary of the modelling method

The main points of our discussion so far have been as follows.
(*i*) We assume that we can model the activities of the brain by multidimensional dynamical systems.
(*ii*) By the C^0-density theorem, we confine our attention to closed orbit attractors, and hence to second order differential equations.
(*iii*) Some of the bifurcations of the latter are modelled by elementary catastrophes.
(*iv*) Neurological artifacts may exhibit diffeomorphic catastrophes and hence provide measurable models for psychological prediction.

We conclude that *some* brain activities *may* be modellable by elementary catastrophes. It would be wrong to deduce any stronger statement, because
(*i*) it may not be possible to model there levant brain activity by a dynamical system;
(*ii*) the C^0-density theorem has not yet been generalised to parametrised systems;
(*iii*) some bifurcations are non-elementary; and
(*iv*) it may not be possible to measure anything that exhibits the relevant bifurcation.

For further discussion see references 9, 10, 15 and 20.

5. Applications

We now give a number of applications to behaviour exhibiting catastrophic jumps. These examples are really not so much predictions as suggestions for designs of experiments.

(i) Sensory inputs

Most sense organs convert amplitude into frequency. For example, brighter lights cause the neurons in the optic tract to fire with the same action potential but more rapidly. Similarly, louder noises, sharper pains, etc., all cause increased firing rates. When the frequency of firing reaches a certain threshold, then the brain will suddenly pay attention. Now the Duffing soft spring provides a simple model in which a frequency threshold causes a sudden jump in amplitude (see Fig. 6). Here the forcing term represents the input message from the sense organ to the brain and the oscillator represents the brain's response.

In Fig. 6 the frequency threshold occurs at ω_3. If the input frequency is now reduced again, then the model makes a prediction about hysteresis, which could be tested experimentally. The response does not switch off again at ω_2, but at a lower frequency ω_1. Indeed between ω_1 and ω_2 the amplitude of the response is in fact slightly enhanced. The hysteresis $\omega_2 - \omega_1$ could be measured.

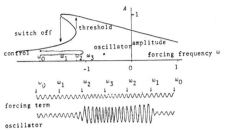

Fig. 6. The Duffing soft spring sketched for $k = F = 1$, $\alpha = -4$. The input frequency is increased to up to ω_3 and down again. The threshold occurs at ω_2, and switch off at ω_1

Another question that could be tested experimentally is the hardening or softening effect due to changes in arousal, fatigue or practice. A second prediction would be an anti-correlation between threshold and hysteresis: if the threshold ω_2 drops (implying softening) then the hysteresis $\omega_2 - \omega_1$ would increase and, conversely, if the threshold increases (implying hardening) then the hysteresis would drop.

(ii) Association

We begin with an idea of Thom.[20] Suppose two memories, as yet unassociated, are each represented by an attracting closed orbit of a dynamical system. The words "as yet unassociated" mean that we can represent both together by the product of the two systems. Therefore when the two memories are stimulated together they are represented by the product of the two attractors, which is a linear flow on an attracting torus. However, the latter is structurally unstable, and an arbitrarily small generic perturbation will, by Peixoto's theorem,[21] furnish a new stable system with a new attracting closed orbit on the torus. (This process is analogous to the locking-on phenomenon in the forced Van der Pol oscillator.) The new attractor represents the new associative memory, associating the two previous memories.

The beauty of this model is that it needs only arbitrarily small random synaptic perturbations in order to work. By contrast most models of associative memory, particularly those used in the design of artificial intelligence machines, need to assume a feedback system of synaptic changes in order to work. For example, a typical machine "learns" by turning up the gain on all those synapses it has just used every time it gives the correct answer; therefore it needs a feedback system from the "answer" to the synapses. And it is unlikely that a real brain could contain the abundance of such feedback systems, that would be necessary to explain the apparent ease with which we make all manner of associations all the time.

(iii) Recall

Consider a stimulus recalling a memory. Represent this by an oscillator driving another oscillator. The simplest model is Duffing's equation, with the forcing term representing the stimulus, and the oscillator the memory. Let us further suppose that when the memory is first laid down the equation is linear, $\alpha = 0$; therefore when the stimulus hits the resonant frequency the memory resonates accordingly (not unlike Thom's model above).

Now suppose that the memory is allowed to lie dormant for some period, before the stimulus is given again. There are two possibilities according as to whether or not the mind has been thinking about closely related thoughts during the intervening period. If it has, then facilitation of the nearby synaptic pathways may make the oscillator easier to drive, in the sense that its amplitude and frequency become correlated, because of the proliferation of short cuts available in the neural pathways. In other words, the oscillator turns into a hard spring, $\alpha > 0$. Therefore when the stimulus is given, represented by increasing the forcing frequency to the original resonant frequency, $\omega = 0$, then the amplitude of the oscillator will increase smoothly to a value lower than the original resonance. Hence the memory will *flow smoothly to mind*, quite unobtrusively.

By contrast, if the mind has not been thinking about any closely related thoughts, then the oscillator may become more difficult to drive, in the sense that amplitude and frequency become anti-correlated. In other words, it has become a soft spring, $\alpha < 0$ (like a simple pendulum). In this case when the stimulus is given then the amplitude will make a sudden jump just before the forcing frequency hits the original resonant frequency (see Fig. 6). In other words, the memory will *suddenly spring to mind*.

The difference between memories that flow-to-mind and spring-to-mind is a well known phenomenon; for instance, consider the ease with which we remember our friends' names, compared with the difficulty of putting a name to a familiar face that has not been seen for some time. It is remarkable how that missing name can sometimes suddenly spring to mind. It would be interesting to try to devise psychological experiments to test this difference between memories that flow-to-mind and spring-to-mind, and to measure the size, s, of the catastrophic jump in the latter case. If this were possible then the model gives a quantitative prediction, as follows.

During the period while the memory is lying dormant and while the oscillator is softening, there is a unique critical moment, at time t_0 say, when α crosses the cusp point. This is the precise moment when the latent memory is "forgotten," in other words is switched from being a flow-to-mind into a spring-to-mind type. In the neighbourhood of the cusp point, the size of the latent catastrophe increases parabolically with time. In other words, we have the quantitative prediction

$$s = \lambda\sqrt{(t-t_0)} + O(t-t_0), \quad \lambda \text{ constant} > 0.$$

(iv) Mood

The influence of environment upon mood and emotion has features that strongly suggest the use of catastrophe models. For example, the persistence of mood, the sudden changes of mood, the delays before those changes, the possibility of different moods under similar circumstances and the inaccessibility of other intermediate moods, all these five properties are typical of the bifurcation of oscillators.[10] How do we measure mood? According to MacLean,[22,23] emotions are probably generated in the limbic system (roughly the middle third of the brain), and so the oscillators involved would be modelling limbic organs, notably the hypothalamus. It is difficult to record directly from the limbic system, but the direct connections from the hypothalamus to the frontal lobes suggest that the latter might provide artifacts, that would echo limbic catastrophes. Similarly, physiological indicators of autonomic nervous activity can provide artifacts, such as the facial expressions of a dog indicating the levels of fear and rage.[24]

Contrary to what might be suspected at first sight, mood may in fact be one of the simplest brain activities to model. Possibly much simpler than, for example, the old favourites of visual perception, aural perception, language and problem solving. For emotionally we tend to be in one mood at a time, whereas intellectually we are able to grasp many things at a time. Perhaps this is because the limbic system tends to oscillate in resonance, due to its three-dimensional interconnectivity, whereas the cortex is able to oscillate out of resonance, differently as different parts, because of its two-dimensionality and its lateral inhibition. If mood *can* be represented by a single attractor, then although the infinite variety of nuances of mood would be difficult to measure because they would be represented by smooth variations of this attractor, nevertheless the noticeable changes of mood might be relatively easy to model using catastrophe models of the attractor's bifurcations, and might be relatively easy to predict using artifactual measurements.

(v) Behaviour

The influence of mood and emotions upon behaviour is the next step. Again, the persistence of behaviour, the sudden behavioural changes, the delays before making those changes, the possibility of divergent behaviour under similar emotions and the inaccessibility of other intermediate behavioural patterns, all suggest catastrophe models. A simple example is the cusp catastrophe model of fear and rage as conflicting factors influencing aggression.[10] There are many such psychological models, for both man and animals, waiting to be tested experimentally against neurological and physiological measurements.

(vi) Anorexia nervosa

Anorexia is a psychological disorder, in which dieting degenerates into obsessive fasting, leading to severe malnutrition and possibly death. It sometimes develops a second phase of alternately fasting and gorging. The psychotherapist, J. Hevesi, and I conjectured that the cause might be an elementary bifurcation of the brain oscillator underlying eating and satiety. This gave rise to a five-dimensional butterfly catastrophe model[10] of both the disorder and its cure under Hevesi's successful technique of trance therapy. This model was effective in permitting a coherent synthesis of a large number of observations that would otherwise appear disconnected. It also made sense of some of the victims' bizarre descriptions of their own disorder. Furthermore, it gave insight into what might be the key operative suggestions of the therapy, which is psychiatrically useful in helping to explain the technique to other therapists.

One of our projects is to extend the model to include sleep, because the disorder interferes with the natural catastrophic jumps of falling asleep and waking up. It is likely that the enlarged model could be a section of the ten-dimensional double cusp catastrophe, of which the mathematics is as yet poorly understood.[25] Even so, the five-dimensional geometry of the butterfly catastrophe is already sufficiently rich to have made some qualitative predictions that have been confirmed by observation. What is now needed, parallel to the theory, is a programme of quantitative testing of the model, for instance the monitoring of EEG and physiological changes in patients during the different states of fasting, gorging, sleeping, dreaming, therapy, etc., and the development of numerical techniques to convert these readings into geometric form, in which higher dimensional catastrophes can be recognised and verified.

(vii) Manic-depression

I am indebted to T. C. Dunn for introducing me both to his patients and to the literature on the subject. In reference 26, I suggested briefly that normality, mania and depression might fit into a cusp catastrophe, as shown in Fig. 7 with some measure of compatibility with the environment as a normal factor and some clinical measurement of the abnormality as splitting factor. The changes between the two pathological states do seem to be catastrophic in the sense that they take place relatively quickly compared with the time spent in either state. I had in mind a bifurcation of some attractor in the limbic brain, similar to the *anorexia* model, of frequency say a few cycles per second. However, M. Schmocker of Tübingen University Nervenklinik suggested that an oscillator with a 24-hour period might be more appropriate. Her studies on sleep deprivation, with results similar to those of Pflug,[27] had shown many indications of the disturbance of circadian rhythms amongst depressive and manic-depressive patients.

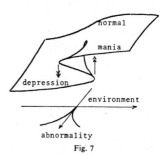

Fig. 7

Therefore let us take the Duffing oscillator as a tentative model, with the forcing term representing the external rhythms of day and night, and the oscillator representing the diurnal variation in the blood content, measured for example by the level of plasma cortisol.[23,28] The limbic brain is directly involved because cortisol production is controlled by the pathway: hypothalamus → pituitary → ACTH secretion → adrenal cortex → cortisol secretion → blood.[23] With this type of control a correlation between frequency and amplitude of the oscillator is plausible, since overstimulation of hypothalamus is liable to increase both. Therefore let us assume the oscillator behaves like a hard spring, $\alpha > 0$.

What we have said so far applies to the normal person. We now turn to the manic-depressive, and take as our main hypothesis that the underlying cause of the abnormality is a *speeding up of the internal circadian rhythm*. Mathematically this is equivalent to decreasing the relative frequency ω of the forcing term. Fig. 8 illustrates the consequences: when ω reaches the threshold ω_1 this causes a catastrophic rise in the amplitude A and a catastrophic drop in the phase-lag ϕ. This is exactly what is observed in some depressives, a substantial increase in plasma cortisol and a forward phase-shift.[28] The fact that the oscillator is now nearly in phase with the forcing term could mean that cortisol secretion is now being paradoxically triggered by the presence, rather than the absence, of cortisol already in the plasma. Moreover, the hysteresis effect locks the system in its abnormal state, even if the internal circadian rhythm slows down again: ω has to reach ω_2 before recovery is possible.

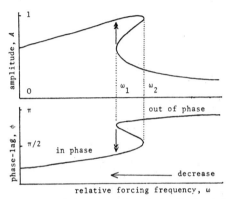

Fig. 8. The Duffing hard spring drawn for $k = F = 1$, $\alpha = 4$. A decrease in the relative forcing frequency ω causes a catastrophic increase in amplitude and forward phase-shift

We now turn to the psychological effects. It is known in Cushing's Syndrome[23] that an overproduction of cortisol due to other causes, for instance an adrenal tumour, is liable to cause mental changes, such as changes of mood ranging from depression to mania. Therefore we can expect these moods to arise from an increase in amplitude due to the Duffing catastrophe. Meanwhile phase-shift symptoms are also observed: sleep patterns are disrupted, manics often enjoy a vigorous night life and depressives are often unable to face the day. If half the body metabolism is out of phase, and sending conflicting messages to the limbic brain, no wonder the latter is liable to generate abnormal moods.

If the underlying cause of the abnormality is so simple, there should be an equally simple cure: just remove the forcing term. Then the abnormal internal circadian rhythm will reassert itself, causing a catastrophic renormalisation of both cortisol level and phase. This is exactly what happens in the sleep deprivation treatment.[27] A depressed patient is kept vigorously awake and active all one night, and in the morning is cured! One 30-year-old manic-depressive patient reported: "After this she felt like a changed person, she could enjoy her breakfast and experience pleasure again. She wanted to make things and read. She said she had not felt so well for 6 months." The snag is that the cure generally lasts only 2 or 3 days. Some fortunate patients stay well (protected by the hysteresis effect) but the more severe depressives revert as soon as the forcing term reasserts itself, *i.e.*, as soon as the patient begins to keep regular hours again.

A simple test of the model would be to remove the forcing term for a longer period, by providing an artificial environment without day and night, or by living in the continuous sunlight of the Arctic summer for instance, so that the patient could revert to his own circadian rhythm. Once the period had been established, then he might be able to devise a life style to suit; admittedly it can be awkward living a non-24-hour day, but worthwhile if it gives freedom from depression. For example, Schulte[27] reports that some victims had independently discovered a method of relieving their own depression by taking sleepless nights periodically.

Another possible cure might be to place a tiny alternating electric field across the patient's bedroom; for Wever[29] has shown that for subjects isolated for a month in an underground electromagnetically shielded bunker, a background field as small as 2·5 volts per metre at 10 cycles per second can have the power of synchronising the endogenous oscillators, and entraining the circadian rhythm. The usual treatment for manic-depression is the drug lithium[30] which, in terms of the model, may interfere with the oscillator sufficiently to cancel the Duffing catastrophe. However, the trouble with lithium is that it can have unpleasant side effects, and an over-accumulation can cause lithium poisoning, possibly resulting in tremor, slurred speech even coma or epileptic seizures.

We now return to our main hypothesis, and ask what happens if the circadian rhythm slows down instead of speeds up. The prediction from Fig. 8 would be a slight decrease in amplitude, and slight phase-shift the opposite way, but since no catastrophes would be involved, this would hardly be detectable, except by statistical analysis of a large number of normal people. However, if the internal rhythm were to slow down exceptionally to a period of more than 36 hours, then the oscillator instead of locking on to the forcing term would lock on to its second harmonic with a similar catastrophe as before. This is exactly what happens in certain rare cases of manic-depressives, who develop a 48-hour cycle.[30] The patient described in reference 30 regularly suffered from 1 day of depression alternating with 1 day of hypomania for 13 years, with the change of state occurring each night during sleep, usually between 2 and 3 a.m. When the patient was put in an artificial 22-hour environment, then, as would be expected from the model, he locked on to a 44-hour psychotic cycle. In normal people free running circadian rhythms of up to 33 hours have been observed.[29]

We conclude by emphasising the tentative nature of the model, and pointing out some reservations. First, it does not explain the difference between mania and depression; perhaps it could be combined with Fig. 7 into a higher dimensional catastrophe model, compatible with the 48-hour syndrome. Second, a closer analysis of cortisol[28] reveals that it is secreted in a series of 7–9 major episodes during the 24 hours. Therefore superimposed upon the circadian rhythm are physiological subrhythms, and any effective model should not only be compatible with the latter, but also perhaps offer an explanation for, and prediction of, the psychological subrhythms of eating/satiety and sleeping/dreaming. One thing is clear: to develop and test such models will need long term collaboration between mathematicians interested in dynamical systems and physiologists and psychiatrists.

References

1. Zeeman, E. C., "Topology of the brain," *in* "Mathematics and Computer Science in Biology and Medicine," Med. Res. Council Publ., 1965.
2. Van der Pol, B., *Phil. Mag.*, 1922, **43** (6), 700–719.
3. Duffing, G., "*Erzwungene Schwingungen bei veränderlicher Eigenfrequenz*," Braunschweig, 1918.
4. Cartwright, M. L., and Littlewood, J. E., *J. Lond. Math. Soc.*, 1945, **20**, 180–189; *Ann. Math.*, 1947, **48** (2), 472–494.
5. Hale, J. K., "Ordinary Differential Equations," Wiley Interscience, 1969.
6. Stoker, J. J., "Non-linear Vibrations," Interscience, 1950.
7. Holmes, P. J., and Rand, D. A., *J. Sound and Vibration*, 1976, **44** (2), 237–253.
8. Zeeman, E. C., *Int. J. Neuroscience*, 1973, **6**, 39–41.
9. Thom, R., "*Stabilité structurelle et Morphogénèse*," Benjamin, New York, 1972; English translation by Fowler, D. H., Benjamin-Addison Wesley, New York, 1975.
10. Zeeman, E. C., *Sci. Amer.*, 1976, **234** (4), 65–83.
11. Hopf, E., *Ber. Verh. Sachs. Akad. Wiss. Leipzig. Math. Phys.*, 1943, **95**, 3–22.
12. Zeeman, E. C., "Differential equations for the heartbeat and nerve impulse," *in* Waddington, C. H., *Editor*, "Towards a Theoretical Biology, Vol. 4," Edinburgh University Press, 1972, 8–67.
13. Penfield, W., and Roberts, L., "Speech and Brain Mechanisms," Princeton University Press, 1959.
14. Adey, W. R., Dunlop, C. W., and Hendrix, C. E., *Arch. Neurol.*, 1960, **3**, 74–90.
15. Zeeman, E. C., "Brain modelling," Symp. on Catastrophe Theory, Seattle, 1975, Springer Lecture Notes in Maths, to be published.
16. Smale, S., "Stability and isotopy in discrete dynamical systems," *in* Peixoto, M. M., *Editor*, "Dynamical Systems," Academic Press, New York, 1973, 527–530.
17. Shub, M., *Bull. AMS*, 1972, **78**, 817–818.
18. de Oliveira, M. M. C., "C^0-density of structurally stable vector fields," Thesis, University of Warwick, 1976.
19. Arrowsmith, D., *Proc. Camb. Phil. Soc.*, 1973, **73**, 301–306.
20. Thom, R., *L'Age de la Science*, 1968, **4**, 219–242.
21. Peixoto, M. M., *Topology*, 1962, **1**, 101–110.
22. MacLean, P. D., "The limbic brain in relation to the psychoses," *in* "Physiological Correlates of Emotion," Academic Press, New York, 1970.
23. Wright, Samson, "Applied Physiology," 12th Edition (Rev.: Keele, C. A., and Neil, E.), Oxford University Press, 1971.
24. Lorenz, K., "On Aggression," Methuen, London, 1966.
25. Zeeman, E. C., "The umbilic bracelet and double-cusp catastrophe," Symp. on Catastrophe Theory, Seattle, 1975, Springer Lecture Notes in Maths, to be published.
26. Zeeman, E. C., "Applications of catastrophe theory," *in* "Manifolds Tokyo," Tokyo University Press, 1973, 11–23.
27. Pflug, B., "Therapeutic aspects of sleep deprivation," 1st Europ. Congr. Sleep Res., Basel, 1972, 185–191.
28. Sachar, E. J., Hellman, L., Roffwary, H. P., Halpern, F. S., Fukushima, D. K., and Gallagher, T. E., *Arch. Gen. Psychiat.*, 1973, **28**, 19–24.
29. Wever, R., *Int. J. Biometeor.*, 1973, **17**, 227–232.
30. Hanna, S. M., Jenner, F. A., Pearson, I. B., Sampson, G. A., and Thompson, E. A., *Brit. J. Psychiatry*, 1972, **121**, 271–280.

ISBN 0-201-09014-7, 0-201-09015-5 (pbk.)

SOCIAL SCIENCES

SOCIAL SCIENCES

Paper.		Page
10.	Some models in the social sciences (with C.A.Isnard).	303
11.	On the unstable behaviour of stock exchanges.	361
12.	Conflicting judgements caused by stress.	373
13.	A model for institutional disturbances (with C.S.Hall P.J. Harrison, G.H. Marriage, P.H.Shapland).	387
14.	Prison disturbances.	403 - 406.

Paper 10, written in collaboration with Carlos Isnard, is an introductory paper describing how catastrophe theory can be used for modelling in the social sciences. The heart of the paper is the discussion on the meaning of "qualitative" in Section 9, pages 319-324. To illustrate the method, we develop a model of the influence of public opinion upon policy, taking as an example the policy of an administration concerning war. Qualitative features of the cusp and butterfly catastrophes are described in some detail. Two further models of opposition and censorship are discussed.

Paper 11 gives a model of stock exchanges. A number of observed qualitative features are translated into local mathematical hypotheses, which determine a cusp catastrophe model with feedback flow, from which the global dynamic behaviour can be deduced, including cycles of growth, boom, recession and recovery.

Paper 12 analyses data of G.C. Drew et al. comparing the effects of alcohol upon the driving skills of introverts and extroverts. The model shows how selectivity of speed-cues by a stressed perceptual mechanism can lead to misestimations of speed, and conflicting judgements. The model gives a good quantitative fit of the observed cusp in the data, and offers a simple explanation of the bimodal behaviour of introverts.

Paper 13, written in collaboration with prison psychologists and a statistician, describes a model of prison disturbances, another example of a feedback flow on a cusp catastrophe. The model is illustrated by applying it to an escalating sequence of events at Gartree Prison in 1972. The choice of variables in the model was based on factor analysis of retrospective data, and the procedures involved in making the model are described in paper 14.

10 *Some models from catastrophe theory in the social sciences*

C. A. ISNARD AND E. C. ZEEMAN*

INTRODUCTION

Phenomena involving sudden large variations traditionally have been assumed to be outside the reach of mathematical treatment, because they lacked what was considered to be an essential precondition, the continuity of the dependence relations between the variables. Recently, a branch of mathematics called catastrophe theory, one of the creations of the French mathematician René Thom (1972), has been applied to such discontinuous phenomena in biology (Thom 1969; 1971a; 1973a; Zeeman 1972a; 1974a) and physics (Fowler 1972; Shulman and Revzon 1972; Thom 1971b; 1973a; Zeeman 1972b; 1973; 1974c). The authors of the present article hope that their modest examples may suggest to specialists in the social sciences the possibility of applying catastrophe theory to similar discontinuous phenomena in their fields. (See also Harrison and Zeeman; Thom 1970; 1973b; Zeeman 1971; 1973; 1974c; 1974d; 1975.) The objective, in each of our examples, is the qualitative characterization of those points where small variations in some variable may cause large variations in a dependent variable, in other words those points where 'catastrophic change' may occur. This is the reason for the name catastrophe theory.

There is also a related phenomenon of 'divergence', where the

---

* Sections 1–11 are the joint work of both authors, and sections 12–16 are by E. C. Zeeman only.

Published in The use of models in the Social Sciences, (Edited L. Collins) Tavistock Publications, London, 1976, pages 44-100.

discontinuity may occur with respect to a variable other than time. For example sharp divisions of opinion can emerge in a population, even though the opinion of each individual may have evolved gradually and smoothly. By contrast the exact sciences are convergent in the sense that small changes in initial data usually cause only small changes in the ensuing motion, and so those sciences displaying divergence have been labelled 'inexact', because again they were thought to be impossible to model.

However, the creation of catastrophe theory has now revealed that sudden change and divergence are not only natural, and interrelated, but also amenable to rigorous mathematical treatment. Our objective is to use the theory to give qualitative understanding and global insight. The next objective, which we do not attempt here, is to provide quantitative models for experimental testing. In this sense catastrophe theory, as Thom himself has pointed out, is not a theory but a method. It is a mathematical tool, like the theory of differential equations, that can be applied to scientific theories, in order to explain and confirm them, or disprove them. The paper is intended to be an introduction to this method.

Contents
1. Influence of public opinion
2. Maxwell and Delay Rules
3. Sociological justification of Delay
4. Catastrophic change
5. Control factors
6. Control-behaviour graph
7. Sociological hypotheses
8. Delay before catastrophe
9. Meaning of 'qualitative'
10. Equivalent graphs
11. Cusp catastrophe
12. Application of the cusp
13. Butterfly catastrophe
14. Compromise opinion
15. Opposition
16. Censorship

1 THE INFLUENCE OF PUBLIC OPINION ON POLICY

In our first example, we consider the influence of public opinion upon an administration, or, more precisely, the effect that changes in the distribution of public opinion have upon the ensuing policy adopted by the administration. We are supposing that this policy is influenced by the opinions of a large group of people that we call the *population* – it may be the total population of a nation, or part of it, for instance the membership of a large party, class, or military group, etc.

In order to illustrate the model more precisely, we shall work with a specific example, although the whole theory that we develop can be applied to almost any situation where there is a continuous spectrum of policies (or several spectra). The specific example we choose to work with is the case of a nation deciding upon its level of action in some war, either a hot war or a cold war. Let the variable x represent the possible alternatives, so that the higher values of x represent stronger military action, and the lower values weaker action, with the lowest values representing withdrawal or surrender. At a given moment, let P(x) be the number of people in the population who would approve of policy x, in other words would approve of the adoption by the administration of the level x of military action. The number P(x) can be weighted, if one wishes, to take into account the relative influence that different segments of the population may have on the administration. Also, the function P can be normalized, if one wishes, in other words scaled down so that there is a unit area under the graph. Hence P can be regarded as the probability distribution of public opinion.

FIGURE 1   Public support for different policies

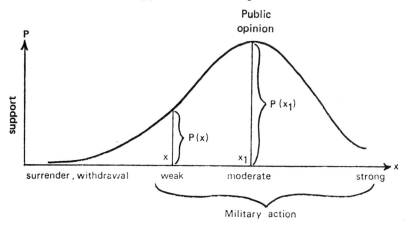

We suppose that *the aim of the administration is to maximize support*. This is, of course, an ideal assumption, made to isolate the particular cause-effect relationship that we want to study below. Also, it is not clear what the word 'maximize' should mean when we come to more complicated distributions, and so we shall need to introduce explicit rules in the next section. But at any rate, given the distribution of public opinion shown in *Figure 1*, then the policy adopted would be x_1. We say that the *behaviour*, at that moment, is x_1.

We choose to use the word 'administration' rather than 'government' in order to emphasize that we are modelling the spirit of the civil servant rather than the elected leader; the obligation of the administrator is to carry out the wishes of others rather than his own, to follow public opinion rather than to lead it.

Remark about maxima and minima This type of mathematics in which a function P is maximized, or minimized, appears in many different fields of application. For example, in economics P might represent profit, or cost, and an agent might choose his behaviour so as to maximize profit, or minimize cost. In psychology, P might represent anxiety level, and an individual might choose the behaviour that is anticipated with least anxiety. In sociology, P might represent the internal tension of a group, and the group might choose the behaviour to minimize this tension. In physics, P might represent the potential energy of some system, and the system would seek a stable equilibrium state, in other words a behaviour that minimized P (see Shulman and Revzon 1972; Zeeman 1972b). In fact, the considerations that follow can be applied to all these phenomena, and many more, and will show how catastrophic changes, and divergence, can occur mathematically in any deterministic model of this nature.

2 MAXWELL'S RULE AND THE DELAY RULE

We return to our original example. Suppose that with time public opinion becomes divided, so that there are now two local maxima at x_2 and x_3, as in *Figure 2*.

FIGURE 2 Split public support

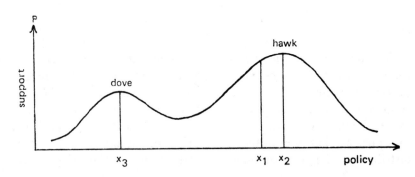

The majority has swung to the right in favour of stronger military action, and a minority has swung to the left in favour of less, or in favour of withdrawal. The contemporary fashion is to label these two groups, or their spokesmen, as *hawks* and *doves*. We suggest that an administration that had been engaged in policy x_1, if it was trying to 'maximize support', would immediately switch to policy x_2, in order to obtain larger support. In this way, as the distribution of public opinion changes with time, so does the adopted policy, or the behaviour. Moreover, if the distribution of *Figure 1* changes smoothly, meaning differentiably with respect to time, into that of *Figure 2*, then the behaviour will change smoothly from x_1 to x_2.

Suppose now that some time later the distribution of opinion changes a little more to *Figure 3*.

FIGURE 3 The Maxwell and Delay rules

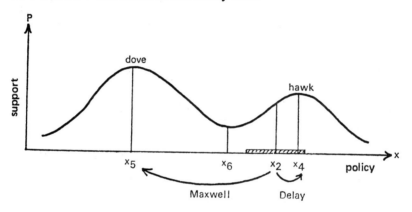

The majority opinion has now swung to the doves (and in so doing may have made both dove and hawk maxima slightly more hawkish). The administration, which had been engaged in policy x_2, now has a problem: should it adapt smoothly to the nearby local maximum x_4 of the hawks, or should it make a 'catastrophic' change of policy to the distant dove maximum x_5, where it would obtain maximum support? One thing is certain: the administration is least likely to try and average everyone's opinion and choose policy x_6, because this would have minimum support, and would incur the criticism of both doves and hawks. Therefore the problem is to choose between the two maxima. We can formalize the procedure by laying down two rules, as follows, and resolve the choice by agreeing to obey one or

other of the two rules. Rule (1) will imply choosing the dove maximum x_5, while rule (2) the hawk maximum x_4.
The rules are:

(1) *Maxwell's rule. Change policy to where support is maximum.*
(2) *Delay rule. Change policy in the direction that locally increases support.*

The directions of the arrows in *Figure 4* illustrate the directions of policy change under the Delay Rule, when this is applied to the distribution in *Figure 3*. At each point the direction of the arrow is determined by the slope of the graph, because this indicates which direction causes the support to increase. The arrows point towards

FIGURE 4 Policy movements under the Delay rule

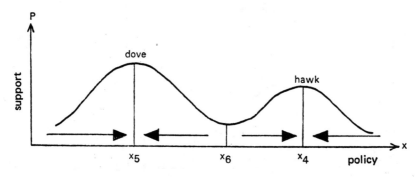

the two local maxima, and therefore under the Delay rule the policy changes until it is at one of the *local maxima*, whereas under the Maxwell rule it always changes to the *over-all maximum*.

Under both rules we assume the policy change happens *instantaneously*. The justification for this is that in practice any change of policy by an administration is usually so much swifter than the slow rate of change of distribution of public opinion, that compared with the latter the policy change can be regarded as instantaneous (see Zeeman 1972a). The reader may well ask the question: if under the Delay rule the policy change is instantaneous, why call it 'Delay'? The answer will become apparent in Sections 4 and 8 below, where we show that (instantaneous) adherence to the Delay rule can cause an administration to delay its response to swings of public opinion. Moreover, in the next section we give several reasons for suggesting that administrations do in general obey the Delay rule. This may help to explain why the delayed response is so common a phenomenon in practice.

Now let us return to *Figure 2*, and verify what happens under the two different rules as *Figure 2* evolves slowly into *Figure 3*. Maxwell's rule would imply a *catastrophic* jump from the old hawk policy x_2 to the new dove policy x_5, as soon as the dove maximum overtook the hawk maximum. Meanwhile, the Delay rule would imply a *smooth* policy change from the old hawk maximum x_2 to the new hawk maximum x_4. In this case, the emerging majority of doves would have been ignored because to change policy in that direction would initially have incurred a decrease of support, until the minimum x_6 had been passed.

Remark about minima There is a small mathematical point here: what if we found ourselves at the minimum x_6 and had to operate the Delay rule? The answer is that such a situation is unstable, and if it did occur, could only be a transient phenomenon, because P is always changing, and as soon as the minimum had moved, then x_6 would no longer be at the minimum, and so we could operate the Delay rule.

Remark about choice of rule At this stage, in any application of catastrophe theory, one must make a choice between rules (1) and (2), because they lead to slightly different models. One can devise other rules, for instance choosing the peak with the largest area beneath might be called the Voting rule. But in each case, the choice of rule must depend upon the particular application, and the one that seems to be best for most applications is the Delay rule. In fact, whenever the behaviour is determined by a differential equation (such as $dx/dt = \partial P/\partial x$) the Delay rule is a theorem, and so holds automatically (Zeeman 1973). For instance, in a biological example like the beating of the heart (Zeeman 1972a) the Delay rule is a consequence of using a differential equation to describe the underlying chemistry. In an economics example about the behaviour of stock exchanges (Zeeman 1974b) the Delay rule is a consequence of using a differential equation to describe the feedback of the market index upon investors. It is only when averaging devices are used that Maxwell's rule applies: in physics, for example, density measures the average packing of molecules in a volume, and consequently Maxwell's rule is used in Van der Waals's equation to describe the catastrophic jump in density as a liquid boils (Fowler 1972; Shulman and Revzon 1972; Thom 1971b; 1972). In fact, it is called

Maxwell's rule after the nineteenth-century Scottish physicist J. C. Maxwell, who formulated an equivalent rule for this particular phenomenon. In most applications to the social sciences, the Delay rule is more appropriate, but nevertheless in each case, at this stage of the argument, its use must be justified.

3 SOCIOLOGICAL JUSTIFICATION OF THE DELAY RULE

We return to our original example. Here we choose to use the Delay rule, and justify our choice as follows. There are broadly speaking five types of reason for preferring the Delay rule to Maxwell's rule: lack of information, intuition, sociological pressures, inertia, and past history.

First, the lack of information. In spite of prolific intelligence services, news media and opinion polls, administrations today are still often in serious doubt as to the weight of public opinion, and may not be able to gauge at all accurately at any given time whether the dove or hawk maximum is higher. Furthermore, it is also difficult to ascertain the precise distant policy that would receive the most support (in other words the position of the point x_5 in *Figure 3*). There is also a serious mathematical point here (which we return to at the end of Section 10 below) because it is possible to alter the relative heights of the two maxima by suitably manipulating the horizontal scale; for example, we could raise the hawk maximum by shrinking the hawk-end of the scale, and simultaneously lower the dove maximum by stretching the dove-end of the scale.

Usually, it is much easier for an administration to gauge the extent of support for nearby alternative policies than for a major shift of policy. It is generally simpler, and may seem relatively safer, to adapt policy smoothly so as to stay at the local maximum, rather than to make the sudden major shift of policy to another maximum which may or may not be higher.

Second, intuition. Compared with the difficulty of determining the global distribution function P, the local determination of which direction brings increased support is relatively easy, and can often be seen by intuition (the analysis required is local, and demands only the ordering by size of the support for nearby alternatives, rather than a full quantification). This becomes important when administrators have to make quick decisions off the cuff, in which intuition plays a greater role.

Third, the sociological pressures. Besides the natural face-saving reluctance to undertake reversals of policy there may be deeper sociological reasons against so doing. It may happen that an administration considers views too distant from its own to be hostile already on a personal level, and therefore the support of the holders of those views would be unconquerable, even by a radical reversal of policy. Such an administration tends to keep a close dialogue restricted to a smaller group of people, whose opinions are closer to its current policy, and amongst whom it tries to maximize support, while dismissing the excluded opinions as extremisms. In *Figure 3*, the shaded interval on either side of x_2 represents the opinions of this restricted group, and therefore the 'viable' near-by alternatives under this view. Within the viable alternatives x_4 commands the maximum support.

Fourth, inertia. It may take a great deal of time, effort, and money to reverse a policy, both in human persuasion and communication, and in the redeployment of resources. Meanwhile, public opinion may be volatile, and if a large proportion of the population is undecided, then the two maxima of doves and hawks may be oscillating slowly up and down like a seesaw – in which case it could be foolish, if not impossible, to indulge in the erratic behaviour of a reversal of policy at every oscillation of the seesaw.

Fifth, past history. The two rules display a significant difference in that Maxwell's rule takes no account of past history, whereas under the Delay rule the recent behaviour plays a crucial role in determining the current behaviour.

Summarizing: the above five reasons all argue in favour of the Delay rule and against Maxwell's rule. They all point towards the administration pursuing a policy that changes smoothly, through the continual successive comparison of the adopted policy with all near-by alternatives, and, at each stage, the adoption of a new policy if it brings larger support – in other words the Delay rule. Exceptions can happen during times of election. For if public opinion, during an election, is divided over some issue as in *Figure 4,* and if there happen to be two main parties that take the two peaks of opinion as their main election planks, then the election itself can perform a simple operation of the Voting rule. However, in our model we are more interested in the behaviour of a single administration during its tenure. Therefore, from now on we assume the Delay rule. In its behaviour the administration will cling to the protection of its local

10. Models in Social Sciences

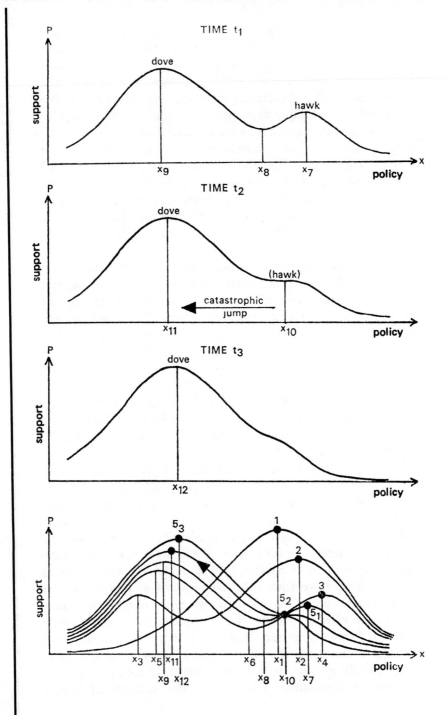

maximum, and will delay making any catastrophic changes in policy until forced to do so, which eventuality we now examine.

4 THE CATASTROPHIC CHANGE

Suppose that the behaviour was in the local hawk maximum x_4 of *Figure 3*. Suppose then that the distribution of public opinion evolves smoothly with time through the three successive distributions shown in *Figure 5*, at times t_1, t_2, t_3. By the time t_1 the behaviour will have changed smoothly from x_4 to x_7. By the time t_2 the hawk maximum x_7 and minimum x_8 have coalesced at x_{10}, and the behaviour is poised momentarily in unstable equilibrium at x_{10}. A moment later, the equilibrium vanishes, and by the Delay rule the behaviour has to make the instantaneous catastrophic jump to the dove maximum x_{11}. By the time t_3, the behaviour has settled stably in the dove policy x_{12}. We can now see why the Delay rule is so called: the catastrophic change is delayed until the last possible moment, when the policy is forced to jump because the local maximum in which the behaviour 'was caught' vanishes completely.

The bottom diagram of *Figure 5* shows all the distributions that we have had so far superimposed, with the numbers referring to the previous figures, and the blobs indicating the adopted policy in each case. Although the population is steadily becoming more dove-like, the track of the adopted policy is quite different. Initially the hawk policy escalates, and continues to escalate for most of the time; then there is a brief de-escalation before the catastrophic jump to the dove policy, after which the latter hardens slightly.

5 CONTROL FACTORS

What influences the changes in public opinion? We begin to elaborate the model by trying to pinpoint certain causes, which we shall call *control factors*. This is not to suggest that we necessarily have any control over them, but merely that this is a convenient terminology. It is actually the dependence of the behaviour x on the control factors that we wish to analyse. Control and behaviour are cause and effect.

We continue with our example of a nation deciding upon its level of action in some war. Consider the two control factors:

a = threat
b = cost.

In other words, *a* is some scale or index that measures how much the population feels their territory is threatened or their security is at stake. Meanwhile, *b* is some scale measuring the cost of the war, in casualties and money. (Possibly cost-per-achievement might be a more significant factor.) The effect of *a* and *b* on public opinion is qualitatively different, because a sense of common danger tends to unify opinion into a fighting mood, while a costly war tends to divide the population. We shall therefore call *a* a *normal factor* and *b* a *splitting factor*.

In order to make these concepts precise, we shall introduce sociological hypotheses, and translate them into mathematics. The deep theorems of catastrophe theory will enable us to synthesize the mathematics. We can then translate the synthesis back into sociological conclusions. It is not immediately apparent, without the use of the intervening mathematics, that the sociological hypotheses imply the sociological conclusions, and that is the purpose of using catastrophe theory. In this manner we shall see how the terms splitting factor and normal factor can be defined abstractly, and so acquire a new depth of significance, that can give insight into a wide variety of applications.

But first we explain how the dependence of the behaviour upon the control factors can be visualized as a graph.

6 THE CONTROL-BEHAVIOUR GRAPH

Let C be a horizontal plane with co-ordinates a,b. We call C the *control space* or *parameter space*. A point c = (a,b) in C is called a *control point*, and represents a particular threat + cost. The control space parametrizes the distribution of public opinion; in other words each control point c determines a particular distribution P_c of opinion x. We can incorporate the parametrized family of distributions into a single *support function* P, by defining

$$P : C \times X \longrightarrow R$$

by the formula $P(c,x) = P_c(x)$.

Let G_c denote the set of maxima of P_c. For example, if P_c is a distribution as in *Figure 1*, with a unique maximum at x_1, then G_c consists of the single point x_1. If, on the other hand, P_c is a distribution as in *Figure 2* with two maxima at x_2, x_3, the G_c consists of the pair of points $\{x_2, x_3\}$.

Now, let X be a vertical line with co-ordinate x. We call X the *behaviour space*, because it represents the possible levels of military action. Then G_c is a function from C to X, which is sometimes single-valued and sometimes double-valued. Our objective is to analyse this function. Now, the best way to visualize a function is to draw its graph. Therefore, let G denote the graph of the function G_c. To be precise, the graph G is contained in three-dimensional space, $C \times X = R^3$, and is defined to be the set of points

$$G = \text{closure} \{(c,x); c \in C, x \in G_c\}.$$

In fact, G will be a surface in R^3, and we claim that it will be equivalent to the folded surface pictured in *Figure 11* of Section 11 below. This is a strong and surprising claim, and to prove it we shall need to define equivalence and assume five hypotheses. Four of these are local sociological hypotheses, local in the sense that they are assumptions about the shape of P_c for particular cases of threat + cost. The other one is a mathematical hypothesis about P as follows.

Hypothesis 1. P is smooth and generic.[1]

This is a technical but harmless mathematical assumption, that enables us to use catastrophe theory in order to weld the four local sociological hypotheses into the single global picture of *Figure 11*, and hence determine G.

7 SOCIOLOGICAL HYPOTHESES

For convenience, let us use the language: *opinion is unified* if the distribution P_c of public opinion has a single maximum as in *Figure 1*, and *opinion is split* if it has two maxima as in *Figures 2–4*.

Hypothesis 2. If the cost of the war is low, then opinion will be unified, and the greater the threat, the greater will be the level of military action called for.

Therefore, keeping b fixed small and letting a vary, we shall have G_c as a single-valued increasing function of a. Moreover, G_c is smooth, since P is smooth by Hypothesis 1. Therefore, we obtain the graph shown in *Figure 6*. We label this G because it is a plane section,

b = constant, of the three-dimensional control-behaviour graph G defined in the last section. This graph, and Hypothesis 2, is an initial formalization of what we mean by saying a is a normal factor.

FIGURE 6   Threat-action graph for low cost

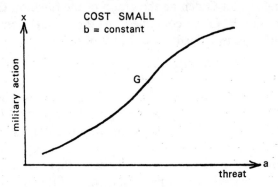

Hypothesis 3.   If the cost is high, and the threat moderate, then opinion will be split between doves and hawks.

Therefore, keeping b fixed large, and letting a vary, the graph of G_c will have two branches called dove and hawk, as shown in *Figure 7* below. This is an initial formalization of what we mean by saying b is a splitting factor.

Hypothesis 4.   If the cost is high, but the threat very great, then opinion will be unified in favour of strong military action.

Therefore, in the graph in *Figure 7*, only the hawk-branch extends over the right-hand end of the a-axis. Therefore, the dove branch must terminate at some point, B. Similarly we have the complementary hypothesis:

Hypothesis 5.   If the cost is high, but the threat very small, then opinion will be unified in favour of withdrawal.

Therefore, only the dove-branch extends over the left-hand end of the a-axis. Therefore, the hawk-branch must terminate at some point, A. The precise fashion in which the hawk branch terminates is illustrated by the sequence of distributions in *Figure 5*. Here we are envisaging a situation in which the sense of threat is diminishing, taking decreasing values $a_1 > a_2 > a_3$ at times $t_1 < t_2 < t_3$, and therefore the hawk-branch will terminate at time t_2 at the point $A = (a_2, x_{10})$.

In *Figure 7* the unbroken line illustrates the graph G of maxima G_c, while the dotted line illustrates the analogous graph of minima. Although the dotted line has no sociological significance (as we pointed out in Section 2, above), nevertheless, it is mathematically interesting that the two graphs continue to give a smooth curve.[2] The two points A, B where the hawk and dove branches terminate are called *fold-points*, because these are the points where the combined graph folds over. In other words, the singularities of the projection G→C.

FIGURE 7 Threat-action graph for high cost

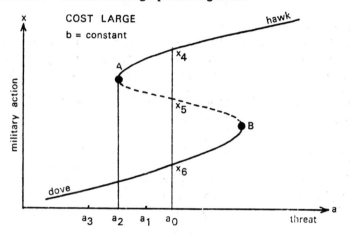

Let $b = b_0$ denote the fixed value of high cost for which *Figure 7* is drawn. Now, also fix $a = a_0$, and let $c = (a_0, b_0)$. Then suppose that the corresponding distribution P_c is illustrated in *Figure 4* with $x_4 =$ hawk maximum, $x_5 =$ dove maximum, $x_6 =$ the minimum. This explains why the vertical line $a = a_0$ in *Figure 7* meets the combined graph in those three values of x.

Summarizing: we have translated the four sociological hypotheses into mathematics by drawing the graphs in *Figures 6* and *7*. These will be the plane sections, $b =$ constant, of the three-dimensional graph G that we are seeking. The problem arises: how does the section shown in *Figure 6* evolve into that in *Figure 7*? The main theorem of catastrophe theory tells us that qualitatively there is only one way for this evolution to occur. It enables us to synthesize the mathematics into a single global picture, and then translate the mathematical properties of that picture back into sociological conclusions. A similar technique was used to study the unstable behaviour of stock exchanges (Zeeman 1974b).

But before we move on to the more complicated three-dimensional geometry in Section 11, let us first illustrate in the next section the type of qualitative conclusion than can already be drawn from *Figure 7*, and then digress in the following section to explain more precisely what the word 'qualitative' means.

8 THE DELAY BEFORE THE CATASTROPHIC CHANGE

We can now begin to put together the two main threads of our model so far:

(i) The variation of public opinion under different situations of cost +threat (described by Hypotheses 1–5 in Sections 7–9).
(ii) The resulting behaviour of an administration, that is seeking to maximize its support by operating the Delay rule (described in Sections 3–6 above).

Suppose we have a war-like situation with high cost (b = constant). The cost may not necessarily be due to military action, but could be due to a cold war, necessitating a high level of military preparedness, and a consequent crippling burden on the economy. Under varying

FIGURE 8 The delay before declaring war or making peace

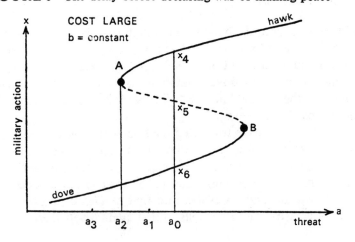

levels of threat, *a*, the maxima of public opinion will be represented by *Figure 7*, which we draw again in *Figure 8* below.

By the Delay rule the administration's policy behaviour will follow the graph G. Therefore, if the threat is small, the administration will pursue a dove policy. If the threat increases the administration will follow the dove graph, changing smoothly to a slightly tougher line, but essentially still pursuing a policy of appeasement. When the critical point B is reached, the administration will suddenly make a catastrophic change in policy to the point B' on the hawk graph, by, for example, declaring war, or launching a military attack. Notice that the point B occurs some time after the majority of opinion has swung in favour of the military action, and so, fortunately for the human race, there is a built-in delay in the process of declaring war. Moreover, the higher the cost the greater the delay, because the S-shape of the graph in *Figure 7* becomes more accentuated as the constant b is increased.

Now, consider the reverse process of what happens if the threat subsequently diminishes. The administration having once adopted the hawk policy will remain on the hawk graph until the critical point A is reached, when it will suddenly make a catastrophic change in policy to the point A' on the dove graph, by, for example, agreeing to a cease-fire, withdrawing, or surrendering. This time, the built-in delay is less fortunate, because the administration will remain entrenched in its war-like policy, possibly causing unnecessary loss of life, some time after the majority of opinion has swung in favour of abandoning that policy. The reason for this delay is illustrated in *Figure 5*. Again, paradoxically this time, the higher the cost the greater the delay. The explanation often given during the delay period for not withdrawing from a costly war is that such withdrawal would make a mockery of the sacrifices already made.

9 DIGRESSION ON THE MEANING OF 'QUALITATIVE'

There are fundamental reasons why the qualitative point of view is important in the social sciences, and therefore it is worthwhile digressing at some length in order to define 'qualitative' and explain the point of view. The reader is also recommended to read Thom's philosophy on the subject (Thom 1969; 1970; 1972; 1973b).

The whole of mathematics rests on three types of structure (i) order, (ii) topological, and (iii) algebraic. In a subject like physics all three types of structure can be given physical meaning, but in the social sciences only the first two types generally have any sociological[3] meaning.

(i) *Order structure*   As soon as one uses comparisons such as 'higher' cost or 'greater' sense of threat, one is giving sociological meaning to concepts of order.

(ii) *Topological structure*   As soon as one uses any kind of continuous scale to describe proximity one is giving sociological meaning to topological concepts. Furthermore, one can begin to use concepts of calculus, like smoothness. At this juncture, it might not be amiss to quote some observations of Tolstoy on calculus from *War and Peace*. Although he is writing in 1869, his words are strangely pertinent today:

> A modern branch of mathematics, having achieved the art of dealing with the infinitely small, can now yield solutions in other more complex problems of motion, which used to appear insoluble.
>
> This modern branch of mathematics, unknown to the ancients, when dealing with problems of motion, admits the conception of the infinitely small, and so conforms to the chief condition of motion (absolute continuity) and thereby corrects the inevitable error which the human mind cannot avoid when dealing with separate elements of motion instead of examining continuous motion.
>
> In seeking the laws of historical movement just the same thing happens. The movement of humanity, arising as it does from innumerable arbitrary human wills, is continuous. To understand the laws of this continuous movement is the aim of history . . . Only by taking an infinitesimally small unit for observation (the differential of history, that is, the individual tendencies of men) and attaining to the art of integrating them (that is, finding the sum of these infinitesimals) can we hope to arrive at the laws of history. (Tolstoy 1869, *War and Peace*, Book XI, ch. 1; see 1970, Vol. III, p. 3; also p. 225.)

Tolstoy puts his finger on exactly what catastrophe theory is trying to do.

(iii) *Algebraic structure*   By contrast, algebraic concepts such as addition and multiplication seldom have any sociological meaning.

For example, one cannot 'add' two senses of threat to give a third, and even 'twice' the cost can become meaningless if one tries to include in the cost of a war the measure of human suffering.

Roughly speaking, in mathematics those properties that depend upon the order and the differential-topological structures are called qualitative, while those that depend upon the algebraic structure are called quantitative. More precisely, we begin by framing the first qualitative definition.

Definition: Let X, X' be two scales describing the same data. We say the changes of scale from X to X', and from X' to X, are qualitative if they are both smooth and order-preserving.[4] *In this case the two scales are said to be qualitatively related.*

Notice the important feature of this definition is that the change of scale can be non-linear; therefore, although the order and topological structure of the scale are preserved under the change, the algebraic structure may not be.

In most mathematical models in the social sciences, if one uses a scale X for the convenience of making experimental measurements, or displaying data, then any qualitatively related scale X' is as valid. Therefore, any conclusion based upon the use of the particular scale X is only valid provided the same conclusion also holds using X'. Such a conclusion is called *qualitatively invariant*, or, more briefly, a *qualitative conclusion*.

There are exceptions: if the scale happens to measure physical quantities like population or raw materials, then it is permissible to restrict attention to *linear* changes of scale, but even then, if other non-physical scales are also involved, it still may only be valid to draw qualitative conclusions. Of course, in the physical sciences one is justified in restricting attention to linear changes of scale, due to the translational symmetries of space and time, and therefore one can expect physical laws to be expressed in quantitative language or formulae, for example like Boyle's law, $PV = RT$. Rutherford's remark that 'qualitative is just poor quantitative' does indeed contain an element of truth in the physical sciences.

However, it would be a fundamental mistake to expect to be able to express the laws of the social sciences in such language, because the language itself is not qualitatively invariant. It is equally mistaken to expect that since we cannot use quantitative language there are no laws. To state the laws of the social sciences we must expect to have to use qualitative language, in other words, mathematical terms that are invariant under qualitative changes of scale.

Of course, to *scientifically prove* any scientific law, social or physical, it is necessary to adopt the classical scientific method of choosing a quantitative model, predicting, and verifying experimentally.[5] To perform an experiment the social scientist must *quantify* each variable, that is to say, select a particular scale, and a particular method of measurement. Two different experimenters may select different scales that are only qualitatively related,[6] and therefore may finish up with graphs that are only qualitatively equivalent. *Therefore, although each experiment may produce a quantitative graph, only the qualitative properties of those graphs can be admissible conclusions.*

Compare again with the situation in physics: two physicists independently checking Boyle's law, for instance, may choose to use different temperature scales, centigrade and Fahrenheit, which have different origins and different sized units, but the essential point is that these two scales are linearly related. Therefore, when one experimenter obtains a straight line graph relating temperature and pressure (under fixed volume), he knows that the other experimenter will also obtain a straight line, and he is therefore entitled to attach significance to the straightness of it, and to express this fact in a formula.

By contrast, consider two social scientists performing experiments to check *Figure 6*, the dependence of action upon threat (under low cost). If one experimenter happened to quantify so as to obtain a straight line graph, then he would *not* be entitled to attach significance to the straightness of it, because he knows that the other experimenter may have chosen only qualitatively related scales, and consequently obtained a curved graph. Now the only qualitative properties possessed by *Figure 6* are single-valuedness, smoothness, and increasingness. Therefore, the only type of law that he could extract from *Figure 6* is that 'action is an increasing function of threat'. In essence this is what we have taken as our Hypothesis 2.

The social scientist may complain, with some justification, that this is a pretty feeble kind of law – and ask if this is the best example that qualitative mathematics has to offer? If we are not allowed to use formulae, what else has qualitative mathematical language to offer? Until recently the mathematician would have had to admit that there were only a few words in this language, because only a few[7] qualitatively invariant terms were known, terms such as 'greater than, increasing, maximum, single-valued, double-valued', etc. It is hardly enough to constitute a language. Not only that, but the terms themselves, although containing hidden subtleties for the mathematician, must have seemed rather unsubtle to the non-mathematician; so translucent, in fact, to our visual intuition, that they had already long been perceived, and incorporated into everyday language, well before the advent of mathematics. As a result, the laws that could be written in qualitative mathematical language were regarded as non-mathematical, because they could be translated into everyday language. For example, in our statement of Hypothesis 2, instead of using the qualitative mathematical language 'action is an increasing function of threat' we found it preferable to use the more familiar translation 'the greater the threat the greater the action'. As a result, qualitative mathematical language was, until recently, too limited and too obvious to be useful in the social sciences, except perhaps for expressing proverbs.[8]

Scientific statements that are written in terms too obvious are generally criticized for being both trivially true and trivially false, in spite of the fact that these criticisms are contradictory. Perhaps this is because of the swiftness with which the mind can leap on to the truth of the statement, and then leap off again to consider all the exceptions. On the other hand, if the statement is more subtle, packs more punch, incorporates more special cases, or more varied phenomena, contains more insight, has the power to arrest the mind with more surprise, then the mind is more ready to dwell upon the statement, sufficiently long perhaps to admit that it might be called a law (or a proverb), and to forgive the exceptions by renaming them as modifications. For example, this is certainly the case with Boyle's law, which is patently false near the critical point of a gas, and therefore needs to be modified as Van der Waals's equation (Fowler 1972); but in spite of this we still call it a 'law' because it still packs the punch.

A scientific law is an intellectual resting point. It is a landing that needs being approached by a staircase, upon which the mind can pause, before climbing further to seek modifications.

Summarizing what we have said so far: the qualitative mathematical language is the natural language for expressing the laws of the social sciences, but until recently it was useless.

Now, with the advent of catastrophe theory, this language has suddenly been unexpectedly enriched in two vital ways. The language has been transformed from being useless to becoming potentially useful for expressing rigorously an unsuspected array of laws in all the social sciences (as well as the biological and physical). First, catastrophe theory has contributed several new qualitatively invariant terms, such as 'fold-point, catastrophic change, cusp catastrophe, divergence, normal factor, splitting factor, bias factor, butterfly catastrophe', etc. Second, these new terms are subtle, and have no familiar translation into everyday language. Therefore, the new scientific statements that can be made using them have new power to synthesize ideas, to lend new insight, and to arrest the mind, perhaps sufficiently to be called laws.

10 EQUIVALENT GRAPHS

So far, we have used two different types of diagrams, probability distributions as in *Figures 1–5*, and cause-effect or control–behaviour graphs as in *Figures 6–8*. In catastrophe theory we are particularly interested in multivalued control–behaviour graphs, as in *Figures 7, 8*, and wish to study their qualitative properties. Therefore, we introduce a definition of equivalence between two such graphs, that is slanted towards catastrophe theory in the sense that it will later enable us to state the theorems precisely.

Definition: Given two planar graphs G, G' we say they are qualitatively equivalent if there is a diffeomorphism[9] of the plane that maps vertical lines to vertical lines and maps G to G'.

Example 1 Figure 9 illustrates two graphs that are qualitatively (but not quantitatively) equivalent.

Example 2 Let C, C' be the two horizontal scales measuring control, and let X, X' be the two vertical scales measuring behaviour. Let $\Psi C: C \to C'$ and $\Psi_X: X \to X'$ be two qualitative changes of scale. Let Ψ denote the product map $\Psi = \Psi C \times \Psi_X$. Let $G' = \Psi G$. Then G is qualitatively equivalent to G' by the diffeomorphism Ψ. Notice that in this example not only are vertical lines mapped to vertical lines, but also horizontal lines to horizontal lines.

Remark 1 In general, the diffeomorphism Ψ in the definition can have an extra freedom over and above that given by a product of two

FIGURE 9   Qualitatively equivalent graphs

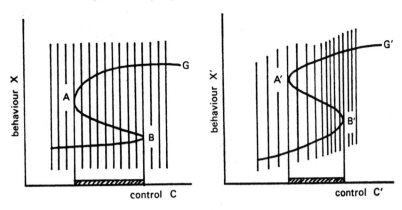

qualitative scale-changes, because the definition is biased in favour of verticals rather than horizontals. There are two reasons for this extra freedom, as we now explain. Since vertical lines are parametrized by the horizontal control axes C, C', and since Ψ maps vertical lines to vertical lines, Ψ induces a unique map $\Psi C: C \to C'$. If, further, ΨC preserves orientation then ΨC is a qualitative scale-change. On the other hand, Ψ does not induce a unique scale-change on the vertical behaviour axis, but rather a smooth family of scale-changes, one for each vertical line. The main reason for allowing the extra freedom of varying vertical scale-changes is that it is mathematically necessary for the main classification theorem of catastrophe theory (see Mather 1969; Thom 1972; Trotman and Zeeman 1974; and Sections 11, 13 below). But also, the extra freedom can be given sociological meaning, because different experimenters may prefer to vary their behaviour scales differently under different conditions of control.

Remark 2   Since qualitative changes of scale are a special case of equivalence, properties that are invariant under equivalence are qualitative properties.[10] *Figure 9* above illustrates two such properties, *fold-points* and *multi-valuedness*. Fold-points of G are where the vertical lines are tangent to G, and are invariant because smooth maps preserve tangency. Therefore, the fold-points A, B of G are mapped to the fold-points A', B' of G'. Consequently, the shaded interval of C, over which G is three-valued, is mapped to the shaded interval of C', over which G' is three-valued; the complement where G is single-valued is mapped to the complement where G' is single-valued. In Section 8 above we have already seen the importance of fold-points in applications, representing thresholds of catastrophic change, while the shaded interval represents the extent of delay.

Remark 3 Fold-points and multi-valuedness are the only qualitative properties of planar graphs that we shall use. We are now ready to pass on to higher dimensions, representing situations where there are two or more control factors influencing one or more modes of behaviour. The crucial question is: what are the higher-dimensional analogues of fold points? Thom (1972) calls these *elementary catastrophes*. His remarkable achievement was to recognize them and classify them for up to five control factors. Siersma (1973) has extended the classification to eight control factors, and higher dimensions are currently being studied.

Remark 4 One advantage of the qualitative point of view is mathematical simplicity. For, when studying qualitative properties of an application, instead of having to work with an analytically awkward graph (such as G in *Figure 9*), which may have been thrown up by the statistical data from experiment, we are at liberty to replace it by any equivalent graph (such as G'). Generally, the most convenient procedure is to choose a graph G' that is algebraically the simplest, given by a polynomial equation: we call this graph a *canonical model*.

Remark 5 Canonical models are particularly useful in higher dimensions. From the quantitative point of view, although it is fairly easy to comprehend and visualize the infinite variety of two-dimensional graphs, in higher dimensions the complexity coupled with the difficulty of visualization might well cause the mathematician to despair. He finds it difficult to comprehend them, let alone clasisfy them. However, the qualitative point of view turns this despair into delight, because the problem is now largely solved. Locally, any graph is equivalent to one of the elementary catastrophes. Therefore, it suffices to look at canonical models of the elemetary catastrophes. There are only a few of these, and they provide the new qualitative language. This language enables us to describe how two or more control factors can interrelate and interfere with one another in influencing behaviour. If the social scientist wishes to acquire the language in order to frame rigorous laws, all he need do is to master the geometry of the elementary catastrophes (Poston and Woodcock;

Thom 1972), the two most important of which, the cusp and the butterfly, we describe in Sections 11 and 13 below.

We must now extend the definition of equivalence from two-dimensional graphs to higher dimensional graphs. First consider the extension to three dimensions. Here we envisage a situation where two control factors are influencing one behaviour mode, similar to the application in Section 6 above of cost + threat influencing action. For convenience we represent the two control factors as horizontal axes, and the behaviour mode as the vertical axis. The graph will be a surface in three-dimensional space, R^3. The definition follows almost word for word the two-dimensional case.

Definition: Given two graphs G, G' in R^3 we say they are qualitatively equivalent if there is a diffeomorphism of R^3 that maps vertical lines to vertical lines, and maps G to G'.

Now, consider the general case, allowing for arbitrarily many controls and modes of behaviour. Suppose there are k scales or factors of control, so that the control spaces C, C' are represented by k-dimensional Euclidean space $C = C' = R^k$. Suppose that $X = R^n$, representing n scales or modes of behaviour, and $X' = R^{n'}$, representing n' modes of behaviour, where n, n' may, or may not, be the same. The control–behaviour graphs G, G' will be k-dimensional manifolds in $C \times X = R^{k+n}$, $C' \times X' = R^{k+n'}$. Call the n-planes, n'-planes parallel to X, X' in $C \times X$, $C' \times X$, *vertical*.

Definition: Given two graphs G, G' in $C \times X$, $C' \times X'$ we say they are qualitatively equivalent if there is a smooth map $C \times X \to C' \times X'$ that maps vertical n-planes to vertical n'-planes, and induces diffeomorphisms $C \to C'$ and $G \to G'$.

The reader may ask why in the general case do we allow n to be different from n'. There are two reasons. First, mathematically it enables the classification theorems below to be stated in a much more powerful form. Second, in applications we may have n > 1, with X representing several behaviour modes simultaneously, while n' = 1 and G' is a canonical model. Therefore, the use of canonical models does not impose any restriction upon the description of behaviour. For an example of this in an application see Section 15 below.

Remark We emphasize that all we have said about equivalence refers only to control–behaviour (or cause-effect) graphs. Before concluding this section, a word of caution needs to be said about probability distributions, which are the other types of diagram that we have been using. The main point to be made is that a probability distribution is a quantitative tool, and part of the experimental procedure leading to the presentation of data in the form of a control-behaviour graph.

$$\text{Experiment} \xrightarrow{\text{quantitative}} \begin{array}{c}\text{Control}\\ \text{behaviour}\\ \text{graph}\end{array} \xrightarrow{\text{qualitative}} \text{Conclusion}$$

Therefore in general it is meaningless to apply qualitative arguments to raw probability distributions. We illustrate this point by stating a sociologically meaningless mathematical lemma:

Given any raw probability distribution P on a scale X, and given any interval I of X, then it is possible, by shrinking I to I', to choose a qualitatively related scale X', such that the new distribution P' has a new maximum in I', as high as we please.

Proof: $\int_I P = \int_{I'} P'$. See *Figure 10*

FIGURE 10 Probability distributions (as opposed to support functions)

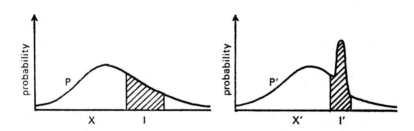

On the other hand, if two experimenters took a scientific hypothesis, tested it by experiment, chose qualitatively related scales, collected statistics giving raw probability distributions, chose methods of weighting and adjusting their data that seemed to them the most accurate way of transforming the raw material into qualitatively meaningful distributions of public support for the various policies, and finally produced the two distributions in *Figure 10*, then this would be another matter. This would be an experimental result that threw doubt upon the scientific hypothesis.

Here in this paper, we are not presenting experimental data, and therefore not giving the quantitative half of the process, but we are concerned with presenting the qualitative half. The ambiguity suggested by *Figure 10* does not enter into our discussion, because in the sociological hypotheses, in Section 7 above, we have assumed that opinion is either unified or split, that is to say, the distribution has one or two maxima, depending upon the control conditions. Therefore, we can proceed rigorously from these hypotheses to the conclusion in Section 12 below. To test the theory, the social scientist would have to test the hypotheses by experiment, or by interpretation of historical data.

11 THE CUSP CATASTROPHE

The simplest elementary catastrophe is the fold-point. The next simplest is the *cusp catastrophe*, and we now describe its canonical model (Thom 1969; 1972). For the benefit of those readers who have not met it before we describe it in some detail, and also pedagogically recommend reading (Zeeman 1971; 1972b).

Let M be the cubic surface in R^3, given by the equation:

$$x^3 = a + bx.$$

Here a,b are horizontal control axes, and x the vertical behaviour axis, as in Section 6 above. The surface is illustrated in *Figure 11*, where, for convenience the control space C is drawn as a horizontal plane below the origin (rather than through the origin). The *fold curve* F is where vertical lines are tangent to M, and is given by differentiating with respect to x:

$$3x^2 = b.$$

The projection of F down onto the control space is called the *bifurcation set* B. Although F is a smooth cubic curve, B has a cusp at the origin, and that is where the name[11] comes from. The equation of B is given by eliminating x from the two equations above:

$$27a^2 = 4b^3.$$

F separates M into two pieces, both of which have F as their common

FIGURE 11 The cusp catastrophe

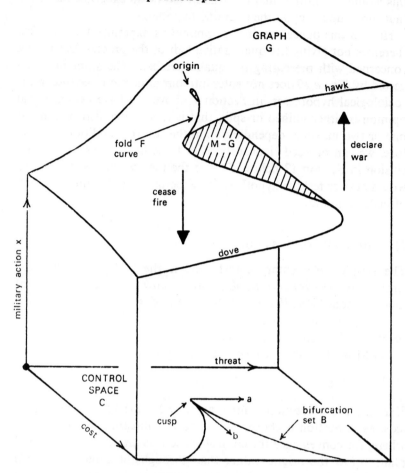

boundary. The larger piece, given by $3x^2 \geq b$, is the *graph* G that we want. G is single-sheeted outside the cusp, and double-sheeted inside the cusp, where the upper and lower sheets, marked hawk and dove, overlap. The smaller piece, M–G, given by $3x^2 < b$, and shown shaded, consists of a single sheet over the inside of the cusp, in between the hawk and dove sheets. The piece M–G is sociologically meaningless like the dotted part of *Figure 7*, and so, from now on, we concentrate on G rather than M.

There are many[12] probability functions P giving rise to this particular graph, for which G is the set of maxima, as defined in Section 7; M–G is the set of minima, and M is given by the equation

$\frac{\partial P}{\partial x} = 0$. However, we are less interested in the probability function than the graph, because the probability function is a quantitative tool, part of the experiment, leading not to the canonical model, but only to an equivalent graph. We are more interested in the qualitative properties of the canonical model, because these properties are automatically shared by both graphs. We have already mentioned some qualitative properties, for example:

(i) The projection B of the boundary F of G has a cusp.
(ii) B is the set of control points over which the qualitative type of distribution changes, or bifurcates.
(iii) G is single-sheeted outside B, and double-sheeted inside.

Now, consider the plane sections of G given by b = constant. For b < 0 the sections are equivalent to *Figure 6*, and for b > 0 to *Figure 7*. Therefore, the cusp catastrophe answers the question how these plane sections can evolve into one another. The deep result is the uniqueness of that evolution, which follows from the main classification theorem below. But first we verify that the evolution can be written in invariant qualitative language.

In the control space define the *splitting factor* to be the direction of the axis of the cusp (at the cusp point). Define a *normal factor* to be any transverse direction, oriented towards a > 0. This gives invariant definitions of normal and splitting factors, that agree with the a,b–axes in the canonical model. In a neighbourhood of the origin any section of G transverse to the splitting factor, b, will be smooth increasing if b < 0, and split if b > 0. The splitting causes the fold curve to appear and hence the resulting catastrophic effects. We are now ready to state the main theorem.

Classification Theorem 1: Let $C = R^2$, $X = R$, P be any smooth generic probability function on $C \times X$, and G the resulting graph in R^3. Then the only singularities of G are fold-curves and cusp-points.

Remark 1. Here, by *singularity*, we mean a singularity of the projection of G onto the control space, C, in other words, a point where the projection is not locally a diffeomorphism. The theorem is so called because it classifies the only two elementary catastrophes that can occur with two control factors. The theorem remains true, word for word, if $X = R^n$, in other words, we allow simultaneously arbitrarily many modes or scales of behaviour. In Section 13 below we state the analogous result for four control factors.

Remark 2   The theorem and conception of proof are due to Thom (1969; 1972) and the details of proof to Mather (1969). See Trotman and Zeeman (1974). The proof is much more complicated than the statement, but the latter can be used without necessarily knowing the proof.

Remark 3   The graph G in the theorem may be many-sheeted, and may have several fold-curves and cusps. But all the fold-curves are alike, and each cusp-point is locally-equivalent[13] to the cusp catastrophe or its dual.[14] Hence, the cusp catastrophe is the most complicated situation that can happen locally. It is stable and persists under small perturbations of P. It is the unique way a sheet can split, and therefore the unique way that continuous sections can evolve into split sections.

FIGURE 12   Normal and splitting factors: conflicting factors

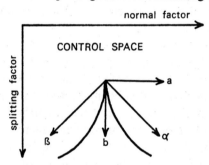

Notation   If there is only one cusp, and no other fold curves other than those associated with the cusp, then the graph is equivalent to the canonical model, and we can choose suitable axes a, b in the control space. We abbreviate the statement of equivalence by saying:

a *is a normal factor and* b *a splitting factor influencing* x.

In some applications one does not use the splitting factor, but rather two normal factors either side of the cusp, such as $\alpha = b + a$, $\beta = b - a$.
In this case we say

α, β *are conflicting factors influencing* x.

The reason for this terminology is that the control α tries to push the behaviour on to the upper surface, while β tries to push it on to the lower, and inside the cusp the two controls conflict. Examples of this nature are 'rage and fear are conflicting factors influencing aggression' (Lorenz 1966: 81; see Zeeman 1971), or 'pressure and temperature are conflicting factors influencing density' (which is Van der Waals's equation (Fowler 1972; 1–7)).

We emphasize that both of the above notational statements are precise qualitative statements that the graph of x as a function of the controls is equivalent to the cusp catastrophe.

Each statement can play several roles: it can serve as:

(i) a *theorem*, provable from simpler hypotheses, as in the next section,
(ii) a *hypothesis* (or conjecture) providing the theoretical framework for design of experiment,
(iii) a *law*, summarizing experimental results.

The statement itself is a synthesis of many ideas, with the following aspects:

Profundity due to the mathematical uniqueness and stability, depending on deep theorems.

Universality In any aspect of nature, or any scientific experiment, where two factors influence behaviour, where splitting and discontinuous effects are observed, and where smooth genericity may be assumed, the graph must contain the cusp catastrophe.

Insight From the model one can explain, predict, and relate a variety of phenomena that previously may not have appeared to be related. As an example we return to the problem on hand.

12 APPLICATION OF THE CUSP CATASTROPHE

Theorem. Threat is a normal factor and cost a splitting factor influencing the level of military action

We are entitled to label this statement as a *theorem* because it is an immediate corollary of the classification theorem and the graphs in *Figures 6, 7* arising from the five hypotheses in Section 8. Alternatively, we could label it as a *hypothesis*, and then the four previous sociological hypotheses would be corollaries, testable by experiment. An 'experiment' in this case might take the form of analysing the behaviour of a particular nation over a period, and would involve quantifying, that is to say inventing methods of measurement from historical date of indices of cost, threat, and action. If sufficiently many experiments were successful, that is to say, the results of different methods of measuring different nations were qualitatively equivalent, and the exceptions were explicable by suitable modifications, then the statement would become a candidate for a *law*.

We now illustrate the use of the statement to explain what happens under varying conditions of cost and threat upon a population, and the resulting behaviour of its administration. Consider the various cost–threat paths in the control space, illustrated in *Figure 13*. As the pressures on the population follow the path in the control space, the administration dutifully follows the path on the graph G, vertically above, obeying the Delay rule.

Path 1 A nation with plenty of resources feels increasingly threatened and, unified, gradually moves into moderate action. The resulting

FIGURE 13 Paths in the control space

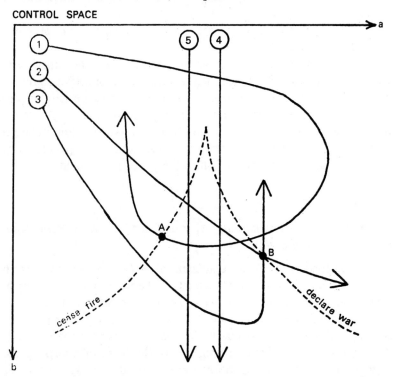

cost escalates, and as a consequence the action also gradually escalates. The threat subsides, and as a result opinion becomes divided between doves and hawks, but the administration's policy is caught entrenched in the hawk opinion, and delays until A before making the (catastrophic) decision to withdraw, after which the cost subsides. The catastrophic change of policy at A is precisely the catastrophe discussed in Sections 5, 9 above, and pictured in *Figures 5, 8*.

Path 2 A nation with more limited resources feels a similar increasing threat, but at the same time feels the escalating cost of military preparedness, and so delays any action until B, when it (catastrophically) declares war.

Path 3 A nation initially finds the cost of war prohibitive, but subsequently, due perhaps to the development or purchase of new defence weapons, may find that war becomes feasible. Therefore, to its enemy's surprise, it may suddenly declare war in the middle of a stable period of constant threat. This is called the *surprise reversal* phenomenon, and tends to appear in many applications, for example in committee behaviour (Zeeman 1973: Example 4).

Paths 4 and 5 Two nations confront each other, and, both feeling threatened, start a costly arms race. As a result, opinion in both nations becomes divided between hawks and doves. Although they may both start in moderately hostile moods that are very close, in one nation the hawks prevail, because path 4 passes just to the right of the cusp, and so this administration gradually adopts a more aggressive posture. Meanwhile, in the other nation the doves prevail, because path 5 passes just to the left of the cusp, and so that administration gradually adopts a posture of appeasement. Both changes are smooth, and this illustrates the phenomenon of divergence.

A similar phenomenon can occur when an international crisis blows up between two nations, or two groups of nations (or for that matter between any institutions where conflict can arise). During the short period of the crisis, lasting perhaps a month, or even as short as a week, the two administrations involved become increasingly aware of the cost of further escalation, and not infrequently the crisis is resolved by one administration standing firm, path 4, while the other backs down path 5. It is well known by diplomats that it is essential for the administration standing firm, path 4, to leave open for their opponents a clear avenue of retreat so that they can follow path 5, and not to succumb to the euphoric temptation to victoriously close all avenues, otherwise, by driving their opponents into a corner, they may marginally increase the feeling of threat experienced by the opposing population. This tiny margin may be sufficient to convert their opponents' path 5 disastrously into path 4. Then as both nations follow path 4, war will probably ensue.

Together, the five paths illustrate the close interrelation between discontinuous and divergent behaviour.

13   THE BUTTERFLY CATASTROPHE

The next sophistication in our model is to investigate the emergence of a compromise opinion, represented by the appearance of a new maximum in the probability distribution as in *Figure 14*.

FIGURE 14   Emergence of compromise opinion

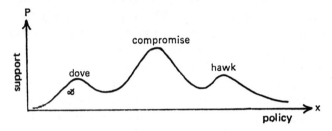

If opinion has become polarized between doves and hawks in a stalemate, then it is of importance to understand the mechanism underlying compromise. This same mechanism applies to nearly all social phenomena where compromise plays a part.

We have seen by Classification Theorem 1 above that mathematically the only way a two-peak distribution can evolve from a one-peak is by means of the cusp catastrophe, which required two control factors. The next theorem states that the only way a three-peak distribution can evolve is by means of the *butterfly catastrophe*, which requires four control factors. Indeed, in everyday experience, to hammer a compromise solution out of a polarized situation generally does need two more factors, some new ingredient plus time.

We now describe the canonical model (Thom 1972) of the butterfly catastrophe. We give it in a form that illustrates it as a generalization of the cusp catastrophe. The control space $C = \mathbf{R}^4$ of the butterfly has 4 control factors:

a = normal factor
b = splitting factor
c = bias factor
d = butterfly factor.

The behaviour space $X = R$, with

x = behaviour mode.

Let M be the four-dimensional manifold in $C \times X = R^5$ given by the equation

$$x^5 = a + bx + cx^2 + dx^3.$$

Let G be the four-dimensional submanifold of M given by

$$5x^4 \geqslant b + 2cx + 3dx^2.$$

Then G is the desired graph of the butterfly catastrophe. Since we cannot draw a five-dimensional picture, the best way to understand the qualitative properties of G is to first draw in *Figure 15* various

FIGURE 15 Sections of the butterfly catastrophe bifurcation set

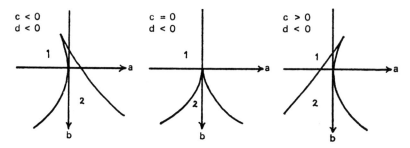

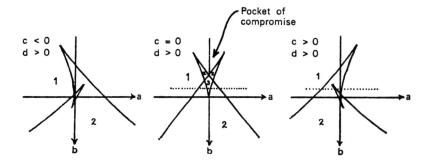

two-dimensional sections of the control space C. Each section is an (a,b)–plane drawn for (c,d) = constant, and illustrates what the bifurcation-set looks like in that section.

Notice the effect of the bias factor, c. When d < 0, the effect of the bias is to bias the position of the cusp, as shown in the top three pictures. When the bias is positive the cusp is biased in the positive direction of the normal factor, and vice versa.

Now, consider the effect of the butterfly factor, d. When this comes into play, d > 0, the effect is to bifurcate the cusp into three cusps, as

FIGURE 16   Section of the butterfly catastrophe

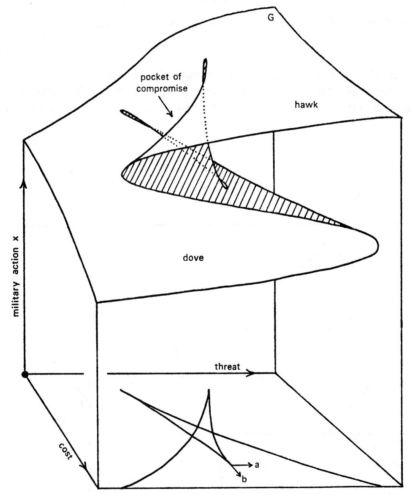

shown in the bottom three pictures. The number in each region indicates the number of peaks in the corresponding probability distribution, and hence the number of sheets of G over that region; as the control point crosses the bifurcation-set the number changes because the distribution bifurcates.

If one turns the central picture upside down it looks a bit like a butterfly, which explains the name. The V-shape in this picture represents the emerging pocket of compromise opinion. There are in fact two Vs, an outer-V with vertex at the cusp at the origin, and an inner-V with vertex at the intersection point above the origin. The compromise maximum exists at all points inside the outer-V, and is the unique maximum at all points inside the inner-V. In between the two Vs, the compromise is competing with the dove and hawk maxima, and inside the little diamond shape all three maxima are competing.

To illustrate this, we now draw the corresponding control–behaviour graph in *Figure 16*. We add the x–axis as a third dimension to the (a,b)–plane, and draw the section of the graph G above this particular section of C. Comparing *Figure 16* with *Figure 11* shows how the pocket of compromise is emerging in between the hawk and dove sheets. Notice that in *Figure 16* each of the two upper cusps is a cusp catastrophe, while the lower cusp is a dual-cusp catastrophe.

The effect of a positive bias, $c > 0$, on *Figure 16* is to distort it and raise it all slightly in the x-direction. This effect is so important in applications that we state it as a lemma and prove it.

Lemma 1: *The effect of more bias is to make behaviour more aggressive.*

Proof Let $c_1 < c_2$ be two values of bias. Fix the values of b, d and consider the two graphs M_1, M_2 of a_1, a_2 as functions of x given by

$$a_1 = x^5 - bx - c_1 x^2 - dx^3$$
$$a_2 = x^5 - bx - c_2 x^2 - dx^3.$$

Therefore

$$a_1 - a_2 = (c_2 - c_1)x^2$$
$$> 0, \text{ for all } x \neq 0.$$

Therefore

$$a_1 > a_2, \text{ except when } x = 0.$$

Now, look at these same two curves M_1, M_2 the other way round as graphs of x_1, x_2 as functions of a given by

$$x_1^5 = a + bx_1 + c_1 x_1^2 + dx_1^3$$
$$x_2^5 = a + bx_2 + c_2 x_2^2 + dx_2^3.$$

Restrict attention to the subgraphs G_1, G_2 given by

$$5x_1^4 \geqslant b + 2c_1 x_1 + 3dx_1^3$$
$$5x_2^4 \geqslant b + 2c_2 x_2 + 3dx_2^3.$$

These are the two control–behaviour graphs corresponding to values of bias c_1, c_2. Since $a_1 > a_2$, we see that $G_1 < G_2$ in the sense that for each point (a, x_1) in G_1 there exists a point (a, x_2) in G_2 such that

FIGURE 17 Sections of the butterfly catastrophe

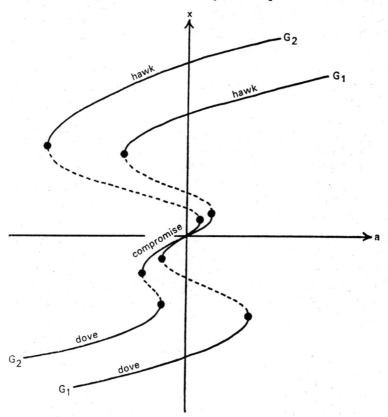

$x_1 < x_2$ (except possibly when $x_1 = x_2 = 0$). Therefore the higher bias c_2 causes behaviour G_2 that is more aggressive than G_1, as required.

Figure 17 illustrates the graphs $G_1 < G_2$ for particular values $b > 0$, $c_1 = 0 < c_2$, and $d > 0$, corresponding to the dotted horizontal lines in the two bottom right pictures of *Figure 15*. If the paper is turned round so that the a-axis if vertical, one can see how the result is caused by the formula

$$a_2 = a_1 - (c_2 - c_1)x^2.$$

Before concluding our description of the butterfly catastrophe we make a small qualitative point about the butterfly factor, d. In many applications d = time, and so at d = 0 the compromise begins, and as time progresses the pocket of compromise grows. In the lower central picture of *Figure 15* the lower cusp is fixed at the origin a = b = 0, and the pocket grows by the two upper cusps moving upwards and outwards. Qualitatively, we could just as well choose to have the pocket growing by the upper cusps moving outwards along the a-axis and the lower cusp moving downwards along the b-axis. In some applications this kind of growth makes more sense because the compromise pocket then gradually takes over the whole plane. In other applications, as in real life, the compromise can grow lopsidedly, or can wax and wane, and each of these growths can be achieved by choosing a suitable time-axis in the canonical model. For example, the downwards growth can be achieved by choosing a curvilinear time-axis along the parabola $(a,b,c,d) = (0, -\frac{9t^2}{20}, 0, t)$.

Qualitatively, there is nothing to choose between a straight or a curved time-axis, and since this parabola has the d-axis as tangent at the origin we are still entitled to refer to d as time.

This completes the description of the butterfly catastrophe sufficient for our applications. More accurate, computer-drawn, quantitative pictures of the canonical model can be found in Poston and Woodcock (1974). The reader who is primarily interested in the applications may now proceed to the next Section; see also Zeeman (1973). But before we leave this section, in order to highlight the fundamental importance of the butterfly, and put it in its correct perspective, we would like to state a four-dimensional version of the Classification Theorem. Thom (1969; 1972) proved that there are seven elementary catastrophes for a four-dimensional control space (together with three duals). However, we maintain that only three of these apply to probability distributions, the fold, the cusp, and the butterfly, and hence these are the most important catastrophes for the social sciences. Therefore, we introduce a definition that singles out the cusp and the butterfly.

Definition: Let G be the graph of maxima of a probability function on $C \times X$. Let Q be a point on G, and let Q' be the projection of Q in the control space C. We say Q is complete *if any neighbourhood of Q in G projects onto a neighbourhood of Q' in C.*

Example (i) Any point not on the boundary of G is complete.

Example (ii) The origin Q of the cusp catastrophe is complete, because although Q lies in the boundary F of G (see *Figure 11*), the projection into C of any neighbourhood of Q in G overlaps itself, and wraps round to form a neighbourhood of the cusp point in C.

Example (iii) The origin of the butterfly catastrophe is complete, for the numbers in *Figure 15* show that there is at least one sheet of G over each region.

Example (iv) The fold-point is incomplete. For instance taking $Q = A$ in *Figure 7*, a small neighbourhood of Q in G projects down to only the left hand side of Q', which is not a neighbourhood of Q' in C.

We have explained, in Sections 5 and 9 above, the importance of fold-points in initiating catastrophic change; therefore, a classification of only the complete catastrophes might at first sight appear inadequate, because it leaves out fold-points. However we suggest that, in graphs derived from probability functions, fold-points do not appear in their own right, as it were, but only as satellites of the cusp and butterfly. We make this argument precise, as follows.

Define a graph G to have a *centre* Q if G is equivalent to arbitrarily small neighbourhoods of Q. Intuitively, this means the global graph G looks like the local graph at Q. For example, both cusp and butterfly have a centre at the origin; in fact all the elementary catastrophes have a unique centre (Thom 1972). In applications the notion of a centre represents being able to capture the heart of the problem in microcosm at one point. If a graph has a complete centre we call the graph complete.

Lemma 2: Suppose G is the graph of maxima of a probability function. If G has a centre then G is complete.

Proof For each control point c, the probability distribution P_c has at least one maximum. Therefore, G has at least one sheet over each control point. Hence, the projection of G covers C. Given any neighbourhood of the centre Q, we can choose a smaller neighbourhood N equivalent to G, since Q is the centre. Since the projection of G covers C, the projection of N covers a neighbourhood of Q'. Therefore Q is complete. Therefore G is complete.

Having justified the concept of completeness we can now state the classification.

Classification Theorem 2: Let $C=R^4, X=R^n$, with n arbitarary, P be any smooth generic probability function on $C \times X$, and G the resulting graph in R^{4+n}. Then G is a four-dimensional manifold and the only complete singularities of G are cusp-surfaces and butterfly-points. For the proof see (Mather 1969; Trotman and Zeeman 1974).

Remark 1 Compare with Classification Theorem 1 in Section 12 above; had we added the word 'complete' to Theorem 1, this would have ruled out fold-curves. But, just as fold-curves appear as satellites of cusp-points, so do swallow-tail-curves appear as satellites of butterfly-points (see Thom 1972).

Remark 2 Folds are intrinsically co-dimension one, and so in Theorem 1 in a two-dimensional graph they appear as fold-curves. Similarly, cusps are intrinsically co-dimension two, and so in Theorem 1 in a two-dimensional graph they appear as cusp-points, while in Theorem 2 in a four-dimensional graph they appear as cusp-surfaces. Meanwhile, butterflies are intrinsically co-dimension four, and so they cannot appear at all in Theorem 1, while in Theorem 2 they appear as butterfly-points. In general, a catastrophe of co-dimension c appears over k controls as a (k–c)-dimensional submanifold of the k-dimensional graph.

Remark 3 For more than four controls a new complete cuspoid appears in each even dimension. In dimension eight another phenomenon appears, the *double-cusp*, which is complete and requires two behaviour modes, and has a wealth of incomplete satellites of lower co-dimension including the umbilics (Siersma 1973; Thom 1972). Although the umbilics have co-dimensions three and four, they do not appear in four-dimensional central probability graphs.

14 COMPROMISE OPINION

Just as our knowledge of the uniqueness of the cusp over two controls gives confidence to postulate the cusp whenever discontinuous or divergent behaviour appears, so the uniqueness of the butterfly over four controls gives confidence to postulate the butterfly whenever compromise opinion appears. We shall now generalize our previous model in this direction. Previously, we deduced the cusp from simpler hypotheses, in order not only to usher our readers more gently into the ideas of catastrophe theory, but also to provide a template for them to manufacture their own applications. This time, we go with confidence straight to the heart of the matter, and take the butterfly as hypothesis.

Hypothesis 6
Threat is a normal factor
Cost is a splitting factor   *influencing the level*
Invulnerability is a bias factor   *of military action.*
Time is a butterfly factor

First, observe that with average vulnerability ($c = 0$), and at the beginning of a war ($d < 0$), we obtain the top central picture of *Figure 15*, which is the ordinary cusp, the same as in *Figures 11 and 13*. Therefore, in these circumstances, our new model reduces to the old model, and so is a strict generalization or modification of the latter.

Next vulnerability. Why choose vulnerability? Vulnerability is in fact one of the strongest influences modifying public opinion. For example, if an industrial nation feels dangerously threatened, its population may call for strong action; however, if its enemy possesses overwhelming air superiority, then strong action may not be feasible due to the vulnerability of its cities. The public will be aware of this vulnerability, and may consequently modify its opinion in favour of appeasement. By contrast a predominantly agricultural community in suitable terrain may feel no such inhibition, due to the feasibility of guerrilla warfare. The more invulnerable the terrain, the more aggressively the population will modify its opinion. In other words, by Lemma 1 of Section 14, invulnerability behaves like a bias factor. Therefore, let c be some index or scale measuring invulnerability.

Now time. Public opinion does modify with time, especially in a stalemate war of moderate threat (a = 0), moderate cost (b=0), and moderate vulnerability (c=0). People are liable to become weary of the war, and a compromise opinion may emerge in favour of a negotiated cease-fire, an opinion in between, and possibly sharply divided from, both hawks, who demand victory, and doves, who counsel appeasement or withdrawal. Therefore, time behaves qualitatively like a butterfly factor, d, and for d > 0 the V-shaped pocket of the lower centre picture of *Figure 15* begins to appear. Moreover, the longer the stalemate, the more of the population who adopt the compromise opinion, the higher grows the centre peak of the probability distribution, and the larger grows the compromise pocket in the threat-cost plane. Eventually the inner-V of the pocket crosses the particular point of that plane representing the current threat+cost, and at that moment the administration fighting the war will make the (catastrophic) switch to the compromise policy, and begin to negotiate for peace. It might save many lives if administrations understood sufficiently to begin negotiation at the earlier moment, when the outer-V of the pocket crossed the point, instead of delaying to the last possible moment.

15 OPPOSITION TO A GOVERNMENT

In our first model, we studied the opinion of a population opposing an external threat. We now study a second model of the opinion of a population opposing its own government, which is quite different. In our first model, we assumed that some local maximum of opinion was actually *realized* by an administration obeying the Delay rule. In the second model, the opinion is not realized, because it only represents what the population would like to do if it had the chance. If an election happened to take place, then the population might be able to realize its majority opinion, by the Voting rule; but most of the time no election is in the offing, either because the election rules may only allow for an election every four or five years, or because elections may have been abolished altogether, and so any antagonistic majority of opinion may only represent a vague threat against the government. Therefore, for perhaps economic reasons, a government may find itself in the position of operating policies against the wishes of most of those governed, and for this reason, in this model, we prefer to use the word *government* rather than administration.

Let x be some index or scale measuring the opposition to the government. In a sample survey of the population, in order to discover the probability distribution of x, one might design a *questionnaire*, showing a *continuous* scale, with varying degrees of support or opposition written against various points on the scale, and ask each person surveyed to position himself approximately on this scale. For instance, the scale might read:

```
         ⎡  Prepared to donate money to re-election fund.
         ⎢  Would canvas for government.
         ⎢  Would vote for government.
         ⎢  Would vote against other parties.
    x_N  ●  Neutral.
         ⎢  Would vote against the government.
         ⎢  Would vote for another party.
         ⎢  Would canvas for another party.
    x_S  ●  Prepared to strike against the government.
         ⎢  Prepared to join a civil disobedience campaign.
         ↓  Prepared to wage guerrilla war against the government.
         x
```

From the government's point of view, the two critical points on this scale are the neutral point, x_N, and the strike point, x_S. For imagine a government that is enjoying a unified opinion in its favour, in other words the distribution of opinion has a single maximum at x_0, where $x_0 < x_N$. Suppose that the government now runs into economic difficulties, which we shall later assume to be a normal factor influencing opposition. Then x_0 will begin to move in the direction of x increasing. As x_0 crosses x_N the government knows that it has lost majority support, and might lose the next election, but nevertheless it may hopefully pursue economic policies in order to remedy the situation. However if x_0 crosses x_S then the majority of the population may be prepared to strike against the government, thereby wrecking its economic plans, and possibly ruining its chances of re-election.

Some readers may protest against the inclusion of industrial action on a political scale, but there are reasons for doing so. First, this is a model, as we shall see, primarily about the government's reactions; and, looking through government eyes, although the industrial action may be directed towards higher pay, or against industrial management, nevertheless the government may see the importance of such action primarily as wrecking its own economic policies. Indeed, for this very reason, many governments in the world today have ruled strike action to be illegal.

Second, admittedly it might seem a more precise procedure to separate industrial and political action as two different modes of behaviour, in other words use a two-dimensional[15] behaviour space X rather than a single scale. But if we did so, then qualitatively we could arrive back at *exactly the same* graph G over C, or more precisely, an equivalent graph, by the definition of equivalence in Section 11, and the statement of Classification Theorem 2 in Section 14. This is one of the remarkable features of catastrophe theory, that enables us to *implicitly* allow for as complicated a modality of behaviour as we wish, and yet *explicitly* worry only about the control factors. We are now ready to state the main hypothesis of the model, after which we explain the meaning of the controls.

Hypothesis 7
Economic malaise is a normal factor
Promise of reform is a splitting factor } influencing the opposition
Failure of reform is a bias factor to a government.
Pressure is a butterfly factor

Economic malaise is caused by such things as chronic inflation, unemployment, depressions, industrial strife, strains of development, shortages and balance of payment problems. If the government fails to cure these ailments, then malaise will erode confidence, and the majority of opinion x_0 will begin to drift towards opposition. Thus, malaise acts as a normal factor influencing opposition.

To stop the drift the government may promise reforms. But such promises have a splitting effect, between those who approve and disapprove of the reforms. For example, promise of land reform may appeal to peasants but not to landlords, and promise of industrial reform may appeal to management but not to unions. Thus, the single maximum of opinion x_0 may split into two maxima x_1, x_2, where $x_1 < x_N < x_2$. Therefore, initially we have the cusp catastrophe by Theorem 1.

If the malaise persists, the government may, as the control point approaches the cusp, desperately promise more and more reforms, in the hope of at least retaining some local maximum of support,

$x_1 < x_N$, staving off the moment when x_1 disappears, and opposition is unified at $x_2 > x_N$, as illustrated by the point B in *Figure 18*. But, alas, the greater the promise of reform, the greater the opposition, and the greater the eventual downfall of government.

Eventually, the population may agree to suspend the electoral rules, so that a unified opposition $x_2 > x_N$ does not require the resignation of a government, and so that at least some government may be given sufficient breathing space to tackle the economic problems on a long-term basis.

FIGURE 18 Government trying to avoid catastrophe

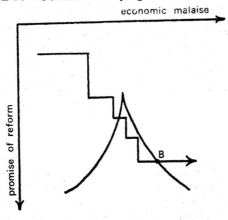

However, this may take a long time. The apparent initial failure of promised reform may act as a bias factor, biasing public opinion yet further against the government. The critical moment occurs when x_2 approaches x_S, because the resulting strike action may have the potential to wreck the long term plans for economic recovery. It is then that the government may feel obliged to exert pressure upon the population, in order to bring the majority opinion back, $x_2 < x_s$. But this in turn may have the effect of splitting off an extremist wing, with opinion centred at a new maximum, $x_3 > x_s$. Thus, we are liable to reach a three-peak probability distribution as shown in *Figure 19*.

By Theorem 2 we must have the butterfly catastrophe, and the butterfly factor that caused the middle peak to appear was the pressure. This time the middle point x_2 does not represent compromise opinion, but rather the opinion of the silent majority, in silent opposition, $x_N < x_2 < x_s$. Eventually, if the economy picks up

FIGURE 19 Public opposition to the government

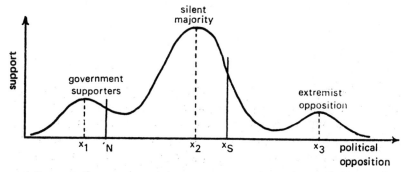

and the malaise subsides, the majority may move back to supporting the government, sufficient for the latter to restore the electoral rules.

Summarizing, the single hypothesis 7, supported by the uniqueness of the butterfly catastrophe, synthesizes a complex behavioural response, and lends itself to the design of many experiments to test it. One of the interesting features of this example is that, while two of the control factors, a and c, depend upon external forces outside the government's jurisdiction, the other two, b and d, are actually under government control, and behave as responses to the stimulus of the first two: a stimulates response b, and c stimulates response d. It is an interesting question as to whether the mathematical ordering of the factors, indicated by the alphabetical ordering, and induced by the Taylor expansion of the canonical probability function, whether or not this ordering can be given a more general interpretation of stimulus and response, beyond the confines of this particular application.

16 CENSORSHIP

To introduce our last example we quote the title of an article by John Wilson. 'The pendulum of censorship or anti-censorship fashions may swing to and fro; but it is not clear that any serious advance in the topic has been made since the days of Plato and Aristotle' (Wilson 1970). In the article Wilson lays his finger on the two criteria that are involved in deciding whether or not to censor a book or a work of art, namely *erotic content* and *aesthetic value*, and he concludes by suggesting: 'But before striking attitudes, and even before collecting more statistics, we need a better conceptual understanding of just how either effect operates; otherwise we shall not know what sort of facts to look for' (Wilson 1970). We suggest that the problem is not how either effect operates by itself, but how the two effects together interfere with one another. By now, the reader will have recognized them as control factors, and have guessed that the cusp catastrophe is going to be involved. But first we must identify the behaviour mode.

The key to the problem is to replace the two states 'censored' and 'uncensored' by a continuous scale of behaviour. Many problems in the social sciences can be unlocked by the same key: the problem may seem to present a finite set of states, giving the appearance of discreteness, but in fact the apparent discreteness may result from discontinuities of an underlying continuous phenomenon, in other words from an elementary catastrophe. The elementary catastrophe may impart to the discrete set a subtlety that escapes description unless one uses the new qualitative language. A typical example is the 'attack' or 'flight' behaviour of an animal (Lorenz 1966; Zeeman 1971).

Here we approach the problem of censorship by designing a simple experiment, which we hope we might tempt one of our readers to perform. Design a *questionnaire* similar to that in the last model. We take the book or work of art into the market-place and ask a sample of passers-by whether it should be censored; we ask each person to place the work on the following continuous scale:

```
       x
       ↑    Definitely censor
       |    Probably should be censored
   O ●      Don't know
       |    Probably should not be censored
       |    Definitely do not censor
```

A probability distribution is obtained for this particular work. Plot the resulting maxima of probability on a three-dimensional graph, as behaviour against control scales measuring erotic content and aesthetic value. Now, do this for many works of art and we obtain a cloud of points, which we conjecture will cluster about a surface equivalent to the cusp catastrophe. More precisely:

Hypothesis 8. Erotic content is a normal factor and aesthetic value a splitting factor influencing opinion about censorship (see Figure 20).

If the work is of little aesthetic value then opinion is likely to be unified, that is to say, that probability distribution is likely to have a single peak, which is likely to give a good judgment of the erotic content relative to the prevailing norm in the market-place at that time. In other words, erotic content acts as a normal factor.

On the other hand, a work of high aesthetic value may well give rise to a double-peak distribution. Some people may feel it should be uncensored because of its high aesthetic value, even though its erotic content may transgress the prevailing norm; others may feel it should be censored because its very aesthetic value endows it with too dangerous an influence, even though its erotic content may be within the prevailing norm. In other words, aesthetic value acts as a splitting factor. The resulting graph is illustrated in the top picture of *Figure 20*; notice that in this picture we only draw the graph G, and omit the middle sheet M − G of *Figure 11*.

For simplicity of discussion let us now assume the graph is not only equivalent to, but actually equal to, the canonical cusp catastrophe. Also, suppose the prevailing censorship-norm is $x = 0$, in other words $x = 0$ is the 'Don't know' position on the *questionnaire* scale. Consider the task of a censor, who is trying to adhere to the prevailing norm.

Define the *type* of a work to be its control point, in other words, the point in the control space that represents its particular erotic content and aesthetic value. Admittedly, it may be difficult to get critics to agree on the type of a work, but for simplicity of discussion suppose that we can. In the control space there is a cusp, as a consequence of Hypothesis 8. If the type of work lies outside the cusp the censor's job is easy; opinion is unified with maximum at x_1, say, and he can censor it if $x_1 > 0$, and uncensor it if $x_1 < 0$. On the other hand, if the type lies inside the cusp he has a problem of choice, because opinion is split, with two maxima x_1, x_2 such that $x_1 < 0 < x_2$, justifying either choice (see *Figure 22* below).

The censorship-norm, $x = 0$, separates the graph G into two regions which we label 'censor', $x > 0$, and 'uncensor', $x < 0$, as shown in the top picture of *Figure 20*. The lower picture shows the projections of these two regions in the control space. This simple diagram helps to explain much of the confusion surrounding censorship. For works of low aesthetic content, $b < 0$, the regions do not overlap, and so their frontier gives a clear delineation of the eroticism-norm, $a = 0$, permitted by the prevailing censorship-norm, $x = 0$. On the other hand, for works of high aesthetic content, $b > 0$, the regions overlap, and so anachronisms can arise, because some works

FIGURE 20 Cusp catastrophe model of censorship

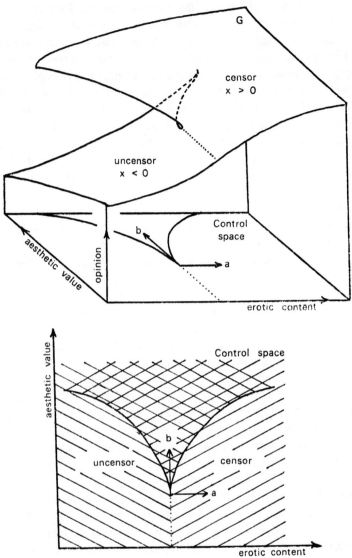

of low eroticism, $a < 0$, can be censored, $x > 0$, while others of high eroticism, $a > 0$, can be uncensored, $x < 0$.

What further confuses the issue is that the phenomena merge smoothly into one another. For example, a new work may 'move up' the aesthetic value scale as it gains recognition amongst the critics.

Two very similar works, on the borderline of the eroticism norm, may experience very close paths as types (like paths 4 and 5 in *Figure 13*), but at the same time, they may undergo divergent treatment at the hands of the censor, one becoming firmly censored and the other firmly uncensored.

Consider now what happens when the market-place becomes more permissive. The censhorship norm moves from $x=0$ to $x=x_N$, say, where $x_N > 0$. The probability distribution of opinion for each work does not change, but rather the origin of the scale is shifted (see *Figure 22*). *Figure 21* shows the resulting change from *Figure 20*.

FIGURE 21   Censorship in a permissive society

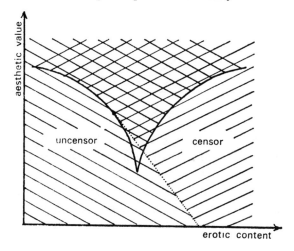

The new frontier between the two regions is the tangent to the cusp at the point $(a,b) = (-2x_N^3, 3x_N^2)$. As a result we should expect the following characteristics of a permissive society.

First, the most noticeable change is a tolerance towards works of low aesthetic value and high erotic content, and the lower the value, the higher the tolerated eroticism. Second, in the neighbourhood of the cusp point, where the debate used to begin, works of modest aesthetic value have all been quietly uncensored. This is not a catastrophic process, because there is no jump from one maximum to the other, but merely a shifting of the norm so that both peaks now lie in the uncensored region. If anything, the debate has moved higher up the aesthetic scale.

Third, there is no qualitative change amongst works of high

aesthetic value, $b > 3x_N^2$, in the sense that types inside the cusp still have one peak either side of the norm as in *Figure 22*, $x_1 < 0 < x_N < x_2$. However, there is a quantitative change in the shift of weight of public opinion which leads to a modified form of the Voting rule as follows. Define the *weight* W by some such formula, as:

$$W = \int_{-\infty}^{x_N} P(x)dx - \int_{x_N}^{\infty} P(x)dx.$$

FIGURE 22   Movement of the norm in a permissive society

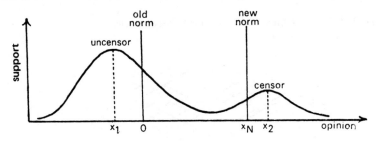

As x_N increases, if W achieves a threshold, $W > W_0$ say, then the weight of opinion should be sufficient to persuade the censor to uncensor a work that had been previously censored. But this time the change is catastrophic, because it is represented by a jump from x_2 to x_1. Consequently, such events are usually accompanied by much ceremony, such as the trial of Lady Chatterley (Rolph 1961).

FIGURE 23   The pendulum of fashion

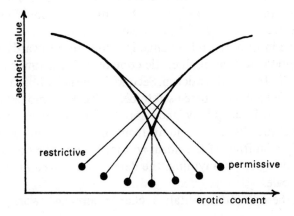

Returning to the metaphor of the 'pendulum of fashion' used in the title of Wilson's article, we can now see this beautifully illustrated by the tangent rolling along the cusp, as shown in *Figure 23*.

Each position of the pendulum gives clear instructions to the censor about the prevailing norm amongst works of low aesthetic value, but the confusion over works of high aesthetic value remains more or less the same, independent of the permissiveness of the particular society or particular age.

Let us look at the metaphor of the pendulum more closely. Thom would probably argue (1970; 1972; 1973b) that since the dynamics of the brain are stable, they are also governed by the same elementary catastrophes, and so intuition would naturally perceive this analogy before reason could explain it; that is why the metaphor appeals to us. Now, a pendulum has not only a swinging bob, but also a fixed pivot. In our case, the fixed pivot manages not only to be a point, the cusp point, but also manages to broaden out so as to include the best of our literature and art, thus providing exactly that fixed backdrop for fashion to swing against.

If, as it seems in Western society today, the pendulum has swung to the permissive side, then our model would predict the presence of a hidden cusp amongst uncensored works of moderate value. It would be very interesting if the experiment described above could reveal so curious a phenomenon. The same model can be applied to a variety of situations where the pendulum of fashion swings.

We conclude with one more short application, to the censorship of facts by governments, because this model links together all three previous models in the paper. The two main reasons for censoring facts are military and political: first, the government is concerned to protect the country against possible external threats, and therefore the military importance of a fact may deem it worthy of censorship. Second, the government is concerned to protect itself against possible internal opposition amongst the population, and therefore the political importance of a fact may deem it worthy of censorship.

Hypothesis 9. Political importance is a normal factor, and military importance a splitting factor, influencing the censorship of facts by governments.

As in the case of pornography, there are variations of censorship amongst different governments and at different times. The most noticeable difference between a permissive and a restrictive regime occurs at low values of the splitting factor, that is to say, in the government attitude towards facts of no military importance, but possibly of political significance. Meanwhile, the attitude of both countries towards facts of high military importance is almost identical. Military weapons, for example, will be either hidden with great secrecy or flamboyantly displayed, and the catastrophic switch from secrecy to display is usually accompanied by much ceremony.

NOTES

1 For the mathematicians: *smooth* means that P, regarded as a function of the three variables a, b, x, is differentiable to all orders. *Generic* means that P, regarded as a map from C to the space of probability distributions on X, is transversal to the natural stratification (Mather 1969; Thom 1972; Trotman and Zeeman 1974); in other words, P is in general position. Smooth generic Ps are open-dense in the space of all smooth Ps, which are dense in the space of all continuous Ps. Therefore, any continuous P can be approximated arbitrarily closely by a smooth generic P, justifying the use of smooth generic models. Furthermore, the latter are stable under sufficiently small, smooth perturbations (Trotman and Zeeman 1974).

2 The combined graph has equation $\frac{\partial P}{\partial x} = 0$, which is a smooth curve by Hypothesis 1. The fold-points are given by the equations
$$\frac{\partial P}{\partial x} = \frac{\partial^2 P}{\partial^2 x} = 0.$$

3 Throughout the paper, and in this section in particular, we use the word 'sociological' to mean pertaining to the social sciences, rather than pertaining just to sociology.

4 In other words, the map $X \to X'$ is an orientation-preserving diffeomorphism (where a *diffeomorphism* is a one-to-one map that is smooth and has smooth inverse).

5 For an application of catastrophe theory in psychology, involving the fitting of experimental data, and leading to predictions, and the design of new experiments, see Zeeman (1974d; 1975).

6 Indeed, in some cases the change of scale may not even be qualitative, and may only be order-preserving up to some tolerance, in which case the conclusions can only be rigorous up to that tolerance. However, this would require additional mathematical techniques that we do not consider in this paper.

7 For the mathematician: it is true that more sophisticated invariants

such as Betti numbers, homology, and homotopy groups, etc., have been known for a long time, but in most cases these appear to be useless for describing graphs of experimental data. Where they have proved useful, they also tend to be translatable into everyday language, such as 'connected', 'single-valued', etc.

8 For example 'more haste, less speed'. A resolution of this paradox can be made using the language of Section 11 below: skill is a normal factor, and haste a splitting factor, for speed. But of course, a paradox is always more attractive than its resolution, because it gives the mind more pause.

9 A *diffeomorphism of the plane* is a one-to-one map of the plane onto itself that is smooth and has smooth inverse.

10 But not conversely. An example of a qualitative property that is *not* invariant under equivalence is *increasingness* in an ordinary single-valued two-dimensional graph. However, more subtly, in the cusp catastrophe, which is the three-dimensional graph described in the next section, near the cusp point increasingness with respect to a normal factor *is* an invariant property.

11 Thom (1972) calls it the *Riemann-Hugoniot catastrophe* in honour of its first application to shock waves in the middle of the nineteenth century. Of course, it was not realized at that time as being part of the family of elementary catastrophes.

12 For example let $P_c(x) = \left[\frac{x^4}{4} - \frac{bx^2}{2} - ax + f(c)\right]^{-1}$, where $c = (a,b)$ and f is the unique function of c such that for each c, the bracket is positive and $\int_{-\infty}^{\infty} P(x)\,dx = 1$.

13 *Locally-equivalent* means there is a neighbourhood that is equivalent.

14 *Dual* means that maxima and minima are interchanged and so the roles of G and S − G are reversed. See *Figure 16*.

15 Mathematically, we could equally well use a 1000-dimensional behaviour space and the graph would still be a two-dimensional surface. However, this surface might then have so many folds and cusps that it could obscure the very phenomenon that we wish to elucidate. Sometimes, an approximate description of behaviour using only a few dimensions can be more effective than a detailed description using many dimensions, in the same way that, in order to study the flow of water over a dam, for instance, it is more effective to pretend that water is a continuous medium, than to analyse the detailed mechanics of all the water molecules.

REFERENCES

FOWLER, D. H. 1972. The Riemann-Hugoniot Catastrophe and Van der Waals's Equation. In C. H. Waddington (ed.), *Towards a Theoretical Biology*, *4*. London: English Universities Press.

HARRISON, P. J. and ZEEMAN, E. C. 1973. Applications of Catastrophe Theory to Macroeconomics. Symposium on Applications of Global Analysis. Utrecht University, Utrecht.

LORENZ, K. 1966. *On Aggression*. London: Methuen.

MATHER, J. N. 1969. Right Equivalence. Warwick University reprint.

POSTON, T. and WOODCOCK, A. E. R. 1974. *The Geometry of the Elementary Catastrophes*. Lecture Notes in Mathematics. Berlin and New York: Springer Verlag.

ROLPH, C. H. 1961. *The Trial of Lady Chatterley*. Harmondsworth: Penguin.

SHULMAN, L. S. and REVZON, M. 1972. Phase Transitions as Catastrophes. *Collective Phenomena*, **1**: 43–7.

SIERSMA, D. 1973. Singularities of $C\infty$ functions of right-codimension smaller or equal than eight. *Indag* **25**: 31–37.

THOM, R. 1972. *Stabilité structurelle et morphogénèse*. New York: Benjamin. English translation by D. H. Fowler forthcoming 1975.

——1969. Topological Models in Biology. *Topology* **8**: 313–35.

——1970. Topologie et linguistique. *Essays on Topology*. Vol. dedicated to G. de Rham. Berlin and New York: Springer Verlag.

——1971a. A Global Dynamical Scheme for Vertebrate Embryology. *AAAS* Some Mathematical Questions in Biology, Symposium V, *Lectures on Mathematics in the Life Sciences* **5**: 3–45. Providence: American Mathematical Society.

——1971b. Phase-transitions as Catastrophes. Conference on Statistical Mechanics. Chicago: Chicago University Press.

——1973a. La théorie des catastrophes: État présent et perspectives. *Manifold* **14** 16–23. Mathematics Institute, University of Warwick.

——1973b. Langage et catastrophes: Eléments pour une sémantique topologique. In M. M. Peixoto (ed.), *Dynamical Systems*. New York and London: Academic Press.

TOLSTOY, L. 1970. *War and Peace*. Vol. III. Trans. by L. and M. Maude, revised and reprinted. London: Oxford University Press.

TROTMAN, D. J. A. and ZEEMAN, E. C. 1974. Classification of Elementary Catastrophes. Warwick University reprint.

WILSON, J. 1970. Censorship. *Guardian*, November 9.

ZEEMAN, E. C. 1971. Geometry of Catastrophe. *Times Literary Supplement*, December 10.

3 — 1972a. Differential Equations for the Heartbeat and Nerve Impulse. In C. H. Waddington (ed.), *Towards a Theoretical Biology*, *4*. London: English Universities Press.

15 — 1972b. A Catastrophe Machine. In C. H. Waddington (ed.), *Towards a Theoretical Biology*, *4*. London: English Universities Press.

— 1973. Applications of Catastrophe Theory. Tokyo International Conference on Manifolds. Tokyo: Tokyo University.

4 — 1974a. Primary and Secondary Waves in Developmental Biology. *AAAS*. Some Mathematical Questions in Biology Symposium VIII. *Lectures on Mathematics in the Life Sciences* **7**. Providence: American Mathematical Society.

11 — 1974b. On the Unstable Behaviour of Stock Exchanges. *Journal of Mathematical Economics* **1**: 39–49.

21 — 1974c. Catastrophe Theory: A reply to Thom. *Manifold* **15**: 4–15. Mathematical Institute, University of Warwick.

12 — 1974d. Conflicting Judgements Caused by Stress. Forthcoming.

13 ZEEMAN, E. C., HALL, C. S., HARRISON, P. J., MARRIAGE, G. H., and SHAPLAND, P. H. 1975. A Model for Prison Riots. Forthcoming.

— Numbering in this book.

11 ON THE UNSTABLE BEHAVIOUR OF STOCK EXCHANGES

Received 29 August 1973

Some of the unstable behaviour of stock exchanges can be explained by a model based on catastrophe theory [Thom (1972), Zeeman (1971)]. A similar model can be applied to currencies, property markets, or any market that admits speculators.

The index. The simplest way of measuring the state of the market in a stock exchange is to choose some index, I, such as the Dow–Jones index. Let $J = \dot{I} = dI/dt$ denote the rate of change of our chosen index. Then $J = 0$ represents a static market, while $J > 0$ represents a bull market and $J < 0$ a bear market. In the case of a currency, let I denote exchange rate of that currency and again let $J = \dot{I}$. Then $J = 0$ represents a stable currency, while $J > 0$ represents a floating revaluation and $J < 0$ a floating devaluation.

The variable J can be regarded as a dependent variable, depending upon the rate of buying and selling of investors. At the same time there is a feedback because the knowledge of J in turn influences the investors, and this is what makes the dynamic interesting. To express this mathematically we need to introduce variables describing the activity of investors.

Main hypothesis. Broadly speaking there are two types of investors, fundamentalists and chartists. Fundamentalists are investors who act on the basis of estimates of large economic factors such as supply and demand, money supply, etc. Before a fundamentalist invests in a firm, he instructs his research team to assess its viability, its growth potential and market potential, etc. Chartists, on the other hand, are investors who base their investment policy upon the behaviour of the market itself, using the charts of recent behaviour to predict future behaviour. Speculators tend to be chartists.

We take as our main hypothesis that we can divide the investors into these two groups, and represent their activities by two variables C, F as follows. Let C be

---

*I am indebted to Alan Kirman for advice about some of the economic hypotheses, to Yung-Chow Wong for discussions about the Hong Kong stock exchange, and to Sharon Hintze for the terms fundamentalist and chartist.

Published in the Journal of Mathematical Economics, Volume 1, 1974, pages 39–49.

the *proportion of the market held by chartists*, or in other words the proportion of speculative money in the market. Let F be the *excess demand for stock by fundamentalists*. In order to keep the model as simple as possible we do not introduce a separate variable for the excess demand by chartists for two reasons. Firstly the total excess demand can probably be represented by some function $f(J)$, and so the excess demand by chartists is merely the difference $f(J)-F$. Secondly the excess demand by chartists is, by definition of chartist, more like part of the internal mechanism of the market, compared with F which is more like an external driving force.

Dynamic flow. We ask the question: how are the variables C, F, J related? Of course the question is too naïve, because they will undoubtedly be affected by many other external factors. However, as a first approximation let us suppose that there is a dynamic relation between them, represented by an ordinary differential equation. If we wanted to add in the effect of external factors, we could, later, superimpose on this differential equation a stochastic noise term.

Take C, F, J are coordinates in 3-dimensional Euclidean space R^3. For intuitive convenience call C and F the horizontal coordinates, and call J the vertical coordinate. The differential equation will be given by a vector field X on R^3. The resulting flow, given by the solution curves of X, will represent the dynamic behaviour of the stock exchange or the currency. It is this flow that we seek to understand qualitatively. Once the flow is understood qualitatively, then it may be possible for economists to use data and build a quantitative model to describe a particular stock exchange or a particular portfolio, or a particular currency, during a particular period. The ultimate objective is prediction, and the design of more effective controls to reduce instability and avoid crashes.

The model. To build up the qualitative picture of the flow, we shall take as hypotheses a number of observed qualitative features of stock exchanges and currencies, and translate each feature into mathematics. The advantage of the translation process is as follows. Whereas it is easy to understand each feature by itself from a local and relatively static point of view, it is more difficult to simultaneously comprehend them all together, and grasp how they interact from a global dynamic point of view. By contrast the mathematical synthesis is relatively easy, because we can use deep theorems. It transpires that all mathematical features can be synthesised into a single mathematical concept, namely *the cusp catastrophe with a slow feedback flow*. This single mathematical entity can be visualised geometrically, and provides an overall grasp of the problem. Not only can the individual features be seen at a glance, but the more complicated patterns of change emerge clearly. Summarising: we insert seven disconnected elementary local hypotheses into the mathematics, and the mathematics then synthesises them for us and hands us back a global dynamic understanding.

Hypothesis 1. J responds to C and F much faster than C and F respond to J.
The main purpose of a stock exchange or money market is to act as a nerve centre so that prices (represented in our elementary model by the rate of change J of the index) can respond as swiftly and as sensitively as possible to supply and demand. Changes in C and F can cause changes in J within minutes, whereas changes in J have a much slower feedback on C and F. The response time for C can be a matter of hours, but is more likely to be days or weeks, while the response time for F can be weeks or months, due to the research involved.

What does this mean mathematically? The speed of response at a point p is represented by the length of the vector $X(p)$, and Hypothesis 1 implies that for most points p the vertical component X_J of X is much larger than the two horizontal components X_C and X_F. Consequently the flow lines are nearly vertical almost everywhere. Another way of saying this is that if we fix C and F, then the forces of supply and demand will cause J to rapidly seek a stable equilibrium position $J = J(C, F)$, where $X_J = 0$. Therefore we obtain a surface S of stable equilibria, given by J as a function of C and F (see fig. 3 below). We call S an *attractor* surface because, if the system starts at a point not on S, then the dynamic will carry the point towards S by a fast flow line that is almost vertical. On S itself the vertical component vanishes; therefore as the point approaches near S, the horizontal components, which previously had been relatively unimportant, now become dominant. Therefore on S, or more precisely near S, there will be a slow flow representing the feedback effect of J on C and F.

Mathematical digression. Some mathematically inclined readers may protest at the apparent imprecision of the words fast and slow, large and small, near and far, and so for their benefit we digress in order to make some precise mathematical statements. But first observe that the difference between large and small is a common phenomenon in applied mathematics, for instance a liquid is treated as a continuum in the large and as particles in the small. In our case the difference between weeks and minutes is so large a quantitative difference that it gives rise to a qualitative difference, which we wish to capture in the mathematics. However, it would be wrong to push this difference to a limit as in classical analysis, because the limiting statement would be both irrelevant and probably false. For instance when we say the flow lines are 'nearly vertical almost everywhere' we do not mean that they become *arbitrarily* close to the vertical *sufficiently* far from S. To state what we do mean precisely, it is necessary to use the differential-topological language of the qualitative theory of differential equations rather than the analytical language of the quantitative theory, as follows.

Let $\bar{S}$ be the subset of R^3 given by $X_J = 0$. Then we have a theorem: given any neighbourhood N of $\bar{S}$, there is a diffeomorphism of R^3 into itself, mapping each horizontal plane onto itself, and mapping all the flow lines outside N onto vertical lines. Therefore from the qualitative point of view we may intuitively think of the flow lines outside N as being vertical. Generically $\bar{S}$ will be a smooth surface without boundary, containing S as a subsurface, and meeting each vertical line in at least one and at most a finite number of points (see fig. 3).[1] Moreover $\bar{S}$ will separate R^3 into two regions, one region 'above' $\bar{S}$ given by $X_J < 0$, where the flow outside N is vertically downwards, and the other region 'below' $\bar{S}$ given by $X_J > 0$, where the flow outside N is vertically upwards.

Flows in 3-dimensions featuring this fast and slow response, and consequently possessing an attractor surface containing a slow feedback flow, have been studied in biological contexts [Zeeman (1971)]. Explicit algebraic examples are given there, which are related to the Van der Pol and Lienard equations, and the corresponding flow lines are illustrated in Zeeman (1971, figs. 7 and 9). This ends the digression.

We return to the surface S given by the stock exchange. The question that interests us here is to study the qualitative shape of S and the qualitative properties of the slow flow on S. In particular we shall investigate whether S has any folds or singularities, corresponding to multivaluedness of the function $J = J(F, C)$. If singularities do occur, then we can call upon the powerful classification theorem of Thom [see Thom (1972) and Zeeman (1971)], stating the only way in which they can occur. Summarising: Hypothesis 1 implies the existence of S, and to study the folds of S and the slow flow on S we need further hypotheses.

Hypothesis 2. If C is small then J is a continuous monotonic increasing function of F passing through the origin. In other words if the speculators are in a minority, and the market is dominated by well-informed investors, then an equal demand for buying and selling by the latter will cause the index to be static, $J = 0$; an excess demand will cause the index to rise, $J > 0$, and an excess supply will cause the index to fall, $J < 0$. Therefore the plane $C =$ constant (small) will intersect S in a graph as in fig. 1a. In particular, since the graph goes through the origin, if we fix $F = 0$ then $J = 0$ is a stable equilibrium for the fast response of J.

Hypothesis 3. If C is large this introduces an instability into the market. What does 'instability' mean mathematically? Suppose for the moment that $F = 0$, representing equal demands for buying and selling by the fundamentalists. Then $J = 0$ will again be an equilibrium position for J, but due to the presence of the large proportion of speculative money it will now become an *unstable* equilibrium. In other words, since J is the rate of change of the index, we are postulating in Hypothesis 3 that it is dynamically unstable for the index to remain constant. Any slight perturbation of the index up or down (by external forces) will at once be amplified by the chartists. If the index begins to rise then it will

[1] See footnote on p. 45.

quickly settle into a steadily rising state as a bull market, in other words J will quickly settle into a stable equilibrium position, $J > 0$. Conversely if the index begins to fall then J will quickly settle into a different stable equilibrium position as a bear market, $J < 0$. We emphasise that the word 'stable' here refers only to the fast response of J assuming C and F are kept fixed; if C and F are allowed to vary then the slow feedback flow on S may slowly return J to zero, or may slowly lead to a sudden change of sign of J, because in the long run a bull or bear market may not be able to sustain itself.

The critical consequence of Hypothesis 3 is that for large C and small F *the function $J(C, F)$ is 2-valued*, and so *the attractor surface S is 2-sheeted*.

Therefore the plane $C = $ constant (large) will intersect S in the graph shown

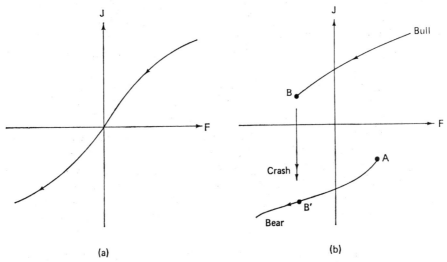

Fig. 1. (a) Small C, (b) large C.

in fig. 1b. Near the origin the graph will consist of the 2 monotonic curves marked bull and bear. If F is large positive, then the demand for buying by fundamentalists will override any external bearish pressures on the market, and guarantee a bull market. Therefore only the bull-graph extends along the positive end of the F-axis, and so the bear-graph must end somehow. Now the bear-graph cannot bend round and go back because this would make S 3-sheeted instead of 2-sheeted, nor can it join onto the bull-graph, because T-shaped junctions cannot occur in a family of sections of a smooth surface. Therefore the bear-graph must have an end point A. Similarly only the bear-graph extends along the negative end of the F-axis, and so the bull graph must have an end point B. Therefore the graph is disconnected.[2]

[2] Alternatively one can use Thom's theorem [Thom (1972), Zeeman (1971)] to deduce the conclusions of the last three sentences from the fact that S is 1-sheeted over some points and 2-sheeted over others.

We can deduce interesting market behaviour as a result of this disconnectedness. Suppose the fundamentalists decide to pull out of a bull market. The resulting slow flow is indicated by arrows on the graphs in fig. 1. If the market contains only a small proportion of speculative money (fig. 1a) then J will slowly fall. Since $\dot{I} = J$, this means that the rising index will gradually flatten as $J \to 0$, and gradually begin to drop as J becomes negative, giving a smooth maximum as in fig. 2a.

If on the other hand the market contains a large proportion of speculative money (fig. 1b), then there will be a delay while the market remains artificially bull until the boundary point B is reached. At this point the vertical fast flow will take over, indicated by the double arrow, causing a sudden drop in J from B to B′ on the bear part of the attractor surface. This will cause a sharp maximum

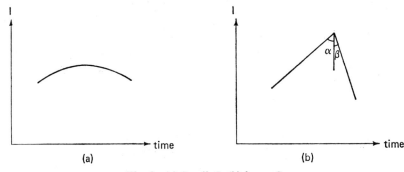

Fig. 2. (a) Small C, (b) large C.

in I as in fig. 2b, with angle $\alpha + \beta$, where $\cot \alpha + \cot \beta = BB'$. Therefore the larger the drop BB′, the smaller the angle β and the steeper the descent of the index. By and large nature encourages us to expect smooth graphs, and so we tend to regard fig. 2a as normal and call fig. 2b a *recession*. If BB′ is very large, causing a very steep descent, then we are liable to call the recession a *crash*.

Conversely fundamentalists often make large profits by quietly investing in a recovering market during the delay period while the market is remaining artificially bear, before it has reached the boundary point A. On the other hand, if the delay is exaggerated by a very high proportion of speculative money, then the fundamentalists' support may be insufficient to reach the boundary point A. This phenomenon occurs for example in a run against a reserve currency, when in spite of positive support by central banks, there is too much speculative money for the support to be effective, and so the currency may have to be devalued, even though in terms of its purchasing power it may then seem to be undervalued.

Synthesis. We now use the deep classification theorem of Thom [see Thom (1972) and Zeeman (1971)] to synthesise the information acquired so far into a

3-dimensional picture of the surface S. The theorem implies amongst other things that:

(i) S is a subsurface of a generic[3] surface $\bar{S}$;
(ii) a generic surface is smooth without boundary;
(iii) when a generic surface is projected orthogonally onto the (C, F)-plane, the only singularities that can occur are fold curves and cusp points;
(iv) the boundary of S equals the fold curves of $\bar{S}$.

Since S has sections as in fig. 1, we can deduce from the theorem that $\bar{S}$ must have a cusp point and therefore must be equivalent[4] to the cusp catastrophe surface, shown in fig. 3.

The cusp catastrophe is described in Thom (1972) and Zeeman (1971) and we now review its properties. The surface $\bar{S}$ shown in fig. 3 is given by the equation

$$J^3 - (C - C_0)J - F = 0.$$

Here C_0 is the value of C at which the discontinuity shown in fig. 1b first begins. The surface S is the subsurface of $\bar{S}$ given by the inequality

$$3J^2 + C_0 \geqq C.$$

The boundary ∂S of S is the fold curve of $\bar{S}$, given by $3J^2 + C_0 = C$. The projection of ∂S onto the (C, F)-plane is the cusp shown in fig. 3. and has the equation

$$4(C - C_0)^3 = 27F^2.$$

Over the outside of the cusp S is single-sheeted and is the same as $\bar{S}$. Over the inside of the cusp S is 2-sheeted and $\bar{S}$ is 3-sheeted, the extra middle sheet being the complement $\bar{S} - S$. The complement $\bar{S} - S$ is given by $3J^2 + C_0 < C$, and represents points of unstable equilibrium; in other words it is a *repellor* surface, the opposite of an attractor surface, and it is shown shaded in fig. 3. It follows from the theorem that the attractor surface S and repellor surface $\bar{S} - S$ together form the smooth surface $\bar{S}$, and are separated by the boundary ∂S. Although $\bar{S}$ is mathematically interesting it is irrelevant from the point of view of the application under consideration, because the system stays only on the attractor subsurface S.

---

[3]*Generic* is a technical mathematical term: generic surfaces arise from generic systems. Here we define X to be generic if $\int_0^r X_J \, dJ, r \in R$, regarded as a map from the (C, F)-plane into $C^\infty(R)$, is transversal to the natural stratification of $C^\infty(R)$. Transversal maps are open dense in the space of all C^∞-maps, endowed with the Whitney C^∞-topology. Therefore any system can be approximated arbitrarily closely by a generic system, and so we are justified in taking as our model a generic surface.

[4]*Equivalent* means the following: two surfaces S_1, S_2 are equivalent if there is a diffeomorphism of R^3 onto itself throwing S_1 onto S_2, and throwing vertical lines to vertical lines. In particular we can incorporate any required scaling of C, F, J into the diffeomorphism.

We now investigate the slow flow on S, and for this we need four more hypotheses describing the feedback effect of J on C and F. The dotted flow lines shown in fig. 3 will be the consequence of these hypotheses.

Recall that $J = \dot{I}$, where $\dot{I}$ denotes the rate of change of the index I.

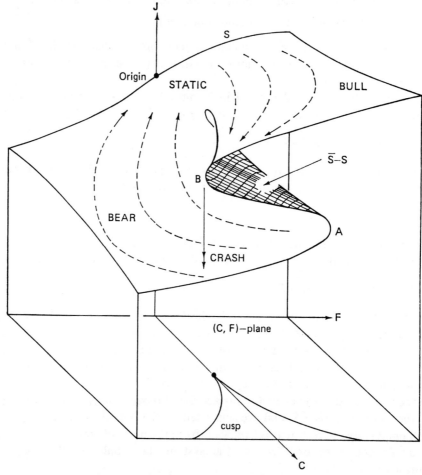

Fig. 3. Note: For convenience we have drawn the (C, F)-plane not through the origin $J = 0$, but below the origin as $J = k$, where k is a negative constant.

Hypothesis 4. *C has the same sign as J.* In other words chartists follow the trend. A bull market attracts the chartists, and so the proportion of speculative money in the market goes up, while a bear market repels chartists, and so the proportion of speculative money goes down.

Hypothesis 5. $\dot{F} < 0$ *after a large rise in I, even though I may still be rising.* In other words fundamentalists, with the experience that a bull market does not last indefinitely, and the knowledge of the industrial capacity behind the market, tend to judge when the market has become overvalued, and tend to cash in while the index is still rising before it has reached its maximum. Therefore if the system has been following a slow curve during which J has been positive and increasing, then $\dot{F}$ will become negative.

Hypothesis 6. $\dot{F} < 0$ *after a short fall in I.* Fundamentalists often pursue a policy of choosing a margin for each stock, and then selling if there is a fall in price of magnitude greater than that margin. In other words the selling price is the maximum price to date minus the margin. This policy of cutting losses would give rise to Hypothesis 6. In particular if J is large negative then $\dot{F} < 0$.

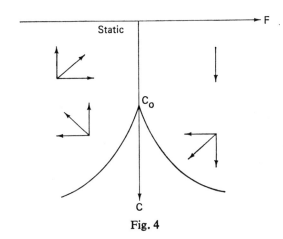

Fig. 4

Hypothesis 7. $\dot{F} > 0$ *if I has been falling for some time and is beginning to flatten out.* In other words a recovering bear market is an attractive investment for fundamentalists because prices are liable to be undervalued, and small losses in the short run are liable to be offset by much greater gains in the long run. In terms of J this occurs when $J < 0$ and $\dot{J} > 0$.

The slow flow on S. Fig. 4 shows the changes in C and F implied by Hypotheses 4–7 using the values of J given by S. The sums of components indicate the direction of the slow flow vectors. The resulting slow flow lines on S are shown dotted in fig. 3.

Global dynamics. We can now deduce a global dynamic behaviour of a stock exchange. If there are no external forces the market will settle into a stable and relatively static equilibrium. Suppose now there is a stochastic noise in the form of an influx of fundamentalist money. This frequently happens to a market which has been static for some time, and often is caused by fundamentalists moving their money away from another market which is experiencing a period of instability, into the stock exchange under consideration, because, due to the latter's recent spell of stability, the fundamentalists have had time to research into its potentialities. The system then becomes caught in the slow flow with the following sequence of consequences. First the rising index attracts the chartists, and the bull market accelerates. If the proportion of chartist money becomes too high then the fundamentalists may begin to cash in. If sufficiently many cash in, then after some delay a recession will occur, causing chartists to withdraw. Finally there is a slow recovery as the fundamentalists begin to reinvest.

The higher the proportion of speculative money before the recession, the longer the delay and the steeper the slope of the recession (the slope = $\cot \beta$ in fig. 2b). If C is very large the slope will be very steep, and the recession will be called a crash. Recall that $C = C_0$ gives the position of the cusp point. As a corollary of Thom's theorem, for $C \geqq C_0$ we have the formulae:

$$\text{delay} = K(C-C_0)^{\frac{1}{2}},$$
$$\text{slope of recession} = L(C-C_0)^{\frac{1}{2}},$$

where K, L are non-vanishing smooth functions of C. In the case of the cusp catastrophe the functions K, L are constant $K = 2/3\sqrt{3}, L = 1/\sqrt{3}$ but in general the surface S is only equivalent to, rather than the same as, that of the cusp catastrophe, and so K, L may not be constant. Nevertheless since K, L do not vanish at C_0, we can as a first approximation take them to be constant near C_0, giving surprisingly precise formulae that might be useful for testing quantitatively. For instance near the cusp $\cot \beta = 2 \cot \alpha$ (see fig. 2).

Of course the above sequence of events could be slowed down or accelerated, or interrupted, by stochastic noise caused by external forces. If the noise is large then its unpredictability may obscure the underlying dynamic, giving the impression that a stock exchange is only stochastic.

If a government wanted to introduce controls to prevent a runaway bull market from crashing, it could impose a temporary heavy tax on the sale of stock proportional to length of time held. This would discourage selling in general, and in particular might persuade fundamentalists to hold on to their long-term investments during the tax period, while the proportion of speculative money slowly subsided. If the model was correct, this would steer the market back to normality by a continuous rather than a discontinuous path.

Note added in proof: A similar economic model, also based on the cusp catastrophe, has been developed by the anthropologist Michael Thompson, to explain the periodic cycles of credit confidence underlying the complex pig-giving ceremonies in the New Guinea Highlands.

References

Thom, R., 1972, Stabilité structurelle et morphogénèse (Benjamin, New York).
Zeeman, E.C., 1971, Differential equations for the heartbeat and nerve impulse, in: C.H. Waddington, Towards a theoretical biology, vol. 4 (E.U.P.) 8–67.

(Number 3 in this book)

12. A MATHEMATICAL MODEL FOR CONFLICTING JUDGEMENTS CAUSED BY STRESS, APPLIED TO POSSIBLE MISESTIMATIONS OF SPEED CAUSED BY ALCOHOL

A mathematical model is proposed, based on catastrophe theory, to describe the qualitative effect of stress upon the neural mechanisms used for making judgements, such as estimating speed. The model is used quantitatively to fit data, and to explain the cusp-shaped results of Drew *et al.* (1959), showing that introverts under alcohol tend to drive either too fast or too slow in a driving simulator. Experiments are suggested in which discontinuous jumps in perception of continuous variables like speed might well appear.

1. Introduction

Some forms of stress impair the integrative capacity of the central nervous system. Such stress may prevent the individual from attending to, and according due weight to, all the data that happen to be entering the sense inputs. When the central nervous system is subjected to information overload, with more incoming sense data than it can integrate, then we suggest that a possible automatic response is *selectivity*; that is to say, the central nervous system will tend to select, and pay more attention to, a coherent subset of the data which it can integrate, and pay less attention to the rest. Therefore the individual will (subconsciously) tend to bias judgements in favour of this coherent subset.

We show, mathematically, how this phenomenon can lead to conflicting judgements. The mathematics offers rational explanations of unexpected behaviour that would otherwise be inexplicable without more complex hypotheses. The model that we give can be applied generally to stress situations, but in order to illustrate it better, we take one particular example, for which there are experimental data available, of a surprising nature. I am indebted to Peter Shapland and Hugh Marriage for drawing my attention to these data, and to John Annett and Stephen Ambler for suggesting improvements to the paper.

2. The Effect of Alcohol on Driving

Drew *et al.* (1959) administered small doses of alcohol to their subjects. They said that: 'Alcohol exerts its effect through the central nervous system, its main pharmacological effect being the depression of central nervous activity . . .'

Published in British Journal of mathematical and statistical Psychology, Volume 29, 1976, pages 19-31.

(p. 1). Therefore it is a reasonable hypothesis that alcohol is the type of stress, described in the Introduction, that would impair integrative capacity.

Their experiment consisted of asking the subject to drive in a simulator; the subject sat in a dummy car, with normal controls, in a darkened room, and watched on a screen the moving picture of a 1-mile stretch of winding road, that responded correctly to the controls. The subjects were asked to drive at normal speed, while sober, and the habitual normal speed of each subject was recorded. Then each subject was asked to drive at normal speed after imbibing alcohol, and it was found that some drove faster and others drove more slowly. The variance in speed was then plotted against an introvert/extravert measure, as shown in Fig. 1 (Drew *et al.* 1959, p. 57). Other results, such as the number of

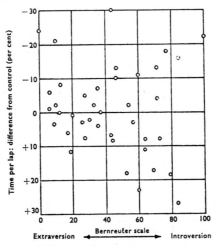

FIG. 1.—Mean time per lap (difference from control) for Dose 4 plotted against extraversion measured by the Bernreuter Scale.

errors, were also recorded. The authors concluded by offering the following explanation of their data: 'In this experiment subjects classified as extraverts appear not to be bothered by the extra stress imposed by alcohol. They drive at much the same speed as before, but they make much more error. Those classified as introverts, on the other hand, appear to be striving to compensate for the alcohol effect, and to be anxious to demonstrate their efficiency. They overreact to the situation by changing their speed markedly. Some slow down, presumably in an attempt to achieve accuracy, although they do not necessarily do so, while others appear to be attempting to demonstrate how quickly they can drive, though not always with a proportionate loss of efficiency' (p. 62).

Now one of the fundamental questions raised by this explanation of the experimental results is: Are there two types of introvert? By using mathematics, we shall be able to answer: 'Not necessarily'. Another question raised by the data in Fig. 1, and not answered by the above explanation, is: Why do the dots fall roughly into the shape of a cusp? Why not, for example, a parabola? We shall show that a cusp is to be expected.

3. Hypotheses

The model is based on certain psychological assumptions, and uses the mathematics of catastrophe theory (Thom, 1972; Isnard & Zeeman, 1975), which, although simple in statement, implicitly makes use of deep theorems. In order to distinguish between the mathematics and the psychological assumptions, we frame the latter as four hypotheses. The hypotheses may appear to lack precision at this stage, because we do not define the terms; however, the terms gain the required precision when translated into mathematics (see Section 8).

Hypothesis 1

Alcohol stress ⇒ impairment of integrative capacity.

Hypothesis 2

Impairment ⇒ selectivity.

Hypothesis 3

Extraverts have greater integrative capacity than introverts. More precisely, under alcohol, those subjects classified as introverts are more selective, either because they are already more selective when sober and so the alcohol merely accentuates the selectivity implied by Hypotheses 1 and 2, or else they are more susceptible to the impairment due to alcohol implied by Hypothesis 1, and consequently more selective by Hypothesis 2.

Hypothesis 4

In the experiment all subjects under alcohol try to drive at what they judge to be their normal speed, to the best of their ability.

The basic idea behind our explanation is that impairment is a *mechanical defect* of the central nervous system, and the selectivity is the *natural efficient response* of a partially defective system. This defect and response take place automatically in the central nervous system while integrating the data and estimating speed. The driver then reacts logically to the situation by adjusting the accelerator according to this estimate.

Summarizing the two main points: firstly, we suggest that alcohol may cause misjudgement of speed, and so the drivers drive at the wrong speed. Secondly, we suggest that this misjudgement is due to selectivity, and as a result we shall be able to prove mathematically in Section 8 below that this implies the cusp illustrated in Fig. 9. There is no need to postulate more complex motives such as 'striving to compensate' or 'anxious to demonstrate'.

4. Estimation of Speed

A driver normally receives a great many cues as to his speed, not only the more obvious visual cues, but also aural cues from the sounds of the engine, tyres and wind, tactile cues from the feel of the wheel, accelerator and brakes, and from the pressures of seat, elbow-rest and floor against body and limbs, muscular cues from the reactions to vibrations, bumps and accelerations, inertial and orientation cues from the intestinal and balance organs, neurological cues as adrenalin enters the bloodstream in response to unexpected events, etc. Each cue may suggest a speed, x, or range of speeds, and, taking into account the weighting of cues acquired by experience, the central nervous system receives from the various sources a weighted probability distribution of speed-cues, as shown in Fig. 2.

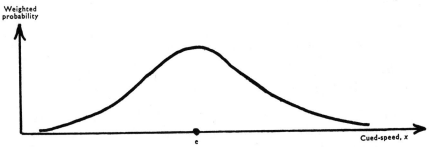

Fig. 2.—Weighted distribution of speed-cues.

Normally the visual cues are dominant (as is shown by the misestimations of speed that can occur whilst driving in fog). One can receive several different visual cues simultaneously; for example, the hedgerows rushing past may suggest a high speed, while the distant mountains moving slowly may suggest a low speed. In the experiment in the simulator there were only visual cues. Nevertheless, to an experienced driver the lack of non-visual cues may, in fact, have contributed negatively to his estimate of speed. Incoherence of cues can in fact induce nausea.

However, in both cases, either in a real car or in a simulator, we may assume that the central nervous system swiftly integrates the material presented to it, and chooses the maximum, e, of the distribution as the *estimated speed*. The point we wish to make is that there are at least two processes going on: (1) the *integration process*, during which the data presented by the sense organs are assembled, and which we have expressed mathematically by the weighted probability distribution; and (2) the *estimation process*, which we have expressed mathematically as the choice of maximum, e.

Suppose that an experienced sober driver is driving under normal conditions at speed s, and the resulting estimated speed is e. Then we translate the words 'experienced driver under normal conditions' into mathematics by postulating the equation $e = s$, giving the trivial diagonal graph in Fig. 3. The purpose of this rather elaborate approach is to distinguish carefully between three variables: firstly, the speed, s; secondly, the cued-speed, x; and thirdly, the estimated speed, e.

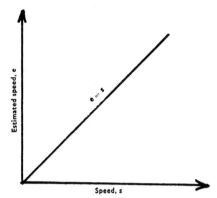

FIG. 3.—The diagonal graph (compare Fig. 7).

5. THE SPLITTING EFFECT OF SELECTIVITY

Suppose now that selectivity enters into the integrative process, due either to personality habit or to stress-impairment, or both. What is the simplest selectivity mechanism? Since the central nervous system cannot integrate all cues, it must select a subset A of the cues that are correlated and sufficiently coherent to be integrated, and to which it can therefore pay more attention, while paying less attention to the complementary subset B. What, then, is the simplest way of splitting the cues into two such subsets, A and B? Before the cues are used for the estimation process (2), they will be initially grouped during the integrative process (1) into clusters, fast cues with fast, slow with slow, etc. The clustering determines a structure† on the set of cues, which is similar to that obtained by ordering them on the x-axis, according to cued-speed. We translate the words 'correlated and sufficiently coherent' into mathematics by requiring that the subset A be connected with respect to the clustering structure. Therefore A corresponds to a connected subset of the x-axis, and the three possibilities are illustrated in Fig. 4.

In other words, Fig. 4 means that the system pays either: (i) more attention to slower cues and less to faster; or (ii) more attention to faster cues and less to slower; or (iii) more attention to moderate cues and less to extreme. Although (iii) may seem the most sensible selectivity process from the conscious reasoning point of view, it is in fact twice as complicated a task for a subconscious mechanism to perform, because the boundary between A and B consists of two

† The expressions 'structure', 'similar' and 'connected with respect to the structure' can be made mathematically precise by using tolerance spaces (Zeeman, 1965), or fuzzy spaces, or comparison spaces, or any topological-combinatorial procedure of cluster analysis.

points rather than one. Therefore (i) and (ii) are simplest subconscious selectivity mechanisms. The net result of such a mechanism will be to bias the weighting given to cues, and hence bias the weighted distribution of Fig. 2 to either one side or the other, as in Fig. 5. The maximum, which previously occurred at

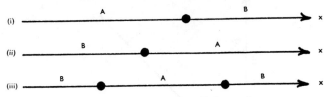

FIG. 4.—Connected subsets of the x-axis.

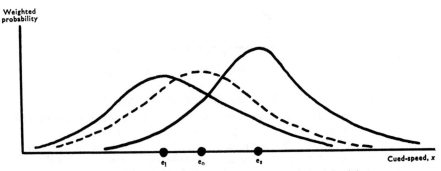

FIG. 5.—Weighted distribution of speed-cues biased by selectivity.

speed $e_0 = s$, is now displaced to e_1 or e_2. Therefore the estimation process (2) will select e_1 or e_2 as the estimated speed.

The above discussion is a somewhat rigid oversimplification of the underlying neural mechanisms, and we now modify this rigidity by introducing a probability distribution P for the estimated speed. Without selectivity we assume that P is unimodal, looking like Fig. 2, and with the peak occurring at the same place. On the other hand, with selectivity our arguments have shown that P becomes bimodal, as in Fig. 6, with peaks occurring at e_1 and e_2 (separated by a minimum at m, say). The end-result of the integration process (1) and the estimation process (2) is to choose one of the two maxima. We explain in the next section why one maximum might be chosen rather than the other. Since

$$e_0 = s,$$
$$e_1 < e_0 < e_2,$$

both maxima will give false estimates of speed. Suppose that a driver normally drives at speed s_N. If he was able to adjust his speed, s, until his underestimated speed, e_1, coincided with his normal speed, $e_1 = s_N$, then he would *think* he was driving at normal speed, but would be in fact be going too fast, $s > s_N$.

Conversely, the driver with overestimated speed, e_2, would be going too slow. We begin to see how Hypotheses 1–4 imply that introverts will drive too fast or too slow.

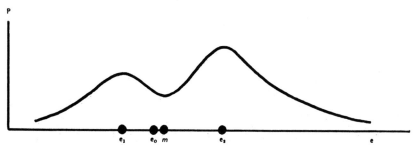

FIG. 6.—Bimodal distribution for e under selectivity.

This phenomenon will be particularly noticeable when the speed itself is nearly normal, because then the cues will be distributed on either side of normal, and so if there is any doubt in the driver's mind as to whether or not his speed is normal, due to impairment of his integrative capacity, then it will be particularly difficult for him to judge whether the speed is faster or slower than normal. On the other hand, if the speed s is very slow then the cues will be predominantly slow, and sufficiently coherent to be integrated without selectivity, in spite of impairment, and so we should expect a unique correct estimated speed. Similarly, if the speed is very fast, then the cues will be coherent in the sense that they are nearly all faster than normal, and so we should again expect a unique estimated speed, although this time with a larger possible margin of error.

Summarizing these observations, we obtain Fig. 7, which is the selectivity analogue of Fig. 3. In Fig. 7 the solid curves are the graph of maxima. Thus, for normal speed s_N, with probability distribution as in Fig. 6, the graph will have two values, e_1, e_2. Therefore the graph is double-valued near s_N; and we have labelled the two branches *underestimate* and *overestimate*. Meanwhile, for very small and very large speeds, the graph is single-valued and approximates the diagonal graph of Fig. 3. The dotted curve represents the minima of probability, such as m in Fig. 6. The reason that the solid and dotted graphs together form a smooth curve is that they are given by the equation $\partial P/\partial e = 0$, where P is regarded as a function of both speed, s, and estimated speed, e, and which we may assume to be differentiable without loss of generality.

6. Driving under Alcohol

Consider now the difficulties experienced by a driver under alcohol, suffering from impairment, and consequently automatically responding with selectivity. He starts slowly and accurately. As his speed increases his estimated speed follows continuously the underestimate graph, causing him to accelerate beyond his normal speed, s_N. When he reaches speed s_2 his estimated speed will be given by the point, Q, which is still too slow. If he accelerates further, $s > s_2$, then the probability graph only has one maximum, given by Q' on the overestimate graph (because the underestimate maximum will by now have coalesced

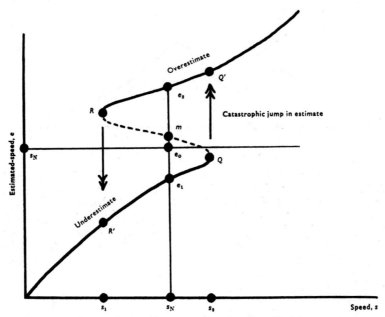

Fig. 7.—Graph of estimated speed under selectivity.

with the minimum, and disappeared). Therefore the subconscious estimation process (2) will present Q' as the unique estimated speed. Therefore, although up to speed s_2 the estimated speed had been increasing continuously† along the underestimate graph, at speed s_2 there is a sudden catastrophic jump from Q to Q', from the underestimate graph to the overestimate graph.

This jump in estimate can be very unnerving to a driver, as he realizes with a sudden shock that he is driving too fast. Notice that this jump is caused by an *automatic continuous* integration mechanism (1) and an *automatic* estimation mechanism (2), both subconscious. Now that the driver is overestimating speed he slows up to speed s_1, where the estimated speed jumps from R to R', and he experiences another slight shock as he suddenly realizes he is driving too slow. And so on. This unnerving hysteresis effect is not uncommon amongst drivers inexperienced under alcohol. The phenomenon that the driver has to face is

† The estimation mechanism follows the graph *continuously* because it depends upon a dynamical system (that of the underlying neurology). Therefore the catastrophic jump is delayed until the last possible moment, that is until the point Q is reached. In other words, the mechanism obeys the Delay Rule (see Thom, 1972; Isnard & Zeeman, 1975).

exhibited by Fig. 7: *at no speed will his estimated speed ever coincide with his normal speed.* s_N, because the horizontal line $e = s_N$ only meets the dotted curve, not the solid curve. (This can be proved mathematically, because in Fig. 8 below, the point R lies above, and Q lies below, the pucker point, which has coordinates $e = s = s_N$.)

With experience under alcohol, however, a driver may learn to cope with this phenomenon by placing himself at Q or R, and driving steadily at speed s_2 or s_1, with his estimated speed hovering on the borderline of catastrophe. His choice of Q or R may depend upon the quantitative properties of the graph, because one of them may represent an estimated speed nearer to the normal speed s_N than the other. For example, Figs. 5, 6, 7 are drawn so that the Q is the nearer. The experienced driver, while driving too fast or too slow because of this automatic and immediate subconscious misestimating, may nevertheless gradually become aware that his speed is abnormal in the sense that his habitual speed under alcohol is different from his habitual speed when sober. Some drivers, for instance, admit that they tend to drive too slowly under alcohol, and too fast when tired. Different types of stress may cause different quantitative graphs, and therefore induce different misestimates and different behaviours in the same person.

7. The Cusp Catastrophe

So far, we have not explained why a cusp should appear in Fig. 1, and for this we need to appeal to the deep classification theorem of catastrophe theory (Thom, 1972; Isnard & Zeeman, 1975). The probability distribution P of the estimated speed, e, depends upon the speed, s, and selectivity, t, and without loss of generality we may assume that P is a differentiable function of all three variables. Then the graph G of e as a function of s, t is given by $\partial P/\partial e = 0$. The classification theorem, in particular, classifies such three-dimensional graphs, and says that, up to qualitative equivalence,† there is a unique graph G with sections as in Figs. 3 and 7, called the *cusp catastrophe*. We can state this result precisely:

Theorem. The graph G of e is a cusp catastrophe, with s as a normal factor, and t as a splitting factor.

The resulting graph G is pictured in Fig. 8. Figure 7 corresponds to the sections $t =$ constant nearer to the eye in Fig. 8. The points Q and R lie in the fold curve of the graph. One of the main properties of the cusp catastrophe—hence the name—is that the fold curve projects down onto a cusp C in the (s, t)-plane. In Section 6 we argued that splitting took place in an interval containing normal speed, s_N. If this is true for all values of t, for which splitting occurs, then s_N is the s-coordinate of the cusp point. Let t_N denote the

† The expressions 'qualitative equivalence', 'normal' and 'splitting factors' are defined in Isnard & Zeeman (1975).

t-coordinate of the cusp point. Then the equation of the cusp near the cusp point is given approximately by the formula

$$(s - s_N)^2 = K(t - t_N)^3,$$

where K is a constant. To obtain a more accurate equation would involve higher powers of $s - s_N$, $t - t_N$, which we shall ignore. One of the remarkable

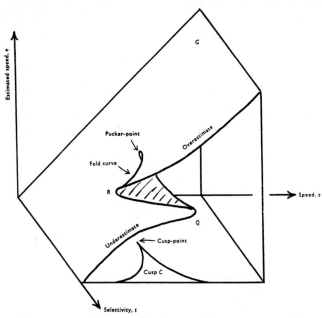

Fig. 8.—The graph G of estimated speed is a cusp catastrophe.

properties of the cusp catastrophe is that this formula is *independent of the particular quantitative properties of the graph G above*. We now utilize this formula to explain the experimental data as follows.

8. Quantification

The two variables used by Drew *et al.* (1959) in plotting the experimental data in Fig. 1 were: B = introvert/extravert measure on the Bernreuter Scale; and L = percentage time per lap difference from control. We must establish a connexion between their plot of individuals with respect to B, L and our formula for the cusp in terms of s, s_N, t, t_N, K.

Now the graph G above refers to a single individual. Therefore each individual has his own formula. However, if we impose the very tough implicit scientific hypothesis that *all human central nervous systems share the same precise subconscious selectivity mechanism*, then we are obliged to use the same constant K

for all individuals. Admittedly, we can choose K to best fit the data, but once chosen it must be fixed.

Recall the psychological hypotheses postulated in Section 3 above. Hypotheses 1, 2 and 3 imply that, for a given dose of alcohol, introverts will be more selective than extraverts. Therefore there is a correlation between t and B. If we assume that all individuals are affected to a greater or lesser extent, and the selectivity tends to zero as $B \to 0$, then the simplest quantification of t is the linear one:

$$t - t_N = B.$$

The relationship between speed and L is straightforward. If τ = time per lap, and τ_N = control, then lap distance = $s\tau = s_N \tau_N$. Hence,

$$\frac{s_N}{s} = \frac{\tau}{\tau_N} = 1 + \frac{L}{100},$$

and therefore

$$s - s_N = -\frac{L}{L+100} s_N.$$

Finally, we use Hypothesis 4. In order to drive at a normal speed to the best of his ability, each individual places himself at Q or R, as we explained in Section 6 above. Therefore his projection on the (s, t)-plane lies on the cusp. Therefore the formula holds. Substituting in the formula

$$\left(\frac{L}{L+100}\right)^2 s_N^2 = KB^3.$$

Since the normal speed s_N is not recorded in the data of Drew *et al.* (1959), let us, as a first approximation, assume that all individuals had the same normal speed, so that we can absorb it into the constant K:

$$\left(\frac{L}{L+100}\right)^2 = KB^3.$$

Now suppose we choose K so as to fit the most introverted individual at $B = 100$, $L = -22$ on to the curve. Therefore

$$K = 100^{-3}\left(\frac{-22}{-22+100}\right)^2 = 0.000000080.$$

We can now plot L as a double-valued function of B:

$$L = \frac{+0.028 B^{\frac{3}{2}}}{1 \mp 0.00028 B^{\frac{3}{2}}}.$$

The curves are shown on Fig. 9.

If, further, we choose a small 'standard deviation' as indicated by the dotted lines, then the region enclosed will include 50 per cent of the individuals (20 out of 40). Trebling this deviation will include over 80 per cent of the individuals.

What is remarkable about the closeness of fit is that we have not merely assumed the formula to be true locally near the cusp point, but have used it *globally* after one single choice for the value of K. Such a fit is strong evidence in

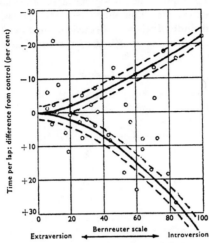

Fig. 9.—Mean time per lap (difference from control) for Dose 4 plotted against extraversion measured by the Bernreuter Scale.

support of our implicit hypothesis about the precision of the subconscious selectivity mechanism.

9. Predictions and Designs of Experiments

The theory we have outlined above suggests a large number of possible experiments and predictions, of which we indicate a sample.

(1) Draw Fig. 9 so as to include the normal speed s_N in the formula, to see if the fit is improved.

(2) Draw Fig. 9 for percentage *speed* difference from control, because this should be more symmetrical about the zero percentage axis.

(3) Draw Fig. 9 for increasing stress (increasing dosages of alcohol) because the cusp should then widen; K will be an increasing function of stress.

(4) The graph of L plotted against stress (alcohol dosage) for a single individual should reveal one branch of a cusp.

(5) An interesting experiment would be to compute the three-dimensional graph G for an individual, by recording the estimated speed. This could be done either directly, by asking the individual to give a running commentary as he drives, saying whether he is going too fast or too slow, or alternatively, indirectly, by keeping a time record of the pressures of his foot upon the controls of accelerator and brake. This would be particularly fascinating if it led to a scientific confirmation of the hidden catastrophes predicted in Section 6 above,

where the estimated speed jumps, and the driver realizes, with a sudden small shock that he is driving too fast or too slow. Once G had been computed for an individual, then these shocks should be predictable, both in magnitude, and the stress and speeds at which they take place.

(6) Similar experiments could be performed with other forms of stress, such as noise, causing conflicting judgements in other forms of skill. This would test whether selectivity, as we have described it, is a general response mechanism of the central nervous system.

REFERENCES

DREW, G. C., COLQUHOUN, W. P. & LONG, H. A. (1959). Effect of small doses of alcohol on a skill resembling driving. *Memo* 38. London: Medical Research Council.

ISNARD, C. A. & ZEEMAN, E. C. (1975). Some models from catastrophe theory in the social sciences. In L. Collins (ed.), *Use of Models in the Social Sciences*. London: Tavistock Publications. [10]

THOM, R. (1972). *Stabilité structurelle et morphogénèse*. New York: Benjamin.

ZEEMAN, E. C. (1965). Topology of the brain. In *Mathematics and Computer Science in Biology and Medicine*. London: Medical Research Council.

— Numbering in this book.

13 A MODEL FOR INSTITUTIONAL DISTURBANCES

By E. C. Zeeman,[†]
University of Warwick

C. S. Hall,
Home Office Prison Dept., Birmingham

P. J. Harrison
University of Warwick

G. H. Marriage,
Home Office Prison Dept., Surbiton

and

P. H. Shapland,
Home Office Prison Dept., Birmingham

Disturbances in institutions are often thought to be due to special local circumstances. This paper outlines a general mathematical model for sudden outbreaks of disorder in an institution. The model is illustrated by applying it to the escalating sequence of events at Gartree Prison during 1972. Although the approach is largely theoretical some suggestions are made about the prediction and handling of disorder. In particular, the model suggests why the policy of 'playing it cool' is generally likely to be successful.

1. The Model

The factors underlying disorder in an institution may be roughly grouped under two headings: (1) tension (frustration, distress) and (2) alienation (division, lack of communication, polarization). This applies not only to prisons, but also to organizations such as universities, hospitals, firms, governments, committees, armies, nations and even to individuals. Setting aside for the moment the problems of definition and measurement, let us first tackle the problem of what ought to be the theoretical shape of the graph of disorder as a function of tension and alienation. Since it is the graph of a function of two variables, it will be three-dimensional and somewhat difficult to describe verbally. So let us begin with a couple of observations describing the effect of each factor separately: (1) The more tension, the more disorder. (2) The more alienation, the more sudden and violent are the outbreaks of disorder.

The reader may well protest that these observations are too crude, and may wish to hedge them about with reservations, and we would agree that they are inadequate. In a sense they are only first approximations to the truth, and are so transparent they they can be accused of being both trivially true and trivially false. For instance, (1) is trivially true in the sense that if one is asked whether disorder is an increasing or decreasing function of tension, then the obvious reply is that it is increasing. On the other hand, (1) is trivially false, because one can easily envisage situations of heightened tension and unnatural quiet, which might be viewed as the calm before a storm. Again, (2) is trivially true in that

Published in the British Journal of mathematical and statistical Psychology, Volume 29, 1976, pages 66-80.

alienated institutions sometimes suffer the worst riots, but trivially false in that at other times they seem to escape disorder.

The reason for making these simplistic observations is that the trivially true part gives us sufficient mathematical information to enable us to deduce the shape of the three-dimensional graph. Then the mathematics in turn is able to hand back to us an explanation of the trivially false part of the observations, which arise from different parts of the graph having different types of section.

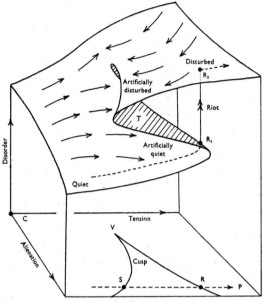

FIG. 1.—Theoretical graph G of disorder as a function of tension and alienation. The dotted path P of continuously rising tension causes a riot at R, where the disorder level jumps catastrophically from the quiet equilibrium at R_1 to the disturbed equilibrium at R_2. The arrows on the surface illustrate the feedback flow.

From (2) we know that the graph must possess discontinuities representing outbreaks of disorder, and that alienation is a 'splitting' factor. The general mathematical theory of discontinuous behaviour arising from continuous underlying forces is called 'catastrophe theory', and for an introduction to the latter the reader is referred to Thom (1972) and Isnard & Zeeman (1975). However, to understand the model it is not necessary to understand the underlying mathematics—it suffices to know that there is a deep theorem giving the qualitative shape of a three-dimensional graph with a splitting factor. This graph is called the *cusp catastrophe* and is the surface shown in Fig. 1. The two horizontal axes are called the *normal* and *splitting* factors; the normal factor is the direction in which the graph is increasing, and the splitting factor the direction in which the fold occurs. Our original observations (1) and (2) enable us to identify these two factors as tension and alienation. Therefore we can now replace those inadequate observations by the more sophisticated hypothesis as follows:

Hypothesis 1. *The graph G of disorder is a cusp catastrophe with tension as normal factor and alienation as splitting factor, as shown in Fig. 1.*

The implicit mathematical philosophy behind Hypothesis 1 is as follows (see Zeeman, 1974). To model the varying state of the institution accurately it would be necessary to use, firstly, a very high dimensional space X to represent all possible states; secondly, a dynamic on X representing the human relationships tending to reach a stable equilibrium state; and thirdly, stochastic noise representing incidents and external influences, continually disturbing that equilibrium. Generally, the dynamic would be able to restore the equilibrium, but exceptionally the noise might carry the system past a threshold, so that the dynamic, instead of carrying it back to the old equilibrium, might carry it forward to a different equilibrium—for example, an incident leading to an outbreak of disorder.

Now the assumptions leading to Hypothesis 1 are as follows. We assume that, for given levels of tension and alienation, provided that the alienation is low, there is a unique† equilibrium state, represented by a single point on the graph G. Therefore for low alienation the graph is single-valued, and we assume that it is a continuous increasing function of tension. Similarly, we assume that if the alienation is high and the tension low, there is a unique equilibrium, with low disorder; for convenience we have labelled this part of the graph 'quiet' in Fig. 1. Similarly, we assume that if both alienation and tension are high there is a unique equilibrium, with high disorder, which we have labelled 'disturbed'.

But if the alienation is high and the tension moderate we assume there are two equilibria, because, with the use of formal controls, it may be possible to hold the institution quiet or by letting it erupt allow the disorder to reach a higher level than might be expected for that level of tension. To emphasize the two equilibria we have labelled them 'artificially quiet' and 'artificially disturbed' in Fig. 1. Therefore over this region the graph becomes two-valued, with the two sheets separated by a third sheet, T, representing the threshold separating the two equilibria.

We now appeal to the classification theory of catastrophe theory (Thom, 1972) and deduce that G and T together form a smooth surface equivalent‡ to

† At first sight, the assumption here of a single equilibrium might appear somewhat sweeping. However, it can be justified mathematically, for when taken in conjunction with Hypothesis 3 below, the resulting behaviour is similar to that arising from many equilibria following some likelihood function.

‡ 'Equivalent' means precisely that there is a projection of X on to the vertical disorder axis throwing $G \cup T$ on to a surface in three dimensions, and then a smooth map of three dimensions on to itself, mapping vertical lines to vertical lines, and throwing $G \cup T$ on to the cusp-catastrophe surface (see Thom, 1972; Isnard & Zeeman, 1975). Since equivalence preserves qualitative features, and since we only want to discuss qualitative features at this stage, it suffices to look only at the canonical surface.

the cusp-catastrophe surface, given by the equation
$$d^3 = t + ad,$$
where d, t and a measure disorder, tension and alienation, respectively (relative to an origin taken at the cusp point). The graph G is the part given by $3d^2 \geq a$, the threshold surface T is the part given by $3d^2 < a$, and the fold curve between them is given by $3d^2 = a$. The fold curve is therefore the boundary of G. The projection of the fold curve in the tension-alienation plane, C, is the cusp, given by the equation
$$27t^2 = 4a^3.$$
The graph G is single-sheeted over the outside of the cusp and double-sheeted over the inside. To see why this represents a 'catastrophe', consider a path P in C representing fixed high alienation and rising tension as shown in Fig. 1. The disorder level is initially quiet, and the dynamic holds the state in the quiet equilibrium until the point R is reached, when the quiet equilibrium at R_1 breaks down, and the dynamic rapidly carries the state into the disturbed equilibrium at R_2; we have indicated the jump in Fig. 1 by a vertical double-headed arrow labelled 'riot'. This sudden outbreak of disorder is a discontinuous change of state caused by gradually rising tension; in other words, a 'catastrophe'. Such discontinuities are one of the characteristic properties of catastrophe models, and the higher the alienation the greater the catastrophe, confirming our original observation (2). Similarly, if we consider the reverse path, with tension subsiding, then the disturbance will continue until the point S, when quiet will suddenly descend. Notice the hysteresis effect over the interval RS, which grows with alienation. This corresponds to the delays that can often be observed: the calm before a storm, and the lingering aftermath of a storm.

It will be seen that the unique feature of this kind of model is its ability to show how a catastrophic discontinuity can be the result of a continuous, even gentle, change in tension. We must now introduce a hypothesis to explain what causes change in tension.

Hypothesis 2. There is a tendency for an institution as a whole to avoid the extremes of 'quiet' and 'disturbance'.

Prisoners can spend literally years in the same institution and we may assume that, however well ordered the regime, many find it monotonous. In this situation it is common for people to seek sensation, that is go out of their way to generate stimulation, and we may assume that it is characteristic of any institution that, if life becomes 'too quiet', there will be a number of people who will spontaneously set about generating stimulation. But if the level of disorder becomes excessive, staff will naturally react and so the institution may be said to have an overall homeostatic tendency to keep within a 'moderate level of disorder': if life becomes too quiet prisoners will begin overtly to seek stimulation, whereas if it becomes too chaotic the staff will start to exercise firmer control.

This tendency can be represented mathematically by a flow on the surface G, as shown by the arrows in Fig. 1. We may regard this flow as a feedback mechanism (see Zeeman, 1974) as follows: firstly, the surface G of Hypothesis 1 represents the influence of tension and alienation upon disorder, and secondly, the flow of Hypothesis 2 represents the feedback influence of disorder upon tension and alienation. The implicit mathematical assumption is that the feedback flow is much slower than the dynamic underlying Hypothesis 1, and therefore the dynamic will hold the state on the surface G while the feedback will move it slowly about on G. In other words, the feedback mechanism uses tension to restore the homeostasis in the disorder level.† It can be seen from Fig. 1 that if the alienation is low and there are good relations between staff and prisoners, then the feedback flow will contribute to maintaining a long-term stable equilibrium in the institution. However, if the alienation is high the result will be catastrophic alternation between periods of quiet and periods of disturbance. For in the alienated quiet state the feedback causes a slow build-up of tension and further alienation (over a matter of weeks or months); eventually the build-up reaches the right-hand side of the cusp, where the fast dynamic takes over, representing the sudden riot. Conversely, in the disturbed state the feedback causes a release of tension, somewhat faster (over a matter of days or weeks); eventually the left-hand side of the cusp is reached, and quiet suddenly descends. Meanwhile, during the disturbance there may be an initial increase in alienation, but possibly a subsequent decrease due to increased communication stimulated by the disturbance. However, while the alienation remains high there will continue to be a hysteresis cycle alternating between quiet and disturbance. Hysteresis is another characteristic property of catastrophe models.

So far, our model is deterministic, and to make it realistic we must add the third and final hypothesis allowing for chance elements.

Hypothesis 3. External events, or internal incidents within the institution, may be represented as stochastic noise.

Stochastic noise is represented mathematically by random displacements in the three variables causing fluctuations on either side of the surface. This is illustrated in Fig. 2, which, for simplicity, is confined to the two variables tension and disorder, representing a section of Fig. 1 at high alienation. The noise is similar to a stationary process and we call the variability of the process the *noise level*.

Noise can be interpreted as covering both external and internal events; for example, external influences such as a strike called by a prisoners' rights group, or internal actions by individuals within the institution. Therefore the noise level will have both a global and a local component; the global component will manifest itself during periods of widespread disturbance throughout several institutions, while the local component will manifest itself in local disruptions.

† The two factors tension and alienation emerged as independent factors after factor analysis of data. In the model they are treated as independent with respect to the fast dynamic, but dependent with respect to feedback.

The local component will depend upon the particular population and environment of the institution, and therefore different institutions will experience different noise levels. It can be shown (Marriage, 1975) that in prisons the noise level is highest amongst populations containing a higher proportion of those younger men who confine themselves to serious offences; therefore this type of population is more volatile and more susceptible to catastrophic oscillation.

Mathematically, if the state is displaced from equilibrium by Hypothesis 3, then it is generally restored to a nearby equilibrium by the dynamic of

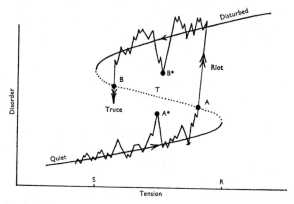

FIG. 2.—Stochastic noise represents incidents that may spark a riot, or moves that may lead to a truce.

Hypothesis 1. Exceptionally, however, a displacement of tension or alienation may cross one of the branches of the cusp and cause a catastrophe. Alternatively, a displacement in disorder may cross the threshold T, causing a catastrophe as illustrated in Fig. 2; for example, the point A might represent an incident during a period of quiet sparking off a series of incidents leading to a riot. Conversely, B might represent a move during a period of disturbance initiating a series of moves, leading to a truce. Therefore catastrophes do not occur exactly on the cusp lines, but earlier inside the cusp, just before the cusp lines are reached. In the centre of the cusp region catastrophes are less likely, as is illustrated by the points A^*, B^*, where noise of the same magnitude as A, B does not reach the threshold T, and therefore does not cause a catastrophe because the disorder level returns to its previous equilibrium surface in each case.

Remark about personality. We have completed our three hypotheses and the theoretical construction of the model, which consists of the surface, the flow thereon and the noise level. Before we go on to analyse data, we make a remark about what to expect from such analysis. Although different institutions may have the same qualitative model, we should not necessarily expect them to have the same quantitative model; in other words, we should not expect them to have exactly the same surface, but only equivalent surfaces (see earlier definition of this term). An institution at any moment will be represented by both its surface (and flow and noise), and a point on the surface; the point represents the behaviour of the institution at that moment, while the surface represents potential behaviour, or, in other words, 'personality' of the institution. The particular personality will depend upon environment, staff, population, traditions, memories and recent past history. Thus as the point wanders back and forth over the surface, the surface itself may respond like a living thing, tightening up after periods of tension or relaxing after periods of quiet. This seemed to occur, for instance, in the data below, illustrated by the shift in the position of the cusp in Fig. 3. The situation is analogous to that of an individual, who behaves in accordance with his personality, but whose personality at the same time is modified by his experience and behaviour.

2. Analysis of Data

We analyse the data arising week by week from Gartree Prison during the year 1972. During this year there was a series of disturbances of varying seriousness, culminating in a riot on the first day of week 48. The build-up of tension and alienation during the year is illustrated in Fig. 3. We shall discuss the nature of the three variables further in Section 3 below. In the mean time they are measured as follows:

(1) *Tension*. We assume that physical sickness, real or apparent, is related to mental frustration, and consequently to the tension within the institution. We therefore add together the five weekly standardized values of (i) men reporting sick, (ii) men reporting special sick (at work), (iii) Governor's applications, and (iv) welfare visits to A wing and (v) to D wing. The reason for adding them was that a factor analysis of data, thought to be relevant to disorder, gave them equal weights in the leading factor. This factor was identified as a sickness measure which fitted in with an *a priori* hypothesis. The insignificant variables were omitted and the simple sum of standardized scores used.

(2) *Alienation*. We assume that the numbers of men in the punishment wing (E wing) is related to the alienation between inmates and staff. We also assume that the numbers of inmates requesting segregation (on Rule 43) is related to alienation amongst inmates. We therefore add these two figures. Initially, the data were transformed but we finally decided that, both for clarity and simplicity, the raw data should be used, since the prison population remained nearly constant.

Remark. It could be argued that segregation is also related to tension. However, adding segregation to sickness gives a similar graph, and so we chose the simpler assumption. The number of disciplinary reports might also be related to alienation, but in this particular case the number of reports was statistically very uneven varying from a weekly average of 12 to an exceptional 143. A square-root transformation was beneficial but in this paper's results we did not use disciplinary reports. We make no claim that the above measurements are the best; they just happened to be the data available in retrospect. We suggest that the model ought to be looked at in other cases with different data, and then methods devised for the best future measurement of the required factors.

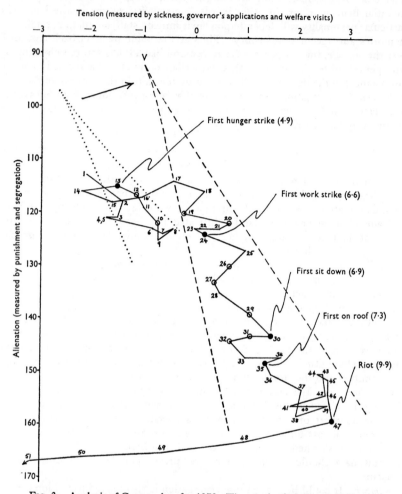

FIG. 3.—Analysis of Gartree data for 1972. Time path of tension and alienation is plotted weekly throughout the year (numbers indicate weeks). The serious incidents are indicated by circles. The solid circles indicate those incidents involving nearly all the inmates in a new form of mass protest; the numbers in brackets indicate an assessment of seriousness (out of 10). A possible initial position of the cusp is shown dotted and a possible subsequent position is shown dashed; the movement of the cusp may represent a higher tolerance level of tension in the institution after the first mass protest.

Smoothing. In order to reveal the underlying trend movement a filter or smoothing is required. Here a one-sided exponential filter was used. More precisely if R_t denotes the raw variable at week t, then the smoothed variable S is given by the formula

$$S_t = (1-\lambda)(R_t + \lambda R_{t-1} + \lambda^2 R_{t-2} + \ldots),$$

where $\lambda = 0 \cdot 8$. The results are shown in Table 1. The advantage of this simple smoothing is that it helps to eliminate irrelevant short-term effects. On the other hand, it has the disadvantage of damping, and therefore possibly hiding the effects of, the faster dynamics. For example, the true rise in tension preceding an outbreak of disorder, indicated by a rightward movement in

Table 1.—Data for Tension and Alienation

(Tension = smoothed standardized sickness + governor's applications + welfare visits.
Alienation = smoothed punishment + segregation)

	Week	Tension		Alienation	Punishment		Segregation	
		Raw	Smooth		Raw	Smooth	Raw	Smooth
	1	−3·3	−2·1	113	38	40	73	73
	2	1·5	−1·4	118	53	43	82	75
	3	−2·0	−1·5	121	56	46	76	75
	4	−2·3	−1·7	121	40	45	80	76
	5	−1·9	−1·7	121	46	45	74	76
	6	2·6	−0·8	123	61	48	71	75
	7	0·0	−0·6	124	53	49	77	75
	8	0·5	0·4	123	42	48	74	75
	9	−2·1	−0·7	125	46	48	84	77
○	10	−0·9	−0·7	122	39	46	72	76
	11	−2·1	−1·0	119	28	42	81	77
○	12	−1·7	−1·1	117	27	39	81	78
●	13	−3·0	−1·5	115	28	37	77	78
	14	−5·1	−2·2	116	41	38	77	78
	15	0·6	−1·6	118	41	39	82	79
	16	1·3	−1·0	117	31	37	85	80
	17	2·2	−0·4	114	24	34	78	80
	18	2·5	0·2	116	42	36	79	80
○	19	−2·0	−0·2	120	46	38	90	82
○	20	4·4	0·7	122	47	40	84	82
	21	−0·4	0·5	123	38	40	89	83
	22	−1·6	0·1	123	34	39	90	84
	23	−0·4	0·0	123	38	39	84	84
●	24	0·8	0·2	124	41	39	87	85
	25	4·2	1·0	127	54	42	87	85
○	26	−0·4	0·7	130	50	44	91	86
○	27	−0·7	0·4	133	51	45	94	88
	28	0·7	0·5	135	51	46	93	89
○	29	3·5	1·1	139	57	48	97	91
●	30	3·3	1·5	143	57	50	100	93

Table 1.—(cont.)

	Week	Tension Raw	Tension Smooth	Alienation	Punishment Raw	Punishment Smooth	Segregation Raw	Segregation Smooth
○	31	−0.7	1.1	143	46	49	99	94
○	32	−0.7	0.7	144	63	52	86	92
	33	2.0	1.0	147	61	54	96	93
	34	4.7	1.7	147	41	51	107	96
●	35	0.4	1.4	148	66	54	86	94
	36	2.0	1.5	150	61	55	97	95
	37	4.3	2.1	153	72	58	97	95
	38	1.8	2.0	158	73	61	103	97
	39	4.8	2.6	156	58	60	94	96
	40	0.4	2.2	156	69	62	84	94
	41	0.5	1.9	156	64	62	94	94
	42	5.2	2.6	154	42	58	103	96
	43	2.3	2.5	150	35	53	103	97
	44	2.1	2.4	150	44	51	108	99
	45	3.2	2.6	151	36	48	118	103
	46	2.4	2.6	154	37	46	126	108
●	47	3.3	2.7	159	49	47	127	112
	48	−5.7	1.0	163	63	50	119	113
	49	−6.9	−0.6	165	65	53	106	112
	50	−8.6	−2.2	166	56	54	110	112
	51	−7.0	−3.2	167	48	53	122	114
	52	−7.5	−4.1	172	55	53	138	119

Fig. 3, and the true drop in tension following incidents of disorder, indicated by a leftward movement, may both be hidden by this smoothing. The value 0.8 was chosen as the best compromise to extract the qualitative behaviour from rough data. It might be possible to use the more powerful but more complicated methods of Mixed Model smoothing to capture the essentials of both dynamics (see Harrison & Stevens, 1971).

Table 2.—Disorder: List of the More Serious Incidents

	Week	Date	Assessment	Description of incident
	5	4 February	2.3	Inmate found in possession of counterfeit £5 note.
○	10	10 March	6.6	Five men involved in assault on officers, then barricaded themselves in recess and smashed it up. All returned to cells in a short time.
○	12	21 March	6.7	Seven men pushed their way into vestibule (location of special pattern key) and refused to move. Let out through outer door and finally returned to main prison when ordered.

Table 2.—(cont.)

	Week	Date	Assessment	Description of incident
		12	24 March	3·9	Assault on an officer.
●	13	29 March	4·9	Hunger strike—total population (*c.* 370). One meal only. Complaint about food.
○	19	8 May	5·1	Refusal to work in No. 5 shop (tailors). Forty involved. Going to shop every day but not working. Complaint about rate of pay. Returned to work 16 May.
○	20	17 May	5·7	Sit-down protest involving 102 men for 12 hours. Complaint about adjudication.
●	24	13 June	6·6	350 men refused lunch, further complaint about food. Work strike (350)—rates of pay. Both ended June 15.
○	26	27 June	5·7	375 men in work strike—complaint about food. Half-day only.
○	27	6 July	5·3	Fire in tailor's shop store. Extensive damage. Fire brigade needed. Cause unknown.
	28	14 July	3·4	Fire in B wing recess. Cause unknown.
○	29	18 July	5·7	364 men in work strike. Complaint about food. Ended same day.
	29	22 July	3·3	Fire in C wing cell. Cause unknown.
●	30	26 July	6·9	378 men in work strike and sit-down in recreation area. Ended July 29.
○	31	4 August	6·6	Small public demonstration outside prison. After dispersal, 208 men sat down on exercise yard and remained until 5 August.
○	32	9 August	6·4	378 men in work strike. Complaint about transfer of a prisoner. Ended 11 August.
○	35	29 August	6·4	364 men in work strike. Complaint about rates of pay and food. Ended 31 August.
●	35	1 September	7·3	350 men in work strike. Complaint about adjudication. 211 sat down on exercise yard and 30 climbed on roof. Ended 4 September.
	46	18 November	3·4	Fire in A wing cell. Fire brigade needed. Cause unknown.
●	47/8	26 November	9·9	Thirteen men in attempted escape. Disturbance developing into violence in C/B dining-hall; three fires; disturbance extended to wings. Fire brigade called; police in reserve. Ended 27 November.

(3) *Disorder*. The list of main incidents during the year is given in Table 2. The seriousness of these incidents was assessed on a scale between 0 and 10 by seven independent assessors, and, since there was a high correlation, this assessment is consistent. The weeks containing an incident of seriousness greater than 5 have been marked with a circle in Table 1 and 2 and in Fig. 3. Those incidents involving nearly all the inmates in some new mode of protest have been marked with a solid circle. Notice that tension tends to be released after a new mode of protest has been tried out for the first time.

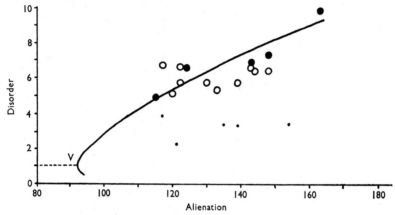

FIG. 4.—Assessment of incidents plotted against alienation. The vertex V of the parabola gives the cusp point.

Positions of the cusp. If the model is to fit the data, then we must justify the position of the cusp drawn in Fig. 3. First, notice the qualitative difference between the two variables: alienation builds up steadily throughout the year whereas tension oscillates to and fro. It could be argued that such oscillations merely reflect rough data, but on the other hand the correlation between the five ingredients suggests that the oscillations are significant. Moreover, the oscillations are precisely what the model leads us to expect within the region inside the cusp. By Fig. 2 catastrophes tend to occur before the cusp lines are reached, and so we have drawn the cusp so as to enclose the oscillations inside.

The next task is to locate the cusp point. The data, by listing the incidents, give us the coordinates of the point R_2 in Fig. 1, for various levels of alienation. In terms of the canonical formulae, R_2 is given by $4a = 3d^2$, $4t = d^3$, where a, t and d are measured relative to the cusp point. Therefore the projection of R_2 in the (a, d)-plane lies on a parabola, with vertex V at the cusp point. Figure 4 shows a plot of disorder against alienation, for all reported incidents. Since the minimum assessment of any reported incident is 2·3, we estimate that the homeostatic level of the institution implied by Hypothesis 2 is approximately at assessment level 1; in other words, this is the disorder level of the cusp point, and of the vertex V of the parabola. Therefore in Fig. 4 we have drawn a parabola through the incidents, paying particular attention to the solid circles, with vertex V at disorder level 1. From Fig. 4 we deduce the alienation value of V is approximately 92. Therefore in Fig. 3 we have located the cusp point at the intersection of alienation level 92 with a line that approximately bisects the oscillations. We emphasize that the above argument is largely guesswork and only gives an accuracy of about 92 ± 10, but it does lend plausibility to the cusp lines drawn in Fig. 3.

The position of the cusp fits the data well for weeks 17–48. However, the initial incidents in weeks 10, 12, 13 do not lie within this cusp, and so we propose a different initial position of the cusp, shown dotted in Fig. 3. We suggest that the first mass protest in week 13 may have caused a change in the personality of the institution, as follows: the awareness by both staff and inmates of the possibility and power of future mass protests may have raised the general tolerance level of tension within the institution, and thereby shifted the entire surface, and hence also the cusp, in the direction of increasing tension, as shown by the arrow in Fig. 3. The personality of the institution then remains static in a state of heightened tension, whilst the behavioural path oscillates down the cusp. During weeks 24–35 the behaviour is mainly on the upper disturbed surface, and during weeks 36–47 on the lower quiet surface. From the graph one might have expected the riot to occur two weeks earlier, but the delay may have been imposed by the planning involved in the attempted escape.

The final release in tension after the riot is very marked, but this part of the graph should also be interpreted as personality change, because the institution temporarily suspended normal activity and many inmates were transferred elsewhere.

3. Discussion

Most of the relevant literature on prison riots looks at the phenomena in sociological terms, although the language used is very similar to the 'catastrophe' presented here. For example, as Vernon Fox (1971) said: 'The way to make a bomb is to build a strong perimeter and generate pressure inside'; or as Shrag (1960) in his review describes: 'Each riot began as a sudden flare-up of violence. In most cases there was no perceptible forewarning, though a state of heightened tension and anxiety was widely recognized'. The problem is one of measurement, as indicated in Fox's (*op. cit.*) opening sentence: 'Finding valid, consistent and reliable information as to why prisoners riot defies most standard methods of gathering data on human behaviour'. The mathematical model presented here, however, provides a theory with which to discuss ideas about the functioning of institutions, and against which to test the measurement of institutional variables. Management information systems for prisons are still in a rudimentary state of development. So far, there has been some speculation about using statistical models, monitoring certain institutional variables, for instance, in order to establish normative data and confidence limits for taking action. This will be appropriate in some areas, but for global institutional processes a more theoretical approach is needed. The advantage of catastrophe theory is to give not only qualitative understanding into phenomena involving catastrophic change but also to provide quantitative models for experimental testing. This may eventually provide experimental confirmation of hypotheses which have so far had only intuitive support. Also eventually it may be possible to design quantitative monitoring that would provide a basis for action. As yet, it is not possible to determine precisely where the cusp lines are. In Fig. 3, for instance, we have drawn cusp lines to provide a plausible fit for retrospective data. In an ongoing situation, however, all that one knows is that the further the time path moves in an easterly direction, the nearer the catastrophe comes and the further it goes in a south easterly direction the larger the disturbance will be. For the further south the wider apart are the upper and lower surfaces in Fig. 1, and so the greater the jump that will occur. Even if effective monitoring became possible, monitoring by itself may be susceptible to interference during times of disturbance. Nevertheless, in such circumstances the underlying theoretical model may still provide a useful framework to fall back upon, a framework that could be adapted to alternative monitoring, or used in conjunction with the intuition of experienced staff. For example, one deduction which may be drawn from the theory concerns the policy of 'playing it cool', as we now explain.

With increasing unrest, and hence increasing tension and alienation, it will be the natural reaction of management to increase staffing levels. Very often staff are sent into the conflict situation with specific instructions to exercise extreme caution. This reaction may be interpreted in the light of the model as management attempting to reduce the 'noise level' and hence reduce the probability of a triggering incident occurring. In the event of confrontation both sides are forced to communicate, with the possible pay-off of a reduction in both tension and alienation resulting from the increased communication. When this situation occurs, the demands made by both sides become explicit and the extent of latitude for negotiation becomes known. Two outcomes are possible from the confrontation situation: (*a*) the increased communication facilitated by the confrontation permits a solution to be negotiated peacefully; or (*b*) the demands of both sides are irreconcilable or, alternatively, communication is impossible. In the latter situation the authorities are faced with alternatives of either waiting to see if negotiation becomes possible and avoiding further incidents by playing it cool, or else attempting to restore order by force. It is in this situation that the model is both helpful and prescriptive, as follows.

If the policy of 'playing it cool' is adopted, then by Hypothesis 2 there will be a reduction in tension, with the time-scale for this reduction being days rather than hours. The net result of this reduction of tension, when taken with the fast dynamic of Hypothesis 1, is to reduce not only tension but alienation as well. This is because the fast dynamic ensures that the process stays on the disorder surface, as long as no incident causes a displacement from it. Thus playing it cool not only prevents the escalation of both tension and alienation to be found in situations of continuing confrontation, but ensures that events move in the desired direction, other things being equal.

This is illustrated in Fig. 2, where B, B^* represent the same move designed to achieve a truce, with B^* made near the beginning of the unrest and B made after tension has subsided somewhat. B^* is unsuccessful because the threshold is too far away but B is successful. It follows that timing is a crucial element in intervention and that there is a minimum period which must elapse before the disturbance can subside; although little systematic evidence exists, support is given to this argument by the fact that highjackers rarely surrender within the first 24 hours, and prison riots seem to take a few days to settle.

The illustration above reveals how Hypothesis 2 is critical to the policy of playing it cool; and it also explains how this policy, by cooperating with, instead of working against, the internal sociological dynamics of the institution, is sometimes able to restore order with minimal use of force.

This study was done retrospectively using what information was available. It seems probable that the number of men on Rule 43 (requesting segregation) and the number of men in E Wing (undergoing punishment) is a reasonable measure of the variable we have called 'alienation'. These measures include some effects of both the staff–inmate and inmate–inmate interfaces. This is not so true, however, of the 'tension' variable. The number of men going sick, special sick, applying to see the Governor and Welfare Officer are intercorrelated. We cannot, however, distinguish with Governor's applications the exact nature of the application. We merely postulate that the applications contain a 'tension' component within the inmate. This means that our tension indicators reflect 'tension' within the inmate population and do not include any measure of tension within staff.

It is hoped shortly to institute an ongoing monitoring system at Gartree, using better measures of the variables. This will be used by management as part of the information on which decisions will be made. Only in this way do we think that the model can be tested for predictive accuracy.

Any opinions expressed are personal and do not necessarily represent the Prison Department's views.

References

Fox, V. (1971). Why prisoners riot. *Fed. Prob.* **35**, 9–14.
Harrison, P. J. & Stevens, C. F. (1971). A Bayesian approach to short-term forecasting. *Opl Res. Q.* **22**, 341–362.
Isnard, C. A. & Zeeman, E. C. (1975). Some models from catastrophe theory in the social sciences. In L. Collins (ed.), *Use of Models in the Social Sciences*. London: Tavistock Publications, pp. 44–100.
Marriage, G. H. (1975). Part of Ph.D. thesis in preparation, University of London.
Schrag, C. (1960). The sociology of prison riots. In *Proceedings of the American Correctional Association*. New York.
Thom, R. (1972). *Stabilité structurelle et morphogénèse*. Benjamin: New York.
Zeeman, E. C. (1974). Levels of structure in catastrophe theory. In *Proceedings of the International Congress of Mathematicians*, University of British Columbia, Vancouver, Canada.

14 PRISON DISTURBANCES

This talk described a catastrophe model for prison riots. Since the material is to be published elsewhere [1], I will comment here mainly upon the <u>procedure</u> adopted. For, although it is easy enough to use the cusp-catastrophe to trot out models of human behaviour by the dozen, which may be both illuminating and entertaining, it is however quite another matter to test such models with data. And if catastrophe theory is going to be of any use in the human sciences, then its models must not only offer qualitative insight, but must also be susceptible to quantitative testing. The prison riot model is one of the first to be fitted to data, and although the results are not conclusive - indeed it is only the first crude test - nevertheless it is perhaps worthwhile discussing the procedure, because this might help in other applications, and perhaps stimulate better procedures.

1. Firstly it takes time; we have been discussing this model on and off for about 5 years.

2. Secondly it needs a partnership between a mathematician and a scientist; in this case I was the mathematician and Peter Shapland was the psychologist, and we meet regularly through our families.

3. Thirdly it needs motivation; Shapland is the chief psychologist of the Midland region of the Prison Department, and has some 25 institutions under his care. He was keen to explore any avenue that might help to explain why an institution sometimes explodes, and at other times remains at peace, be

---

Published in Structural stability, the theory of catastrophes and applications in the sciences, Springer Lecture Notes in Mathematics, Volume 525, 1976, pages 402-406.

it a prison, a university, or what have you. An explosion is a catastrophic jump in the level of disorder, and so we began to look for a catastrophe model. What are the causal factors?

4. It needs patience to collect data. Shapland decided to look at Gartree prison, which experienced an escalating sequence of incidents during 1972. He asked one of his staff, Chris Hall, to collect as much likely-looking retrospective data as possible. Hall found about 25 indices, such as the numbers of prisoners reporting sick, etc., and collected the 52 weekly totals of each during the year, standardised them, and did a computerised factor analysis. The factor analysis threw up two main factors, which it seemed reasonable to call "tension" and "alienation". The tension factor arose from correlations between sickness, numbers of welfare visits, and numbers of applications to see the governor. The alienation factor arose from a correlation between the numbers of prisoners in the punishment wing, and those who requested segregation (to avoid other prisoners).

5. It needs statistical skill to handle rough data, knowing what weight to attach to various analyses, knowing when to smooth, and when not to despair; for this we called in the expertise of a statistician, Jeff Harrison, who quickly detected the oscillatory nature of the tension compared with the steady growth of alienation (see Figure 2).

6. Then came the building of the model. For this purpose we all assembled together for a day, joined by a third prison psychologist, Hugh Marriage, who had experience of different types of prison populations in other regions. It was very important for us all to meet together for a whole day, because a great many ideas were tried out and rejected, and relatively few passed the stringent test of being acceptable to all the disciplines represented. The psychologists were essential because they could judge from

their expertise which hypotheses were plausible, and which unrealistic, which conclusions were valuable, and which surprising, and they chose carefully the right words to use. The mathematician was essential because frequent questions arose as to what exactly the classification theorem implied, what restrictions it placed, what freedom it permitted, and what predictions would be expected from it - difficult questions to answer without discussing at some length the context in each case. The conclusion of this session was the somewhat innocent looking cusp-catastrophe shown in Figure 1.

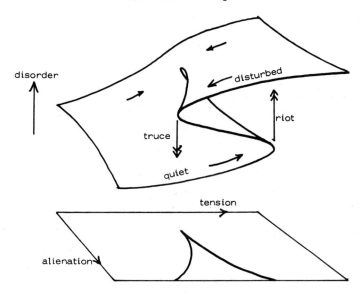

Tension was chosen as a normal factor because more tension normally means more disorder. Alienation was chosen as a splitting factor, because the more alienation, the more sudden and violent are likely to be the outbreaks of disorder.
7.   The data was plotted in the context of the model, and somewhat to our surprise appeared quite suggestive : see Figure 2*. For an analysis see [1]. Let me hasten to add that the cusps drawn in Figure 2 are largely guesswork, because there is insufficient retrospective data to plot a

---

* See Figure 3 on page 394.

3-dimensional graph. Indeed the only measurements of disorder available were the catastrophic jumps, in the escalating sequence of incidents during the year.

8. The rest can be imagined : drafts circulating and being tinkered about with. Another long day together, questioning, reassessing, and polishing What are the conclusions?

9. The model helps to explain the internal dynamics of an institution; and understanding helps one to cooperate with, rather than fight against, the dynamics. For example it suggests why the policy of playing it cool is generally likely to be successful, because a disturbance may cause a gradual release of tension, and therefore suddenly become spent.

10. An on-going monitoring system has been instituted at Gartree, using better measures of the variables, and the information will be used to assist in decision making. Only in this way can the model be tested for predictive accuracy.

REFERENCE

1. E.C. Zeeman, C.S. Hall, P.J. Harrison, G.H. Marriage and P.H. Shapland, A model for institutional disturbances, British Jour. Math. and Stat. Psychology (to appear).

(Number 13 in this book)

ISBN 0-201-09014-7, 0-201-09015-5 (pbk.)

PHYSICAL SCIENCES

PHYSICAL SCIENCES

Paper		Page
15.	A catastrophe machine.	409
16.	Euler buckling.	417
17.	Stability of ships.	441 - 493.

Paper 15 describes a pedagogical device that is easy to make and intuitively rewarding.

Paper 16 presents some classical results from elasticity in the language of catastrophe theory, concerning the buckling of beams. Both the free beam and the pinned beam are analysed. The last section on globalisation gives the new result that the bifurcation set of a pinned beam is the evolute of an ellipse.

Paper 17 introduces a global geometric approach to the stability of ships, complementary to the classical analytic approach. The paper begins with an elementary exposition of the classical linear theory of ship motions. The non-linear statics is then described by taking the position of the centre of gravity of the ship as a parameter, and analysing the manifold of equilibrium states; the manifold contains a variety of elementary catastrophes associated with heeling, capsizing and the point of inflextion on the lever arm curve. The non-linear dynamics is described by the parametrised system of damped Hamiltonian flows, and various unsolved problems are discussed. The paper is exploratory, in order to see if this approach might eventually lead to a better understanding of non-linear coupling, non-linear ship motions in a seaway, handling procedures in heavy weather, and design criteria for stability.

15 A catastrophe machine

This is a small toy made out of elastic bands designed to illustrate catastrophe theory, and in particular the cusp (or Riemann-Hugoniot) catastrophe [1-4]. I designed it for pedagogical reasons, in order to provide a concrete example in which all the variables were obvious and measurable, and the relationship between them differentiable and computable. But then I found it so illuminating to experiment with, giving such a powerful intuition of how a continuous change of control can cause discontinuous jumps in behaviour, that I strongly recommend the reader to make one for himself or herself.

Materials needed: Two elastic bands, three drawing-pins, a piece of cardboard and a piece of wood (the desk-top will do if you don't mind sticking drawing-pins into it).

Take as unit length the length of the unstretched elastic bands. With this unit the wood needs to be about 2×6. Cut a cardboard disk of unit diameter. Let X be the first drawing-pin. Use X to pin the centre of the disk to the centre of the wood, so that it spins freely. Make sure that it does spin freely by enlarging the hole if necessary—sometimes a small washer under the disk helps. Stick drawing-pin Y into the wood a distance 2 from X. Now fix the elastic bands to the disk at a point near the circumference; the easiest way to do this is to stick drawing-pin Z *upwards* through the disk near the circumference, tie the two elastic bands together in a reef knot, slip the knot over the point of Z and pull it tight. (Of course the knot does have a tendency to slip off occasionally, but the enthusiast no doubt can replace drawing-pin Z by a suitable nut and bolt.)

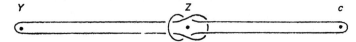

FIGURE 1

Now slip the other end of one of the elastic bands over Y, and the machine is complete, ready to go.

Hold the other end c of the other elastic band so that both elastics are taut, and move c smoothly and slowly about the plane. The disk will respond by moving smoothly and slowly most of the time, but occasionally it will make a sudden

---

Published in Towards a theoretical biology (Edited C.H. Waddington) Edinburgh University Press, Volume 4, 1972, pages 276-282.

jump. This jump is called a *catastrophe*. Now put a pencil dot where *c* was each time the machine makes a jump. Very soon the dots will build up a concave diamond-shaped curve with cusps *P, P', Q, Q'* (see Figures 2 and 6). This curve is called the *bifurcation set B*. The curve is symmetrical about the line *XY*. The cusps *P, P'* lie on the line *XY* at distances approximately $XP = 1.41$, $XP' = 2.46$.

The reader will observe that the machine does not jump when *c* enters *B*, but only where *c* exits from *B*, and then only provided *c* has previously entered *B* from the opposite side. If *c* makes a complete circle round *B*, then the disk executes a smooth circle. If *c* makes little circles round *P* or *P'* then the disk hiccoughs once each time round. By contrast if *c* makes little circles round *Q* or *Q'* then the disk makes smooth oscillations. All this qualitative behaviour is surprising at first, but becomes easy to understand with the pictures of catastrophe theory. Therefore let us put the mathematics of the machine into the framework of catastrophe theory. Define

Control point $= c =$ the held end of the elastic.
Control space, $C =$ plane $=$ all possible positions of *c*.
State, $\theta =$ angle $Y\dot{X}Z =$ position of disk.
State space, $S =$ circle $=$ all possible positions of disk.
Potential $V_c(\theta) =$ potential energy in elastic bands with control held at *c*, and disk held at θ.

Therefore

$V_c : S \to R$ is a smooth function from the circle *S* into the reals *R*, and
$V : C \times S \to R$ is a smooth function $C \times S$ into the reals.

If *c* is held fixed and the disk is released then the disk will jump into a state θ_c, that is a local minimum of V_c, and stay there. Friction damps out any oscillations, so that the system is dissipative. Therefore the machine obeys the fundamental requirement of catastrophe theory, that the state θ seeks a local minimum θ_c of the potential V_c. It transpires that if *c* lies *outside B* then V_c has one minimum (and one maximum). Therefore the position of the disk is determined uniquely. However if *c* lies *inside B* then V_c has two minima (separated by two maxima). The choice of which of the two minima that potential V_c will take, and therefore which of the two positions the disk will seek, is determined by the past history of *c*, as follows. If *c* moves smoothly then a minimum of V_c will move smoothly, unless it happens to be annihilated by coalescing with a maximum as *c* exits from *B*. In this case, if the disk were in the minimum that was annihilated, then it will have to jump into the other minimum. This is illustrated by Figure 3 which shows a sequence of potentials V_c as *c* runs along the dotted line in Figure 2, near the

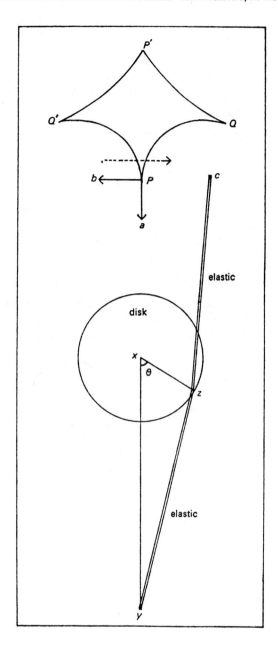

FIGURE 2

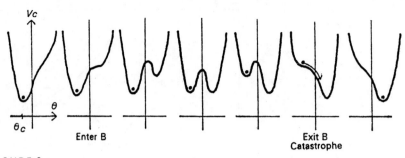

FIGURE 3

cusp P, parallel to the b-axis. In each case the graph is drawn for small value of θ.

The next diagram, Figure 4, shows the graph of θ_c as a function of c, drawn for small values of control coordinates $c=(a, b)$ with origin at the cusp P, and for small values of θ. More precisely the graph is given by $\frac{\partial V}{\partial \theta}=0$, because the graph includes not only the minima of V_c but also the maxima of V_c near $\theta=0$, shown shaded. The other single valued surface of maxima near $\theta=\pi$ is not shown in Figure 4 (but is shown in Figure 5).

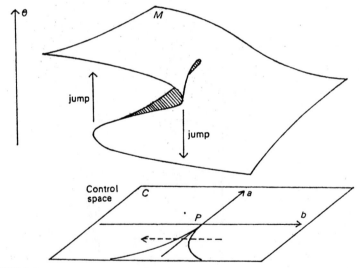

FIGURE 4

Using Hooke's law for the energy in the elastic bands one can compute the approximate equation for M:
$$\theta^3 - 1\cdot 3b\theta^2 + 1\cdot 8a\theta + 1\cdot 3b = 0 \ ,$$
which is equivalent (in sense of [4, 1.10]) to the canonical cusp catastrophe surface
$$\theta^3 + a\theta + b = 0 \ .$$
As c varies the state θ_c stays on the surface M of minima. Therefore as c moves along the dotted line the state has to jump from the lower branch of the surface on to the upper branch at the second crossing of the cusp. If c moves to and fro the state performs hysteresis action.

We may summarize Figures 3 and 4 by saying that the machine obeys the *delay convention* which says 'delay jumping until necessary'. Delay is a consequence of the fact that the dynamics of the damped disk obeys a differential equation in θ with the property that $\dot\theta$ and $\frac{\partial V}{\partial \theta}$ have opposite signs. Any appli-

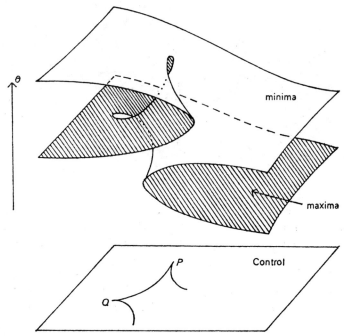

FIGURE 5

cation of catastrophe theory satisfying a differential equation with this property will exhibit delay—for example the applications to heartbeat and nerve impulse [4]. By contrast the application to phase-transition [1] does not exhibit delay. The reason is that the variables temperature, pressure and density in phase-transition do not obey a differential equation direct, but are averaging devices, exhibiting the average behaviour of very many differential equations obeyed by particles of the substrate.

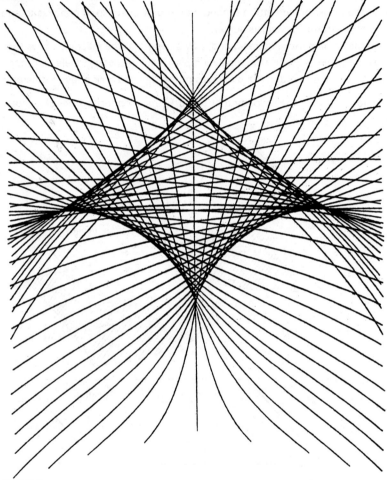

FIGURE 6

If we now turn our attention to the cusp Q we find a surface similar to Figure 4, but dual in the sense that maxima and minima are interchanged. If we put the pictures for P and Q together we obtain Figure 5. It is difficult to combine pictures for all four cusps, because this would involve bending the θ-axis round into a circle. We leave this as an amusement for the reader.

The precise shape of the bifurcation set B has been studied by Poston and Woodcock [2]. They have drawn it accurately by computer, and I am grateful to them for allowing me to reproduce it in Figure 6. Here the lines represent the intersections of M with the planes $\theta =$ constant, and hence B is the envelope for all such lines.

References

1. D. H. Fowler, The Riemann–Hugoniot Catastrophe and van der Waals Equation, in (C. H. Waddington, ed.) *Towards a Theoretical Biology 4: Essays* pp. 1-7 (Edinburgh University Press 1972).

2. T. Poston and A. E. R. Woodcock, Zeeman's Catastrophe Machine, *Proc. Camb. Phil. Soc.* (In press).

3. R. Thom, Topological models in biology, in (C. H. Waddington, ed.) *Towards a Theoretical Biology 3: Drafts* pp. 89-116 (Edinburgh University Press 1970).

4. E. C. Zeeman, Differential equations for the heartbeat and nerve impulse, in (C. H. Waddington, ed.) *Towards a Theoretical Biology 4: Essays* pp. 8-67 (Edinburgh University Press 1972).

(Number 3 in this book)

16 EULER BUCKLING

There has recently been a frutiful interaction between elasticity theory and catastrophe theory. This paper offers an introduction, by giving an elementary exposition of the classical buckling beam in terms of cusp catastrophes. The last section on globalisation contains new material. See also [1,2,4,8]. Many other interesting examples can be found in the book [6] of Michael Thompson and Giles Hunt.

The study of instability of structures leads naturally to the analysis of the first few terms of Taylor expansions of energy functions of several variables, in order to find the equilibria. Thus several authors, including W.T. Koiter [11] as early as 1945, and in the 1960's M.J. Sewell [12], Thompson and Hunt, had independently discovered some of the elementary catastrophes. Note particularly Hunt's elegant computer drawings of the elliptic and hyperbolic umbilics [7,8], made before he had heard of Thom's classification theorem. However these discoveries were computations of particular approximations, rather than the recognition that the elementary catastrophes were diffeomorphism invariants. What catastrophe theory has to offer to elasticity theory is theorems, proofs, and higher dimensional singularities for handling compound buckling. Meanwhile, in return, what elasticity theory has to offer to catastrophe theory is an abundance of examples, fresh insights and problems. I am indebted to David Chillingworth, Maurice Dodson, Tim Poston and Michael Thompson for discussions.

Contents.

1. The simple Euler arch.
2. The Euler strut.
3. The pinned Euler strut.
4. Globalisation.

Publsihed in Structural stability, the theory of catastrophes and applications in the sciences, Springer Lecture Notes in Mathematics, Volume 525, 1976, 373-395.

1. THE SIMPLE EULER ARCH.

We begin with a simple example consisting two rigid arms supported at the ends and pivoted together at the centre, with a spring tending to keep them at 180°, as illustrated in Figure 1.

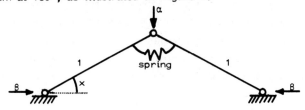

Figure 1.   The simple Euler arch.

If the ends are compressed with a gradually increasing horizontal force β then the arms will remain horizontal until β reaches a critical value, when they will begin to buckle upwards (or downwards). If β is now fixed, and a gradually increasing vertical load α is applied to the pivot, as in Figure 1, then the arch will support the load until α reaches a critical value, when it will suddenly snap catastrophically into the downwards position. It is this behaviour that we shall explain by our first cusp catastrophe.

Suppose that the arms each have length 1, and let μ denote the modulus of elasticity of the spring. Initially we assume $\alpha = 0$.

<u>Theorem 1.</u>   The arch buckles when $\beta = 2\mu$.

Before proving theorem 1, we go on to describe what happens in the neighbourhood of the buckling point. Let $\beta = 2\mu + b$, and let x denote the angle of the arms to the horizontal. We assume α, b, x are small. In 3-dimensions choose the (α, β)-axes horizontal, and the x-axis vertical. Call the horizontal (α, β)-plane the <u>control plane</u>, C. Let M be the graph of x as a function of α, β.

Theorem 2. M is a cusp catastrophe with ($-\alpha$) as normal factor and β as splitting factor (Figure 2).

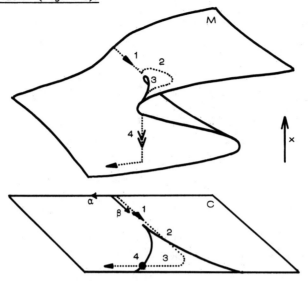

Figure 2. The dotted path shows the arch (1) compressed, (2) buckling upwards, (3) loaded, and (4) snapping downwards.

Proof of Theorem 1.

$$\text{Energy in spring} = \frac{1}{2}\mu(2x)^2.$$

$$\text{Energy gained by load} = \alpha \sin x.$$

$$\text{Energy lost by compression} = -2\beta(1-\cos x)$$

$$\therefore \text{total energy, } V = 2\mu x^2 + \alpha \sin x - 2\beta(1-\cos x).$$

The surface M of equilibria is given by $V' = 0$, where the prime denotes $\partial/\partial x$. The fold lines are given by $V'' = 0$, and the cusp point by $V''' = 0$.

$$V' = 4\mu x + \alpha \cos x - 2\beta \sin x = 0$$

$$V'' = 4\mu - \alpha \sin x - 2\beta \cos x = 0$$

$$V''' = \quad\quad - \alpha \cos x + 2\beta \sin x = 0.$$

Add the first and last: $4\mu x = 0$. But $\mu \neq 0$. $\therefore x = 0$.

Substitute in the first: $\alpha \cos x = 0$. $\therefore \alpha = 0$.

Substitute in the second: $4\mu - 2\beta = 0$. $\therefore \beta = 2\mu$.

This completes the proof of Theorem 1.

<u>Proof of Theorem 2.</u> Put $\beta = 2\mu + b$. Let O^5 denote $O(x^5)$. Therefore

$$V = 2\mu x^2 + \alpha(x - \frac{x^3}{6}) + 2(2\mu+b)(-\frac{x^2}{2} + \frac{x^4}{24}) + O^5$$

$$= \alpha x - b x^2 - \frac{\alpha}{6} x^3 + \frac{2\mu+b}{12} x^4 + O^5 .$$

When $\alpha = b = 0$, $V = \frac{\mu}{6} x^4 + O^5$.

Hence x obeys a cusp catastrophe since $\mu > 0$. We can eliminate the x^3-term by the translation of coordinates $x = x_1 + \frac{\alpha}{2(2\mu+b)}$, and then eliminate the tail of the Taylor series by a non-linear change of coordinates by [9, Theorem 2.9]. Therefore

$$V \sim \frac{\mu}{6} x^4 + \alpha x - b x^2 ,$$

by the isomorphism of unfoldings [9, Theorem 6.9]. This is the potential for a cusp catastrophe with $(-\alpha)$ as normal factor and b (or β) as splitting factor, thus completing the proof of Theorem 2 and Figure 2.

2. THE EULER STRUT.

We now turn from the discrete to the continuous, from the simple pivot to the elastic strut*, compressed under force β, as shown in Figure 3.

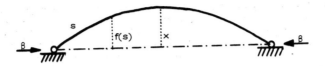

Figure 3. The Euler strut.

Let λ denote the length of the strut, and μ the modulus of elasticity per unit length.

<u>Theorem 3 (Euler [3] 1744).</u> <u>The strut buckles when $\beta = \mu(\pi/\lambda)^2$. The buckled shape is a sine-curve, to second order.</u>

<u>Proof.</u> Let s be a parameter for arc-length, $0 \le s \le \lambda$. Let f(s) denote the vertical displacement of the point s, which we assume is small. The shape of the strut is therefore given by the function $f:[0,\lambda] \to R$. We assume f is a C^∞-function satisfying the boundary conditions :

$f = 0$ at the ends (since the ends are supported),

$f'' = 0$ at the ends (since there is no bending moment there),

where primes denote $\partial/\partial s$. Let $\theta(s)$ be the inclination of the strut to the horizontal at s. Then

---

* The Battelle Research Centre conveniently provides plastic Euler struts for stirring coffee, which when held between thumb and forefinger make excellent experimental material; otherwise try using a 1" x 4" piece of thin cardboard. See Figure 7.

$$f' = \sin\theta$$

$$f'' = \cos\theta \cdot \theta'$$

$$\therefore \text{curvature} = \theta' = \frac{f''}{\cos\theta}$$

$$= f'', \text{ neglecting fourth order terms.}$$

$$\therefore \text{energy in increment ds of strut} = \tfrac{1}{2}(\mu ds)(f'')^2$$

$$\therefore \text{energy in strut} = \tfrac{1}{2}\int_0^\lambda \mu(f'')^2 ds$$

The contraction between the ends $= \int_0^\lambda (1-\cos\theta)ds$

$$= \tfrac{1}{2}\int (f')^2 ds$$

$$\therefore \text{energy lost by compression force} = -\tfrac{1}{2}\int \beta(f')^2 ds$$

$$\therefore \text{total energy}, V = \int F ds, \text{ where } F = \tfrac{1}{2}[\mu(f'')^2 - \beta(f')^2].$$

By the calculus of variations, using the boundary condition, the requirement for equilibrium is given by the Euler equation

$$\left(\frac{\partial F}{\partial f''}\right)'' - \left(\frac{\partial F}{\partial f'}\right)' = 0$$

$$\therefore \mu f'''' + \beta f'' = 0.$$

Solving this equation, using the boundary conditions,

$$f(s) = x \sin(s\sqrt{\beta/\mu}),$$

where x = constant, and $\lambda\sqrt{\beta/\mu}$ is a multiple of π. Therefore, if $\beta < \mu(\pi/\lambda)^2$, only the zero solution is possible. Buckling first begins when $\beta = \mu(\pi/\lambda)^2$ and then, to second order,

$$f(s) = x\sin\frac{\pi s}{\lambda}, \; x \text{ small constant}.$$

The solution is correct to second order because the next term is of order x^3 (see below). This completes the proof of Theorem 3.

Harmonics.

Write f as the Fourier series $f(s) = \sum_1^\infty x_n \sin\frac{n\pi s}{\lambda}$. We call x_n the nth harmonic of f.

Meaning of x.

The constant x occuring in the proof of Theorem 3 above can be interpreted in three ways, which agree to first order, but differ in higher orders :

    (i) x is the vertical displacement of the centre of the strut.

    (ii) x is the first harmonic of f.

    (iii) x is a perturbation parameter.

Using x in the last sense, one can expand f by perturbation theory; see for example [6, pages 28-34] :

$$f = x - \frac{1}{64}(\pi/\lambda)^2 [\sin(\pi s/\lambda) + (3\pi s/\lambda)] x^3 + O^5$$

$$\beta = \mu(\pi/\lambda)^2 + \tfrac{1}{4}\mu(\pi/\lambda)^4 x^2 + O^4$$

Qualitative approach.

The disadvantage of perturbation theory is that it tends to carry us away from the conceptual point of view of regarding the compression force β as the "cause" and the shape f as the "effect". What we really want to do is to use the perturbation expansion to draw the graph G of f as a function of β, as in Figure 4. (Alternatively we could use Theorem 5 below to deduce the shape of G.)

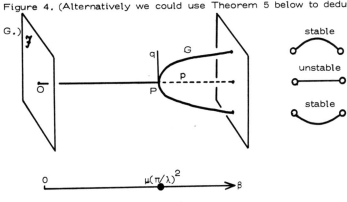

Figure 4. The graph G of shape f as a function of compression β.

The graph is stably constant $f = 0$ up to the Euler buckling force, where it bifurcates parabolically at the point P into two stable branches, representing buckling upwards and downwards, while $f = 0$ becomes an unstable branch.

There are two technical difficulties about this graph: firstly G has a singularity at P, and secondly the space $\mathcal{F}$ in which f lies (which we have drawn as 2-dimensional in Figure 4) is in fact ∞-dimensional. Catastrophe theory helps us to get rid of the first difficulty as follows. Observe that G is equivalent to the section $\alpha = 0$ of the surface in Figure 2. Therefore if we introduce a small vertical load α on the centre of the strut, then the β-axis will be embedded in a 2-dimensional control plane C, with coordinates (α, β), and G will be embedded in a smooth equilibrium surface $M \subset C \times \mathcal{F}$, such that the projection $M \to C$ is a cusp catastrophe.

Meanwhile the second difficulty of ∞-dimensionality can be either met or avoided; Chillingworth [1] shows how to meet it by embedding $\mathcal{F}$ in a Hilbert space, but here we shall avoid it, as follows. We capture the qualitative behaviour by selecting the significant harmonic, which in this case is the first harmonic, and computing it to first order. To study the quantitative behaviour one could equally well compute a finite number of harmonics up to the required accuracy. More precisely, let $h: \mathcal{F} \to R$ be the function mapping f to its first harmonic, $h(f) = x$. Then, although $1 \times h: C \times \mathcal{F} \to C \times R$ crushes (2+∞)-dimensions down onto 3-dimensions, it nevertheless does not crush the equilibrium surface M that interests us, but embeds M smoothly into $C \times R$, enabling us to visualise it and compute it.

Lemma 4. $1 \times h$ maps a neighbourhood of P in M diffeomorphically into $C \times R$.

Proof. Let p,q denote the two tangent lines to G at P. In Figure 3 p coincides with the line $f = 0$, and is horizontal parallel to the β-axis, while q is tangent to the parabola, and vertical in the sense of being parallel to $\mathcal{F}$. By the proof of Theorem 3 the derivative Dh maps q isomorphically onto R, and hence D(1xh) maps the plane spanned by {p,q} isomorphically into CxR. But this is the tangent plane to M at P. Therefore 1xh maps a neighbourhood of P in M diffeomorphically into CxR.

Let $M' = (1 \times h)M$, $P' = (1 \times h)P$. Then we have a commutative diagram

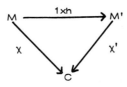

where χ, χ' are induced by projection.

Corollary. The singularity of χ at P is equivalent to that of χ' at P'.

Therefore to prove that M is a cusp catastrophe it suffices to show that M' is. Thus we have avoided the ∞-dimensional problem, because we work with $M' \subset C \times R$, which is 3-dimensional.

For convenience choose units so that the length of the strut, $\lambda = \pi$, and the modulus of elasticity per unit length, $\mu = 1/\pi$. Then the Euler buckling force $= \mu = 1/\pi$. Let $\beta = (1+b)/\pi$, and assume x, α, b are small.

Theorem 5. M' is a cusp catastrophe with $(-\alpha)$ as normal factor and β as splitting factor. In other words the Euler strut behaves exactly as the simple Euler arch in Theorem 2 and Figure 2.

Proof.   By Theorem 3, $f = x\sin s + O^3$.

$$\therefore \text{energy in strut} = \frac{1}{2}\int_0^\pi \mu(f'')^2(1-(f')^2)^{-1}ds.$$
$$= \frac{1}{2\pi}\int x^2\sin^2 s(1+x^2\cos^2 s)ds + O^6$$
$$= \tfrac{1}{4}(x^2 + \tfrac{1}{4}x^4) + O^6,$$

because the other O^4-terms disappear in the integration.

$$\text{Energy lost by compression} = -\int \beta[1-\sqrt{1-(f')^2}]ds$$
$$= -\tfrac{\beta}{2}\int(x^2\cos^2 s + \tfrac{1}{4}x^4\cos^4 s)ds + O^6$$
$$= -\tfrac{1}{4}(1+b)(x^2 + \tfrac{3x^4}{16}) + O^6.$$

Energy gained by load $= \alpha(x+O^3)$

$$\therefore \text{total energy, } V = \tfrac{1}{64}x^4 + \alpha(x+O^3) - b(\tfrac{1}{4}x^2 + O^4) + O^6$$
$$\sim \tfrac{1}{64}x^4 + \alpha x - \tfrac{1}{4}bx^2 \ .$$

This completes the proof of Theorem 4.

3. THE PINNED EULER STRUT

We now turn to a much more interesting property of the Euler strut, namely its load bearing capacity and imperfection sensitivity, when the ends are pinned as in Figure 5.

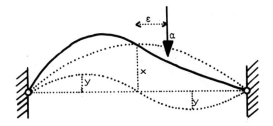

Figure 5. In the pinned Euler strut with an offset load the second harmonic y is significant.

Let r be the first harmonic when the strut is unloaded; r is a constant related to the difference between the length of the strut and the distance between the pins. We now apply a vertical load α off-set from the centre by a distance ε, as shown in Figure 5. We call ε the imperfection, because this model sometimes simulates a manufacturing imperfection. The question is : when will the strut snap catastrophically into the downwards position ? This time the second harmonic will be the significant one, as shown in Figure 5.

A nice realisation in stone of Figure 5 can be seen in the western arch of Clare College bridge over the river Cam, in which the keystone is about 10° to the vertical.

Theorem 6. The graph of y as a function of (α,ε) is a dual cusp catastrophe, with ε as normal factor and $(-\alpha)$ as splitting factor.

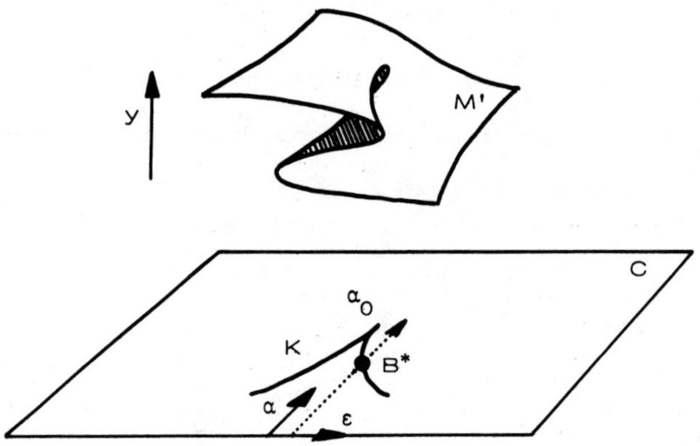

16. Euler Buckling

Figure 6. In the pinned Euler strut the second harmonic y obeys a dual cusp catastrophe. The cusp K illustrates the load carrying capacity and imperfection sensitivity.

Remark 1. Here the word <u>dual</u> means that the middle sheet represents stable equilibria, while the upper and lower sheets represent unstable equilibria. We call ε the normal factor, rather than $(-\varepsilon)$, because on the stable middle sheet y increases with ε.

Remark 2. Theorem 6 is a local statement and therefore we have only drawn the graph over a neighbourhood of the cusp point. In particular the result does not extend to $\alpha = 0$. We shall treat $\alpha = 0$ in the globalisation below.

<u>Load carrying capacity.</u> The cusp K in Figure 6 is the bifurcation set. If the load α is gradually increased keeping the imperfection

ε fixed, as shown on the dotted path, then the stability of the upward position of the strut breaks down when the path crosses K at the point B^*, and the strut will snap catastrophically into the downward position. The latter is represented by another sheet of stable equilibria, which is not shown in Figure 6, but is explained in Figure 8 and Theorem 9 below.

The cusp K is therefore the graph in C of the load carrying capacity α as a function of the imperfection ε. The sharp reduction in load carrying capacity away from the maximum α_0 by the ⅔-power law of the cusp is called <u>imperfection-sensitivity</u>. Obviously this can be critical in designing structures. Roorda [5, 6 Figure 44] has tested struts of high-tensile steel, and obtained accurate experimental confirmation of the cusp. Before proving Theorem 6 we compute the maximum load.

<u>Lemma 7.</u> <u>The first harmonic x and the second harmonic y are related by the equation $x^2 + 4y^2 = r^2$.</u>

<u>Proof.</u> As before suppose $\lambda = \pi$, $\mu = 1/\pi$. We shall work to O^2, and ignore higher harmonics. Therefore

$$f = x\sin s + y\sin 2s.$$

Distance between pins $= \int_0^\pi \sqrt{1-(f')^2}\,ds$

$\qquad\qquad\qquad\qquad = \int (1-\tfrac{1}{2}(x\cos s + 2y\cos 2s)^2)\,ds$

$\qquad\qquad\qquad\qquad = \pi(1-\tfrac{1}{4}x^2 - y^2)$

$\qquad\qquad\qquad\qquad = \pi(1-\tfrac{1}{4}r^2)\quad$ when the strut is unloaded.

$\therefore x^2 + 4y^2 = r^2.$

<u>Lemma 8.</u> The maximum load is $\alpha_0 = \dfrac{3r}{2}$.

Proof.   Energy in strut $= \frac{1}{2} \int \mu(f'')^2 ds$

$= \frac{1}{4}(x^2 + 16y^2)$

$= \frac{1}{4}(r^2 + 12y^2)$, by Lemma 7.

Energy of load $= \alpha f(\pi/2 + \varepsilon)$

$= \alpha(x - 2\varepsilon y)$, ignoring ε^2

$= \alpha(r - \frac{2y^2}{r} - 2\varepsilon y)$, by Lemma 7.

$\therefore$ total energy, $V = $ constant $- 2\alpha\varepsilon y + (3 - \frac{2\alpha}{r})y^2$.

The cusp point occurs when $V' = V'' = 0$, where prime denotes $\partial/\partial y$. Therefore $\varepsilon = 0$ and $\alpha = \frac{3r}{2}$.

Proof of Theorem 6. We can either work locally with higher orders or globally with order 2, and since both approaches shed light we shall do both. First we work locally, and show that y satisfies a dual cusp catastrophe by computing the coefficient of y^4 in the expansion of V, and verifying that it is negative. Since we are working near the cusp point we may assume $y \ll r$. Therefore we shall suppose $r, x = O^1$, $y = O^2$, and ignore higher harmonics. Then working to order O^6 gives a refinement of Lemma 7 :

$x^2 + 4y^2 + \frac{3}{2}x^2 y^2 = r^2$

$\therefore x = r - \frac{2y^2}{r} - \frac{3ry^2}{4} - \frac{2y^4}{r^3}$

Energy in strut $= $ constant $+ (3 + \frac{13r^2}{8})y^2$

Energy in load $= \alpha(x - 2\varepsilon y)$, ignoring ε^2

$= \alpha[r - 2\varepsilon y - (\frac{2}{r} + \frac{3r}{4})y^2 - \frac{2y^4}{r^3}]$

$\therefore$ total energy $= $ const $- 2\alpha\varepsilon y + [(3 + \frac{13r^2}{8}) - \alpha(\frac{2}{r} + \frac{3r}{4})]y^2 - \frac{2\alpha y^4}{r^3}$

$\therefore \alpha_0 = \frac{3r}{2} - \frac{r^3}{4}$.

Put $\alpha = \alpha_0 + a$.

$\therefore V \sim \frac{3}{r^2}y^4 - 3r\varepsilon y - \frac{2}{r}ay^2$

$\sim -(y^4 + \varepsilon y + ay^2)$, absorbing the constants in the variables. This is the potential of a dual cusp with normal factor ε and splitting factor $(-a)$, completing the proof of Theorem 6.

A simple constraint experiment.

We can obtain the shape of Figure 5 by holding a plastic or cardboard strut between thumb and forefinger as shown in Figure 7(a), and constraining it by pushing slowly down, off centre.

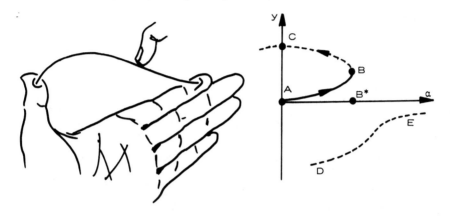

Figure 7. (a) Experiment with a plastic or cardboard strut, and (b) the resulting graph of load α and second harmonic y.

However this experiment is slightly different from that described by Theorem 6 because the push down is a <u>constraint</u> rather than a load. It is true that the constraint will exert a downward load equal to the upward resistance of the strut, which is exactly what is needed to keep it in equilibrium, but the nature of the stability changes : for example an equilibrium point on the upper sheet of Figure 6, which would be <u>unstable</u> with respect to a fixed load, becomes <u>stable</u> with respect to a fixed constraint. We can test this as follows.

Figure 7(b) shows the section of Figure 6 over the dotted path through B^*, extended back to $\alpha = 0$. The black curve AB is the section of the stable middle sheet, and the dotted curves BC, DE are the sections of the unstable upper and lower sheets. The point A represents the initial unloaded shape with maximum first harmonic and zero second harmonic. As the strut is slowly pushed down it follows smoothly the curve AB; a steadily increasing load is required, and the second harmonic grows noticeably. B is the fold point over B^*, and would be the point where strut would have snapped into the downward position in the loading experiment of Theorem 6. But here we are constraining rather than loading, and the constraint does not continue to exert the critical load B^*; instead the actual load applied by the constraint begins to decrease again, and the behaviour, instead of snapping, continues to follow smoothly round the curve BC. The change in the nature of the stability can be detected by observing that for points on AB a push down increases the resistance of the strut, whereas for points on BC a push down decreases the resistance. Eventually the point C is reached, which represents maximum second harmonic and zero first harmonic; here the resistance changes sign, and so the constraint no longer constrains, allowing the strut at last to snap into the downward position.

4. GLOBALISATION

The disadvantage of Theorem 6 is that it only gives a local analysis near the cusp point. Similarly the disadvantage of Figure 6 is that does not show the other sheet of stable equilibria, representing the strut in the downward positions, $x < 0$. Although the cusp K tells us where the catastrophic jumps will occur, and although the fold curves above tell us where they jump <u>from</u>, we cannot see where they will jump <u>to</u>, because this receiving sheet is missing. And indeed, were we to include the missing sheet on Figure 6, this would be even more confusing because it would intersect the other sheets already there.

To clarify matters we must use the first harmonic as well as the second. Let H be the plane with coordinates (x,y) representing both harmonics, and let $h: \mathcal{F} \to H$ denote the map $f \mapsto (x,y)$. We are interested in the equilibrium surface $M \subset C \times \mathcal{F}$. The relation $x^2 + 4y^2 = r^2$ of Lemma 7 determines an ellipse $E \subset H$. Therefore the product map

$$1 \times h : C \times \mathcal{F} \to C \times H$$

maps M onto a surface M' inside the solid torus $C \times E$. The questions we want to answer are :

What is the geometry of M'?

How does M' lie in $C \times E$?

What is the singular set of the projection $\chi' : M' \to C$?

What is the bifurcation set, Bif $\chi' = \chi'(\text{Sing } \chi')$?

If we can show M' is smooth, then $1 \times h$ must have mapped M diffeomorphically on to M', and so the commutative diagram

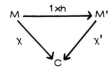

ensures that Bif χ = Bif χ'. Therefore an analysis of M' will reveal the global catastrophe set, Bif χ, that we are looking for.

Torque.

The moment of the load α about the mid-point between the pins is $\alpha\varepsilon$. Define the <u>torque</u> τ = ½$\alpha\varepsilon$. (The factor ½ is included for technical convenience in the proof of Theorem 9 below.) <u>Locally</u> Theorem 6 and Figure 6 remain the same if the imperfection ε is replaced by the torque τ, because, for α bounded away from zero (as it is in Theorem 6), the change of variable $\varepsilon \to \tau$ is a diffeomorphism and the cusp catastrophe is invariant under diffeomorphism. Globally τ is more convenient than ε, and so from now on we use τ. Therefore the control plane C will now have coordinates (α,τ). In particular the control point $\alpha = 0$, $\tau \neq 0$ means apply torque, without load, to the centre of the strut.

Construction.

We make a mysterious construction. Let j:H $\to$ C be the linear isomorphism given by

$$\begin{cases} \alpha = -x/2 \\ \tau = 2y \end{cases}$$

Recall that E is the ellipse $x^2+4y^2=r^2$. Therefore jE is the **ellipse** $4\alpha^2+\tau^2=r^2$. Let W denote the <u>evolute</u> of jE, in other words the envelope of normals which is the concave diamond with four cusps shown in Figure 8. Define the <u>normal bundle</u> M" of jE to be the subset of C×jE given by

M" = {(c,e); c$\in$C, e$\in$jE, c lies on the normal at e}.

Now comes the surprise that justifies the construction.

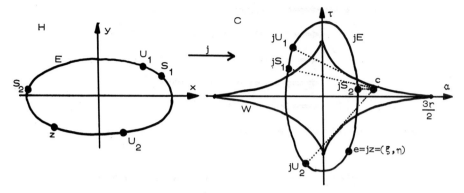

Figure 8. W is the evolute of jE. The harmonics of the stable and unstable equilibria corresponding to control point c are given by the inverse images S_1, S_2, U_1, U_2 of the feet of the normals from c to jE.

Theorem 9. $1 \times j : C \times E \to C \times jE$ maps M' diffeomorphically onto M".

Proof. Let $z = (x,y) \in H$, $c = (\alpha, \tau) \in C$. From the proof of Lemma 8, the total energy

$$V = \tfrac{1}{4}(x^2+16y^2) + \alpha(x-2\varepsilon y)$$
$$= \tfrac{1}{4}x^2 + 4y^2 + \alpha x - 4\tau y.$$

The surface $M' \subset C \times E$ is given by the stationary values of V with respect to x,y, subject to the constraint $z \in E$. Let $e = jz \in C$, and suppose e has coordinates $e = (\xi, \eta)$. Then

$$x = -2\xi$$
$$y = \eta/2.$$
$$V = \xi^2 + \eta^2 - 2\alpha\xi - 2\tau y$$
$$= (\xi-\alpha)^2 + (\eta-\tau)^2 - (\alpha^2+\tau^2)$$
$$= \delta^2 - (\alpha^2+\tau^2)$$

where δ is the distance between the points c,e. The constraint becomes $e \in jE$. Therefore keeping c fixed and varying z,

$(c,z) \in M' \iff V$ stationary, subject to $z \in E$

$\iff \delta$ stationary, subject to $e \in jE$

$\iff e$ is the foot of a normal from c

$\iff c$ lies on the normal at e

$\iff (c,e) \in M''$

Therefore $1 \times j$ maps M' to M'', as required.

Problem. In the proof of Theorem 9 **we** used the second order approximation for V, and therefore the result is not proved over the whole of C, but only over an open subset U of C. However if r is small, then U contains the evolute W, which is the set we are really concerned with. It would be interesting to compute lower bounds for the size of U in terms of r. The subtlety of the problem is that Theorem 9 is strictly neither local or global, but semi-global.

Remark. The mysterious construction of j and the factor ½ in the definition of τ were merely devices to identify the intrinsic Riemannian structure of the problem with that of the Euclidean plane.

The evolute W. We have reduced the geometry of M' to that of M'', which is well known. Topologically the normal bundle is a cylinder, showing that the equilibrium states are connected, and contain an essential cycle*. Let c_e denote the centre of curvature of jE at e. Then the singular set of the projection $\chi'': M'' \to C$ is

$$\text{Sing } \chi'' = \{(c_e, e); e \in jE\}.$$

---

* Hence a stiffened elastic panel can become locked in a <u>global</u> buckling mode, which cannot be localised and is difficult to push out. If you try to push it out then it will tend to slide around. The only way to get rid of it is to hold down the two second harmonics, and then press with sufficient force to call into play the third harmonic.

Therefore the bifurcation set

$$\text{Bif } \chi'' = \{c_e; e \in jE\} = \text{the evolute, W.}$$

Since the bifurcation sets of $\chi, \chi' \chi''$ are all equal, we have achieved the desired result :

Corollary. Bif χ = W.

This answers a question of Sewell [4].

The evolute can be written parametrically :

$$\begin{cases} \alpha = \frac{3r}{2} \cos^3 \theta \\ \tau = \frac{3r}{4} \sin^3 \theta, \ 0 \leq \theta < 2\pi. \end{cases}$$

The four cusp points are $(\pm\frac{3r}{2}, 0), (0, \pm\frac{3r}{4})$. In particular $(\frac{3r}{2}, 0)$ is the cusp point of Theorem 6.

<u>Geometric interpretation.</u> Not only does the evolute give us the bifurcation set, in other words the loads and torques in C at which the catastrophic jumps take place, but the map j also enables us to identify the shapes of the strut, in other words the points in the harmonic plane H where the jumps begin and end.

Consider a point c in the interior of W, as in Figure 8. The four tangents from c to W are the same as the normals from c to jE. Let jS_1, jS_2, jU_1, jU_2 be the feet of those normals, where S stands for stable (the distance from c to jE being a local minimum) and U stands for unstable (when it is a local maximum). Let S_1, S_2, U_1, U_2 denote their inverse images under j. The latter points determine the shapes of the stable and unstable equilibria of the strut, corresponding to control c. For example Figure 5 illustrates the stable point S_1, with x,y > 0.

Now increase the load, α. As c crosses W the points S_1 and U_1 coalesce, causing the stability of S_1 to break down. Therefore the strut snaps into position S_2, with $x < 0$, $y > 0$, as shown in Figure 9. Conversely we can snap the strut up again by smoothly changing the sign of α until c hits the left hand side of W, and then return the strut smoothly to its original position S_1 by

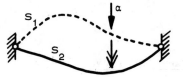

Figure 9. The snap $S_1 \to S_2$.

increasing α again. Alternatively we could return the strut to S_1 along an entirely smooth path by making c encircle the top cusp of W.

Note that when c is at the top cusp the points S_1, U_1, S_2 coalesce, confirming that it is an ordinary cusp catastrophe, with an unstable sheet in between two stables; similarly for the bottom cusp. Meanwhile when c is at the right hand cusp the points U_1, S_1, U_2 coalesce, confirming that it is a dual cusp catastrophe, with a stable sheet in between two unstables, as we proved in Theorem 6 and Figure 6. Similarly with the left hand cusp. The global configuration is diffeomorphic to that of the **catastrophe machine** in [10].

Summarising, Figure 8 gives us a global comprehension of both the qualitative and the quantitative behaviour of the pinned Euler strut.

REFERENCES

1. D. Chillingworth, The catastrophe of a bucking beam, Dynamical Systems - Warwick 1974, Springer Lecture Notes in Maths., 468, 86-91.

2. C.T.J. Dodson & M.M. Dodson, Simple non-linear systems and the cusp catastrophe, Lancaster & York Universities preprint, 1975.

3. L. Euler, Methodus inveniendi lineas curvas maximi minimive proprietate gaudentes (Appendix, De curvis elasticis), Marcum Michaelum Bousquet, Lausanne & Geneva, 1744.

4. M.J. Sewell, Some mechanical examples of catastrophe theory, Bull. Inst. Math. and Appl. (to appear).

5. J. Roorda, Stability of structures with small imperfections, J. Engng. Mech. Div. Am. Soc. civ. Engrs. 91, (1965) 87-106.

6. J.M.T. Thompson & G.W. Hunt, A general theory of elastic stability, Wiley, London, 1973.

7. J.M.T. Thompson & G.W. Hunt, Towards a unified bifurcation theory, J. Appl. Math. & Physics 26 (1975).

8. J.M.T. Thompson, Experiments in catastrophe, Nature, 254, 5499 (1975) 392-395.

9. D.J.A. Trotman & E.C. Zeeman, Classification of elementary catastrophes of codimension ≤ 5, this volume. *[18]*

10. E.C. Zeeman, A catastrophe machine, Towards a theoretical biology, 4 (Ed. C.H. Waddington), Edinburgh Univ. Press (1972), 276-282. *[15]*

11. W.T. Koiter, On the stability of elastic equilibrium, Dissertation, Delft, Holland, 1945.

12. M.J. Sewell, The static perturbation technique in buckling problems, J. Mech. Phys, Solids, 13, 247 (1965).

Numbering in this book.

17

A CATASTROPHE MODEL FOR THE STABILITY OF SHIPS

1. INTRODUCTION.

Catastrophe theory provides a new way of looking at the statics of a ship, and this in turn lends a new simplicity to the global non-linear dynamics. The weight of the ship and the position of the centre of gravity are taken as parameters. Then the set of equilibrium positions form a smooth manifold that maps onto the parameter space. It is the singularities of this map that are recognisable as elementary catastrophes.

For example heeling and capsizing are fold catastrophes. At the metacentre there is a cusp catastrophe. The point of inflexion of the lever arm curve is caused by another cusp catastrophe. The increased likelihood of capsizing when overloaded, or when the crest of a wave is amidships, is due to a swallowtail catastrophe. The evolution of hull shape from canoe to modern ship is characterised by a butterfly catastrophe. On the metacentric locus there are also hyperbolic umbilic catastrophes. The sudden onset of heavy rolling due to non-linear resonance with the wave is a dynamic fold catastrophe.

For the naval architect this approach in terms of canonical forms offers a qualitative geometry that is complementary to the classical approach [2,3,9,11]. Whether or not such formulation will be of any use remains to be seen, because to make quantitative predictions it is still necessary to choose coordinates and approximations as in the classical theory. However as a general principle it is always advantageous to retain the dynamics in a conceptually simple form for as long as possible, so that the important qualitative features can be kept in the forefront of the

---

Published in the Proceedings of III Escola Latino Americana de Matematica, I.M.P.A., Rio de Janeiro, Brazil, 1976.

mind unobscured by detail, allowing the eventual approximation to be tailored to the job on hand. Catastrophe theory should be used like the zoom lens on a microscope, for gaining a global view of the problem, and thus enhancing the discrimination with which one selects the necessary tool from classical mathematics to zoom in and solve it.

For the mathematician the interest of this example lies in the rich variety of the mathematics that it brings together; it is a prototype revealing catastrophe theory as a natural generalisation of Hamiltonian dynamics. Besides having a parameter space, an elementary catastrophe model has a state space, a potential, and a dynamic minimising the potential. Here the state space is the usual phase space of Hamiltonian dynamics, namely the cotangent bundle of the configuration space of the ship, while the potential is none other than the Hamiltonian, which is a Lyapunov function for, and therefore locally minimised by, the Hamiltonian flow with any type of damping. The subtlety comes from the bifurcations of the dynamic over the parameter space, which are governed by the elementary catastrophes in the evolute of the buoyancy locus.

We begin the paper by giving in Sections 2 - 6 a brief elementary sketch of the classical linear theory of ship motions, partly for the benefit of readers unfamiliar with the topic, and partly to see later how the model generalises it. The reader familiar with the linear theory is recommended to proceed at once to Section 7, where we describe the significance of the geometry. The model itself is introduced in Section 10, at first for rolling only, and then successively enlarged to include pitching, heaving and loading. At the end in Section 14 we place the treatment in its proper group-theoretical setting.

An unsolved problem is how best to incorporate into the model the forcing terms of wind and wave, so as to prove global existence theorems concerning induced oscillations, resonance and capsizing. The model already sheds some light on capsizing. In the conclusion we tentatively suggest two areas for further exploration.

CONTENTS

1. Introduction.
2. Linear theory of rolling.
3. Quantitative estimates.
4. Rolling in a seaway.
5. Resonance.
6. Pitching and heaving.
7. Cusp catastrophe at the metacentre.
8. Global metacentric locus.
9. Lever arm curve.
10. Catastrophe model for rolling.
11. Global dynamics.
12. Model including pitching.
13. Model including heaving.
14. Model including loading.
15. Conclusion.

2. LINEAR THEORY OF ROLLING.

The classical linear theory of how a ship behaves like a pendulum hanging from the metacentre dates back to Euler [7] in 1737 and Bouguer [4] in 1746. For simplicity we begin by confining ourselves to the artificial 2-dimensional problem of rolling only. When we come to the catastrophe model, the notation will allow us to generalise with ease to the full 3-dimensional situation.

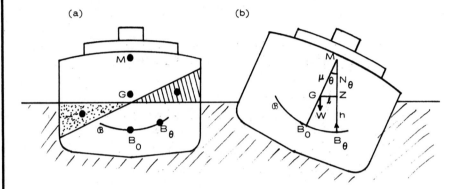

Figure 1. The buoyancy locus and metacentre.

Let G = centre of gravity of ship.

B_0 = centre of buoyancy when ship floats vertically
    = centre of gravity of water displaced.

B_θ = centre of buoyancy when ship is at angle θ.

$\mathcal{B}$ = <u>buoyancy locus</u> = $\{B_\theta; -\pi < \theta \leq \pi\}$

N_θ = normal to $\mathcal{B}$ at B_θ.

M = <u>metacentre</u> = centre of curvature of $\mathcal{B}$ at B_0.

μ = GM = <u>metacentric height.</u>

Lemma 1. $\mathcal{B}$ is a convex closed curve. When the ship is heeled at angle θ, the normal N_θ is vertical.

Proof. Let A_θ denote the water displaced at angle θ. The area of A_θ is independent of θ, since by Archimedes principle the weight of water displaced equals that of the ship. A_θ is obtained from A_0 by adding the immersed wedge, shown shaded in Figure 1, and subtracting the emerged wedge, shown dotted. Since the wedges have equal area, $B_0 B_\theta$ is parallel to the line joining the centres of gravity of the two wedges. As $\theta \to 0$ this line tends to the horizontal in Figure 1(a), and hence the tangent to $\mathcal{B}$ at B_0 is horizontal in Figure 1(a). The above argument is independent of the symmetry at $\theta = 0$, and hence applies to any θ. Therefore the tangent to $\mathcal{B}$ at B_θ is horizontal in Figure 1(b), and hence the normal N_θ is vertical.

For small θ, the inclination of $B_0 B_\theta$ to the horizontal has the same sign as θ, and hence the curvature of $\mathcal{B}$ at B_0 is upwards. The same argument applies to any θ, and hence $\mathcal{B}$ is convex.

Corollary. For small θ, the buoyancy force passes approximately through the metacentre M.

Proof. The buoyancy force acts vertically upwards at B_θ, and therefore along the normal N_θ. For small θ the normals pass approximately through the centre of curvature, M. In fact since B_0 is a point of symmetry of $\mathcal{B}$, the distance from M to N_θ is of order θ^3.

Remark. For the 3-dimensional problem the same result holds, with the proviso that $\mathcal{B}$ is now a convex closed surface; the proof is the same. Note that the result is independent of the shape of hull, and holds not only for ordinary ships but also for catamarans and icebergs, for example.

The <u>righting couple</u> in Figure 1(b) consists of the weight W of the ship acting downwards at G, and the buoyancy force W acting upwards at B_θ. Let ℓ denote that <u>lever arm</u> of this couple :

ℓ = distance from G to N_θ

= GZ, in Figure 1(b) ,

where Z is the foot of the perpendicular from G to N_θ. Newton's law of motion gives :

$$I\ddot{\theta} = -W\ell \quad \ldots\ldots\ldots\ldots\ldots\ldots (1)$$

where I is the moment of inertia of the ship (and the entrained water) about G. From the Corollary to Lemma 1 and Figure 1(b),

$\ell = \mu \sin \theta$, to second order in θ,

$= \mu\theta$, again to second order.

Hence the approximate linear equation is

$$I\ddot{\theta} = -W\mu\theta. \quad \ldots\ldots\ldots\ldots\ldots\ldots (2)$$

This has rolling solution

$$\theta = \theta_o \cos \frac{2\pi t}{T} \quad \ldots\ldots\ldots\ldots\ldots (3)$$

where θ_o is the amplitude and T the period of the roll. The amplitude is arbitrary, and the period is given by

$$T = 2\pi \sqrt{\frac{I}{W\mu}} . \quad \ldots\ldots\ldots\ldots\ldots (4)$$

The viscosity of the water has a damping effect, and the simplest way to put this into the dynamics is to add a small damping term $2\varepsilon\dot{\theta}$ to equation (2), where ε is a small positive constant :

$$\ddot{\theta} + 2\varepsilon\dot{\theta} + \frac{W\mu}{I} \theta = 0 . \quad \ldots\ldots\ldots\ldots (5)$$

This has the effect of multiplying the solution (3) by a decay factor $e^{-\varepsilon t}$, and lengthening the period by a factor of order ε^2, which we can ignore.

3. QUANTITATIVE ESTIMATES.

By (2) the larger the metacentric height the more stable is the ship. However by (4) the larger the metacentric height the shorter is the period of roll, and the more uncomfortable is the ship; therefore choice of μ is an important feature of ship design.

What is at first surprising is how small μ can be compared with the size of the ship : for example a liner 300 metres long may have a metacentric height of only half a metre. Therefore we shall digress briefly to make some very rough estimates of μ and T, in order to give a quantitative feel for the problem complementary to the qualitative feel given by the subsequent catastrophe theory. We shall see in Section 7 that if G is too near M small alterations in the position of G may seriously increase the danger of capsizing.

Let $2a$ = beam of ship = width at water line,
A = area below water-line,
(x,y) = coordinates of B_θ relative to B_0,
$\rho = B_0 M$ = radius curvature of $\mathcal{B}$ at B_0.

Lemma 2. $\rho = \dfrac{2a^3}{3A}$.

Proof. Let $t = \tan\theta$.
Let 0^n denote order t^n.
Each wedge has area $\frac{1}{2}a^2 t + 0^2$, and the coordinates of the centres of gravity of the wedges relative to the mid point of the water line are :

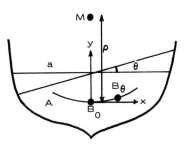

Figure 2. Computing ρ.

$$(\pm \tfrac{2a}{3} + 0^1, \pm \tfrac{at}{3} + 0^2).$$

The coordinates (x,y) of B_θ are given by taking moments of the wedges about B_0 :

$$Ax = (\tfrac{1}{2}a^2 t)\tfrac{4a}{3} + 0^2 = \tfrac{2a^3 t}{3} + 0^2$$

$$Ay = (\tfrac{1}{2}a^2 t)\tfrac{2at}{3} + 0^3 = \tfrac{a^3 t^2}{3} + 0^3$$

Therefore putting $\rho = \dfrac{2a^3}{3A}$, $\quad x = \rho t + 0^2$
$\quad\quad\quad\quad\quad\quad\quad\quad\quad\quad y = \tfrac{1}{2}\rho t^2 + 0^3$.

Therefore the equation of ⓑ is
$$y = \frac{x^2}{2\rho} + O(x^3),$$
which has radius of curve ρ at the origin, as required.

Define a ship to be <u>wall-sided</u> if the sides are parallel at the water-line, as in Figure 1. Let β denote the maximum angle of heel for which the water-line still meets the parallel part of the sides (in Figures 1 and 3, θ is about $25°$).

<u>Lemma 3.</u> In a wall-sided ship the buoyancy locus for $|\theta| < \beta$ is precisely the parabola $x^2 = 2\rho y$.

<u>Proof.</u> In the proof of Lemma 2, leave out all the O^n's.

In order to estimate the metacentric height and period of roll we now make a couple of very rough quantitative assumptions. Assume (i) that A is approximately a rectangle with the draught equal to a third of the beam, as in Figure 3. Therefore

$$A = 2a \times \frac{2a}{3} = \frac{4a^2}{3}$$

$$B_0 M = \rho = \frac{\frac{2a^3}{3}}{3A} \text{ (by Lemma 2)}$$

$$= \frac{a}{2}.$$

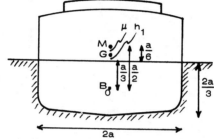

Figure 3. Estimate of μ.

Meanwhile B_0 is $\frac{a}{3}$ below the water-line, and so M is $\frac{a}{6}$ above. Therefore if h_1 = height of G above the water-line,

$$\mu = \frac{a}{6} - h_1 \quad \ldots\ldots\ldots\ldots\ldots (6)$$

Assume (ii) that the moment of inertia I is the same as that of a solid disk of radius a:

$$I = \frac{W}{g} \frac{a^2}{2} \quad \ldots\ldots\ldots\ldots\ldots (7)$$

where g = gravitational acceleration = 9.81 m/sec^2. We can now compute the period of roll:

$$T = 2\pi\sqrt{\frac{I}{W\mu}} \quad \text{by (4)}$$

$$= 2\pi\sqrt{\frac{a^2}{2g\mu}} \quad \text{by (7)}$$

$$= 1.42\,\frac{a}{\sqrt{\mu}} \quad \ldots\ldots\ldots\ldots\ldots \quad (8)$$

Although the above assumptions are crude the resulting orders of magnitude are not unreasonable for both naval and merchant ships [2, pages 107,335]. Where ships tend to differ is in the height of G above the water-line, and so let us work out a couple of examples of a destroyer and a liner in order to illustrate the contrast. In each case we assume typical values for the beam and position of G, and deduce the resulting metacentric height and period of roll.

	Table 1	Destroyer	Liner
Assume	Beam, 2a	10m	30m
	Height of G above water-line, h_1	0	2m
Deduce	Metacentric height, μ, by (6)	0.8m	0.5m
	Period of roll, T, by (8)	8 secs	30 secs

Notice that the greater metacentric height of the destroyer gives a greater righting couple, and hence makes her more stable, so that she can perform tighter manoeuvres, as well as causing a faster roll. By contrast the lesser metacentric height of the liner makes her less stable, although this does not matter so much since she does not have to indulge in manoeuvres; meanwhile she benefits from the increased comfort of the slower roll and smaller accelerations, which ensure that the passengers are less likely to be seasick.

4.   ROLLING IN A SEAWAY.

The forced oscillations induced by periodic waves, and the resonance that occurs when the period of the waves coincides with the natural period T of rolling, were first studied by Bernoulli [3] in 1759.

Typical Atlantic ocean waves usually have a wave-length between 50 and 100 metres [2, page 177]. In deep water the wave-length determines both the speed and the period of the wave. For example a 60 metre wave has a speed of 35 Km/hour (= 19 knots) and a period of 6 seconds. The height of the wave from trough to crest is independent of the wave-length, and in the Atlantic, for instance, wave heights of over 6m occur on an average of 55 days in the year. In a 60m wave of height 6m the maximum inclination of water surface to the horizontal is approximately $18°$.

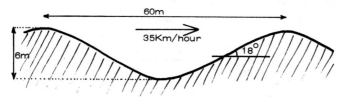

Figure 4. Typical Atlantic wave of period 6 seconds.

What is the effect of the waves on the ship? Suppose that in our 2-dimensional example the water surface is inclined at an angle α to the horizontal, as in Figure 5. The buoyancy force is the sum of all the pressures on the hull that would have sufficed to keep the displaced water in equilibrium with its surface inclined at angle α; hence the buoyancy force acts at $B_{\theta+\alpha}$ perpendicular to the water surface, in other words through M (approximately). Therefore the equation (2) for rolling must be modified to

$$I\ddot{\theta} = -W\mu(\theta+\alpha) \qquad \ldots\ldots\ldots\ldots (9)$$

Suppose now that we have a beam sea, in other words waves are coming from the side with period τ and maximum inclination α_o to the horizontal. The inclination α will be a periodic function of time, t, approximately equal to

$$\alpha = \alpha_o \cos \frac{2\pi t}{\tau} \qquad \ldots\ldots\ldots\ldots (10)$$

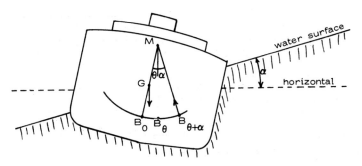

Figure 5. Inclined water surface.

Substituting (4) and (10) in (9) we obtain

$$\left(\frac{T}{2\pi}\right)^2 \ddot{\theta} + \theta = -\alpha = -\alpha_o \cos\frac{2\pi t}{\tau} . \quad \ldots\ldots (11)$$

The particular integral of (11) of period τ will give the rolling induced by the waves. The general solution of (11) will in fact be the sum of this particular integral together with a natural roll (3) of period T; however if we were to add a damping factor as in (5), then the latter would decay exponentially leaving only the particular integral, which is therefore an attractor of the dynamic. The particular integral will in fact be an amplification of the wave

$$\theta = \lambda\alpha = \lambda\alpha_o \cos\frac{2\pi t}{\tau} . \quad \ldots\ldots\ldots (12)$$

where λ is a constant amplifying factor. We can compute λ by substituting (12) into (11) :

$$\left[-\left(\frac{T}{\tau}\right)^2 + 1\right]\lambda\alpha = -\alpha .$$

Therefore
$$\lambda = \left[\left(\frac{T}{\tau}\right)^2 - 1\right]^{-1} \quad \ldots\ldots\ldots\ldots (13)$$

We can now compute the effect of a typical Atlantic beam sea upon our destroyer and liner. Assume τ = 6 secs and α = 18° as in Figure 4.

Table 2	Destroyer	Liner
Period of roll T, by Table 1	8 secs	30 secs
Amplifying factor λ, by (13)	$\frac{9}{7}$	$\frac{1}{24}$
Amplitude of induced roll, by (12)	$23°$	$1°$

Thus while the destroyer will be wallowing in the seaway the liner will hardly notice it. As Barnaby observes [2,p.337].

> "This is the great paradox of naval architecture — that the more stable the vessel really is, the more unstable she appears in a seaway."

He reinforces his observation with a revealing anecdote [2,p.355] that gives life to our computations :

> "This can be illustrated by the case of two yachts that were virtual sister ships, differing only in metacentric height. The captain of the first yacht reported her to be a magnificent sea-boat, extremely dry, and "stiff as a church" in a sudden squall [like our destroyer]. Her owner thought she was much too quick and lively in a seaway. As yachts have to be built for owners rather than for captains, the second vessel was given less metacentric height [like our liner]. The same captain transferred to the new yacht, and, in accordance with the best seafaring tradition, greatly preferred his old ship. His new command was wetter and more sluggish in a seaway, and, for these reasons, he considered her a worse sea-boat. The new owner was delighted, and said his yacht was "stiff as a church" in a seaway."

5. RESONANCE.

Before leaving the seaway we make some remarks concerning resonance. Suppose as before that the ship encounters waves of fixed period.

(i) If the wave-period, τ, happens to coincide with the natural roll-period, T, then the amplifying factor reaches a maximum. Because of damping this maximum is finite. Formula (13) gives $\lambda = \infty$, but this is wrong because damping has been ignored. If, as in (5), a damping term of $2\varepsilon\dot{\theta}$ is added to (9), then for $\tau = T$ the resonant solution has a $\frac{\pi}{2}$-phase-shift,

$$\theta = -\lambda\alpha_o \sin\frac{2\pi t}{T}$$

with amplifying factor $\lambda = \frac{\pi}{\varepsilon T}$. However the induced roll in this case may be so large that the linear theory is no longer valid. Indeed in heavy seas resonance may cause capsizing.

(ii) Let V, v denote the speeds of ship and wave. If the ship's course is such as to cause resonance we call it a <u>sensitive course</u>.

<u>Lemma 4.</u> <u>If $V > v(1+\frac{\tau}{T})$ there are 4 sensitive courses.</u>

Proof. Figure 6 shows how the period with which the ship encounters the waves depends upon the course. The blob represents the velocity of the waves, and the circle represents the different courses of the ship at speed V. The dashed line indicates the two courses perpendicular to the velocity of the waves (beam sea with period τ). The dotted lines indicate the two courses on which the ship will ride with the waves (period ∞). The other four lines indicate the sensitive courses. In order to encounter the waves with period T, the relative velocity of the ship must have a component $\pm\frac{v\tau}{T}$ in the direction of the waves. Therefore the ship's velocity is $v(1\pm\frac{\tau}{T})$ in this direction. The condition for the four

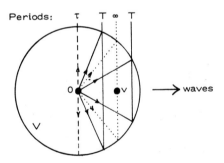

Figure 6. Sensitive courses.

directions to exist is that this component be less than V. This completes the proof. On the two sensitive courses closest to the wave direction the ship overtakes the waves, while on the other two ship is overtaken by the waves.

(iii) The non-linearity of the accurate rolling equation makes a wall-sided ship behave like a hard spring for small angles of roll, and a soft spring for large angles of roll (see Section 9 and Figure 19 below). The latter is liable to produce a Duffing effect [8,15] near resonance, as shown in Figure 7. If there is a Duffing effect, then when the ship reduces speed on a sensitive course overtaking the waves the amplitude of induced roll may at first increase slightly, and then drop suddenly when a critical speed V_1 is reached. Conversely, if the ship increases speed again a hysteresis delay will occur, during which the roll will remain deceptively small, until a higher critical speed V_2 is reached,

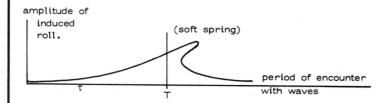

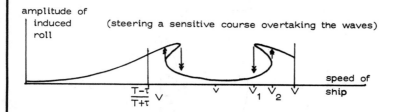

Figure 7. Duffing effects.

when the amplitude will suddenly increase again. Such a catastrophic jump could be dangerous because the dynamic stability of the resulting resonance might lead to capsizing before remedial action had time to take effect. Similar catastrophes occur near $(\frac{T-\tau}{T+\tau})V$, which in the case of the liner above equals $\tfrac{3}{8}V$. Note these catastrophes are different from those in the main model in Sections 7 - 14 below.

6. PITCHING AND HEAVING.

Passing from the 2-dimensional problem to the 3-dimensional problem, the buoyancy locus $\mathcal{B}$ becomes a convex closed surface rather than a convex closed curve. Therefore $\mathcal{B}$ has two principle directions of curvature at B_0, and two centres of curvature. One is the metacentre M for rolling about the longitudinal axis that we have already discussed, and the other is the metacentre M^* for pitching about the transverse axis. We now estimate the period of pitching, using the same notation as before, only with asterisks.

For simplicity assume (i) that the area below water is a rectangle with length L and draught D. Then by Lemma 2 the radius of curvature is

$$\rho^* = \frac{2(L/2)^3}{3LD} = \frac{L^2}{12D}.$$

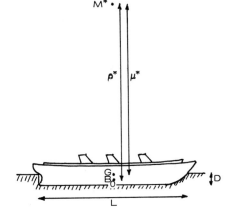

Figure 8. Pitching metacentre.

Assuming (ii) that $L > 24D$, then $\rho^* > 2L \gg B_0G$, and therefore B_0G can be ignored in the estimate of μ^*:

$$\mu^* = \rho^* = \frac{L^2}{12D} \quad \ldots\ldots\ldots\ldots\ldots (14)$$

17. Stability of Ships

Assume (iii) that the moment of inertia about the transverse axis is the same as that of a rod of length L. Therefore

$$I^* = \frac{W}{g} \frac{(L/2)^2}{3} = \frac{WL^2}{12g} \quad \ldots\ldots\ldots (15)$$

By (4) the period of pitching is

$$T^* = 2\pi \sqrt{\frac{I^*}{W\mu^*}}$$

$$= 2\pi \sqrt{\frac{D}{g}} \quad \text{by (14) and (15)} \quad \ldots (16)$$

$$= 2\sqrt{D}, \text{ approximately.}$$

Applying this to our two ships:

		Destroyer	Liner
Assume	Length, L	100m	300m
	Draught, D	3.3m	10m
Deduce	Pitching metcentric height μ^*, by (14)	250m	750m
	Period of pitching T^*, by (16)	4 secs	6 secs

Since μ^* is several hundred times larger than μ, the ship is far more stable with respect to pitching than to rolling, and hence the period is short, and amplitude kept small. The accelerations involved may be greater, and hence pitching is sometimes more uncomfortable than rolling.

Heaving refers to oscillations up and down. Let q denote the height of the ship above the equilibrium position. For a wall-sided ship of draught D, the volume of displaced water is reduced by a factor of approximately $\frac{q}{D}$.

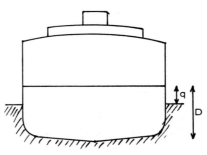

Figure 9. _Heaving._

Therefore the buoyancy force is reduced by the same factor, and so there is a net downward restoring force of $\frac{Wq}{D}$. Therefore by Newton's law

$$\frac{W}{g}\ddot{q} = -\frac{Wq}{D}.$$

Therefore
$$\ddot{q} = -\frac{g}{D}q \quad \ldots\ldots\ldots\ldots\ldots\ldots (17)$$

Hence the period for heaving is the same as that for pitching (16). In practice of course the periods differ slightly, because our assumptions are too imprecise. However the proximity of the periods implies that pitching and heaving will be coupled, and the classical theory of the coupling originated with Krilov in 1893 (see [3]).

The other 3 normal modes of oscillation, yawing (about the vertical axis), swaying (from side to side) and surging (fore and aft) differ from rolling, pitching and heaving in that buoyancy does not provide a natural restoring force; therefore these modes tend to occur only as induced, or secondary effects. We express the difference in group-theoretic terms in Lemma 12 below. This completes our elementary sketch of the classical linear theory, and we now begin the catastrophe theory, which is the main business of the paper.

7. CUSP CATASTROPHE AT THE METACENTRE.

For simplicity return to the 2-dimensional problem of rolling. For large angles the linear theory is no longer valid because the buoyancy force no longer goes through M. We need to look at not just one centre of curvature, but all of them. Therefore define the <u>metacentric locus</u> $\mathfrak{M}$ of the ship to be the locus of centres of curvature of the buoyancy locus; in other words $\mathfrak{M}$ is the evolute of $\mathfrak{B}$.

Now B_0 is a point of symmetry of $\mathfrak{B}$ where the radius of curvature is stationary, and hence M is a cusp point of $\mathfrak{M}$. The geometric question arises : Which way does the cusp branch, upwards or downwards? In Figure 10 we show two cases arising from different shapes of hull, on the left a modern wall sided ship, and on the right an old-fashioned canoe, shaped like an ellipse with major axis horizontal.

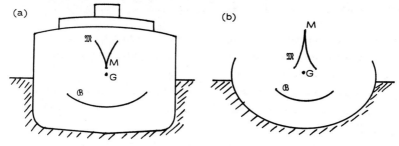

Figure 10. Cusps in ship (a) and canoe (b).

Lemma 5. The cusp branches upwards in the ship (a) and downwards in the canoe (b).

Proof. In a wall-sided ship $\mathcal{B}$ is locally a parabola $x^2 = 2\rho y$, by Lemma 3. This has evolute
$$27\rho x^2 = 8(y-\rho)^3 ,$$
which is a cusp branching upwards. The result for an elliptical-shaped hull follows from the Corollary to Lemma 7 below.

We will now explain the physical significance of which way the cusp branches. Suppose that the position of the centre of gravity of the ship changes for some reason : for example the canoeist might lean over the side, or stand up, or put up a mast. In a liner the passengers might crowd to one side to see something interesting. A cargo boat might load or unload, or restow its cargo (see Figures 27 and 28 below). In heavy weather the cargo might shift* by itself, or slosh about if liquid, or the ship might

---

* Capsizing and the shifting of cargo are still hazards. During 1975, according to Lloyds Casualty Return [10], 125 merchant ships foundered, mostly in heavy weather (not to mention another 211 lost, missing, burnt, wrecked, or in collison). Of the 125 foundered, 13 are known to have capsized and sunk, 14 others developed a list before sinking, and in 9 cases the list was known to be due to the cargo shifting.

accumulate ice to windward, or sea-water on deck. Fishing vessels may be tempted by a good catch to take on more than is advisable. For **simplicity** assume for the moment that the position of G changes without altering the total weight, so that the buoyancy locus remains the same (in Section 14 we allow for change of weight).

<u>Question</u> : given the position of G, at what angles can the ship float in equilibrium?

<u>Answer</u> : by Lemma 1 it will be those angles θ, such that G lies on the normal N_θ to $\mathcal{B}$ at B_θ. But the normals to $\mathcal{B}$ are tangents to $\mathcal{M}$; therefore the angles are obtained by drawing tangents from G to the cusp.

In Figure 11 we plot the graph of θ as a function of G, for the two boats pictured in Figure 10. In each case the position of G is represented by a point in the horizontal plane, C. The value of θ is represented by a point on the vertical axis $\mathbb{R}$, and for simplicity we assume $|\theta| < \beta$, where β is some suitable bound (in the full model in Section 10 below, we allow θ to be arbitrary). On the vertical line above each position of G we plot the corresponding equilibrium values of θ, and as G varies these points trace out a smooth surface, which we call the <u>equilibrium surface</u>, E. We shall prove in Theorem 2 below that in each case E is a cusp-catastrophe. By definition

$$E = \{(G,\theta); G \in N_\theta, |\theta| < \beta\}$$
$$= \{N_\theta \times \theta; |\theta| < \beta\}, \subset C \times \mathbb{R}.$$

Therefore E is a smooth ruled surface, consisting of horizontal lines parallel to the normals, one for each θ. In other words E is the normal bundle of $\mathcal{B}$. Over the outside of the cusp E is single-sheeted because if G lies outside (as in Figure 11(a)) there is only one tangent from G to the cusp. On the other hand over the inside of the cusp E is triple-sheeted, because if G lies inside (as in Figure 11(b)), there are three tangents from G to the cusp.

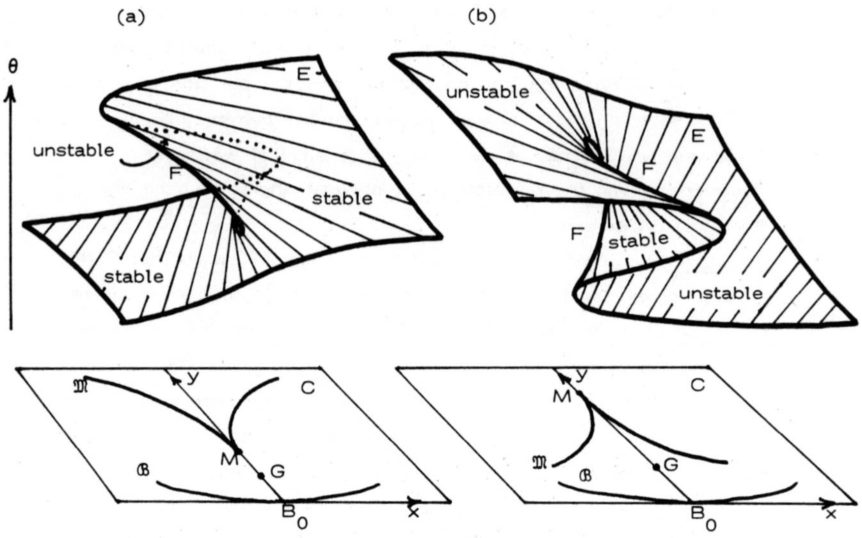

17. Stability of Ships

Figure 11. The cusp-catastrophe in the ship (a) and the canoe (b).

If E is projected down onto C it becomes folded along a curve, called the <u>fold curve</u> F, which projects onto the cusp. Hence the cusp is a <u>bifurcation set</u>. F separates E into two components, one representing stable equilibria and the other unstable equilibria. For example if G lies on the axis of symmetry (the y-axis) then by equation (2) in Section 1 the equilibrium position $\theta = 0$ is stable or unstable according as to whether the metacentric height μ is positive or negative, in other words whether G is below or above M.

If, further, G lies inside the cusp then the $\theta = 0$ equilibrium is represented by a point on the middle sheet of E, whilst the other two equilibria are represented by points on the upper and lower sheets of E. We call these <u>heeling angles</u> if the cusp branches upwards, as in case (a), and <u>capsizing angles</u> if the cusp branches downwards, as in case (b).

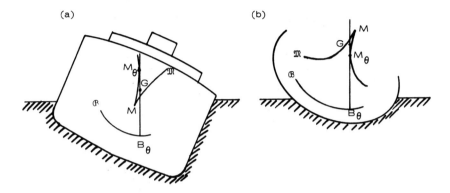

Figure 12 (a) Heeling angle. (b) Capsizing angle.

Here the difference between the two boats becomes apparent because:

Lemma 6. <u>Heeling angles are stable, whereas capsizing angles are unstable.</u>

<u>Proof.</u> Let θ be a heeling or capsizing angle. If M_θ denotes the corresponding metacentre, the centre of curvature of $\mathcal{B}$ at B_θ, then the tangent from G touches the cusp at M_θ. Figure 12 shows that in case (a) G lies below M_θ, and so the ship behaves stably like a pendulum hanging from M_θ. By contrast in case (b) G lies above M_θ, and so the canoe behaves unstably, balanced precariously over M_θ; any perturbation reducing θ will produce a righting couple that returns the canoe upright, whereas any perturbation the opposite

way will produce an opposite couple, that will cause the canoe to turn turtle.

Therefore in Figure 11(a) the upper and lower sheets of E are stable, while the middle sheet over the inside of the cusp is unstable. In Figure 11(b) it is the other way round, and so in this case E is called a <u>dual</u> cusp-catastrophe. The difference is emphasised in Figure 13 which shows in each case the section of E over the y-axis, rotated through 90° so as to make the y-axis vertical. The section in each case consists of the line θ = 0 bisecting a curve (which is a parabola modulo θ^4). The stable equilibria are shown firm, and the unstable dotted. In case (a) the curve is stable and rising and represents heeling angles, while in case (b) it is unstable and falling and represents capsizing angles. The specific angles are given by the intersection of the curve with the horizontal line through G.

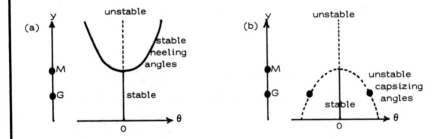

Figure 13. Sections of E over the y-axis for (a) ship and (b) canoe.

Now imagine an experiment in which the centre of gravity is raised up the y-axis past M. In case (a) when G passes M the ship will heel over to one side or the other, and will settle into stable equilibrium at the heeling angle. This happens for instance in certain old cargo boats when they unload, because they used to be designed with negative metacentric height when empty (see

[2, p.71] and Figures 27, 28 below). It also tends to happen in toys like model gondolas for the same reason, and sometimes the heel can be corrected by loading the model with a little ballast.

Suppose that G is on the y-axis above M with the ship heeled to the right, represented by a point on the upper sheet of Figure 11(a). If G is now moved to the left the ship will stay heeled to the right until G crosses the left side of the cusp, when it will suddenly heel over to the left; this is represented in Figure 11(a) by the point crossing the fold curve and jumping catastrophically onto the lower sheet. Conversely if G is now moved to the right then the ship will delay until G crosses the right side of the cusp when it will suddenly heel back again. Therefore in case (a) the cusp is a heeling bifurcation set.

By contrast case (b) is more dangerous because if G is raised up the y-axis past M the canoe will suddenly capsize. Worse still, if G happens to be off-centre when it is raised then it will cross the cusp sooner because the cusp branches downwards, and hence capsize sooner. This imperfection-sensitivity explains why when standing up in a canoe it is advisable to keep perfectly in the centre. Similarly if G is moved sideways, then as soon as G crosses either side of the cusp the canoe will capsize. A graphic description of such an event is given by Gerald Durrell [6, p.163].

"Peter nodded, braced himself, clasped the mast firmly in both hands, and plunged it into the socket. Then he stood back, dusted his hands, and the Bootle-Bumtrinket, with a speed remarkable for a craft of her circumference, turned turtle."

Evidently raising the mast raised G close to the metacentre where the cusp was very narrow, and stepping back was sufficient to cause G to cross the cusp - or maybe only just to reach the cusp, and it was actually the dusting of the hands that gave the final perturbation across it. Durrell goes on to explain how the problem was solved :

"For the rest of the morning he kept sawing

bits off the mast until she eventually floated upright, but by then the mast was only about three feet high."

Summarising :

Theorem 1. <u>An upward branching cusp is a heeling bifurcation set, and a downward branching cusp is a capsizing bifurcation set.</u>

Notice that this theorem only refers to the statics, because although we have used local righting couples to determine the local nature of the stability, the global dynamics has been ignored. We shall return to the dynamics again in Section 10.

8. GLOBAL METACENTRIC LOCUS.

We have yet to prove that the cusp in a canoe branches downwards. The easiest way to tackle this is to investigate the global metacentric locus of a completely elliptical ship, (like a submarine before it submerges).

Lemma 7. <u>The buoyancy locus of an ellipse is a similar ellipse.</u>

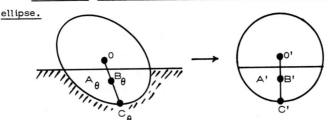

Figure 14.

Proof. Let O denote the centre of the ellipse, and C_θ the lowest point when heeled at angle θ. The line OC_θ bisects all horizontal chords, therefore bisects the region A_θ below the water line, and hence contains the centre of buoyancy B_θ. Map the ellipse onto a circle by an affine area-preserving map, and let A',B',C',O' denote the images of $A_\theta, B_\theta, C_\theta, O$. Then B' is the centre of gravity of A', and since the area of A' is independent of θ

the ratio $O'B'/O'C'$ = constant, k say, independent of θ. Since the map is affine $OB_\theta/OC_\theta = k$. Therefore, as C_θ traces out the ellipse, B_θ traces out a similar ellipse k times the size. This completes the proof. Notice that the result is independent of the weight (or density) of the ship.

Corollary. The metacentric locus of an ellipse has 4 cusps as shown in Figure 15.

For it is merely the evolute of the buoyancy locus, which by the lemma is a similar ellipse. In particular this completes the proof of Lemma 5, for in an elliptical shaped canoe with major axis horizontal, the

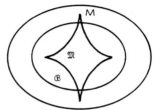

Figure 15.

metacentre M is the topmost cusp, and hence the cusp branches downwards.

Remark. In the 3-dimensional situation exactly the same proof shows that the buoyancy locus of an ellipsoid is a similar ellipsoid. The evolute of an ellipsoid, however, is more difficult to visualise because it consists of two sheets corresponding to the two metacentres, one for rolling and one for pitching. It can be regarded as 2 spheres, pinched along 3 elliptical cusped edges, one of which contains 4 hyperbolic umbilics [5].

Theorem 2. The equilibrium surface E has a cusp catastrophe at the metacentre M.

Proof. If the buoyancy locus ℬ is generic, then from general theory its evolute 𝔐 will have a generic cusp at M, and its normal bundle E will have a cusp catastrophe. However we cannot be sure that the curves in question are generic without checking the explicit formulae for wall-sided and elliptical ships.

The equation of E in (x,y,θ)-space is formally the same as that of the normal N_θ in (x,y)-space, with the proviso that θ is reinterpreted as a coordinate rather than a parameter. In case (a) of the wall-sided ship $\mathcal{B}$ is locally a parabola

$$x^2 = 2\rho y,$$

by Lemma 3. The normal N_θ is given by

$$x + (y-\rho)\tan\theta - \frac{1}{2}\rho \tan^3\theta = 0.$$

As a surface this is differentially equivalent (in the sense of [13]) to

$$x + (y-\rho)\theta - \theta^3 = 0,$$

which is a canonical cusp catastrophe at $(0,\rho)$ with x,y as normal and splitting factors.

In case (b) of the canoe, $\mathcal{B}$ is an ellipse by Lemma 7, and the equation of the ellipse with radius of curvature ρ at the origin and eccentricity e (where e is the ratio of the vertical axis to horizontal axis) is :

$$x^2 + \left(\frac{y}{e}\right)^2 = 2\rho y.$$

The normal N_θ is given by

$$x + (y-\rho)\tan\theta + \rho(1-e^2)\tan\theta\,[1-e(e^2+\tan^2\theta)^{-\frac{1}{2}}] = 0$$

Since $e < 1$, this is differentially equivalent, as a surface, to

$$x + (y-\rho)\theta + \theta^3 = 0,$$

which is a canonical cusp-catastrophe at $(0,\rho)$ with $-x,-y$ as normal and splitting factors. This completes the proof.

<u>Remark.</u> If in case (b) the eccentricity e is increased until $e > 1$, this converts the horizontal axis of the ellipse into the minor axis, and changes the sign of θ^3, converting the cusp from downwards branching into upwards branching, as in case (a).

The question now arises : what is the complete metacentric locus for a modern wall-sided ship? Where does $\mathfrak{M}$ go to after the initial upward branching of the cusp? How does $\mathfrak{M}$ compare globally with the 4-cusp evolute of an elliptical hull shown in Figure 15? We start by looking at a rectangle :

Theorem 3. In a rectangular hull of density ½ the buoyancy locus is the union of 4 pieces of parabolas, and the metacentric locus has 8 cusps.

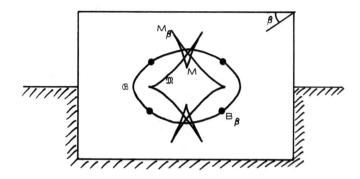

Figure 16. The metacentric locus of a rectangle.

Proof. The rectangle is a wall-sided ship for $|\theta| < \beta$, where β is the inclination of the diagonal to the horizontal. Therefore for $|\theta| < \beta$, $\mathcal{B}$ is a piece of a parabola by Lemma 3, and contributes an upwards branching cusp to $\mathfrak{M}$ by Lemma 5. There are 4 pieces, corresponding to the 4 sides of the rectangle. Two pieces of $\mathcal{B}$ join at B_β, and here the two parabolas have the same tangent by Lemma 1, the same radius of curvature by Lemma 2, and hence the same centre of curvature M_β. Therefore two pieces of $\mathfrak{M}$ touch at M_β, producing a parabolic cusp. Therefore $\mathfrak{M}$ is continuous, containing 4 generic cusps (of index $\frac{3}{2}$) separated by 4 non-generic parabolic cusps (of index 2), as shown in Figure 16.

Remark 1. If the density is reduced (or increased) the 4 parabolas in $\mathcal{B}$ are separated by 4 pieces of rectangular hyperbolas. When the density reaches ½tanβ then 4 swallowtails appear giving raise to another 8 cusps in $\mathfrak{M}$, making 16 in all (see Figures 25 and 28 below).

Remark 2. The non-genericity of the 4 parabolic cusps is due to the non-smoothness of the corners of the rectangle. If the corners are rounded-off in a C^∞-fashion, then the 4 parabolic cusps become generic, and the qualitative shape of $\mathfrak{M}$ is stable under small perturbations. Now the cross-section of a large modern ship can be regarded as a perturbation of a rounded-off rectangle. Therefore :

Conjecture. Large modern ships have metacentric locus $\mathfrak{M}$ similar to that in Figure 16. Detailed computations for individual ships show the top three cusps [11, p.135].

Remark 3. The evolution of hull shape from ellipse to rounded rectangle will cause a bifurcation of $\mathfrak{M}$ from the 4 cusps in Figure 15 to the 8 cusps in Figure 16. The reader familar with the elementary catastrophes will recognise immediately canonical sections of the butterfly catastrophe [12].

Problem. Prove, for an explicit isotopy of hull shape, that the bifurcation is an <u>unbiased butterfly</u>, in other words is equivalent to the symmetry section of the butterfly catastrophe given by putting the bias factor (the coefficient of θ^3) equal to zero. The governing potential at the bifurcation point should be

$$-k\theta^6 + i\theta^4 + \tfrac{1}{2}(\rho-y)\theta^2 - x\theta,$$

where x,y are coordinates of G, ρ the radius of curvature, $k > 0$, and i the isotopy parameter running from $i < 0$ for the canoe to $i > 0$ for the ship. The unbias is due to the symmetry of the ship, and a full butterfly should be obtained by introducing a bias factor measuring lop-sidedness of hull.

Remark 4. Globally the bifuraction set of the modern ship is not so very different from that of a canoe, and therefore the ship is not so safe as Theorem 1 would at first sight suggest. Therefore it is necessary to take another qualitative look at the heeling and capsizing. We assume the ship has metacentric locus similar to Figure 16, as conjectured above.

Theorem 4. The only heeling part of the metacentric locus 𝔐 is the cusp at M, shown dotted in Figure 17(a); the rest is capsizing. The equilibrium surface E is a section of a dual* butterfly catastrophe, as shown in Figure 17(b). The stable equilibria are shown shaded. Therefore for stability G must lie below 𝔐.

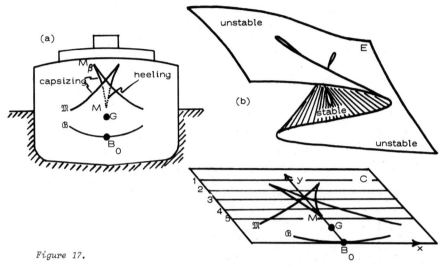

Figure 17.
(a) *Metacentric locus* 𝔐 *is part heeling (dotted) and part capsizing (firm).*
(b) *Equilibrium surface* E *is a section of a butterfly catastrophe.*

Proof. The bifurcation set in Figure 16 determines that E is a butterfly section, as shown in Figure 17(b) (see [12,13]). The identification of stable and unstable components of E is deduced from Figure 11(a), which is a subset of Figure 17(b). Hence E is a dual butterfly. The heeling and capsizing parts of 𝔐 are determined by consideration of the 5 sections of E over the 5 lines

---

* The <u>dual</u> butterfly [13] has germ $-\theta^6$, as opposed to the butterfly which has germ $+\theta^6$. This is the only application I know of the dual butterfly.

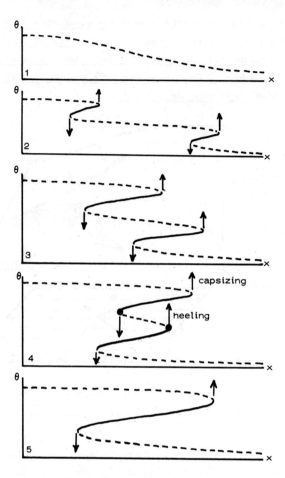

Figure 18. Sections of E.

parallel to the x-axis in Figure 17(b), as follows. The 5 sections are shown in Figure 18, with firm lines indicating stable equilibria, and dashed lines unstable equilibria. The nature of the stability determines which way the couple acts upon θ, and hence determines the direction of the catastrophic jump at each fold point, as indicated by the arrows. The catastrophe is heeling if the arrow tip stands on another stable sheet, and this only occurs for the middle two arrows of the fourth section. Hence the roots of those arrows (indicated by blobs) are the only two heeling parts of $\mathfrak{M}$. All the other parts of $\mathfrak{M}$ induce capsizing catastrophes. This completes the proof of Theorem 4.

9. LEVER ARM CURVE.

Assume G fixed. Recall that ℓ denotes the lever arm of the righting couple (see Figure 1(b)). The graph of ℓ for $0 \leq \theta \leq \frac{\pi}{2}$ is called the <u>lever arm curve</u>, and is illustrated in Figure 19 for the two boats shown in Figure 10. The slope of the lever arm curve at the origin is equal to the metacentric height, μ, because for small θ the linear approximation of the lever arm is $\ell = \mu\theta$.

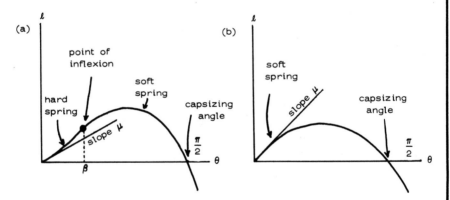

Figure 19. Lever arm curves for (a) ship and (b) canoe.

Lemma 8. In a wall-sided ship (a) the curvature of the lever arm curve is initially positive (like a hard spring), whereas in a canoe (b) it is negative (like a soft spring).

Proof. In the case of a wall-sided ship the normal N_θ is given by
$$x \cos\theta + (y-\rho)\sin\theta - \frac{1}{2}\rho \tan^2\theta \sin\theta = 0$$
by the proof of Theorem 2. Hence the distance ℓ from $G = (0, \rho-\mu)$ to N_θ, choosing the sign to be positive, is
$$\ell = \mu \sin\theta + \frac{1}{2}\rho \tan^2\theta \sin\theta$$
$$= \mu\theta + (\frac{\rho}{2} - \frac{\mu}{6})\theta^3 + o(\theta^5) \ .$$
Since $\mu < \rho$ the coefficient of θ^3 is positive, and hence for θ positive and small, the curvature is positive. In the canoe the normal N_θ is given by
$$x \cos\theta + (y-\rho)\sin\theta + \rho(1-e^2)\sin\theta[1-e(e^2+\tan^2\theta)^{-\frac{1}{2}}] = 0 \ .$$
Hence
$$\ell = \mu \sin\theta - \rho(1-e^2)\sin\theta[1-e(e^2+\tan^2\theta)^{-\frac{1}{2}}]$$
$$= \mu\theta - \theta^3[\frac{\rho}{2}e^{-2}(1-e^2) + \frac{\mu}{6}] + o(\theta^5) \ .$$
Since $e < 1$ the coefficient of θ^3 is negative, and hence, for θ small and positive, the curvature is negative.

Remark 1. In Figure 19 the difference in sign between the initial curvatures can be intuitively explained by which way the cusps branch in Figure 10. For in case (a) the upward branching causes the hard spring, while in case (b) the downward branching causes the soft spring. To be precise the conditions are slightly different : the cusp branches up or down as $e \gtrless 1$, whereas the curvature is hard or soft as $e \gtrless (1-\mu/3\rho)^{-\frac{1}{2}}$, and since $\mu < \rho$ this constant lies between 1 and $\sqrt{3/2}$.

Remark 2. In Figure 19(a) the change of curvature from hard to soft at the point of inflexion can be intuitively explained by the butterfly section in Figure 16. In the case of a rectangular hull, the initial hard spring is caused by the upward branch MM_β of $\mathfrak{M}$, the point of inflexion occurs at the angle β of the cusp M_β, and the subsequent soft spring is caused by the subsequent downward branch of $\mathfrak{M}$, to the capsizing angle, as indicated in Figure 12(b). In the wall-sided ship Figure 17 the same holds, except that the smoothness of hull causes the angle at which the inflexion occurs to be displaced slightly below that at which the cusp occurs.

<u>Remark 3.</u> The dynamical importance of the lever arm curve was first recognised by Atwood in 1796. Its use for judging stability in the design of ships was first proposed by Reed in 1868, and today various key features of the curve are widely used as stability criteria by naval architects and marine authorities [9]. What is new in this paper is the relationship between seemingly ad hoc features of the curve and generic properties of the metacentric locus arising from canonical sections of the butterfly catastrophe.

10. CATASTROPHE MODEL FOR ROLLING.

In Sections 2 - 6 we have discussed the local linear dynamics, and in Sections 7 - 9 the non-linear statics. We now weld the two together in order to study the global non-linear dynamics.

<u>Definition</u> : An <u>elementary catastrophe model</u>* is a parametrised system of gradient-like differential equations, specified by four things :

     (i) a parameter space C,
     (ii) a state space X,
     (iii) an energy function $H: C \times X \to \mathbb{R}$, and
     (iv) a dynamic D on X, parametrised by C, that locally minimises H.

---

* In the language of [14] this is at structure level 2.

The function H determines the equilibrium manifold, $E \subset C \times X$, by the equation $\nabla_X H = 0$. The catastrophe map $\chi: E \to C$ is induced by projection. The bifurcation set is the image in C of the singularities of χ. If H is generic then E has the same dimension as C, and the only singularities of χ are elementary catastrophes, by the classification theorem [12,13].

We first construct the model for the 2-dimensional rolling problem only, where it is easy to understand and visualise, and then in subsequent sections extend it to 3-dimensions to include pitching, heaving and loading.

(i) Define the <u>parameter space</u> to be the plane C containing our 2-dimensional ship. The parameter $G \in C$ is the position of the centre of gravity of the ship (relative to the hull).

(ii) Define the configuration space to be the unit circle, S. The configuration of the ship is uniquely determined by the angle $\theta \in S$. Define the <u>state space</u>, $X = T^*S$, to be the cotangent bundle* of S. The state of the ship is given by $(\theta, \omega) \in T^*S$, where ω is the angular momentum. As before let

W = weight of ship

I = moment of inertia of ship and entrained water.

$h = h(G, \theta)$ = height of G above B_θ, at angle θ

$= ZB_\theta$, in Figure 1(b)

Lemma 9.

<u>The potential energy of the system</u> is: $\quad P = P(G, \theta) = Wh$.

<u>The kinetic energy of the system</u> is: $\quad K = K(\omega) = \dfrac{\omega^2}{2I}$.

Proof.

Let $\quad h_1$ = height of G above the water line,

$\quad h_2$ = depth of B_θ below the water line.

---

*For a general treatment of Hamiltonians on cotangent bundles see [1].

Then taking the water line as zero potential,

Wh_1 = potential energy of the ship,

Wh_2 = potential energy of the displaced water.

Therefore P = total potential energy = $Wh_1 + Wh_2 = Wh$.

The angular momentum $\omega = I\dot\theta$, and therefore as required

$$K = \text{kinetic energy} = \frac{1}{2}I\dot\theta^2 = \frac{\omega^2}{2I}.$$

(iii) Define the energy function of the model to be the Hamiltonian $H: C \times T^*S \to \mathbb{R}$ given by

$$H = P + K = Wh + \frac{\omega^2}{2I}$$

We can now deduce the equilibrium surface and bifurcation set from H, as follows.

<u>Lemma 10.</u> $\quad \frac{\partial h}{\partial \theta} = \ell$.

<u>Proof.</u> Let M_θ, ρ_θ be the centre and radius of curvature of $\mathcal{B}$ at B_θ. Let $\mu_\theta = GM_\theta$, $\alpha = G\hat{M}_\theta B_\theta$. Then

$h(G, \theta) = \rho_\theta - \mu_\theta \cos\alpha$

$h(G, \theta+\varphi) = \rho_\theta - \mu_\theta \cos(\alpha+\varphi) + O(\varphi^2)$

$\frac{\partial h}{\partial \theta} = [\frac{\partial}{\partial \varphi} h(G, \theta+\varphi)]_{\varphi=0}$

$\qquad = [\mu_\theta \sin(\alpha+\varphi) + O(\varphi)]_{\varphi=0}$

$\qquad = \mu_\theta \sin\alpha = \ell$.

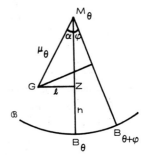

Figure 20.

The buoyancy locus $\mathcal{B}$ has coordinate θ, and is contained in the ambient space C. Therefore the <u>normal bundle</u> $N\mathcal{B}$ of $\mathcal{B}$ is defined by

$$N\mathcal{B} = \{N_\theta \times \theta\} \subset C \times S.$$

The <u>geodesic spray</u> is the natural map $N\mathcal{B} \to C$ of the normal bundle into the ambient space induced by projection $C \times S \to C$. The image of the singularities of the geodesic spray is the evolute of $\mathcal{B}$, which we have called the metacentric locus $\mathfrak{M}$.

<u>Theorem 5.</u> The equilibrium surface E is the normal bundle $N\mathcal{B}$ the buoyancy locus. The catastrophe map $\chi: E \to C$ is the geodesic spray. The bifurcation set is the metacentric locus, $\mathfrak{M}$.

<u>Proof.</u> E is given by $\nabla_X H = 0$, in other words by the equations
$$\frac{\partial H}{\partial \theta} = \frac{\partial H}{\partial \omega} = 0.$$
Now $\frac{\partial H}{\partial \omega} = \frac{\omega}{I}$, and hence $\omega = 0$. Therefore $E \subset C \times S \subset C \times T^*S$, where S is identified with the zero-section of T^*S. Also $\frac{\partial H}{\partial \theta} = W \frac{\partial h}{\partial \theta} = W\ell$, by Lemma 10, and hence $\ell = 0$. Therefore $G \in N_\theta$, the normal to $\mathcal{B}$ at B_θ. Therefore
$$E = \{(G, \theta, 0); G \in N_\theta\} = \{N_\theta \times \theta\} = N\mathcal{B}.$$
The catastrophe map χ merely says "forget θ", mapping each normal to itself, and giving the geodesic spray. Hence the bifurcation set, which is defined to be the image of the singularities of χ, equals $\mathfrak{M}$. This completes the proof of Theorem 5. To complete the model there remains to define the dynamic.

Assuming G fixed, the <u>Hamiltonian dynamic</u> on T^*S is uniquely determined from H by the intrinsic symplectic structure of the cotangent bundle (Newton's law of motion is built-in [1]). Explicitly the dynamic is given by the Hamiltonian equations
$$\dot{\theta} = \frac{\partial H}{\partial \omega} = \frac{\omega}{I}$$
$$\dot{\omega} = -\frac{\partial H}{\partial \theta} = -W\frac{\partial H}{\partial \theta} = -W\ell.$$
Therefore
$$I\ddot{\theta} = \dot{\omega} = -W\ell.$$
which is the same as equation (1) in Section 2. The resulting Hamiltonian flow is the accurate global non-linear generalisation of the approximate local simple harmonic rolling solution (3). However as yet we have not included any friction, because the Hamiltonian flow is conservative, conserving the energy H.

(iv) Define the <u>dynamic D</u> of the catastrophe model to be the Hamiltonian dynamic with non-zero damping. There is no need to be any more specific about the nature of the damping other than saying that energy is dissipated, because this ensures that H decreases along the orbits of D. Therefore H is a Lyapunov function for D. Therefore D locally minimises H, and depends upon the parameter G, as required. The model is complete.

11. GLOBAL DYNAMICS.

In order to understand the catastrophe dynamic D we fix G and draw the phase portrait of the resulting flow in Figure 21(c). The phase portrait is the family of orbits on T^*S. Since T^*S is a cylinder, we cut the cylinder along the generator $\theta = \pi = -\pi$, and lay it out flat, with the understanding that the two sides should be identified. In Figure 21 the dotted parts to the right of $\theta = \pi$ are merely the periodic repeats of the left hand sides of the portraits.

Before drawing the portrait of the damped flow, we first draw two portraits of the undamped Hamiltonian flow for two different positions of G in Figure 21(a) and (b). The latter is easier to understand because the Hamiltonian orbits are contained in the energy levels of H, which are themselves 1-dimensional since T^*S is 2-dimensional.

Figure 21(a) shows the Hamiltonian flow for G on the axis of symmetry below M, as in the case of the ship or the canoe in Figure 10. The 4 equilibria are given by $\omega = 0$ and

$\theta_2 = 0$, stable vertical

θ_1, θ_3, unstable capsizing angles (see Figure 12(b))

$\theta_4 = \pi$, stable turned turtle.

We are assuming that the ship does not sink if it capsizes, but is capable of floating upside down in stable equilibrium. The other three equilibria, $\theta_1, \theta_2, \theta_3$ correspond to the three points above G on the three sheets of E in Figure 11(b) for the canoe and

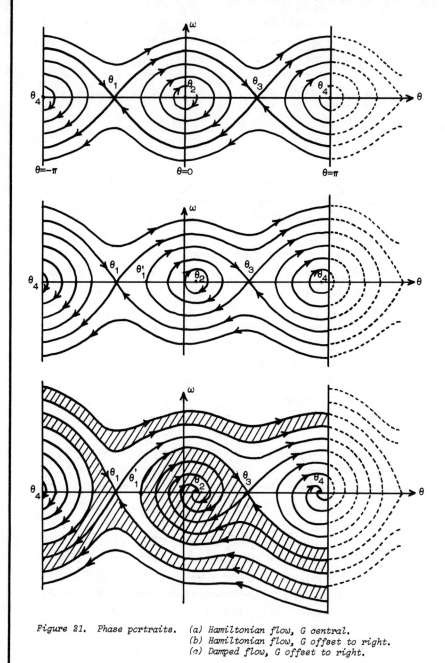

*Figure 21. Phase portraits. (a) Hamiltonian flow, G central.
(b) Hamiltonian flow, G offset to right.
(c) Damped flow, G offset to right.*

Figure 17(b) for the ship. The closed orbits round θ_2 represent rolling, as in the solution (3) of Section 2. The closed orbits round θ_4 represent rolling while turned turtle. The uppermost orbit going from left to right (which is closed since $\pi = -\pi$) represent the ship rolling over and over clockwise; the lowermost orbit represents the same but anticlockwise.

Figure 21(b) shows the Hamiltonian flow when G has been displaced slightly to the right. As a result, in stable equilibrium the ship will heel slightly to the right : $\theta_2 > 0$. Figures 11(b) and 17(b) show that the capsizing angles will also change slightly, θ_3 decreasing and $|\theta_1|$ increasing. Therefore it takes less energy to capsize to the right than to the left, because the θ_3-energy-level lies inside that of θ_1.

There is also a dynamic capsizing phenomenon, as follows. Let θ_1' be the intersection of the θ-axis with the θ_3-energy-level, such that $\theta_1 < \theta_1' < 0$. We call θ_1' the <u>dynamic capsizing angle.</u> If the ship is displaced to angle θ such that $\theta_1 < \theta < \theta_1'$, then it will not capsize to the left, but on the return roll to the right it will roll right over and then capsize, because the orbit will lie outside the θ_3-energy-level. Consequently, from the point of view of resonance, the dynamic capsizing angle is more dangerous than the static capsizing angle, because it is smaller

$$|\theta_1'| < \theta_3 < |\theta_1|.$$

On the other hand, if the ship rolls to the right almost as far as θ_3, then the recovery will take much longer, because, for small ε, the time to roll back from $\theta_3-\varepsilon$ upright again will vary as $|\log \varepsilon|$, as can be seen by considering the linear approximation at θ_3. Therefore the ship will tend to hang periously in the brink of capsizing for a long time, vulnerable to the chance wave or squall that might tilt her over the brink.

Figure 21(c) shows the damped flow for the same position of G. Since the dynamic is now dissipative the two stable equilibria

θ_2 and θ_4 are now attractors. If this figure was superimposed upon the figure above, all the damped orbits would cross the energy levels towards the attractors. The basin of attraction of the upright attractor θ_2 is shown shaded, and the complement is the basin of attraction of the turned turtle attractor θ_4. The capsizing angles are qualitatively unchanged since, being saddle-points of the Hamiltonian flow, they were already structurally stable. Similarly all the remarks about capsizing hold good.

   Problem. What is the best way to introduce the effect of wind and wave into the model, so as to generalise the induced linear rolling of Section 4 to the non-linear situation? The effect of periodic waves can be simulated by introducing a new cyclic parameter that translates the phase-portrait to and fro parallel to the θ-axis, corresponding to adding a forcing term representing the varying water surface, as in Figure 5 and equation (9). The induced roll is then an attracting closed orbit lying over the parameter-cycle. Perhaps the effect of the wind could be similarly modelled by translating the phase-portrait parallel to the ω-axis, to simulate the impulse transmitted by the wind into angular momentum.

12. MODEL INCLUDING PITCHING.

   We first enlarge the model to 3-dimensions to incorporate pitching as well as rolling, as follows.

   (i) The parameter space for G is now 3-dimensional, $C = \mathbb{R}^3$.

   (ii) The configuration space is now the unit sphere, $S = S^2$. The configuration of the ship is uniquely determined by the spherical coordinate $\theta \in S$. The state space is the cotangent bundle, $X = T^*S$, which is now a non-trivial 4-dimensional bundle. The state is given by $(\theta, \omega) \in T^*S$, where $\omega \in T^*_\theta S$ now represents the horizontal component of angular momentum at θ (the vertical yawing component of angular momentum is automatically excluded - see Lemma 12 below).

(iii) The energy is the Hamiltonian $H = P+K$, where the potential energy $P = Wh$, exactly as before. Meanwhile the kinetic energy $K = K_\theta(\omega)$ is the usual quadratic form on $T^*_\theta S$, defined as follows. If $(\omega_1, \omega_2, \omega_3)$ are coordinates of ω relative to the principal axes of inertia of the ship (which are tilted by θ), and (I_1, I_2, I_3) are the corresponding moments of inertia, then

$$K_\theta(\omega) = \Sigma \, \omega_i^2 / 2I_i \, .$$

(iv) As before, the dynamic is the damped Hamiltonian dynamic, and this completes the 3-dimension model.

Theorem 6. Theorem 5 holds for the 3-dimensional model.
Proof. E is defined by $\nabla_\theta H = \nabla_\omega H = 0$. Since the kinetic energy $K_\theta(\omega)$ is positive definite in ω, $\nabla_\omega H = \nabla_\omega K = 0$ implies $\omega = 0$. Therefore $E \subset C \times S$ as before.

Let Figure 20 represent the vertical plane through G and B_θ. Although θ is now a spherical coordinate, we can still define the angle α, and give meaning to $\theta+\varphi$, where φ is an angle. As in the proof of Lemma 10,

$$[\tfrac{\partial}{\partial \varphi} h(G, \theta+\varphi)]_{\varphi=0} = \ell \, .$$

$\nabla_\theta H = 0 \implies \nabla_\theta h = 0$
$\implies$ left-hand side vanishes
$\implies \ell = 0$
$\implies G \in N_\theta$
$\implies E = N\mathfrak{B}$, as required.

The rest of the Theorem follows naturally, as before.

The geometry of the 3-dimensional model is more complicated, for this time E is a 3-manifold folded over $\mathfrak{M}$, and $\mathfrak{M}$ is a surface with cusped edges and singular points. In particular, as the ship pitches, the rolling metacentre M traces out a curve on $\mathfrak{M}$, which we call the pitching curve, $\mathfrak{P}$, and which, due to the bilateral symmetry of the ship, is a cusped edge of $\mathfrak{M}$ containing various singular points. In an ellipsoidal ship $\mathfrak{B}$ is a similar ellipsoid, and $\mathfrak{P}$ is an ellipse containing 4 hyperbolic umbilics, and these are the only singular points of $\mathfrak{M}$ (see [5]). For any floating shape, $\mathfrak{M}$ always has at least 4 hyperbolic umbilics, because, by counting indices, the number of hyperbolic umbilics minus the number of elliptic umbilics is twice the Euler characteristic of $\mathfrak{B}$, which is 2.

In a wall-sided ship there are 4 other singular points on $\mathfrak{P}$, namely 4 unbiased-butterflies, where the butterfly sections (shown in Figures 16 and 17) bifurcate back into single cusps fore and aft. These butterfly points explain why large pitching enhances the danger of capsizing, because if the pitching angle passes beyond them then the rolling-capsizing-angle decreases. In other words pitching has the same qualitative effect as isotoping a ship-shaped section into a canoe-shaped section (see the Problem after Theorem 3 Remark 3), and the resulting decrease in capsizing angle can be seen by comparing Figures 12(b) and 17(a). Moreover large pitching can occur at the same time as resonant rolling, if the ship is less than half the wave-length and steers a sensitive course (see Section 5 and Figure 7). Therefore the geometry of the butterfly may be relevant to the study of capsizing of small vessels in heavy seas. Of the 13 merchant ships that were known to have capsized and sunk during 1975, 11 were small cargo or fishing vessels of 350 tons or less [10]. In Section 14 below we discuss the effect on the capsizing angles of heaving, or having the crest of a wave amidships.

13. MODEL INCLUDING HEAVING.

Next we enlarge the model to incorporate heaving. The enlarged model automatically contains all the coupling between the three modes of oscillation, rolling, pitching and heaving.

(i) As before, the parameter space is $C = \mathbb{R}^3$.

(ii) The configuration of the ship is given by $(\theta, q) \in S \times \mathbb{R}$ where θ is the spherical coordinate, and q the vertical coordinate of heaving. Let $B_{\theta,q}$ denote the resulting centre of buoyancy. The state is given by
$$(\theta, q, \omega, p) \in T^*(S \times \mathbb{R}),$$
where p denotes the vertical linear momentum and ω is as before.

(iii) The potential energy is given by
$$P = p(G, \theta, q) = Wh_1 + Vh_2,$$
where
h_1 = height of G above water line
h_2 = depth of $B_{\theta,q}$ below water line
V = weight of water displaced

} all functions of θ, q.

The kinetic energy is given by
$$K = K_\theta(\omega) + K(p) = \Sigma\, \omega_i^2/2I_i + gp^2/2W.$$
As before the energy
$$H : C \times T^*(S \times \mathbb{R}) \longrightarrow \mathbb{R}$$
is the Hamiltonian, $H = P+K$, and the dynamic is the damped Hamiltonian dynamic. We define the buoyancy locus to be the same as before, consisting only of those centres of buoyancy for which there is no heaving,
$$B = \{B_\theta\} = \{B_{\theta,q};\ q=0\}.$$

Theorem 7. *Theorem 5 holds.*

Proof. E is given by $\nabla_\theta H = \nabla_\omega H = \dfrac{\partial H}{\partial q} = \dfrac{\partial H}{\partial p} = 0$. Now $\dfrac{\partial H}{\partial p} = \dfrac{gp}{W}$, and so $p = 0$. We shall show $\dfrac{\partial H}{\partial q} = 0$ implies $q = 0$. Hence the problem reduces to the previous case, $E \subset C \times S$, and the result follows from Theorem 6. Let

$k_i = h_i(\theta, 0)$
$U = W - V$.
u = height of centre of emerged slice above water line.

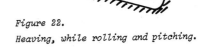

Figure 22.
Heaving, while rolling and pitching.

Then
$$h_1 = k_1 + q.$$

By taking moments of the displaced water about the water-line,
$$W(k_2 - q) = Vh_2 - Uu.$$

Therefore
$$P = Wh_1 + Vh_2 = W(k_1+q) + W(k_2-q) + Uu$$
$$= W(k_1+k_2) + Uu.$$

Therefore
$$\frac{\partial H}{\partial q} = \frac{\partial P}{\partial q} = \frac{\partial U}{\partial q}u + U\frac{\partial u}{\partial q}, \text{ since } k_1 + k_2 \text{ independent of } q.$$

From Figure 22,
$$U, u \gtreqless 0 \text{ as } q \gtreqless 0, \text{ and } \frac{\partial U}{\partial q} \geq 0, \frac{\partial u}{\partial q} > 0.$$

Therefore
$$\frac{\partial H}{\partial q} \gtreqless 0 \text{ as } q \gtreqless 0.$$

Therefore
$$\frac{\partial H}{\partial q} = 0 \text{ implies } q = 0, \text{ as required.}$$

Lemma 11. *For small heaving while upright the Hamiltonian dynamic reduces to the linear theory.*

Proof. In the case of a wall-sided ship of draught D we have approximately
$$U = \frac{Wq}{D}, \quad u = \frac{q}{2}, \text{ and therefore } Uu = \frac{Wq^2}{2D}.$$

The Hamiltonian equations give
$$\dot{q} = \frac{\partial H}{\partial p} = \frac{gp}{W}, \quad \dot{p} = -\frac{\partial H}{\partial q} = -\frac{\partial}{\partial q}(Uu) = \frac{Wq}{D}$$

Therefore $\ddot{q} = \frac{g}{D}q$, as in equation (17) of Section 6.

14. MODEL INCLUDING LOADING.

Finally we enlarge the model to incorporate loading by allowing the weight of the ship to act as another parameter. At the same time we place the treatment in a more elegant group-theoretical setting.

(i) The parameter is (W,G), where W is the weight and G the centre of gravity. The condition for the ship to float is $W \in \overline{W}$, where $\overline{W}$ denotes the open internal $0 < W < \mathscr{W}$, and $\mathscr{W}$ is the weight that would just sink it. Therefore the parameter space is 4-dimensional, $C = \overline{W} \times \mathbb{R}^3$.

(ii) Let $\mathcal{G}$ be the 6-dimensional group of Euclidean motions in $\mathbb{R}^3$. Choose an arbitrary reference position of the ship, and let $\mathcal{G}$ act on this reference position by right action. The potential energy of the water displaced is determined by $g \in \mathcal{G}$, and the potential energy of the ship determined by g and the parameter (W,G), as in the last section. The kinetic energy is the classical positive definite quadratic on $T^*\mathcal{G}$ for a rigid body. The sum gives the Hamiltonian
$$H : C \times T^*\mathcal{G} \longrightarrow \mathbb{R}.$$

Remark 1. The kinetic energy depends not only upon the weight but also upon the inertia tensor I of the ship. Therefore both H and the dynamic depend on I, which we could take as another 6-dimensional parameter. However in equilibrium the kinetic energy vanishes, and so I does not affect the equilibrium manifold. Therefore to identify the latter it is not necessary to specify I, which can be arbitrary.

Remark 2. The above H is no good for a catastrophe potential because it is not generic, and would give a 7-dimensional equilibrium manifold, which is 3 dimensions too big. We therefore need to factor out by the symmetries of H, as follows. $\mathcal{G}$ acts on the left of $T^*\mathcal{G}$ by :
$$h(g,t) = (hg, (T_g^*h)^{-1}t), \quad h \in \mathcal{G}, \; g \in \mathcal{G}, \; t \in T_g^*\mathcal{G},$$
and hence on the right of H by :
$$(Hh)(c,g,t) = (c, h(g,t)).$$
Define the symmetry group
$$\text{Sym}(H) = \{h \in \mathcal{G}; \; Hh = H\}.$$

Let $\mathfrak{H}$ be the 3-dimensional subgroup of $\mathfrak{G}$ preserving the horizontal (consisting of horizontal translations and rotations about vertical axes).

Lemma 12. Sym(H) = $\mathfrak{H}$.

Proof. The action of $\mathfrak{G}$ does not affect kinetic energy, and so we are only concerned with potential energy. $\mathfrak{H}$ does not alter potential energy and so $\mathfrak{H} \subset \text{Sym}(H)$. Conversely $\text{Sym}(H) \subset \mathfrak{H}$, because $\mathfrak{G}/\mathfrak{H}$ is spanned by the 1-parameter subgroups of rolling, pitching and heaving, which do alter H by the proof of Theorems 6 and 7.

We can now construct the model. Define the state space to be $(T^*\mathfrak{G})/\mathfrak{H}$, which is a 6-vector bundle over the 3-manifold $\mathfrak{G}/\mathfrak{H}$. In other words the state of the ship is determined by its 6 coordinates of momentum and its 3 coordinates of roll, pitch and heave (while its other 3 coordinates of latitude, longitude and course are automatically ignored). Define the potential to be the induced Hamiltonian

$$H : C \times (T^*\mathfrak{G})/\mathfrak{H} \longrightarrow \mathbb{R} .$$

Define the dynamic to be the damped Hamiltonian dynamic. This completes the model, and we now proceed to identify the equilibrium manifold E.

As before let S denote the unit sphere. Let

$$\theta : \mathfrak{G}/\mathfrak{H} \longrightarrow S$$

denote the projection given by defining $\theta(\mathfrak{H}g)$ to be the direction of the image of the vertical under g. (This is well-defined since $\mathfrak{H}$ preserves the horizontal, and hence also the vertical).

Each $W \in \overline{W}$ determines a buoyancy locus $\mathfrak{B}^W \subset \mathbb{R}^3$. As W varies from 0 to W, $\{\mathfrak{B}^W\}$ is a decreasing nested family of convex closed surfaces filling the convex hull of the ship's hull, running from the boundary $\mathfrak{B}^0$ to the single point $\mathfrak{B}^W$ at the submerged centre of buoyancy.

Figure 23. Buoyancy loci $\{\mathfrak{B}^W\}$.

For each W, $\mathcal{B}^W$ has evolute $\mathfrak{M}^W$, normal bundle $N\mathcal{B}^W$, and geodesic spray

$$N\mathcal{B}^W = \{N^W_\theta \times \theta\} \xrightarrow{\subset} \mathbb{R}^3 \times S \xrightarrow{proj} \mathbb{R}^3 .$$

Define the <u>complete</u> buoyancy locus to be the 3-manifold

$$\mathcal{B} = \{W \times \mathcal{B}^W\} \subset \overline{W} \times \mathbb{R}^3 = C ,$$

with complete metacentric locus,

$$\mathfrak{M} = \{W \times \mathfrak{M}^W\} \subset C$$

and complete normal bundle and geodesic spray

$$N\mathcal{B} = \{W \times N\mathcal{B}^W\} \xrightarrow{\subset} C \times S \xrightarrow{proj} C .$$

Theorem 8. θ <u>induces a diffeomorphism from the equilibrium</u> <u>manifold E onto the complete normal bundle</u> $N\mathcal{B}$ <u>such that the</u> <u>diagram is commutative.</u>

$$\begin{array}{ccc} E & \xrightarrow{\cong} & N\mathcal{B} \\ \cap \downarrow & & \downarrow \cap \\ C \times \mathcal{Q}/\mathcal{D} & \xrightarrow{1 \times \theta} & C \times S \end{array}$$

Therefore the catastrophe map $\chi : E \to C$ is equivalent to the geodesic spray $N\mathcal{B} \to C$, and the bifurcation set is the complete metacentric locus $\mathfrak{M}$.

<u>Proof.</u> Since kinetic energy is positive definite, equilibria lie in the zero section $\mathcal{Q}/\mathcal{D}$ of the state space $(T^*\mathcal{Q})/\mathcal{D}$. Fix W, and let E^W denote the corresponding section of the equilibrium manifold. Let φ^W denote the diffeomorphism

$$\varphi^W : \mathbb{R}^3 \times \mathcal{Q}/\mathcal{D} \longrightarrow \mathbb{R}^3 \times S \times \mathbb{R}$$
$$(G, \mathcal{D}g) \longmapsto (G, \theta, q) ,$$

where $\theta = \theta(\mathcal{D}g)$ and q is the height of Gg above the level that G would be at angle θ, were a weight W of water displaced. (This is well defined since $\mathcal{D}$ preserves that level). Then $\varphi^W E^W = N\mathcal{B}^W \times 0$ by Theorem 6. Therefore there are diffeomorphisms :

$$\begin{array}{ccccc} E^W & \xrightarrow{\cong} & N\mathcal{B}^W \times 0 & \xrightarrow{\cong} & N\mathcal{B}^W \\ \cap \downarrow & & \downarrow \cap & & \downarrow \cap \\ \mathbb{R}^3 \times \mathcal{Q}/\mathcal{D} & \xrightarrow[\cong]{\varphi^W} & \mathbb{R}^3 \times S \times \mathbb{R} & \xrightarrow{project} & \mathbb{R}^3 \times S \end{array}$$

Premultiplying by W, and taking the union for all $W \in \overline{W}$, give the result. Projecting onto C yields the required equivalence.

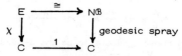

Both sides have the same image of singularities, and hence the bifurcation set is $\mathfrak{M}$. This completes the proof of Theorem 8.

<u>Theorem 9.</u> The metacentric loci $\mathfrak{M}^W$ and $\mathfrak{M}^{\omega-W}$ are similar. $\mathfrak{M}^W$ is obtained by reflecting $\mathfrak{M}^{\omega-W}$ in the submerged buoyancy centre $\mathcal{B}^\omega$ and scaling by $\frac{\omega-W}{W}$.

<u>Proof.</u> Let V_θ^W denote the volume displaced by weight W at angle θ. Then the volume of the ship can be written as a disjoint union

$$V^\omega = V_\theta^W \cup V_{T\theta}^{\omega-W}$$

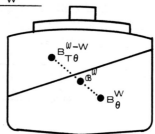

Figure 24.

where T is the antipodal map of S. Therefore the buoyancy centre B_θ^W is obtained by reflecting $B_{T\theta}^{\omega-W}$ in B^ω and scaling by $\frac{\omega-W}{W}$. Therefore this similarity maps B^W to $B^{\omega-W}$, and hence $\mathfrak{M}^W$ to $\mathfrak{M}^{\omega-W}$.

<u>Corollary 1.</u> $\mathfrak{M}^{\omega/2}$ is symmetrical about $\mathcal{B}^\omega$.
<u>Corollary 2.</u> $\mathfrak{M}^{3\omega/4}$ is one third the size of $\mathfrak{M}^{\omega/4}$.

The Corollaries are illustrated in Figure 25 for a typical 2-dimensional section of a ship having beam:draught ratio = 3:1 when $W = \omega/2$. The position of G is fixed, and the properties are the same as those of the liner in Section 2. The metacentric locus in Figure 25(b) has 8 cusps, and when the weight increases or decreases 4 swallowtails appear creating 8 more cusps, 16 in all.

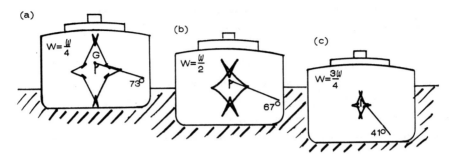

Figure 25. Decrease in size of $\mathfrak{M}^W$ and capsizing angle, with weight.

As the weight increases the ship becomes more susceptible to capsizing, not only because the bifurcation set shrinks closer to G, but also because the capsizing angle drops dramatically. The latter is caused by the appearance of the swallowtails. Figure 26 shows graphs of metacentric height and capsizing angle as functions of W, assuming G fixed. The dynamic capsizing angle can drop even further if there is a shift of cargo (see Figure 21).

This geometry is also relevant to the capsizing of small ships in following seas (small means the ship length is less than the wave-length), for when the crest of the wave is amidships, the buoyancy amidships carries more of the weight.

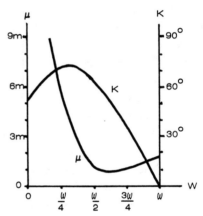

Figure 26. Metacentric height μ and capsizing angle K as functions of weight W.

Therefore although the ship may be designed as in Figure 25(b), she may unwittingly find herself riding on the crest of a wave as in Figure 25(c). The usual approach of linear theory is to emphasise the loss of stability due to the loss of metacentric height [9], but Figure 26 suggests that the non-linear drop in capsizing angle may in fact be more dangerous. A similar phenomenon may occur in a head sea, if the high frequency of encountered waves happens to resonate with heaving.

We conclude by considering the loading of cargo onto the ship. As W increases G may move, and so the parameter (W,G) follows a path p in the parameter space C. If p crosses $\mathfrak{M}$ the ship may heel, or right itself, or capsize. To illustrate this consider two such loading paths shown in Figure 27. The weight of the ship is $W/6$ when empty and $5W/6$ when full. During path p_1 the cargo is stowed so as to keep G fixed at the centre $\mathbb{B}^W$ of the ship, whereas during p_2 the cargo is loaded from the bottom upwards, as in a tanker. Meanwhile the dotted line shows how the height of the metacentre M^W depends upon W. The latter is drawn for a narrow ship, (with beam:draught ratio = 2:1 when $W = W/2$), because the **narrowness lowers** the path of M^W so as to cut p_1. Therefore p_1 suffers the two catastrophes of heeling and righting, whereas p_2 suffers none because it skirts below the dotted line.

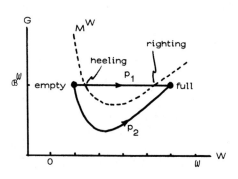

Figure 27. *Two loading paths.*

The corresponding behaviour of the ship is illustrated in Figure 28. In each case the metacentric locus $\mathfrak{M}^W$ is sketched,

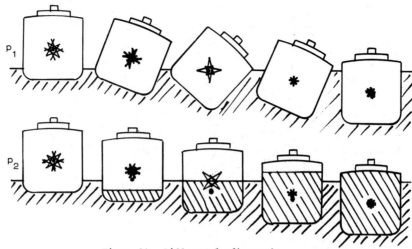

Figure 28. Different loading paths.

and G is indicated by a blob. Certain heavy narrow old cargo boats have a loading path like p_2, but the heaviness of the empty boat displaces the empty position in Figure 27 to the right across the dotted line; therefore such a boat heels with negative metacentric height when empty, and then suddenly rights itself during loading. If there is not enough cargo then the boat must take on ballast before sailing.

15. CONCLUSION

Could this geometrical approach to be **of any practical use**? It is difficult to say at this stage, but we tentatively suggest two areas for further exploration. Firstly the singularities involved are stable and qualitatively simple phenomena, sitting halfway between the geometric complexity of the shape of the ship and the dynamic complexity of its behaviour in the seaway. If details of a number of individual ships were analysed in the light of these singularities, it might give some further insight into how design can influence performance, and eventually might even suggest more sophisticated stability criteria.

Secondly there is at present a lack in the theory of non-linear coupling. Consequently there is a lack of mathematical language in which to express and communicate the intuition, which experienced pilots possess, of how to handle a ship in heavy weather. It is not enough to say that capsizing is probably due to that one-in-a-million freak wave, because capsizings do occur more frequently than we would wish [10], and it might be that the intuitive knowledge of one experienced pilot could have saved another. A case in point is Lindemann's discovery of how to get out of a flat spin in an aircraft, by pushing the stick fully forward and kicking hard on the opposite rudder. What was at one time a situation dreaded by all fliers, is now a routine recovery procedure taught to all beginners. Analogously in ship stability, the qualitative simplicity of the singularities that we have been discussing might eventually lead to a better understanding of the routine procedures for handling that freak wave.

REFERENCES.

1. R. Abraham & J.E. Marsden, Foundations of mechanics, Benjamin, New York, 1967.

2. K.C. Barnaby, Basic naval architecture, Hutchinson, London, 1963.

3. S.N. Blagoveshchensky, Theory of ship motions (English trans.), Dover, New York, 1962.

4. P. Bouguer, Traite du Navire, de sa construction et de ses mouvements, Paris, 1746.

5. A. Cayley, On the centro-surface of an ellipsoid, Trans. Camb. Phil. Soc. 12 (1873) 319-365.

6. G. Durrell, My family and other animals, Penguin, England, 1956.

7. L. Euler, Scientia Navalis, St. Petersburg, 1749.

8. P.J. Holmes & D.A. Rand, The bifurcations of Duffing's equation : and application of catastrophe theory, J. Sound & Vib, 44 (1976) 237-253.

9. C. Kuo, International Conference on Stability of Ships and Ocean vehicles, University of Strathclyde, Glasgow 1975.

10. Lloyd's Register of Shipping, Casualty Return, Lloyd's, London 1975.

11. A.M. Robb, Theory of Naval Architecture, Griffin, London, 1952.

12. R. Thom, Structural stability and morphogenesis, (trans : D D.H. Fowler), Benjamin, New York, 1975, (French edition 1972)

13. D.J.A. Trotman & E.C. Zeeman, The classification of elementary catastrophes of codimension ≤ 5, Structural stability, the Theory of catastrophes, and Applications in the Sciences, Springer Lecture Notes in Math. 525 (1976) 263-327.

14. E.C. Zeeman, Levels of structure in catastrophe theory, Proc. Int. Cong. Math. Vancouver (1974) 2, 533-546.

15. E.C. Zeeman, Duffing's equation in brain modelling, Bull. Inst. Math. & Appl., 12 (1976), 207-214.

Numbering in this book.

MATHEMATICS

MATHEMATICS

Paper.		Page
18.	The classification of elementary catastrophes of codimension ≤ 5 (with D.J.A. Trotman).	497
19.	The umbilic bracelet and the double-cusp catastrophe.	563 – 601.

Paper 18 gives in 65 pages a complete minimal proof of the classification theorem. The precise theorem that we prove is stated on page 502. Our statement differs slightly from that presented in most of the other mathematical expositions of the subject, because we are interested in the applied point of view, as opposed to the point of view of pure mathematicians; our objective is to provide a global theorem that is useful for applications, whereas most presentations prefer to stay within the more convenient local framework of singularity theory. Consequently we prove a theorem about whole functions, whereas most presentations confine themselves to germs; to do this we have to globalise in Chapter 8, using transversality theory. Also we highlight the catastrophe map, which tends to be neglected in most presentations. The importance of the catastrophe map can be seen from the many applications in this book, because it is the singularities of the catastrophe map that give rise to the discontinuities and the other characteristic features of catastrophe models.

Paper 19 gives a partial description of the double-cusp catastrophe, which is as yet not fully understood. The double-cusp is the most important elementary catastrophe beyond Thom's original seven, and will become increasingly significant in the future as more and more complex phenomena are modelled. We draw the reader's attention to the elegant definition of unfolding on page 574.

18 THE CLASSIFICATION OF ELEMENTARY CATASTROPHES OF CODIMENSION* ≤ 5.

(Notes written and revised by David Trotman)

INTRODUCTION.

These lecture notes are an attempt to give a minimal complete proof of the classification theorem from first principles. All results which are not standard theorems of differential topology are proved. The theorem is stated in Chapter 1 in a form that is useful for applications [12].

The elementary catastrophes are certain singularities of smooth maps $\mathbb{R}^r \to \mathbb{R}^r$. They arise generically from considering the stationary values of r-dimensional families of functions on a manifold, or from considering the fixed points of r-dimensional families of gradient dynamical systems on a manifold. Therefore they are of central importance in the bifurcation theory of ordinary differential equations. In particular the case $r = 4$ is important for applications parametrised by space-time.

The concept of elementary catastrophes, and the recognition of their importance, is due to René Thom [10]. He realized as early as about 1963 that they could be finitely classified for $r \leq 4$, by unfolding certain polynomial germs $(x^3, x^4, x^5, x^6, x^3 \pm xy^2, x^2y+y^4)$. Thom's sources of inspiration were fourfold: firstly Whitney's paper [11] on stable-singularities for $r = 2$, secondly his own work extending these results to $r > 2$, thirdly light caustics, and fourthly biological morphogenesis.

---

*This paper, giving a complete proof of Thom's classification theorem, seems not to be readily available. In response to many requests from conference participants, Zeeman and his collaborator, David Trotman, agreed to make a revised version of the paper (July, 1975) available for the conference proceedings. I would like to express my appreciation to both Christopher Zeeman and David Trotman.

Peter Hilton

Lecture Notes of a course given at Warwick University in Spring 1973.
Published in Structural stability, the theory of catastrophes and applications in the sciences, Springer Lecture Notes in Mathematics, Volume 525, 1976, pages 263-327.

However although Thom had conjectured the classification, it was some years before the conjecture could be proved, because several branches of mathematics had to be developed in order to provide the necessary tools. Indeed the greatest achievement of catastrophe theory to date is to have stimulated these developments in mathematics, notably in the areas of bifurcation, singularities, unfoldings and stratifications. In particular the heart of the proof lies in the concept of unfoldings, which is due to Thom. The key result is that two transversal unfoldings are isomorphic, and for this Thom needed a C^∞ version of the Weierstrass preparation theorem. He persuaded Malgrange [3] to prove this around 1965. Since then several mathematicians, notably Mather, have contributed to giving simpler alternative proofs [4,5,7,8] and the proof we give in Chapter 5 is mainly taken from [1].

The preparation theorem is a way of synthesising the analysis into an algebraic tool; then with this algebraic tool it is possible to construct the geometric diffeomorphism required to prove two unfoldings equivalent. The first person to write down an explicit construction, and therefore a rigorous proof of the classification theorem, was John Mather, in about 1967. The essence of the proof is contained in his published papers [4,5] about more general singularities. However the particular theorem that we need is somewhat buried in these papers, and so in 1967 Mather wrote a delightful unpublished manuscript [6] giving an explicit minimal proof of the classification of the germs of functions that give rise to the elementary catastrophes. The basic idea is to localise functions to germs, and then by determinacy reduce germs to jets, thereby reducing the ∞-dimensional problem in analysis to a finite dimensional problem in algebraic geometry. Regrettably Mather's manuscript was never quite finished, ahtough copies of it have circulated widely. We base Chapters 2, 3, 4, 6 primarily upon his exposition.

Mather's paper is confined to the local problem of classifying germs of functions. To put the theory in a usable form for applications three further steps are necessary. Firstly we need to globalise from germs back to functions again, in order to obtain an open-dense set of functions, that can be used for modeling. For this we need the Thom transversality lemma, and Chapter 8 is based on Levine's exposition [2].

Secondly we have to relate the function germs, as classified by Mather, to the induced elementary catastrophes, which are needed for the applications. For instance the elliptic umbilic starts as an unstable germ $\mathbb{R}^2 \to \mathbb{R}$, which then unfolds to a stable-germ $\mathbb{R}^2 \times \mathbb{R}^3 \to \mathbb{R} \times \mathbb{R}^3$, or equivalently to a germ $f: \mathbb{R}^2 \times \mathbb{R}^3 \to \mathbb{R}$, and eventually induces the elementary catastrophe germ $\chi_f: \mathbb{R}^3 \to \mathbb{R}^3$. The relation between these is explained in Chapter 7.

Finally in Chapter 9 we verify the stability of the elementary catastrophes, in other words the stability of χ_f under perturbations of f. A word of warning here: although the elementary catastrophes are singularities, and are stable, they are different from the classical stable-singularities [1,2,4,5,11]. The unfolded germ is indeed a stable-singularity, but the induced catastrophe germ may not be. The difference can be explained as follows. Let M denote the space of all C^∞ maps $\mathbb{R}^r \to \mathbb{R}^r$, and C the subspace of catastrophe maps. Then $C \neq M$ because not all maps can be induced by a function. Therefore a stable-singularity, such as Σ_2, may appear in M, but not in C, and therefore will not occur as an elementary catastrophe. Conversely an elementary catastrophe, such as an umbilic, may appear in C, and be stable in C, but become unstable if perturbations in M are allowed, and therefore will not occur as a stable-singularity. For $r = 2$ the two concepts accidentally coincide, because Whitney [11] showed that the only two stable-singularities were the fold and cusp, and these are the two elementary catastrophes. However for $r = 3$ the concepts diverge, and for $r = 4$, for instance, there are 6 stable-singularities and 7 elementary catastrophes, as follows:

stable-singularities   elementary catastrophes

2   4   3
Σ_2's   cuspoids   umbilics

We are grateful to Mario De Oliveira and Peter Stefan for their helpful comments: these have led to several corrections in the text.*

---

*As editor, I would also like to express my gratitude to Sandra Smith for adapting the original manuscript to a form suitable for the Lecture Notes, and to Sarah Rosenberg for her skillful reproduction of the diagrams.

                                        Peter Hilton

CONTENTS

Chapter 1. Stating Thom's Theorem

Chapter 2. Determinacy

Chapter 3. Codimension

Chapter 4. Classification

Chapter 5. The Preparation Theorem

Chapter 6. Unfoldings

Chapter 7. Catastrophe Germs

Chapter 8. Globalisation

Chapter 9. Stability

CHAPTER 1. STATING THOM's THEOREM.

Let $f: \mathbb{R}^n \times \mathbb{R}^r \to \mathbb{R}$ be a smooth function. Define $M_f \subset \mathbb{R}^{n+r}$ to be given by $(\frac{\partial f}{\partial x_1}, \ldots, \frac{\partial f}{\partial x_n}) = \text{grad}_x f = 0$, where $x_1, \ldots, x_n$ are coordinates for $\mathbb{R}^n$, and $y_1, \ldots, y_r$ are coordinates for $\mathbb{R}^r$. Generically M_f is an r-manifold because it is codimension n, given by n equations. Let $X_f: M_f^r \to \mathbb{R}^r$ be the map induced by the projection $\mathbb{R}^{n+r} \to \mathbb{R}^r$. We call X_f the <u>catastrophe map</u> of f. Let F denote the space of C^∞-functions on $\mathbb{R}^{n+r}$, with the Whitney C^∞-topology. We can now state Thom's theorem.

<u>Theorem.</u> If $r \le 5$, there is an open dense set $F_* \subset F$ which we call <u>generic</u> functions. If f is generic, then

(1) M_f is an r-manifold.

(2) Any singularity of X_f is equivalent to one of a finite number of types called <u>elementary catastrophes</u>.

(c) X_f is locally stable at all points of M_f with respect to small perturbations of f.

The number of elementary catastrophes depends only upon r, as follows:

r	1	2	3	4	5	6	7
elem. cats.	1	2	5	7	11	∞	∞

Here equivalence means the following: two maps $X: M \to N$ and $X': M' \to N'$ are equivalent if $\exists$ diffeomorphisms h, k such that the following diagram commutes:

Now suppose the maps X, X' have singularities at x, x' respectively. Then the singularities are equivalent if the above definition holds locally, with $hx = x'$.

Remarks. The reason for keeping $r \leq 5$ is that for $r > 5$ the classification becomes infinite, because there are equivalence classes of singularities depending upon a continuous parameter. One can obtain a finite classification under topological equivalence, but for applications the smooth classification in low dimensions is more important. The theorem remains true when $\mathbb{R}^{n+r}$ is replaced by a bundle over an arbitrary r-manifold, with fibre an arbitrary n-manifold.

The theorem stated above is a classification theorem: we classify the types of singularity that 'most' X_f can have. We find that if X_f has a singularity at $(x,y) \in \mathbb{R}^{n+r} \cap M_f$, and if η is the germ at (x,y) of $f|\mathbb{R}^n \times y$, then the equivalence class of X_f at (x,y) depends only upon the (right) equivalence class of η (Theorem 7.8). This result is hard and requires an application of the Malgrange Preparation Theorem, itself a consequence of the Division Theorem (Chapter 5), and study of the category of unfoldings of a germ η (Chapter 6).

To use it we have first to classify germs η of C^∞ functions $\mathbb{R}^n, 0 \to \mathbb{R}, 0$. We use two related integer invariants, determinacy and codimension, and the Jacobian ideal $\Delta(\eta)$ (the ideal spanned by $\frac{\partial \eta}{\partial x_1}, \ldots, \frac{\partial \eta}{\partial x_n}$ in the local ring E of germs at 0 of C^∞ functions $\mathbb{R}^n \to \mathbb{R}$). The determinacy of a germ η is the least integer k such that if any germ ξ has the same k-jet as η then ξ is right equivalent to η. Theorem 2.9 gives necessary and sufficient conditions for k-determinacy in terms of Δ. Defining the codimension of η as the dimension of m/Δ, where m is the unique maximal ideal of E, we use this theorem to show that $\det \eta - 2 \leq \text{cod } \eta$ in Lemma 3.1. If $r \leq 5$ and $f \in F_*$ then if $\eta = f|\mathbb{R}^n \times y$, for any $y \in \mathbb{R}^r$, we have $\text{cod } \eta \leq r$. Hence since we can restrict to $\text{cod } \eta \leq 5$ we need only look at 7-determined germs in the vector space J^7 of 7-jets. We must restrict to $r \leq 5$, for if $\text{cod } \eta \geq 7$ there are equivalence classes depending upon a continuous parameter, and the definition of F_* ensures that if $r = 6$ then each of these equivalence classes contains an $f|\mathbb{R}^n \times y$ for some $y \in \mathbb{R}^r$ and $f \in F_*$.

The 7-jets of codimension ≥ 6 form a closed algebraic variety Σ in J^7, and the partition by codimension of $J^7-\Sigma$ forms a regular stratification (Chapters 3 and 8). We in fact use a condition implied by a-regularity (Definition 8.2). This is necessary to show that F_* is open in F. That it is dense follows from Thom's transversality lemma; and transversality gives that M_f is an r-manifold for $f \in F_*$ (Chapter 8).

The classification of germs of codimension ≤ 5 is completed in Chapter 4 and in Chapter 7 the connection is made with catastrophe germs. Finally in Chapter 9 we show the local stability of X_f.

CHAPTER 2. DETERMINACY.

Definition. Suppose $C^\infty(M,Q)$ is the space of C^∞ maps $M \to Q$, where M and Q are C^∞ manifolds. If $x \in M$ and f and $g \in C^\infty(M,Q)$ let $f \sim g$ if $\exists$ a neighborhood N of x such that $f|N = g|N$. The equivalence class $[f]$ is called a _germ_, the germ of f at x.

Let E_n be the set of germs at 0 of C^∞ functions $\mathbb{R}^n \to \mathbb{R}$. It is a real vector space of infinite dimension, and a ring with a 1, the 1 being the germ at 0 of the constant function taking the value $1 \in \mathbb{R}$. Addition, multiplication, and scalar multiplication are induced pointwise from the structure in $\mathbb{R}$.

Definition. A _local ring_ is a commutative ring with a 1 with a unique maximal ideal.

We shall show that E_n is a local ring with maximal ideal m_n being the set of germs at 0 of C^∞ functions vanishing at 0 (written as functions $\mathbb{R}^n, 0 \to \mathbb{R}, 0$).

Lemma 2.1. m_n is a maximal ideal of E_n.

Proof. Suppose $\eta \in E_n$ and $\eta \notin m_n$. We claim that the ideal generated by m_n and η, $(m_n, \eta)_{E_n}$, is equal to E_n.

Let the function $e \in \eta$, i.e, η is the germ at 0 of e, and choose a neighborhood U of 0 in $\mathbb{R}^n$ such that $e \neq 0$ on U. Then $1/e$ exists on U. Let ξ be the germ $[1/e]$, then $\xi\eta = [1/e]\cdot[e] = [1/e\cdot e] = [1] = 1$. Also $\xi\eta \in (m_n,\eta)_{E_n}$. Thus $(m_n,\eta)_{E_n} = E_n$.

Lemma 2.2. m_n is the unique maximal ideal of E_n.

Proof. Given $I \subsetneq E_n$, we claim $I \subset m_n$. If not $\exists \eta \in I - m_n$, and then as in Lemma 2.1 an inverse exists in E_n. $1 = 1/\eta \cdot \eta \in I$, and so $I = E_n$.

Lemma 2.1 and Lemma 2.2 show that E_n is a local ring.

Let G_n be the set of germs at 0 of C^∞ diffeomorphisms $\mathbb{R}^n, 0 \to \mathbb{R}^n, 0$. G_n is a group with multiplication induced by composition. We shall drop suffices and use E, m and G, when referring to E_n, m_n and G_n rather than E_s when $n \neq s$, etc. Given $\alpha_1, \ldots, \alpha_r \in E$, we let $(\alpha_1,\ldots,\alpha_r)_E$ be the ideal generated by $\{\alpha_i\} = \{\sum_{i=1}^{r} \varepsilon_i \alpha_i : \varepsilon_i \in E\}$, and drop the suffix if there is no risk of confusion. Choose coordinates $x_1, \ldots, x_n$ in $\mathbb{R}^n$ (linear or curvilinear). The symbol 'x_i' will be used ambiguously as:

(i) coordinate of $x = (x_1,\ldots,x_n)$, $x_i \in \mathbb{R}$.

(ii) function $x_i: \mathbb{R}^n, 0 \to \mathbb{R}, 0$.

(iii) the germ at 0 of this function in $m \subset E$.

(iv) the k-jet of that germ (see below).

Lemma 2.3. $m = (x_1,\ldots,x_n)_E$

$\quad$ = ideal of E generated by the germs x_i.

Proof. Given $\eta \in m$, represent η by $e: \mathbb{R}^n, 0 \to \mathbb{R}, 0$. $\forall x \in \mathbb{R}^n$,

$$e(x) = \int_0^1 \frac{\partial e}{\partial t}(tx)dt$$
$$= \int_0^1 \sum_{i=1}^{n} \frac{\partial e}{\partial x_i}(tx) x_i(x) dt$$
$$= \sum_{i=1}^{n} e_i(x) x_i(x).$$

$e = \sum_{i=1}^{n} e_i x_i$ as functions and so $\eta = \sum_{i=1}^{n} e_i x_i$ as germs. Thus $m \subset (x_1,\ldots,x_n)$. $(x_1,\ldots,x_n) \subset m$ because each $x_i \in m$.

Corollary 2.4. m^k is the ideal generated by all monomials in x_i of degree k.

Corollary 2.5. m^k is a finitely generated E-module.

We let J^k be the quotient E/m^{k+1}, and let $\mathcal{J}^k$ be m/m^{k+1}. j^k denotes the canonical projection $E \to J^k$.

Lemma 2.6. J^k is 1) a local ring with maximal ideal $\mathcal{J}^k$,

2) a finite-dimensional real vector space (generated by monomials in $\{x_i\}$, of degree $\leq k$).

Proof. 1) J^k is a quotient ring of E and thus is a commutative ring with a 1. There is a 1-1 correspondence between ideals:

$$\begin{array}{ccc} E & & E/m^{k+1} = J^k \\ \cup & & \cup \\ I & \longleftrightarrow & I/m^{k+1} \\ \cup & & \\ m^{k+1} & & \end{array}$$

So J^k is a local ring.

2) J^k is a quotient vector space of E and is finite-dimensional. For given $\eta \in E$, the Taylor expansion at 0 is,

$$\eta = \eta_0 + \eta_1 + \ldots + \eta_k + \rho_{k+1},$$

where η_j is a homogeneous polynomial in $\{x_i\}$ of degree j, with coefficients the corresponding partial derivatives at 0, and $\rho_{k+1} \in m^{k+1}$.

Definition. The **k-jet** of $\eta = j^k \eta = \eta_0 + \ldots + \eta_k$

= Taylor series cut off at k.

J^k and $\mathcal{J}^k$ are spaces of k-jets, or **jet spaces**.

Definition. If $\eta, \xi \in E$ we say they are **right equivalent** ($\sim$) if they belong to the same G-orbit. $\eta \sim \xi \Leftrightarrow \exists \gamma \in G$ such that $\eta = \xi\gamma$.

Definition. If $\eta, \xi \in E$ we say they are __k-equivalent__ ($\overset{k}{\sim}$) if they have the same k-jet. $\eta \overset{k}{\sim} \xi \Leftrightarrow j^k \eta = j^k \xi$.

Definition. $\eta \in E$ is __k-determinate__ if $\forall \xi \in E$, $\eta \overset{k}{\sim} \xi \Rightarrow \eta \sim \xi$. Clearly η k-determinate $\Rightarrow \eta$ i-determinate $\forall i \geq k$. The __determinacy__ of η is the least k such that η is k-determinate. We write $\det \eta$.

Lemma 2.7. If η is k-determinate then

1) $\eta \overset{k}{\sim} \xi \Rightarrow \xi$ k-determinate,

2) $\eta \sim \xi \Rightarrow \xi$ k-determinate.

Proof. 1) follows at once from 2), which we shall prove. Assume $\eta \sim \xi$, i.e. $\eta = \xi\gamma_1$, some $\gamma_1 \in G$. Suppose $\xi \overset{k}{\sim} \nu$, i.e. $j^k\xi = j^k\nu$, i.e. $j^k(\eta\gamma_1^{-1}) = j^k\nu$.

Then $j^k\eta = j^k(\eta\gamma_1^{-1}\gamma_1) = j^k(\eta\gamma_1^{-1}) \cdot j^k(\gamma_1) = j^k\nu \cdot j^k\gamma_1 = j^k(\nu\gamma_1)$. So $\eta \overset{k}{\sim} \nu\gamma_1$, which $\Rightarrow \eta \sim \nu\gamma_1$, i.e. $\eta = \nu\gamma_1\gamma_2$ some $\gamma_2 \in G$. Then $\xi\gamma_1 = \nu\gamma_1\gamma_2$, and $\xi = \nu\gamma_1\gamma_2\gamma_1^{-1}$, i.e. $\xi \sim \nu$. So 2) is proved.

Definition. If $\eta \in E$, choose coordinates $\{x_i\}$ for $\mathbb{R}^n$, and let $\Delta = \Delta(\eta) = (\frac{\partial\eta}{\partial x_1}, \ldots, \frac{\partial\eta}{\partial x_n})_E$. Δ is independent of the choice of coordinates. For if $\Delta_x = (\frac{\partial\eta}{\partial x_i})$ and $\Delta_y = (\frac{\partial\eta}{\partial y_j})$, $\frac{\partial\eta}{\partial y_j} = \sum_{i=1}^n \frac{\partial\eta}{\partial x_i}\frac{\partial x_i}{\partial y_j} \in \Delta_x$ and so $\Delta_y \subset \Delta_x$. ($\frac{\partial\eta}{\partial x_i} \in \Delta_x$, each i, and $\frac{\partial x_i}{\partial y_j} \in E$, each i, j). Similarly $\Delta_x \subset \Delta_y$, so $\Delta_x = \Delta_y$.

Lemma 2.8. If $\eta \in E - m$, and $\eta' = \eta - \eta(0) \in m$, then $\Delta(\eta) = \Delta(\eta')$, and η is k-determinate $\Leftrightarrow \eta'$ is k-determinate.

Proof. $\Delta(\eta) = \Delta(\eta')$ is trivial. $\eta \overset{k}{\sim} \xi \Leftrightarrow \begin{cases} \eta' \overset{k}{\sim} \xi', \text{ trivially.} \\ \eta(0) = \xi(0). \end{cases}$

Also $\eta = \xi\gamma \Leftrightarrow \begin{cases} \eta' = \xi'\gamma, \gamma \in G \\ \eta(0) = \xi(0). \end{cases}$

Thus $\eta \sim \xi \Leftrightarrow \begin{cases} \eta' \sim \xi' \\ \eta(0) = \xi(0). \end{cases}$

So from now on we shall suppose $\eta \in m$.

Theorem 2.9. If $\eta \in m$ and $\Delta = \Delta(\eta)$, then

$$m^{k+1} \subset m^2\Delta \Rightarrow \eta \text{ is k-determinate} \Rightarrow m^{k+1} \subset m\Delta.$$

Proof. We shall use the following form of Nakayama's Lemma:

Lemma 2.10. If A is a local ring, a its maximal ideal, and M, N are A-modules (contained in some larger A-module) with M finitely generated over A, then $M \subset N + aM \Rightarrow M \subset N$.

Sublemma. $\lambda \in A$, $\lambda \notin a \Rightarrow \lambda^{-1} \in A$.

Proof. λA is an ideal $\not\subset a$. So $\lambda A = A \ni 1$, $\exists \mu$ such that $\lambda\mu = 1$.

Proof of Lemma 2.10. We shall first prove the special case of $N = 0$, i.e., $M \subset aM \Rightarrow M = 0$. Let $v_1, \ldots, v_r$ generate M. $v_i \in aM$ by hypothesis,

$$\text{so } v_i = \sum_{j=1}^{r} \lambda_{ij} v_j \quad (\lambda_{ij} \in a)$$

or $\sum_{j=1}^{r} (\delta_{ij} - \lambda_{ij}) v_j = 0$, i.e. $(I-\Lambda)v = 0$, where Λ is an $(r \times r)$-matrix (λ_{ij}), and $v = \begin{pmatrix} v_1 \\ \vdots \\ v_r \end{pmatrix}$. The determinant $|I-\Lambda| = 1 + \lambda$, some $\lambda \in a$. Now $1 + \lambda \notin a$, else $1 \in a$ and $a = A$. So $(1+\lambda)^{-1}$ exists by the sublemma. Then $(I-\Lambda)^{-1}$ exists, giving $v = 0$ and $M = 0$.

To prove the general case consider the quotient by N, $(M+N)/N \subset N/N + (aM+N)/N$. We claim the R.H.S. $= a(M+N)/N$. (*) Then by the special case, $(M+N)/N = 0$, giving $M \subset N$. Q.E.D.

The A-module structure on $(M+N)/N$ is induced by that on $M + N$ by $\lambda(v+N) = \lambda v + N$.

$$a(M+N)/N = \{\lambda(v+N) : \lambda \in a, v \in M\}$$
$$= \{\lambda v + N : \lambda \in a, v \in M\}$$
$$= (aM+N)/N, \text{ proving (*).}$$

Continuing the proof of Theorem 2.9, we assume $m^{k+1} \subset m^2\Delta$, and must show that $\eta \overset{k}{\sim} \xi \Rightarrow \eta \sim \xi$. The idea of the proof is to change η into ξ continuously with the assumption $\eta \overset{k}{\sim} \xi$. Let Φ denote the germ at $0 \times \mathbb{R}$ of a function $\mathbb{R}^n \times \mathbb{R} \to \mathbb{R}$ given by $\Phi(x,t) = (1-t)\eta(x) + t\xi(x)$, $x \in \mathbb{R}^n$, $t \in \mathbb{R}$. Let

$$\phi^t(x) = \Phi(x,t) = \begin{cases} \eta(x) & t = 0 \\ \xi(x) & t = 1. \end{cases}$$

Lemma 1. Fixing t_0, $0 \leq t_0 \leq 1$, $\exists$ a family $\Gamma^t \in G$ defined for t in a neighborhood of t_0 in $\mathbb{R}$ such that 1) Γ^{t_0} = identity

2) $\phi^t \Gamma^t = \phi^{t_0}$.

Lemma 1 will give $\eta \sim \xi$: Using compactness and connectedness of $[0,1]$, cover by a finite number of neighborhoods as in Lemma 1, then pick $\{t_i\}$ in the overlaps, and construct γ satisfying $\eta = \xi\gamma$ by a finite composition of $\{\Gamma^{t_i}\}$, i.e. $\eta = \phi^0 \sim \ldots \sim \phi^1 = \xi$.

Lemma 2. For $0 \leq t_0 \leq 1$, $\exists$ a germ Γ at (p,t_0) of C^∞ maps $\mathbb{R}^n \times \mathbb{R} \to \mathbb{R}^n$ satisfying (a) $\Gamma(x,t_0) = x$,

(b) $\Gamma(0,t) = 0$,

(c) $\Phi(\Gamma(x,t),t) = \Phi(x,t_0)$,

for all (x,t) in some neighborhood of $(0,t_0)$.

Lemma 2 will give Lemma 1: Define $\Gamma^t(x) = \Gamma(x,t)$ from a neighborhood of 0 in $\mathbb{R}^n$ to $\mathbb{R}^n$; Γ^t is a germ of C^∞ maps $\mathbb{R}^n, 0 \to \mathbb{R}^n, 0$ by (b); Γ^{t_0} is the identity by (a). C^∞ diffeomorphisms are open in the space of C^∞ maps $\mathbb{R}^n, 0 \to \mathbb{R}^n, 0$ (because they correspond to maps with Jacobian of maximal rank, i.e. to the non-vanishing of a certain determinant), and so $\exists$ a neighborhood of t_0 such that Γ^t is a germ of diffeomorphisms for t in that neighborhood, i.e. $\Gamma^t \in G$.

Lemma 3. (c) in Lemma 2 is equivalent to,

(c') $\sum_{i=1}^n \frac{\partial \Phi}{\partial x_i}(\Gamma(x,t),t) \frac{\partial \Gamma_i}{\partial t}(x,t) + \frac{\partial \Phi}{\partial t}(\Gamma(x,t),t) = 0$.

(c) $\to$ (c'): by differentiation with respect to t.

(c') $\to$ (c): $0 = \int_{t_0}^t (c') dt = \Phi(\Gamma(x,t),t) - \Phi(\Gamma(x,t_0),t_0)$

$= \Phi(\Gamma(x,t),t) - \Phi(x,t_0)$ by (a) in Lemma 2.

Thus we have (c).

Lemma 4. For $0 \leq t_0 \leq 1$, $\exists$ a germ Υ at $(0,t_0)$ of a C^∞ map $\mathbb{R}^n \times \mathbb{R} \to \mathbb{R}^n$

satisfying (d) $\Psi(0,t) = 0$,

(e) $\sum_{i=1}^{n} \frac{\partial \Phi}{\partial x_i}(x,t)\Psi_i(x,t) + \frac{\partial \Phi}{\partial t}(x,t) = 0$,

for all (x,t) in some neighborhood of $(0,t_o)$.

Lemma 4 → Lemmas 3 and 2: The existence theorem for ordinary differential equations gives a solution $\Gamma(x,t)$ of $\frac{\partial \Gamma}{\partial t} = \Psi(\Gamma,t)$, with initial condition $\Gamma(x,t_o) = x$ (i.e. (a) of Lemma 2). In (e) put $x = \Gamma(x,t)$ to give (c'). (d) → $\Gamma = 0$ is a solution, i.e. $\Gamma(0,t) = 0$ for all t in some neighborhood of t_o, which is (b).

Let A denote the ring of germs at $(0,t_o)$ of C^∞ functions $\mathbb{R}^n \times \mathbb{R} \to \mathbb{R}$. Projection $\mathbb{R}^n \times \mathbb{R} \to \mathbb{R}^n$ induces an embedding $E \subset A$ by composition. Let $\Omega = (\frac{\partial \Phi}{\partial x_1}, \ldots, \frac{\partial \Phi}{\partial x_n})A$.

<u>Lemma 5.</u> $m^{k+1} \subset m^2\Delta \to m^{k+1} \subset m^2\Omega$.

Lemma 5 → Lemma 4 as follows:

$$\frac{\partial \Phi}{\partial t} = \xi - \eta \in m^{k+1} \qquad (\eta \overset{k}{\sim} \xi)$$
$$\subset m^2\Omega.$$

Thus $\frac{\partial \Phi}{\partial t} = \Sigma_j \mu_j \omega_j$, $\mu_j \in m^2$, $\omega_j \in \Omega$. (finite sum)

$= \Sigma_{ij} \mu_j a_{ij} \frac{\partial \Phi}{\partial x_i}$, where $\omega_j = \Sigma_i a_{ij} \frac{\partial \Phi}{\partial x_i}$, $a_{ij} \in A$.

$= -\Sigma_i \Psi_i \frac{\partial \Phi}{\partial x_i}$, setting $\Psi_i = -\Sigma_j \mu_j a_{ij} \in A$.

This gives (e).

Now $\mu_j = \mu_j(x)$ and $a_{ij} = a_{ij}(x,t)$. $\Psi = \{\Psi_i\}$ is a germ at $(0,t_o)$ of a map $\mathbb{R}^n \times \mathbb{R} \to \mathbb{R}^n$, and $\Psi_i(0,t) = 0$ as each $\mu_j(0) = 0$, so (d) holds for Ψ.

<u>Proof of Lemma 5.</u> (and hence the completion of the proof of a sufficient condition for k-determinacy)

$$\frac{\partial \Phi}{\partial x_i} = \frac{\partial \eta}{\partial x_i} + t\frac{\partial}{\partial x_i}(\xi-\eta)$$
$$\in \frac{\partial \eta}{\partial x_i} + Am^k \qquad (t \in A, \xi - \eta \in m^{k+1})$$

i.e. $\frac{\partial \eta}{\partial x_i} \in \frac{\partial \Phi}{\partial x_i} + Am^k \subset \Omega + Am^k$.

So $\Delta \subset \Omega + Am^k$.

Denote the maximal ideal of A by a, i.e. those germs vanishing at $(0,t_0)$. Then $m \subset a$. Now $Am^{k+1} \subset Am^2$ (hypothesis)
$$\subset Am^2(\Omega + Am^k)$$
$$= m^2\Omega + Am^{k+2}$$
$$\subset m^2\Omega + aAm^{k+1}.$$

Now apply Nakayama's Lemma 2.10 for A, a, M, N where $M = Am^{k+1}$ is finitely generated by monomials in $\{x_i\}$ of degree $k+1$ by Corollary 2.4, and $N = m^2\Omega$. This gives $Am^{k+1} \subset m^2\Omega$. In particular $m^{k+1} \subset m^2\Omega$, completing Lemma 5.

Now we prove that $m^{k+1} \subset m\Delta$ is a necessary condition of k-determinacy. $\exists$ a natural map $m \xrightarrow{\pi} J^{k+1} \longrightarrow J^k$, $\pi = j^{k+1}/m$.
$$\eta \longmapsto j^{k+1}\eta \longmapsto j^k\eta$$

Let $P = \{\xi \in m: \eta \overset{k}{\sim} \xi\}$, and $Q = \{\xi \in m: \eta \sim \xi\}$ = orbit ηG.

Assuming that η is k-determinate then $P \subset Q$, so that $\pi P \subset \pi Q$. $\quad (*)$

$P = \eta + m^{k+1}$, so $\pi P = z + m^{k+1}/m^{k+2} = z + \pi m^{k+1}$. (Letting $z = j^{k+1}\eta$). The tangent plane to πP at z, $T_z(\pi P) = \pi m^{k+1}$.

Let G^k denote the k-jets of germs belonging to G; G^k is a finite-dimensional Lie group. Now $j^{k+1}(\eta\gamma) = j^{k+1}(\eta)j^{k+1}(\gamma)$ for $\gamma \in G$, i.e. π is equivariant with respect to G, G^{k+1}. So $\pi Q = \pi(\eta G) = zG^{k+1}$, an orbit under a Lie group, and hence is a manifold. In particular $T_z(\pi Q)$ exists.

Lemma 2.11. $T_z(\pi Q) = \pi(m\Delta)$.

Now $(*)$ gives $T_z(\pi P) \subset T_z(\pi Q)$. Then Lemma 2.11 gives $\pi m^{k+1} \subset \pi(m\Delta)$, i.e. $m^{k+1} \subset m\Delta + m^{k+2}$. Apply Nakayama's Lemma 2.10 with $A = E$, $a = m$, $M = m^{k+1}$, $N = m\Delta$, using Lemmas 2.1, 2.2 and Corollary 2.4, to yield $m^{k+1} \subset m\Delta$.

Proof (of 2.11). Suppose $\gamma \in G$. As $\mathbb{R}^n$ is additive we can write $\gamma = 1 + \delta$, where 1 is the germ of the identity map, and δ is the germ at 0 of a C^∞ map $\mathbb{R}^n, 0 \to \mathbb{R}^n, 0$. Join 1 to γ by a continuous path of map-germs, $\gamma^t = 1 + t\delta$,

$0 \leq t \leq 1$. When $t = 0$ or 1, γ^t is a diffeomorphism-germ. Diffeomorphisms are open in the space of C^∞ maps, and so $\exists\, t_o > 0$ such that $\gamma^t \in G$, $0 \leq t \leq t_o$.

Then $\{\gamma^t\}$ is a path in G starting at 1,

$\{\eta\gamma^t\}$ is a path in Q starting at η,

$\{\pi\eta\gamma^t\}$ is a path in πQ starting at z.

The tangent to the path at $t = 0$ is given by

$$\frac{d}{dt}(\pi\eta\gamma^t)\Big|_{t=0} = \pi\Big[\frac{d}{dt}\eta(1+t\delta)\Big|_{\eta\gamma^t}\Big]_{t=0}$$

Now $\delta = (\delta_1,\ldots,\delta_n)$ where δ_i is a germ of a C^∞ function $\mathbb{R}^n, 0 \to \mathbb{R}, 0$. (Remember m is a ring and a vector space so we can define differentiation).

So $\frac{d}{dt}(\pi\eta\gamma^t)_{t=0} = \pi[\sum_{i=1}^{n} \frac{\partial\eta}{\partial x_i}(1+t\delta)\cdot\delta_i\Big|_{t=0}]$

$= \pi[\sum_{i=1}^{n} \frac{\partial\eta}{\partial x_i}\cdot\delta_i]$

$\in \pi(m\Delta)$. $(\delta_i \in m, \frac{\partial\eta}{\partial x_i} \in \Delta)$.

This tangent is in $T_z(\pi Q)$; moreover any tangent in $T_z(\pi Q)$ arises from a path in πQ, so from a path in G^{k+1}, so from a path in G starting at 1. Allowing δ to vary in G gives all such paths. Hence $T_z(\pi Q) \subset \pi(m\Delta)$. Given $\xi \in m\Delta$, we can write $\xi = \sum_{i=1}^{n} \frac{\partial\eta}{\partial x_i}\delta_i$, $\delta_i \in m$. The δ_i assemble into δ determining a path in G.

Hence $\pi(m\Delta) \subset T_z(\pi Q)$, and we have $T_z(\pi Q) = \pi(m\Delta)$.

Corollary 2.12. η is finitely determinate $\Leftrightarrow m^k \subset \Delta$, some k.

Proof. '$\Rightarrow$' follows as η k-determinate $\Rightarrow m^{k+1} \subset m\Delta \subset \Delta$.

'$\Leftarrow$'. $m^k \subset \Delta$, so $m^{k+2} \subset m^2\Delta$, and η is $(k+1)$-determinate.

Corollary 2.13. $\eta \in m - m^2 \Rightarrow \eta$ is 1-determinate.

Proof. $\eta'(0) \neq 0$, i.e. some $\frac{\partial\eta}{\partial x_i} \notin m$, so $\Delta = E$. $m^2\Delta = m^2$ and then η is 1-determinate by Theorem 2.9.

So we may effectively assume $\eta \in m^2$ from now on.

Definition. With chosen coordinates $\{x_i\}$, the __essence__ of η (with respect to this coordinate system) is the least k for which $j^k\eta$ contains all the x_i. We write __ess__ η.

Corollary 2.14. det $\eta \geq$ ess η (with respect to any coordinate system).

Proof. $k <$ ess $\eta \Rightarrow j^k\eta$ does not contain x_i, some i. Let $\xi = j^k\eta$ as a germ. So $\Delta(\xi) \not\ni$ any power of x_i,
$$\not\supset m^k, \forall k.$$

Thus ξ is not finitely determinate (Corollary 2.12). But $\eta \stackrel{k}{\sim} \xi$, so if η were k-determinate, Lemma 2.7 would give a contradiction, i.e. $k <$ det η.

Counterexample 1. Let $\eta = x^{k+1}$, $n = 1$. Then $\Delta = (x^k) = m^k$, and $m\Delta = m^{k+1}$. Det $\eta \geq$ ess $\eta = k + 1$, (Corollary 2.14). η is not k-determinate and so the implication, η k-determinate $\Rightarrow m^{k+1} \subset m\Delta$ in Theorem 2.9 is not reversible.

Counterexample 2. D. Siersma found $\eta = \frac{x^3}{3} + xy^3$, $n = 2$. Here $\Delta = (x^2+y^3, xy^2)$. $m^2 = (x^2, xy, y^2)$, so $m^2\Delta = x^4 + x^2y^3, x^3y + xy^4, x^2y^2 + y^5, x^3y^2, x^2y^3, xy^4,$
$$\ni x^3y^2 + xy^5, x^2y^3 + y^6, x^2y^4, x^3y^3, x^4y^2, x^5y, x^6,$$
$$\supset m^6 \text{ (5-determinate)}$$
$$\not\ni y^5, \not\supset m^5.$$

Det $\eta \geq$ ess $\eta = 4$. By computation it __is__ 4-determinate, and so the implication $m^{k+1} \subset m^2\Delta \Rightarrow \eta$ k-determinate is not reversible.

CHAPTER 3. CODIMENSION.

Remember that we work in m^2 using Corollary 2.13.

Definition. The codimension of $\eta = \dim_R m/\Delta(\eta)$. We write cod η. The definition makes sense because if $\eta \in m^2$, each $\frac{\partial \eta}{\partial x_i} \in m$ and so $\Delta(\eta) \subset m$. If η were in $m - m^2$, $\Delta(\eta) = E$ and by convention cod $\eta = 0$.

Lemma 3.1. Either both cod η and det η are infinite, or both are finite and det $\eta - 2 \leq$ cod η.

Proof. $\eta \in m^2 \Rightarrow \Delta \subset m.$ ($\Delta = \Delta(\eta)$)

We have a descending sequence of vector subspaces of m,

$$m = m + \Delta \supset m^2 + \Delta \supset m^3 + \Delta \supset \ldots \supset m^k + \Delta \supset \ldots \quad (3.2)$$

Either (i) $\exists\, k$ such that $m^{k-1} + \Delta = m^k + \Delta$, and k is the least such, or

(ii) $\not\exists$ such a k.

Case (i): $m^{k-1} \subset m^k + \Delta$, and we may apply Nakayama's Lemma 2.10 yielding $m^{k-1} \subset \Delta$, so $m^{k+1} \subset m^2 \Delta$. By Theorem 2.9 η is k-determinate, so $\det \eta \leq k$, i.e. $\det \eta$ is finite. Now $\operatorname{cod} \eta = \dim m/\Delta \leq \dim m/m^{k-1}$, and m/m^{k-1} is finitely generated, by monomials in $\{x_i\}$ of degree ≥ 1 and $< k - 1$. So $\operatorname{cod} \eta$ is finite. Now $m^{k-1} + \Delta = \Delta$, and so the above sequence (3.2) descends strictly to the $m^{k-1} + \Delta$ term, and we have,

$$m/\Delta \supsetneq (m^2 + \Delta)/\Delta \supsetneq \ldots \supsetneq (m^{k-1} + \Delta)/\Delta = 0$$

$$\longleftarrow k-2 \text{ steps} \longrightarrow$$

Hence $\operatorname{cod} \eta = \dim m/\Delta \geq k - 2 \geq \det \eta - 2$, as required.

Case (ii): If $\det \eta$ is finite, then $m^k \subset \Delta$ for some k (Corollary 2.12). Then $m^k + \Delta = \Delta = m^{k+1} + \Delta$, and we are in Case (i). So $\det \eta$ is infinite. $m/\Delta \supset (m^2 + \Delta)/\Delta \supset \ldots$ is a strictly decreasing sequence and so $\operatorname{cod} \eta$ ($= \dim m/\Delta$) is infinity.

Let $\Gamma_c = \{\eta \in m^2 : \operatorname{cod} \eta = c\}$ (a 'c-stratum' of m^2), and let $\Omega_c = \{\eta \in m^2 : \operatorname{cod} \eta \leq c$, and $\Sigma_c = \{\eta \in m^2 : \operatorname{cod} \eta \geq c\}$, so that

$$m^2 = \Gamma_0 \cup \Gamma_1 \cup \Gamma_2 \cup \ldots \cup \Gamma_c \cup \ldots \cup \Gamma_\infty. \quad \text{(disjoint union)}$$

Let $\Gamma_c^k, \Omega_c^k, \Sigma_c^k$ be the images of $\Gamma_c, \Omega_c, \Sigma_c$ under the map $\pi : m^2 \to I^k$ ($\pi = j^k|m^2$), where I^k is defined as m^2/m^{k+1} just as J^k is m/m^{k+1}.

Theorem 3.3. If $0 \leq c \leq k - 2$, then $I^k = \Omega_c^k \cup \Sigma_{c+1}^k$ (disjoint union), and Σ_{c+1}^k is a (closed) real algebraic variety.

Remark. Both statements are false for $c > k - 2$.

Lemma 3.4. $\operatorname{Dim} E/m^{k+1} = \dfrac{(n+k)!}{n!k!}$, $\forall\, n, k \geq 0$.

Proof. If $n = 0$, $E = \mathbb{R}$, $m = 0$; L.H.S. $= 1 =$ R.H.S. $\forall k$. If $k = 0$, $E/m = \mathbb{R}$; L.H.S. $= 1 =$ R.H.S. $\forall n$. Use induction on $n + k$.

Then $E/m^{k+1} =$ polynomials of degree $\leq k$ in $x_1, \ldots, x_n$
$=$ (polynomials of degree k in $x_1, \ldots, x_{n-1}$)
$+ x_n$ (polynomials of degree $k - 1$ in $x_1, \ldots, x_n$)

So $\dim E/m^{k+1} = \frac{(n+k-1)!}{(n-k)!k!} + \frac{(n+k-1)!}{n!(k-1)!}$ (by induction)

$= \frac{(n+k)!}{n!k!}$

Proof of Theorem 3.3. We define an invariant $\tau(z)$ for $z \in I^k = m^2/m^{k+1}$. Choose $\eta \in \pi^{-1}z$. $\eta \in m^2$, so $\Delta(\eta) = \Delta \subset m$. Define $\tau(z) = \dim m/(\Delta + m^k)$. We claim that $\tau(z)$ is independent of the choice of η. Let η' be another choice, $\Delta(\eta') = \Delta'$. Then $\eta - \eta' \in m^{k+1}$, so $\frac{\partial \eta}{\partial x_1} - \frac{\partial \eta'}{\partial x_1} \in m^k$, and $\frac{\partial \eta}{\partial x_1} \in \Delta' + m^k$. Hence $\Delta \subset \Delta' + m^k$ and $\Delta + m^k \subset \Delta' + m^k$.
$\Delta' + m^k \subset \Delta + m^k$ by symmetry.

Hence $\Delta + m^k = \Delta' + m^k$ and $\tau(z)$ is well defined.

We claim that,

(3.5) $\begin{cases} \text{(i)} \quad \tau(z) \leq c \Rightarrow \text{cod } \eta = \tau(z), \text{ so } z \in \Omega_c^k. \\ \text{(ii)} \quad \tau(z) > c \Rightarrow \text{cod } \eta > c, \text{ so } z \in \Sigma_{c+1}^k. \quad (\text{cod } \eta \text{ perhaps } \neq \tau(z)) \end{cases}$

Because (i) and (ii) are disjoint, I^k is the disjoint union of Ω_c^k and Σ_{c+1}^k, once we have shown (i) and (ii) hold.

We have

(Lemma 3.4)

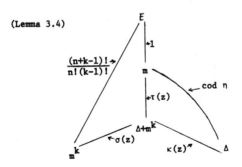

Note that $\tau(z)$ is finite, although cod η may be infinite.

Case (ii): $\text{cod } \eta \geq \tau(z)$ (from the diagram) ⎫
$> c$ (hypothesis of (ii)) ⎬ Thus (ii) holds.
⎭

Case (i): $k - 2 \geq c$ (hypotheses of the theorem)
$\geq \tau(z)$ (hypothesis of (i))

We have a sequence,

$$0 = m/m = m/\Delta + m \leftarrow m/\Delta + m^2 \leftarrow \ldots \leftarrow m/\Delta + m^k$$
$$\longleftarrow k-1 \text{ steps} \longrightarrow$$

$k - 2 \geq \tau(z) = \dim m/\Delta + m^k$, so one step must collapse, i.e. $\Delta + m^{i-1} = \Delta + m^i$, for some $i \leq k$, i.e. $m^{i-1} \subset \Delta + m^i$. Nakayama's Lemma 2.10 $\Rightarrow (m^k c) m^{i-1} \subset \Delta$. Therefore $\Delta + m^k = \Delta$, and so $\kappa(z) = 0$ where $\kappa(z) = \dim (\Delta + m^k)/\Delta$ as in the diagram. We observe that $\tau(z) = \text{cod } \eta$, and so (i) holds.

Now $\sigma(z) = \frac{(n+k-1)!}{n!(k-1)!} - 1 - \tau(z)$, from the diagram. If $\tau(z) > c$, then

$\sigma(z) < \frac{(n+k-1)!}{n!(k-1)!} - 1 - c = K$, say. $\Sigma_{c+1}^k = \{z \in I^k: \text{cod } \eta > c\}$
$= \{z \in I^k: \tau(z) > c\}$ $(c \leq k - 2)$
$= \{z \in I^k: \sigma(z) < K\}$, which we shall

show is an algebraic variety (real).

If $x_1, \ldots, x_n$ are coordinates for $\mathbb{R}^n$, let the monomials of degree $\leq k$ in $\{x_i\}$ be $\{X_j\}$ as below:

$X_1 \ X_2 \ X_3 \ \ldots \ X_{n+1} \ X_{n+2} \ X_{n+3} \ \ldots \ X_\beta$ $(\beta = \frac{(n+k)!}{n!k!})$

$1 \ x_1 \ x_2 \ \ldots \ x_n \ x_1^2 \ x_1 x_2 \ \ldots \ x_n^k$

Now J^k is the space of polynomials in $\{x_i\}$ of degree $\leq k$ with coefficients in $\mathbb{R}$ and no constant term. $z \in I^k$ can be written $z = \sum_{j=n+2}^{\beta} a_j X_j$ $(a_j \in \mathbb{R})$. Because $\frac{\partial z}{\partial x_i}$ is a polynomial of degree $k - 1$ with no constant term it belongs to J^{k-1}, so $\frac{\partial z}{\partial x_i} = \sum_{j=2}^{\bar{\beta}} a_{ij} X_j$, $(\bar{\beta} = \frac{(n+k-1)!}{n!(k-1)!})$, where each a_{ij} is an integer multiplied by some a_k.

Just as Δ is the ideal of E generated by $\{\frac{\partial \eta}{\partial x_i}\}$, so $(\Delta + m^k)/m^k$ is the ideal of J^{k-1} generated by $\{\frac{\partial z}{\partial x_i}\}$. Now J^k as a vector space has a basis $X_2, \ldots, X_\beta$. $(\Delta + m^k)/m^k$ is now the vector subspace of J^{k-1} spanned by

$\{\frac{\partial z}{\partial x_i} X_j\}$. Let each $\frac{\partial z}{\partial x_i} X_j = \sum_{k=2}^{\bar{\beta}} a_{ij,k} X_k$, where each $a_{ij,k}$ is some $a_{\ell m}$.

We put M = the matrix $(a_{ij,k})$

= the coordinates of vectors spanning $(\Delta + m^k)/m^k$.

Now $\sigma(z) < K \Leftrightarrow \dim (\Delta + m^k)/m^k < K$

$\Leftrightarrow$ rank of $M < K$

$\Leftrightarrow$ all K-minors of M vanish.

And so Σ_{c+1}^k is given by polynomials in the $\{a_{ij,k}\}$, k.e. by polynomials in the $\{a_i\}$, each $a_i \in \mathbb{R}$. Hence Σ_{c+1}^k is a real algebraic variety in the real vector space I^k of dimension $\frac{(n+k)!}{n!k!} - n - 1$, itself a subspace of J^k which is $(\frac{(n+k)!}{n!k!} - 1)$-dimensional.

Corollary. I^k is the disjoint union $\Gamma_0^k \cup \Gamma_1^k \cup \ldots \cup \Gamma_{k-2}^k \cup \Sigma_{k-1}^k$, and each Γ_c^k is the difference $\Sigma_c^k - \Sigma_{c+1}^k$ between 2 algebraic varieties.

Recall that the map $\pi: m^2 \to I^k$ is equivariant with respect to G, G^k;

$$\eta \longmapsto z$$

also the image of the orbit ηG is zG^k, a submanifold of I^k, as in the proof of Theorem 2.9.

Theorem 3.7. Let $\eta \in m^2$ and cod $\eta = c$ where $0 \leq c \leq k - 2$. Then zG^k is a submanifold of I^k of codimension c.

Proof. By Lemma 2.11, $T_z(zG^k) = \pi(m\Delta)$. $(\Delta = \Delta(\eta))$

By Lemma 3.1, det $\eta - 2 \leq $ cod $\eta = c \leq k - 2$, by the hypotheses. So det $\eta \leq k$, i.e. η is k-determinate. By Theorem 2.9, $m^{k+1} \subset m\Delta$.

The codimension of zG^k in $I^k = \dim I^k - \dim \pi(m\Delta)$

$= \dim m^2/m^{k+1} - \dim m\Delta/m^{k+1}$

$= \dim m^2/m\Delta$.

Now $m/m\Delta = m/m^2 + m^2/m\Delta$, so $\dim m^2/m\Delta = \dim m/m\Delta - \dim m/m^2$. So the codimension of zG^k in $I^k = \dim m/m\Delta - \dim m/m^2$

$= \dim m/\Delta + \dim \Delta/m\Delta - \dim m/m^2$

$= c + n - n$,

using the following lemma.

Lemma 3.8. If $\eta \in m^2$ and $\text{cod } \eta < \infty$, then $\dim \Delta/m\Delta = n$.

This completes the proof of the theorem.

Proof of Lemma 3.8. Since Δ is the ideal of E generated by $\{\frac{\partial \eta}{\partial x_i}\}$, every $\xi \in \Delta$ can be written as $\xi = \sum_{i=1}^{n} \alpha_i \frac{\partial \eta}{\partial x_i}$ where $\alpha_i \in E$, $\alpha_i = a_i + \mu_i$, $\mu_i \in m$, $a_i \in \mathbb{R}$. Then $\xi = \sum_{i=1}^{n} a_i \frac{\partial \eta}{\partial x_i}$ mod $m\Delta$. So $\{\frac{\partial \eta}{\partial x_i}\}$ span Δ over $\mathbb{R}$, mod $m\Delta$, and $\dim \Delta/m\Delta \leq n$. It remains to prove $\dim \Delta/m\Delta \geq n$.

Suppose not, i.e. that $\dim \Delta/m\Delta < n$. Then $\{\frac{\partial \eta}{\partial x_i}\}$ are linearly dependent mod $m\Delta$. $\exists a_1, \ldots, a_n \in \mathbb{R}$, not all zero, such that

$$\sum_{i=1}^{n} a_i \frac{\partial \eta}{\partial x_i} = \sum_{i=1}^{n} \mu_i \frac{\partial \eta}{\partial x_i} \in m\Delta, \text{ some } \{\mu_i\} \in m.$$

Then $X\eta = \sum_{i=1}^{n} (a_i - \mu_i) \frac{\partial \eta}{\partial x_i} = 0$ where $X = \sum_{i=1}^{n} (a_i - \mu_i) \frac{\partial}{\partial x_i}$ is a vector field on a neighborhood of 0 in $\mathbb{R}^n$. X is nonzero at 0 because $\{\mu_i\} \in m$ and so vanish at 0 and $\{a_i\}$ are not all zero.

Change local coordinates so that $X = \frac{\partial}{\partial y_1}$ where $\{y_i\}$ are the new coordinates. Then $\frac{\partial \eta}{\partial y_1} = 0$. So $\eta = \eta(y_2, \ldots, y_n)$. Ess $\eta = \infty$ with respect to $\{y_i\}$. But $\det \eta \geq \text{ess } \eta$, by Corollary 2.14. By Lemma 3.1., $\text{cod } \eta = \infty$, . We have shown that $\dim \Delta/m\Delta = n$.

Theorem 3.7 justifies the notation $\text{cod } \eta$, as an abbreviation for codimension.

CHAPTER 4. CLASSIFICATION

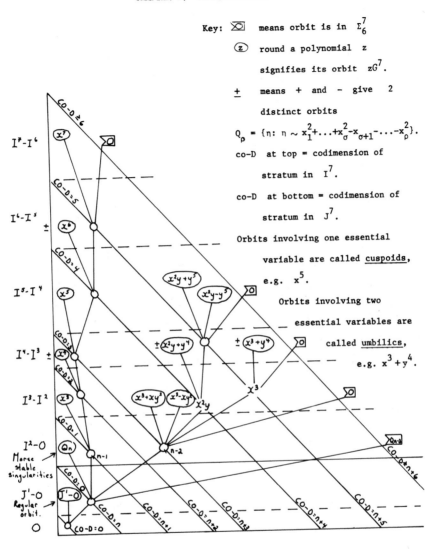

Diagram 4.1. Classification of I^7 (and J^7).

This chapter will complete the above classification of I^7 as in Diagram 4.1. Supposing we already have our classification it follows that:

Theorem 4.2. In $I^7 - \Sigma_6^7$ there are exactly $16n - 7$ orbits under G^7.

Proof. We merely add the orbits in Diagram 4.1.

$$\begin{array}{ll} \text{Stable singularities} & n + 1 \\ \text{Cuspoids} & 7n \\ \text{Umbilics} & 8(n-1) \\ \hline & 16n - 7 \end{array}$$

Corollary 4.3. If $0 \leq c \leq 5$, Γ_c^7 is a submanifold of I^7 of codimension c.

Proof. Γ_c^7 is the union of a finite number of orbits by Theorem 4.2. By Theorem 3.7, each of these is a submanifold of codimension c.

We note that our classification also gives that Σ_6^7 is the union of a finite number of parts each of codimension ≥ 6 in I^7. (See Diagram 4.1)

Theorem 4.4. $I^7 = \Gamma_0^7 \cup \Gamma_1^7 \cup \Gamma_2^7 \cup \Gamma_3^7 \cup \Gamma_4^7 \cup \Gamma_5^7 \cup \Gamma_6^7$ (disjoint) and each Γ_c^7 is of codimension c in I^7 and Σ_6^7 is of codimension 6 in I^7.

We shall now proceed with the classification.

Lemma 4.5. Let $\pi: m \to J^1 = m/m^2 \cong \mathbb{R}^n$, where $\pi = j^1|m$. Then $\pi^{-1}(J^1-0)$ is the orbit of regular germs.

Proof. Given $\eta \in m$, if $j^1\eta \neq 0$, then $\eta = \eta_1 + $ higher terms, where η_1 is a nonzero linear term. Then $\Delta = E$, as in Lemma 2.2. $m^2\Delta = m^2$ and by Theorem 2.9, η is 1-determinate. So $\eta \sim \eta_1 = \sum_{i=1}^{n} a_i x_i = y_j$ for some linear change of coordinates. Thus $\eta \sim x_1$ by the linear map sending y_j to x_1, and $\eta \in$ orbit of x_1 which as a function is regular.

These regular germs are precisely those with no singularity, or rather are not singularities. We observe that $J^7 = J^1 \times J^7/J^1$. Lemma 4.5 tells us that $(J^1-0) \times J^7/J^1$ is the regular orbit. The remainder, $0 \times J^7/J^1 = m^2/m^8 = I^7$, are the irregular orbits which we must classify.

$\eta \in m^2 \Rightarrow \eta = q + $ higher terms, where q is a quadratic form in

18. Classification of Elementary Catastrophes

$\{x_i\}$, say

$$q(x_1,\ldots,x_n) = \sum_{ij} a_{ij} x_i x_j \qquad (a_{ij} = a_{ji})$$

Write A for the matrix (a_{ij}), which is symmetric, and define rank η, the **Hermitian rank** of η or of $j^k\eta$ $(k \geq 2)$ to be rank A. Then $0 \leq \text{rank } \eta \leq n$.

Lemma 4.6. Let rank $\eta = \rho$. By an elementary theorem of linear algebra there is a linear change of coordinates such that $q = y_1^2 + y_2^2 + \ldots + y_\sigma^2 - y_{\sigma+1}^2 - \ldots - y_\rho^2$.

Corollary 4.7. $\eta \sim (x_1^2 + \ldots - x_\rho^2) +$ higher terms, if rank $\eta = \rho$.

Let $Q_\rho = \{q: \text{Hermitian rank of } q = \rho\}$, in $I^2 = m^2/m^3$ which is diffeomorphic to $R^{\frac{1}{2}n(n+1)}$ because it is the linear space of all quadratic forms with coordinates $\{a_{ij}\}$, $i \leq j$. Then $I^2 = Q_n \cup Q_{n-1} \cup \ldots \cup Q_0$.

Lemma 4.8. $Q_{n-\lambda}$ is a submanifold of I^2 of codimension $\frac{1}{2}\lambda(\lambda+1)$.

Proof. Each Q_ρ is a submanifold because each component is an orbit under the action of the general linear group.

Choose $q \in Q_\rho$. By Lemma 4.6. we may assume that $q = x_1^2 + \ldots - x_\rho^2$. Then the associated matrix is $\begin{pmatrix} E & 0 \\ 0 & 0 \\ \underline{\rho} & \underline{\lambda} \end{pmatrix}$ $(\rho = n-\lambda)$, where $E = \begin{pmatrix} 1 & & 0 \\ & \ddots & \\ 0 & & -1 \end{pmatrix}$. Suppose q' has matrix $\begin{pmatrix} A & B \\ B' & C \\ \underline{\rho} & \underline{\lambda} \end{pmatrix}$. $\exists$ a neighborhood N of q in I^2 such that if $q' \in N$, then $|A| \neq 0$. There rank $q' = \text{rank} \begin{pmatrix} A & B \\ B' & C \end{pmatrix}$

$= \text{rank} \begin{pmatrix} A^{-1} & 0 \\ -B'A^{-1} & I \end{pmatrix} \begin{pmatrix} A & B \\ B' & C \end{pmatrix}$

$= \text{rank} \begin{pmatrix} I & A^{-1}B \\ 0 & C-B'A^{-1}B \end{pmatrix}$

Thus rank $q' = \rho \Leftrightarrow C = B'A^{-1}B$, i.e. the entries of C are determined by the entries of A and B. Then $Q_\rho \cap N$ only has the freedom of the entries of A and B.

So the codimension of $Q_\rho = Q_{n-\lambda}$ is $\frac{1}{2}\lambda(\lambda+1)$, which is the number of free entries in symmetric C.

Now $I^7 = I^2 \times m^3/m^8 = (Q_n \times m^3/m^8) \cup \{(Q_\rho \times m^3/m^8)\}_{\rho<n}$. $Q_n \times m^3/m^8$ is the union of orbits of stable singularities (studied in Morse theory) and by Lemma 4.8. is an open set in I^7. It is in fact Γ_0^7 (clear).

Suppose now that rank $\eta = \rho$ and that as in Lemma 4.6. we have chosen coordinates $x_1, \ldots, x_n$ so that $\eta \doteq x_1^2 + \ldots - x_\rho^2 +$ higher terms. We call $x_1, \ldots, x_\rho$ the <u>dummy</u> variables and $x_{\rho+1}, \ldots, x_n$ the <u>essential</u> variables. The following lemma justifies these terms.

<u>Lemma 4.9.</u> (Reduction Lemma) Let $\eta \in m^2$ and $j^2\eta = q = x_1^2 + \ldots - x_\rho^2$. Then $\forall k, \exists \eta' \in m^2$ such that $\eta \sim \eta'$, and $j^k\eta' = q + p(x_{\rho+1},\ldots,x_n)$ where p is a polynomial in only the essential variables with $3 \leq$ degree of monomials of $p \leq k$.

<u>Proof.</u> Use induction on k. The lemma is true for $k = 2$. Suppose it is true for $k - 1$. In I^k, $j^k\eta' = q + p(x_{+1},\ldots,x_n) + \eta_k(x_1,\ldots,x_n)$, where $3 \leq$ degree of monomials of $p \leq k - 1$, and η_k is homogeneous of degree k. Write $\eta_k = 2x_1P_1$ (all terms containing x_1; P_1 a homogeneous polynomial
in $x_1, \ldots, x_n$ of degree $k - 1$)
$+ 2x_2P_2$ (all terms containing x_2, not x_1)
$+ 2x_3P_3$ (all terms containing x_3, not x_1, x_2)
$+ \ldots - 2x_{\sigma+1}P_{\sigma+1} - \ldots - 2x_\rho P_\rho$
$+ P_1(x_{\rho+1},\ldots,x_n)$ (all terms not containing dummy variables).

First incorporate the 2's and -'s into the $\{P_i\}$. Then let $y_i = \begin{cases} x_i + P_i & i \leq \rho \\ x_i & i > \rho \end{cases}$

If $i \leq \rho$, $y_i^2 = (x_i + P_i)^2 = x_i^2 + 2x_iP_i$ because monomials of degree $> k$ vanish in I^k. So $j^k\eta' = y_1^2 + \ldots - y_\rho^2 + p(y_{\rho+1}, \ldots, y_n) + P_1(y_{\rho+1}, \ldots, y_n)$, completing the lemma.

<u>Addendum 4.10.</u> The function $\eta \mapsto p$ is well-defined because the construction is explicit.

Lemma 4.11. If rank $\eta \geq n - 3$, then cod $\eta \geq 6$.

Proof. Either η is not finitely determinate, in which case cod $\eta = \infty$, (Lemma 3.1), or η is k-determinate, some k, i.e. $\eta \sim j^k\eta$, and $j^k\eta \sim q + p$ (essentials), by Lemma 4.9. Then cod $\eta = $ cod $(q+p)$. $\Delta(q+p) = (2x_1,\ldots,-2x_\rho, \frac{\partial p}{\partial x_{\rho+1}},\ldots,\frac{\partial p}{\partial x_n})$.

So cod $\eta = \dim m/\Delta(q+p)$

$\geq \dim m/(\Delta(q+p)+m^3)$

= number of the missing linear and quadratic terms in the essentials.

If $n - $ rank $\eta = \lambda$, all λ linear terms are missing, as too are at least all but λ of the $\frac{1}{2}\lambda(\lambda+1)$ quadratic terms. So cod $\eta \geq \lambda + \frac{1}{2}\lambda(\lambda+1) - \lambda = \frac{1}{2}\lambda(\lambda+1)$.

If rank $\eta \leq n - 3$, then $\lambda \geq 3$ and cod $\eta \geq 6$.

We have that $\bigcup_{\rho \leq n-3} \{Q_\rho \times m^3/m^8\}$ consists of η with cod $\eta \geq 6$. By Lemma 4.8., this subspace has codimension 6 in I^7. It remains to investigate $Q_{n-1} \times m^3/m^8$ and $Q_{n-2} \times m^3/m^8$.

Lemma 4.12. (Classifying cuspoids) If rank $\eta = n - 1$, then $\eta \sim q + x_n^k$, $3 \leq k \leq 7$, or cod $\eta \geq 6$.

Proof. By the reduction lemma 4.9., $\eta \sim \eta'$ where $j^7\eta' = q$ and p is a polynomial $p(x_n)$ with $3 \leq$ degree of monomials of $p \leq 7$. Let k be the least degree appearing, so that $p = a_k x_n^k + \ldots$. Then $j^k\eta'$ is k-determinate, because $\Delta(j^k\eta') = (x_1,\ldots,x_{n-1},x_n^{k-1})$ and so $m^2\Delta \supset m^{k+1}$, and we can use Theorem 2.9. Thus $\eta' \sim j^k\eta' = q + a_k x_n^k$

$= q + y_n^k$, changing coordinates so that $|a_k|^{1/k} x_n = y_n$.

If k is odd changing coordinates $y_n \to -y_n$ makes $\eta' = q - y_n^k \sim q + y_n^k$.

Classify as $q + p$ where $p = x_n^3 \quad \pm x_n^4 \quad x_n^5 \quad \pm x_n^6 \quad x_n^7 \quad 0$

and cod$(q+p)$ = 1 $\quad$ 2 $\quad$ 3 $\quad$ 4 $\quad$ 5

Lemma 4.13. The cuspoids η with cod $\eta \geq 6$ form a submanifold of I^7 of codimension 6.

Proof. If η is a cuspoid, $j^2\eta = q = x_1^2 + \ldots - x_{n-1}^2$.

Write $m^3/m^8 = R \times S$, where R is the set of polynomials involving one of $x_1, \ldots, x_{n-1}$ and such that $3 \leq$ degree of monomials in $r \in R \leq 7$, and S is the set of polynomials in x_n only, so that $S \cong \mathbb{R}^5$. Then $j^7\eta = q + r + s$, $r \in R$, $s \in S$. The reduction lemma 4.9. gave a (unique algebraic) map $\theta: R \to S$ such that $\eta \sim \eta'$, and $j^7\eta' = q + 0 + (\theta r + s)$.

$\qquad$ cod $\eta \geq 6 \Leftrightarrow$ cod $\eta' \geq 6$

$\qquad\qquad\quad \Leftrightarrow \theta r + s = 0$

$\qquad\qquad\quad \Leftrightarrow s = -\theta r$

$\qquad\qquad\quad \Leftrightarrow (r,s) \in M_\theta$, where M_θ is the graph of $-\theta$, and is a submanifold of $R \times S$ of codimension 5. (θ is algebraic and so graph $\theta \cong$ source of θ.) As q varies through Q_{n-1} we find that the required set of cuspoids η with cod $\eta \geq 6$ form a bundle over Q_{n-1} (of codimension 1 in m^2/m^3 by Lemma 4.8) with fibre M_θ which has codimension 5 in m^3/m^8. Thus the bundle has codimension 6 in $m^2/m^8 = I^7$.

Now we classify the umbilics, $Q_{n-2} \times m^3/m^8$. Let $\eta \in m^2$ be such that $j^2\eta = q$, and $q = x_1^2 + \ldots - x_{n-2}^2$. By the reduction lemma 4.9., $\eta \sim \eta'$ where $j^3\eta' = q + p$ and p is a homogeneous cubic in x_{n-1}, x_n.

In place of x_{n-1}, x_n we shall use x, y respectively, for clarity. Note that Lemma 4.12., which classifies cuspoids, has been interpreted in this way in Diagram 4.1 with x replacing x_n.

Let $(x,y) \in \mathbb{R}^2$. The space of cubic forms in x, y is, $\{(a_1 x^3 + a_2 x^2 y + a_3 xy^2 + a_4 y^3): a_1, a_2, a_3, a_4 \in \mathbb{R}\} = \mathbb{R}^4$. The action of $GL(2,\mathbb{R})$ on $\mathbb{R}^2$ induces an action on $\mathbb{R}^4$.

Lemma 4.14. There are 5 $GL(2,\mathbb{R})$-orbits in $\mathbb{R}^4$, and so each $p \in \mathbb{R}^4$ is equivalent to one of 5 forms:

			dimension	codimension
(1)	$x^3 + y^3$	hyperbolic umbilic	4	0
(2)	$x^3 - xy^2$	elliptic umbilic	4	0
(3)	$x^2 y$	parabolic umbilic	3	1
(4)	x^3	symbolic umbilic	2	2
(5)	0		0	4

Proof. Consider the roots x, y of $p(x,y) = 0$, $p \in \mathbb{R}^4$.

There are 5 cases (1) 2 complex, 1 real
                (2) 3 real distinct
                (3) 3 real, 2 same
                (4) 3 real equal
                (5) 3 equal to zero

Case (4): $p = (a_1x+a_2y)^3 = u^3$ by changing coordinates, $\begin{cases} u = a_1x + a_2y \\ v = \text{independent.} \end{cases}$
$\sim x^3$

Case (3): $p = u^2 v$ where u, v are independent linear forms in x, y.
$\sim x^2 y$

Case (2): $p = d_1 d_2 d_3$, product of 3 linear forms, $d_i = a_i x + b_i y$. We have
$k_1 = \begin{vmatrix} a_2 & b_2 \\ a_3 & b_3 \end{vmatrix} \neq 0$ because the root of $d_2 \neq$ the root of d_3. Let

$\left.\begin{array}{l} u + v = k_1 d_1 = u' \\ u - v = k_2 d_2 = v' \end{array}\right\}$ (*). We claim this is a nonsingular coordinate change.

$u,v \mapsto u', v'$ has a change of basis matrix with determinant $= \begin{vmatrix} 1 & 1 \\ 1 & -1 \end{vmatrix} = -2$.

$x,y \mapsto u', v'$ has a change of basis matrix with determinant $= k_1 k_2 \begin{vmatrix} a_1 & b_1 \\ a_2 & b_2 \end{vmatrix}$

$= k_1 k_2 k_3 \neq 0$

Adding (*), $2u = k_1 d_1 + k_2 d_2$
$= (a_2 b_3 - a_3 b_2)(a_1 x + b_1 y) + (a_3 b_1 - a_1 b_3)(a_2 x + b_2 y)$
$= x(a_1 a_2 b_3 - a_1 a_3 b_2 + a_2 a_3 b_1 - a_1 a_2 b_3) + y(\ldots)$
$= a_3 x(a_2 b_1 - a_1 b_2) + b_3 y(a_2 b_1 - a_1 b_2)$
$= -k_3 (a_3 x + b_3 y)$
$= -k_3 d_3$.

So $u^3 - uv^2 \sim 2u(u^2-v^2) = -k_1 k_2 k_3 d_1 d_2 d_3 \sim p$. Thus $p \sim x^3 - xy^2$.

Case (1): This is the same as Case (2) except that $a_2 = \bar{a}_1$, $b_2 = \bar{b}_1$ and a_3, b_3 are real. $d_2 = \bar{d}_1$, $k_1 = \begin{vmatrix} a_2 & b_2 \\ a_3 & b_3 \end{vmatrix} = \begin{vmatrix} \bar{a}_1 & \bar{b}_1 \\ a_3 & b_3 \end{vmatrix} = -\bar{k}_2$,

$k_3 = \begin{vmatrix} a_1 & b_1 \\ a_2 & b_2 \end{vmatrix} = a_1\bar{b}_1 - \bar{a}_1 b_1 = it$, $t \in \mathbb{R}$. Change coordinates, $\left.\begin{array}{l} iu + v = k_1 d_1 \\ iu - v = k_2 d_2 \end{array}\right\}$ $\binom{*}{*}$.

We claim this is a real change. Adding, $2iu = k_3 d_3 = itd_3$ and td_3 is real. Subtracting, $2v = k_1 d_1 - k_2 d_2 = k_1 d_1 + \bar{k}_1 \bar{d}_1$, which is real. So both u and v are real. It is a non-singular change because $\begin{vmatrix} 1 & 1 \\ 1 & -1 \end{vmatrix} = -2i \neq 0$. The product of $\binom{*}{*}$ is $2u(-u^2-v^2) = k_1 k_2 tp \sim p$. So $p \sim 2(u^3+uv^2)$, absorbing $-$ into the u-coordinate. $2(u^3+uv^2) \sim 2(u^3+3uv^2)$ absorbing $3^{\frac{1}{2}}$ into v.

$$= u'^3 + v'^3 \text{ with } \left.\begin{array}{l} u' = u+v \\ v' = u-v \end{array}\right\}$$
$$\sim x^3 + y^3.$$

By calculation $x^3 + y^3$ and $x^3 - xy^2$ are both 3-determinate and both $\text{cod}(x^3+y^3)$ and $\text{cod}(x^3-xy^2)$ equal 3. Thus the orbits corresponding to these are of codimension 3 in I^7 by Theorem 3.7.

Lemma 4.15. If $\eta = q + p$, $q \in Q_{n-2}$, $p = x^2 y$ + higher terms, then either

(1) $\eta \sim q + (x^2 y \underline{+} y^4)$ and $\text{cod } \eta = 4$ (the parabolic umbilic)

or (2) $\eta \sim q + (x^2 y \underline{+} y^5)$ and $\text{cod } \eta = 5$.

or (3) η belongs to Σ_6^7.

Proof. If $k \geq 4$, then if $p = x^2 y \pm y^k$, $\text{cod } p = k = \det p$.

Lemma 4.16. If $k \geq 4$ and $j^{k-1}p = x^2 y$ then $p \sim x^2 y \pm y^k$, or $p \sim p'$ and $j^k p' = x^2 y$.

Lemma 4.16. clearly gives Lemma 4.15.

Proof of Lemma 4.16. $j^k p = x^2 y$ + a polynomial of degree k
$$= x^2 y + ax^k + 2xyP + by^k,$$

where P is a homogeneous polynomial of degree $k - 2 \geq 2$.

$(x+P)^2 (y+ax^{k-2}) = (x^2+2xP)(y+ax^{k-2}) = x^2 y + 2xyP + ax^k$ in I^k. Put $u = (x+P)$ and $v = y + ax^{k-2}$; $v^k = y^k$ in I^k. So $j^k p = u^2 v + bv^k$. There are two cases. $b \neq 0$: $j^k p \sim u^2 v \pm v^k$ absorbing $|b|^{1/k}$ into v, and absorbing $1/|b|^{1/2k}$ into u. $b = 0$: $j^k p = u^2 v \sim x^2 y$.

Lemma 4.17. If $\eta = q + p$, $p \in Q_{n-2}$ and $p = x^3 +$ higher terms in x, y, then either (1) $\eta \sim q + x^3 \pm y^4$ and cod $\eta = 5$

or (2) $\eta \in \Sigma_6^7$.

Proof. Calculation shows that $x^3 \pm y^4 = p'$ is 4-determinate and cod $p' = 5$. $j^4 p = x^3 + a_0 x^4 + a_1 x^3 y + a_2 x^2 y^2 + a_3 xy^3 + a_4 y^4$. $a_4 \neq 0$: Put $v = y + \frac{a_3 x}{4 a_4}$. Then $j^4 p = x^3 + 3x^2 P + a_4 v^4$, where P is a homogeneous polynomial of degree 2 in x, v. In I^4 $j^4 p = (x+P)^3 + a_4 v^4$

$\sim u^3 \pm v^4$, putting $u = x + P$ and absorbing $|a_4|^{\frac{1}{4}}$ into v.

$a_4 = 0$: As above we find that $j^4 p \sim x^3 + xy^3$, which is 4-determinate as stated in Chapter 2. (This is Siersma's germ) In any case a short calculation gives cod $\eta = \text{cod}(x^3 + xy^3) = 6$, so $\eta \in \Sigma_6^7$.

Lemma 4.14 and a straightforward calculation produce the following facts. The symbolic umbilic (S) is a twisted cubic curve of dimension 1 in $\mathbb{R}^3$. The parabolic umbilic (P) is a quartic surface with a cusp edge along S. The elliptic umbilic (E) is inside the cusp. The hyperbolic umbilic (H) is outside the cusp. (4.18)

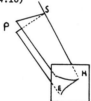

CHAPTER 5. THE PREPARATION THEOREM.

This chapter is self-contained and is devoted to proving a major result, the Preparation Theorem, which we need for Chapter 6.

The words "near 0" will always be understood to mean "in some neighborhood of 0."

Theorem 5.1. (Division Theorem) Let D be a C^∞ function defined near 0, from $\mathbb{R} \times \mathbb{R}^n$ to $\mathbb{R}$, such that $D(t,0) = d(t) t^k$ where $d(0) \neq 0$ and d is

C^∞ near 0 in $\mathbb{R}$. Then given any $C^\infty E: \mathbb{R} \times \mathbb{R}^n \to \mathbb{R}$ defined near 0, $\exists$ C^∞ functions q and r such that: (1) $E = qD + r$ near 0 in $\mathbb{R} \times \mathbb{R}^n$,

where (2) $r(t,x) = \sum_{i=0}^{k-1} r_i(x) t^i$ for $(t,x) \in \mathbb{R} \times \mathbb{R}^n$ near 0.

Notation. Let $P_k: \mathbb{R} \times \mathbb{R}^k \to \mathbb{R}$ be the polynomial $P_k(t,\lambda) = t^k + \sum_{i=0}^{k-1} \lambda_i t^i$.

Theorem 5.2. (Polynomial Division Theorem) Let $E(t,x)$ be a $\mathbb{C}$-valued C^∞ function defined near 0 in $\mathbb{R} \times \mathbb{R}^n$. Then $\exists$ $\mathbb{C}$-valued C^∞ functions $q(t,x,\lambda)$ and $r(t,x,\lambda)$ defined near 0 in $\mathbb{R} \times \mathbb{R}^n \times \mathbb{R}^k$ satisfying:

(1) $E(t,x) = q(t,x,\lambda) P_k(t,\lambda) + r(t,x,\lambda)$, and

(2) $r(t,x,\lambda) = \sum_{i=0}^{k-1} r_i(x,\lambda) t^i$,

where each r_i is a C^∞ function defined near 0 in $\mathbb{R}^n \times \mathbb{R}^k$. Moreover if E is $\mathbb{R}$-valued, then q and r may be chosen $\mathbb{R}$-valued.

Note that if E is $\mathbb{R}$-valued we merely equate real parts of (1) in Theorem 5.2 to give the last part.

Proof of Theorem 5.1 using Theorem 5.2. Given D, E we can apply Theorem 5.2 to find q_D, r_D, q_E, r_E such that $D = q_D P_k + r_D$ and $E = q_E P_k + r_E$; let now

$$r_D(t,x,\lambda) = \sum_{i=0}^{k-1} r_i^D(x,\lambda) t^i \quad (*).$$

Now $t^k d(t) = D(t,0) = q_D(t,0) P_k(t,0) + r_D(t,0) \quad (\lambda = 0)$
$= q_D(t,0) t^k + \sum_{i=0}^{k-1} r_i^D(0) t^i.$

Comparing coefficients of powers of t, $r_i^D(0) = 0$ and $q_D(0) \neq 0$ ($d(0) \neq 0$). Write $s_i(\lambda) = r_i^D(0,\lambda)$. We claim that $\left| \dfrac{\partial s_i(0)}{\partial \lambda_j} \right| \neq 0$.

$t^k d(t) = D(t,0) = q_D(t,0,\lambda)(t^k + \sum_{i=0}^{k-1} \lambda_i t^i) + \sum_{i=0}^{k-1} s_i(\lambda) t^i.$ Differentiating with respect to λ_j and setting $\lambda = 0$, $0 = \dfrac{\partial q_D}{\partial \lambda_j}(t,0) t^k + q_D(t,0) t^j + \sum_{i=0}^{k-1} \dfrac{\partial s_i}{\partial \lambda_j}(0) t^i.$

Thus $\dfrac{\partial s_i}{\partial \lambda_j}(0) = 0$ if $i < j$ and $\dfrac{\partial s_j}{\partial \lambda_j}(0) = -q_D(0)$. So $\left(\dfrac{\partial s_i}{\partial \lambda_j}(0) \right)$ is a lower triangular matrix, and as $q_D(0) \neq 0$, $\left| \dfrac{\partial s_i}{\partial \lambda_j} \right|(0) \neq 0$.

By the implicit function theorem, $\exists$ C^∞ functions $\theta_i(x)$ $(0 \le i \le k-1)$ such that (a) $r_j^D(x,\theta) \equiv 0$, and (b) $\theta(0) = 0$ (recall $r_j^D(0) = 0$). Let $\bar{q}(t,x) = q_D(t,x,\theta)$ and $P(t,x) = P_k(t,\theta)$. Then $D(t,x) = \bar{q}(t,x)P(t,x)$ (as $r_D(t,x,\theta) \equiv 0$ by (a).) As $\bar{q}(0) = q_D(0) \ne 0$, $P(t,x) = \dfrac{D(t,x)}{\bar{q}(t,x)}$ near 0 in $\mathbb{R} \times \mathbb{R}^n$.

By (*), $E(t,x) = q_E(t,x,\theta)P_k(t,\theta) + r_E(t,x,\theta) = q(t,x)D(t,x) + r(t,x)$, where $q(t,x) = \dfrac{q_E(t,x,\theta)}{\bar{q}(t,x)}$ and $r(t,x) = r_E(t,x,\theta) = \sum_{i=0}^{k-1} r_i^E(x,\theta)t^i$. Finally let $r_i(x) = r_i^E(x,\theta)$.

Suppose $f: \mathbb{C} \to \mathbb{C}$, $f = u + iv$ and $u, v: \mathbb{C} \to \mathbb{R}$. If $z = x + iy$, then $\dfrac{\partial u}{\partial \bar{z}} = \dfrac{\partial u}{\partial x} \cdot \dfrac{\partial x}{\partial \bar{z}} + \dfrac{\partial u}{\partial y} \cdot \dfrac{\partial y}{\partial \bar{z}} = \dfrac{1}{2}[\dfrac{\partial u}{\partial x} + \dfrac{i\partial u}{\partial y}]$. A similar result for v gives us that

$$\dfrac{\partial f}{\partial \bar{z}} = \dfrac{1}{2}[(\dfrac{\partial u}{\partial x} - \dfrac{\partial v}{\partial y}) + i(\dfrac{\partial u}{\partial y} + \dfrac{\partial v}{\partial x})] \tag{5.3}$$

Lemma 5.4. Let $f: \mathbb{C} \to \mathbb{C}$ be C^∞ as a function $\mathbb{R}^2 \to \mathbb{R}^2$. Let γ be a simple closed curve in $\mathbb{C}$ whose interior is U. Then for $w \in U$,

$$f(w) = \dfrac{1}{2\pi i} \int_\gamma \dfrac{f(z)}{z-w} dz + \dfrac{1}{2\pi i} \iint_U \dfrac{\partial f}{\partial \bar{z}}(z) \dfrac{dz \wedge d\bar{z}}{z-w}.$$

(If f is holomorphic this reduces to the Cauchy Integral Formula since f is holomorphic $\Leftrightarrow \dfrac{\partial f}{\partial \bar{z}} \equiv 0$.)

Proof. Let $w \in U$ and choose $\varepsilon < \min\{|w-z|: z \in \gamma\}$. Let $U_\varepsilon = U -$ (disc radius ε about w), and $\gamma_\varepsilon = \partial U_\varepsilon$.

Recall Green's Theorem for $\mathbb{R}^2$. If $M, N: U_\varepsilon \to \mathbb{R}$ are C^∞ on γ_ε, then

$$\int_{\gamma_\varepsilon} (Mdx + Ndy) = \iint_{U_\varepsilon} (\dfrac{\partial N}{\partial x} - \dfrac{\partial M}{\partial y}) dx \wedge dy.$$

Green's Theorem and (5.3) for $f = u + iv$ give

$$\int_{\gamma_\varepsilon} f\, dz = \int_{\gamma_\varepsilon} (u+iv)(dx+idy) = 2i \iint_{U_\varepsilon} \dfrac{\partial f}{\partial \bar{z}} dx \wedge dy.$$

$2i\, dx \wedge dy = -dz \wedge d\bar{z}$, so $\int_{\gamma_\varepsilon} f\, dz = -\iint_{U_\varepsilon} \dfrac{\partial f}{\partial \bar{z}} dz \wedge d\bar{z}$ \hfill (*)

Apply (*) to $\frac{f(z)}{z-w}$, noting that $\frac{1}{z-w}$ is holomorphic on U.

$$-\iint_{U_\varepsilon} \frac{\partial f(z)}{\partial \bar{z}} \frac{dz \wedge d\bar{z}}{z-w} = \int_{\gamma_\varepsilon} \frac{f(z)}{z-w} dz = \int_\gamma \frac{f(z)}{z-w} dz - \int_{C_\varepsilon} \frac{f(z)}{z-w} dz, \qquad (\substack{* \\ *})$$

where C_ε is the circle, radius ε, centre w.

With polar coordinates at w, $\int_{C_\varepsilon} \frac{f(z)}{z-w} dz = \int_0^{2\pi} f(w+\varepsilon e^{i\theta}) i d\theta$. As $\varepsilon \to 0$, R.H.S. of $(\substack{* \\ *}) \to \int_\gamma \frac{f(z)}{z-w} dz - 2\pi i f(w)$, and L.H.S. of $(\substack{* \\ *}) \to -\iint_U \frac{\partial f(z)}{\partial \bar{z}} \frac{dz \wedge d\bar{z}}{z-w}$.

(The limit exists because $\frac{\partial}{\partial \bar{z}}$ is bounded on U, and $\frac{1}{z-w}$ is integrable over U.)

Proof of Theorem 5.2. Let $\tilde{E}(z,x,\lambda)$ be a C^∞ function defined near 0 in $\mathbb{C} \times \mathbb{R}^n \times \mathbb{C}^k$ such that $\tilde{E}(t,x,\lambda) = E(t,x) \ \forall \ t \ \mathbb{R}$, i.e. $\tilde{E}$ is an __extension__ of E. Then $\tilde{E}(w,x,\lambda) = \frac{1}{2\pi i} \int_\gamma \frac{\tilde{E}(z)}{z-w} dz + \frac{1}{2\pi i} \iint_U \frac{\partial \tilde{E}(z)}{\partial \bar{z}} \frac{dz \wedge d\bar{z}}{z-w}$, by Lemma 5.4. Let $P_k(z,\lambda) - P_k(w,\lambda) = (z-w) \sum_{i=0}^{k-1} p_i(z,\lambda) w^i$, i.e. $\frac{P_k(z,\lambda)}{z-w} = \frac{P_k(w,\lambda)}{z-w} + \sum_{i=0}^{k-1} p_i(z,\lambda) w^i$.

In the expression for $\tilde{E}(w,x,\lambda)$ multiply top and bottom inside the integrals by $P_k(z,\lambda)$ and expand $\frac{P_k(z,\lambda)}{z-w}$ giving $\tilde{E} = qP_k + r$ on $\mathbb{C} \times \mathbb{R}^n \times \mathbb{C}^k$ where

$$q(w,x,\lambda) = \frac{1}{2\pi i} \int_\gamma \frac{\tilde{E}(z,x,\lambda)}{P_k(z,\lambda)} \cdot \frac{dz}{(z-w)} + \frac{1}{2\pi i} \iint_U \frac{\partial \tilde{E}(z,x,\lambda)}{\partial \bar{z}} \cdot \frac{1 \cdot dz \wedge d\bar{z}}{P_k(z,\lambda)(z-w)}$$

and $r_i(x,\lambda) = \frac{1}{2\pi i} \int_\gamma \frac{\tilde{E}(z,x,\lambda)}{P_k(z,\lambda)} \cdot p_i(z,\lambda) \cdot dz + \frac{1}{2\pi i} \iint_U \frac{\partial \tilde{E}}{\partial \bar{z}}(z,x,\lambda) \cdot \frac{p_i(z,\lambda)}{P_k(z,\lambda)} \cdot dz \wedge d\bar{z}$,

so long as these integrals are well defined and yield C^∞ functions.

The first integral in the definition of both q and r is well-defined and C as long as the zeros of $P_k(z,\lambda)$ do not occur on the curve γ for λ near 0 in $\mathbb{C}^k$. Such a γ is easily chosen.

But U may contain zeros of P_k. So we need $\tilde{E}$ such that $\frac{\partial \tilde{E}}{\partial \bar{z}}$ vanishes on zeros of P_k and for real z to ensure q, r well-defined. As the integrands are bounded we need C^∞ $\tilde{E}$ such that $\frac{\partial \tilde{E}}{\partial \bar{z}}$ vanishes to infinite order on zeros of P_k and for real z to ensure q and r C^∞.

Lemma 1. (Nirenberg Extension Lemma) Let $E(t,x)$ be a C^∞ $\mathbb{C}$-valued function defined near 0 in $\mathbb{R} \times \mathbb{R}^n$. Then $\exists$ a C^∞ $\mathbb{C}$-valued function $\tilde{E}(z,x,\lambda)$ defined near 0 in $\mathbb{C} \times \mathbb{R}^n \times \mathbb{C}^k$ such that,

(1) $\tilde{E}(t,x,\lambda) = E(t,x) \ \forall \ t \in \mathbb{R}$.

(2) $\dfrac{\partial \tilde{E}}{\partial \bar{z}}$ vanishes to infinite order on $\{\text{Im } z = 0\}$.

(3) $\dfrac{\partial \tilde{E}}{\partial \bar{z}}$ vanishes to infinite order on $\{P_k(z,\lambda) = 0\}$.

Lemma 2. (E. Borel's Theorem) Let $f_0, f_1, \ldots$ be a sequence of C^∞ functions on a given neighborhood N of 0 in $\mathbb{R}^n$. Then $\exists$ a C^∞ function $F(t,x)$ on a neighborhood of 0 in $\mathbb{R} \times \mathbb{R}^n$ such that $\dfrac{\partial^i F}{\partial t^i}(0,x) = f_i(x) \ \forall \ i$.

Proof. Let $\rho: \mathbb{R} \xrightarrow{C^\infty} \mathbb{R}$ be such that $\rho(t) = \begin{cases} 1 & |t| \leq \frac{1}{2} \\ 0 & |t| \geq 1 \end{cases}$

Let $F(t,x) = \sum\limits_{i=0}^{\infty} \dfrac{t^i}{i!} \rho(\mu_i t) f_i(x)$, where $\{\mu_i\}$ is a rapidly increasing sequence of real numbers tending to ∞, so that F is C^∞ near 0.

(Lemma 2 may be used to show that for any power series about 0 in $\mathbb{R}^n$ $\exists$ a C^∞ real-valued function with its Taylor series at 0 the given power series.)

Lemma 3. Let V, W be complementary subspaces of $\mathbb{R}^n$ ($= V+W$). Let g, h be C^∞ functions near 0 in $\mathbb{R}^n$, such that for all multi-indices α,
$\dfrac{\partial^{|\alpha|} g(x)}{\partial x^\alpha} = \dfrac{\partial^{|\alpha|} h(x)}{\partial x^\alpha} \ \forall \ x \in V \cap W$. Then $\exists \ C^\infty \ F$ near 0 in $\mathbb{R}^n$, such that

$\forall \alpha, \ \dfrac{\partial^{|\alpha|} F(x)}{\partial x^\alpha} = \begin{cases} \dfrac{\partial^{|\alpha|} g(x)}{\partial x^\alpha} & x \in V \\ \dfrac{\partial^{|\alpha|} h(x)}{\partial x^\alpha} & x \in W \end{cases}$ (A **multi-index** $\alpha = (a_1,\ldots,a_n)$

and $|\alpha| = a_1 + \ldots + a_n$ so that

$$\dfrac{\partial^{|\alpha|} g(x)}{\partial x^\alpha} = \dfrac{\partial^{a_1+\ldots+a_n} g(x)}{\partial x_1^{a_1} \ldots \partial x_n^{a_n}}.)$$

Proof. Without loss of generality $h \equiv 0$, for if F_1 is the required extension for $(g-h)$ and 0, then $F = F_1 + h$ is the required extension for g and h.

Choose coordinates $y_1, \ldots, y_n$ so that $V \equiv y_1 = \ldots = y_j = 0$

and $W \equiv y_{j+1} = \ldots = y_k = 0$. Let

$$F(y) = \sum_{|\alpha|=0}^{\infty} \frac{y^\alpha}{\alpha!} \frac{\partial^{|\alpha|} g}{\partial y^\alpha}(0,\ldots,0,y_{j+1},\ldots,y_n)\rho(\mu_{|\alpha|} \sum_{i=1}^{j} y_i^2),$$ where ρ is as in

$\alpha = (a_1,\ldots,a_j,0,\ldots,0)$

Lemma 2 and $\{\mu_i\}$ increases to ∞ rapidly enough so that F is C^∞ near 0.
If $y \in W$, each term of $\frac{\partial^{|\beta|} F(y)}{\partial y^\beta}$ contains a factor $\frac{\partial^{|\gamma|} g}{\partial y^\gamma}(0,\ldots,0,y_{k+1},\ldots,y_n)$.
Since $(0,\ldots,0,y_{k+1},\ldots,y_n) \in V \cap W$, this factor $= 0$ ($h \equiv 0$). So $\frac{\partial^{|\beta|} F(y)}{\partial y^\beta} = 0$.

If $y \in V$, note that $\frac{\partial^{|\gamma|}}{\partial y^\gamma} \rho(\mu_{|\alpha|} \sum_{i=1}^{j} y_i^2)\bigg|_{y_1=\ldots=y_j=0} = \begin{cases} 1 & \gamma = 0, \\ 0 & \gamma \neq 0 \end{cases}$

and then $\frac{\partial^{|\beta|} F(y)}{\partial y^\beta} = \sum_{|\alpha|=0}^{\infty} \frac{\partial^{|\beta|}}{\partial y^\beta}\left[\frac{y^\alpha}{\alpha!} \frac{\partial^{|\alpha|} g(y)}{\partial y^\alpha}\right]\bigg|_{y_1=\ldots=y_j=0}$.

If $b_i \neq a_i$ some $i \leq j$, then this term is 0. In fact the only nonzero term is $\frac{\partial^{|\beta|} g(y)}{\partial y^\beta}$.

<u>Lemma 4.</u> Let f be a C^∞ $\mathbb{C}$-valued function near 0 in $\mathbb{R}^n$ and let X be a vector field on $\mathbb{R}^n$ with $\mathbb{C}$ coefficients. Then $\exists$ C^∞ $\mathbb{C}$-valued F near 0 in $\mathbb{R} \times \mathbb{R}^n$ so that

(a) $F(0,x) = f(x) \quad \forall x \in \mathbb{R}^n$.

(b) $\frac{\partial F}{\partial t}$ agrees to infinite order with XF at all $(0,x) \in \mathbb{R} \times \mathbb{R}^n$.

<u>Proof.</u> Try $\overline{F}(t,x) = e^{tX}f = \sum_{k=0}^{\infty} \frac{t^k}{k!} X^k f$. Differentiating termwise at $t = 0$ gives (b). Clearly (a) holds. To ensure that $\overline{F}$ is C^∞ use Lemma 2 to choose C^∞ F such that $F = \sum_{k=0}^{\infty} \frac{t^k}{k!} X^k f \rho(\mu_k t)$.

<u>Proof of Lemma 1.</u> We use induction on k. If $k = 0$, $P_k(z,\lambda) \equiv 1$, so we need C^∞ $\tilde{E}(z,x)$ such that $\tilde{E}(t,x) = E(t,x) \ \forall t \in \mathbb{R}$ and $\frac{\partial \tilde{E}}{\partial \bar{z}}(t,x)$ vanishes to infinite order $\forall t \in \mathbb{R}$. Let $z = s + it$, $2\frac{\partial}{\partial \bar{z}} = \frac{\partial}{\partial s} + i\frac{\partial}{\partial t}$. (Compare 5.3) Then Lemma 4 with $X = -i\frac{\partial}{\partial s}$ gives such an $\tilde{E}$.

Suppose Lemma 1 is proved for $k - 1$. We show $\exists$ $C^\infty F(z,x,\lambda)$ and $G(z,x,\lambda)$ such that

(1)' F and G agree to infinite order on $\{P_k(z,\lambda) = 0\}$

(2)' F is an extension of E.

(3)' $\frac{\partial F}{\partial \bar{z}}$ vanishes to infinite order on $\{\text{Im } z = 0\}$.

(4)' Let $M = F|\{P_k(z,\lambda) = 0\}$. $\frac{\partial M}{\partial \bar{z}}$ vanishes to infinite order on $\{\frac{\partial P_k}{\partial \bar{z}}(z,\lambda) = 0\}$.

(5)' $\frac{\partial G}{\partial \bar{z}}$ vanishes to infinite order on $\{P_k(z,\lambda) = 0\}$.

Existence of F and G proves Lemma 1. Let $u = P(z,\lambda) \equiv P_k(z,\lambda)$ and $\lambda' = (\lambda_1,\ldots,\lambda_{k-1})$. Consider $(z,\lambda_0,\lambda') \to (z,u,\lambda')$ on $\mathbb{C} \times \mathbb{C} \times \mathbb{C}^{k-1}$. This is a valid coordinate change because $\frac{\partial u}{\partial \lambda_0} \equiv 1$. In the new coordinates, $\{P_k(z,\lambda) = 0\}$ is given by $u = 0$. By Lemma 3 $\exists \tilde{E}$ agreeing to infinite order with G on $u = 0$ and to infinite order with F on $\text{Im } z = 0$. ($u = 0$ and $\text{Im } z = 0$ intersect transversally in $\mathbb{R}^{2k+2}$.) (2)', (3)' and (5)' now imply $\tilde{E}$ is the desired extension of E.

Existence of F and G. Suppose we have that F exists. In (z,u,λ')-coordinates, $\frac{\partial}{\partial z}$ becomes $\frac{\partial}{\partial z} + \frac{\partial P}{\partial z}\frac{\partial}{\partial u}$, and $\frac{\partial}{\partial \bar{z}}$ becomes $\frac{\partial}{\partial \bar{z}} + \frac{\overline{\partial P}}{\partial z}\frac{\partial}{\partial \bar{u}}$. So in these coordinates we need $G(z,x,u,\lambda')$ such that

(a) $F = G$ to infinite order on $\{u = 0\}$, and

(b) $(\frac{\partial}{\partial \bar{z}} + \frac{\overline{\partial P}}{\partial z}\frac{\partial}{\partial \bar{u}})G = 0$ to infinite order on $\{u = 0\}$.

Let $X = -(\frac{\overline{\partial P}}{\partial z})^{-1}\frac{\partial}{\partial \bar{z}}$. As in Lemma 4 we must find C^∞ G satisfying (a) and

(b') $\frac{\partial G}{\partial \bar{z}} = XG$ to infinite order on $\{u = 0\}$. The formal solution is,

$$G = \sum_{i=0}^{\infty} \frac{(\bar{u})^i}{i!} X^i M(z,x,\lambda') \rho(\mu_i |\bar{u}|^2) \qquad (*)$$

As $\frac{\partial M}{\partial \bar{z}} = 0$ to infinite order on $\{\frac{\partial P}{\partial \bar{z}}(z,\lambda') = 0\}$ by (4)', $X^i M$ is C^∞ in (z,x,λ') $\forall i$, so we can choose $\{\mu_i\}$ to increase quickly enough to make G C^∞.

We need only a C^∞ F so that in (z,x,u,λ')-coordinates,

(2)' $F(t,x,u,\lambda') = E(t,x)$ $\forall t \in \mathbb{R}$

(3)' $\frac{\partial F}{\partial \bar{z}} = XF$ to infinite order on $\{\text{Im } z = 0\}$

(4)' If $M = F|\{u = 0\}$, $\frac{\partial M}{\partial \bar{z}} = 0$ to infinite order on $\{\frac{\partial P_k}{\partial \bar{z}} = 0\}$.

Consider $u = 0$ and the coordinate change $\lambda' = (\lambda_1,\ldots,\lambda_{k-1}) \longmapsto (\frac{\lambda_1}{1},\ldots,\frac{\lambda_{k-1}}{k-1}) = \lambda''$. The conditions are now that we find C^∞ $M(z,x,\lambda'')$ such that,

(I) $M(t,x,\lambda'') = E(t,x) \; \forall \; t \in \mathbb{R}$
(II) $\frac{\partial M}{\partial \bar{z}}$ vanishes to infinite order on $\{\text{Im } z = 0\}$, and
(III) "   "   "   "   "   " $\{P_{k-1}(z,\lambda'') = 0\}$.

The induction hypothesis gives such a C^∞ $M(z,x,\lambda'')$, and we can view M as a C^∞ function of (z,x,λ').

Let $F(z,x,u,\lambda') = \sum_{i=0}^{\infty} \frac{(\bar{u})^i}{i!} X^i M(z,x,\lambda') \rho(\mu_i|\bar{u}|^2)$. Compare (*). By (III), $X^i M$ is C^∞ in z, x, λ', and so the $\{\mu_i\}$ may be chosen so that F is a C^∞ function satisfying (2)', (3)'. Also, on $u = 0$, $F = M$ and (III) gives (4)'.

The completes the proof of Lemma 1.

The remarks before Lemma 1 state that this suffices to prove the (Polynomial Division Theorem) Theorem 5.2.

Let π be projection $\mathbb{R}^{n+s} \to \mathbb{R}^s$. π induces $\pi^*\colon E_s \to E_{n+s}$, where E_s is the set of germs at 0 of C^∞ functions $\mathbb{R}^s \to \mathbb{R}$, as usual. Let M be an E_{n+s}-module, and let $\underline{M}$ denote the same set regarded as an E_s-module with structure induced by π^*.

Theorem 5.5. (Preparation Theorem) Suppose that

(1) M is a finitely generated E_{n+s}-module,
(2) $M/(\pi^* m_s)M$ is a finite-dimensional real vector space.

Then $\underline{M}$ is finitely generated as an E_s-module.

Proof. There are 2 steps.

Step 1. Let $\pi_1\colon \mathbb{R}^s \times \mathbb{R} \to \mathbb{R}^s$ and $t\colon \mathbb{R}^s \times \mathbb{R} \to \mathbb{R}$ denote the projections. We prove the theorem for $n = 1$, $\pi = \pi_1$. Let $v_1, \ldots, v_p$ be elements of M generating M as an E_{s+1}-module, whose images in $M/(\pi^* m_s)M$ span this vector space. Then any $v \in M$ can be written $v = \sum_{i=1}^{p} a_i v_i + \sum_{i=1}^{p} \alpha_i v_i$ where $a_i \in \mathbb{R}$, and $\alpha_i \in (\pi^* m_s) E_{s+1}$. In particular $\exists\ a_{ij} \in R$, $\alpha_{ij} \in (\pi^* m_s) E_{s+1}$ ($1 \leq i,\ j \leq p$), such that $tv_i = \sum_{j=1}^{p} (a_{ij} + \alpha_{ij}) v_j$. Let D be the determinant $|t\delta_{ij} - a_{ij} - \alpha_{ij}|$; by Cramer's rule $Dv_i = 0$, $i = 1, \ldots, p$. Expanding the determinant we see that D is regular of order k, some $k \leq p$, since $D|(0 \times R, \theta)$ is a monic polynomial in t of order p ($\alpha_{ij} = 0$ on $0 \times R$). Since $D.M = 0$, M is an $(E_{s+1}/D.E_{s+1})$-module.

Now D is regular of order k (i.e. $D(t,0) = d(t) t^k$, where $d(0) \neq 0$ and d is C^∞ near 0, and D is C^∞ defined near 0 in $R^s \times R$) and so using the Division Theorem 5.1., $E_{s+1}/D.E_{s+1}$ is finitely generated as an E_s-module.

Since M is finitely generated as an $(E_{s+1}/D.E_{s+1})$-module, it follows that $\underline{M}$ is finitely generated as an E_s-module.

Step 2. We complete the proof of the theorem. Factor π as follows:

$$R^s \times R^n \xrightarrow{\pi_n} \ldots \xrightarrow{\pi_2} R^s \quad R \xrightarrow{\pi_1} R^s,$$

where $\pi_i : R^s \times R^i \to R^s \times R^{i-1}$ is the germ of the projection,

$$(y, a_1, \ldots, a_i) \longmapsto (y, a_1, \ldots, a_{i-1}).$$

For each i, $0 \leq i \leq n + s$, we give M the E_{s+i}-module structure induced by $(\pi_{i+1} \circ \ldots \circ \pi_n)^*$. If $i = 1$ this is the E_s-module structure of $\underline{M}$ since $\pi = \pi_1 \circ \ldots \circ \pi_n$.

Now we prove by decreasing induction on i that M is finitely generated as an E_{s+i}-module $\forall\ i$, $0 \leq i \leq n$. By hypothesis, it is true for $i = n$, so it suffices to carry out the inductive step. Assume M is finitely generated as an E_{s+i+1}-module.

$(\pi^* m_s) M = (\pi_1 \circ \ldots \circ \pi_{i+1})^* (m_s) M$. (On the L.H.S. M is regarded as an E_{n+s}-module, and on the R.H.S. as an E_{s+i+1}-module.) So $(\pi^* m_s) M \subset (\pi_{i+1}^* m_{s+1}) M$. In particular $M/(\pi_{i+1}^* m_{s+1}) M$ is finitely generated as a real vector space. In particular the hypotheses of the theorem are satisfied for π_{i+1} in place of π. Thus we may apply Step 1 to see that M is finitely generated as an E_{s+i}-module.

This completes the inductive step and also the proof as $i = 0$ is the statement of the theorem.

Definition. Let π be projection $\mathbb{R}^{n+s} \to \mathbb{R}^s$. A <u>mixed homomorphism over</u> $\pi*$ <u>of finite type</u> (a <u>mixture</u>) is a diagram:

$$\begin{array}{ccc} & & B \\ & & \downarrow \beta \\ A & \xrightarrow{\alpha} & C \\ \downarrow & & \downarrow \\ E_s & \xrightarrow{\pi*} & E_{n+s} \end{array}$$

where A is a finitely generated E_s-module,

B is an E_{n+s}-module,

C is a finitely generated E_{n+s}-module;

α is a module homomorphism over $\pi*$, i.e. $\alpha(\eta a) = (\pi*\eta)(\alpha a)$, $\eta \in E_s$ and $a \in A$; β is an E_{n+s}-module homomorphism.

Corollary 5.6. $C = \alpha A + \beta B + (\pi* m_s)C \Rightarrow C = \alpha A + \beta B$.

Proof. Let $C' = C/\beta B$ and $\rho: C \to C'$ be the projection. As C is a finitely generated E_{n+s}-module so is C'. \hfill (1)

$(\pi* m_s)C' = m_s \underline{C}'$, so $C'/(\pi* m_s)C' = \underline{C}'/m_s\underline{C}'$. \hfill (2)

Our hypothesis $\Rightarrow C' = \rho\alpha A + (\pi* m_s)C' \Rightarrow \underline{C}' = \rho\alpha A + m_s\underline{C}'$ \hfill (3)

and this $\Rightarrow \underline{C}'/m_s\underline{C}'$ is a finitely generated E_s-module. Choose now a finite base $\{c_i\}$ for $\underline{C}'$ mod $m_s\underline{C}'$ as an E_s-module. Any $c \in \underline{C}'$ can be written,

$$c = \sum_i \eta_i c_i \text{ mod } m_s\underline{C}' \quad \text{(finite sum)} \quad \eta_i \in E_s.$$

Now $\eta_i = \eta_i(0) + \eta_i'$, $\eta_i(0) \in \mathbb{R}$, $\eta_i' \in m_s$ in the notation of Lemma 2.8. So $c = \sum_i \eta_i(0) c_i$ mod $m_s\underline{C}'$. Because c was arbitrary we have shown that $\underline{C}'/m_s\underline{C}'$ is a finite-dimensional vector space over $\mathbb{R}$, and hence by (2) so is $C'/(\pi* m_s)C'$. \hfill (4)

(1) and (4) for C' are the two hypotheses of the Preparation Theorem 5.5, and so $\underline{C}'$ is a finitely generated E_s-module. We can now apply Nakayama's Lemma 2.10 with $A = E_s$, $a = m_s$, $M = \underline{C}'$ and $N = \underline{\rho\alpha A}$ to (3). Therefore $\underline{C}' = \underline{\rho\alpha A}$.

And so $C' = \rho\alpha A$, i.e. $C = \alpha A + \beta B$.

CHAPTER 6. UNFOLDINGS

We defint the category of unfoldings of η, for fixed $\eta \in m^2$. An **object** (r,f) is a germ $f: \mathbb{R}^n \times \mathbb{R}^r, 0 \to \mathbb{R}, 0$ (shorthand for "is a germ f of a C^∞ function $\mathbb{R}^n \times \mathbb{R}^r, 0 \to \mathbb{R}, 0$"), such that $f|\mathbb{R}^n \times 0 = \eta$, i.e.

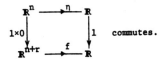

commutes.

A **morphism** $(\phi, \bar{\phi}, \varepsilon): (s,g) \to (r,f)$ is a germ $\phi: \mathbb{R}^{n+s}, 0 \to \mathbb{R}^{n+r}, 0$,

a germ $\bar{\phi}: \mathbb{R}^s, 0 \to \mathbb{R}^r, 0$,

a sheer germ $\varepsilon: \mathbb{R}^s, 0 \to \mathbb{R}, 0$, (6.1)

such that $\phi|\mathbb{R}^n \times 0 = 1$, and

if $\pi_r: \mathbb{R}^{n+r} \to \mathbb{R}^r$ is projection,

$\pi_r \phi = \bar{\phi} \pi_s$ and $g = f\phi + \varepsilon \pi_s$.

Definition. (r,f) is said to be **universal** if, $\forall (s,g) \exists$ a morphism, $(s,g) \to (r,f)$.

Definition. $(\phi, \bar{\phi}, \varepsilon)$ is an **isomorphism** if it has an inverse. Note that this requires $r = s$, and ϕ and $\bar{\phi}$ are diffeomorphism-germs, so $(\phi^{-1}, \bar{\phi}^{-1}, -\varepsilon\bar{\phi}^{-1})$ will do.

Prolongation of a germ. Given $\eta \in m^2$, let $z = j^k \eta$. Choose a representative function of η, $e: \mathbb{R}^n, 0 \to \mathbb{R}, 0$. $\mathbb{R}^n$ operates on e by translation as follows. Given $w \in \mathbb{R}^n$, define $w(e): \mathbb{R}^n, 0 \to \mathbb{R}, 0$

$$x \mapsto e(w+x) - e(w).$$

Graph $w(e)$ = graph e with origin moved to $(w, e(w))$.
Denote by $j_1 e$ the map obtained: $\mathbb{R}^n, 0 \to m, \eta$

$$w \mapsto \text{germ at } 0 \text{ of } w(e).$$

Let $j_1 \eta$ denote the germ at 0 of $j_1 e$ (we shall show this is unambiguous). $j_1 \eta$ is called the **natural germ prolongation** of η. $j_1^k \eta = \pi \circ j_1 \eta$ is called the **natural k-jet prolongation** of η, where π is the usual projection $m \to J^k$.

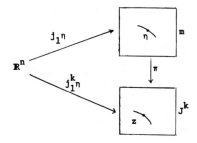

Lemma 6.2. (1) $j_1 \eta$ and $j_1^k \eta$ are uniquely determined by η (not by z, necessarily), i.e. they are independent of the choice of e.

(2) If η is $(k+1)$-determinate, $j_1^k \eta$ is the germ of an embedding $\mathbb{R}^n, 0 \to J^k, z$.

(3) The tangent plane $T_z(\text{im } j_1^k \eta)$ lies in $\pi\Delta(\Delta=\Delta(\eta))$ transverse to $\pi(m\Delta)$, and is spanned by $\{j^k \frac{\partial \eta}{\partial x_i}\}$.

Proof. If e and e' are 2 representatives of η, then $e = e'$ on N, some neighborhood of 0 in $\mathbb{R}^n$. $w(e) = w(e')$ if $w + x$, $w \in N$. So $j_1 \eta$ is well defined (and clearly $j_1^k \eta$ is too). (1) is proved. (2) follows using (3) and the definition of determinacy. (3). Clearly $T_z(\text{im } j_1^k \eta)$ is spanned by $j^k\{\frac{\partial \eta}{\partial x_i}\}$ which are in $\pi\Delta$. By the definition of Δ (the ideal generated by $\{\frac{\partial \eta}{\partial x_i}\}$), $m\Delta$ and the space spanned by $\{\frac{\partial \eta}{\partial x_i}\}$ are transversal in Δ (use Lemma 3.8). Quotient out by m^{k+1}. Hence $T_z(\text{im } j_1^k \eta)$ is transverse to $\pi(m\Delta)$ in $\pi\Delta$.

We define the k-jet prolongation of an unfolding (r,f) of a germ $\eta \in m^2$ in a similar way. Represent f by a function $\tilde{f} \colon \mathbb{R}^{n+r}, 0 \to \mathbb{R}, 0$. Let F be the germ at 0 of the map $\mathbb{R}^{n+r}, 0 \to J^k, z$

$$(x',y') \mapsto \text{k-jet at } 0 \text{ of the function}$$
$$\mathbb{R}^n, 0 \to \mathbb{R}, 0$$
$$x \mapsto \tilde{f}(x'+x, y') - \tilde{f}(x', y')$$

F is the <u>k-jet prolongation of the unfolding</u> (r,f).

Definition. We say the unfolding (r,f) is **k-transversal** if the germ F is transversal to the orbit zG^k in J^k.

Let $x_1, \ldots, x_n$ be coordinates for $\mathbb{R}^n$ and $y_1, \ldots, y_r$ be coordinates for $\mathbb{R}^r$. Choose $\tilde{f} \in f$, and then for each $j = 1, \ldots, r$ we have a function $\frac{\partial \tilde{f}}{\partial y_j}$ from $\mathbb{R}^{n+r}, 0$ to $\mathbb{R}$, $\frac{\partial \tilde{f}}{\partial y_j}(0,0)$. Let $\partial_j f$ be the germ at p of $\frac{\partial \tilde{f}}{\partial y_j} | \mathbb{R}^n \times 0 - \frac{\partial \tilde{f}}{\partial y_j}(0,0)$. $\partial_j f$ is in m. V_f will denote the vector subspace of m spanned by $\partial_1 f, \ldots, \partial_r f$. (6.3)

Lemma 6.4. An unfolding (r,f) of a germ η is k-transversal
$\Leftrightarrow m = \Delta + V_f + m^{k+1}$.

Proof. In J^k, i.e. mod m^{k+1}, the tangent to the orbit zG^k is $m\Delta$ (Lemma 2.11), the tangent to the x-direction of $F(=j_1^k \eta)$ is $T_z(\text{im } j_1^k \eta)$, and these two are transverse in Δ by Lemma 6.2 (3). The tangent to the y-direction of F is V_f (6.3). So F is transversal to $zG^k \Leftrightarrow \Delta + V_f$ span m mod m^{k+1}.

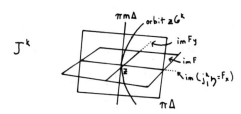

Corollary 6.5. Let η have finite determinacy and $c = \text{cod } \eta$, then $\exists$ an unfolding (c,f), which is k-transversal $\forall k > 0$.

Proof. Because det η is finite, so is $\text{cod } \eta = c$ finite by Lemma 3.1; by definition $\text{cod } \eta = \dim m/\Delta$ ($\Delta = \Delta(\eta)$). Choose $u_1, \ldots, u_c \in m$ such that their images in m/Δ form a basis for m/Δ. Define an unfolding (c,f) by,

$f: \mathbb{R}^n \times \mathbb{R}^c \to \mathbb{R}$ Then $\frac{\partial f}{\partial y_j} = u_j(x)$, so $\partial_j f = \frac{\partial f}{\partial y_j} | \mathbb{R}^n \times 0 - \frac{\partial f}{\partial y_j}(0,0)$
$(x,y) \to \eta(x) + \sum_{j=1}^{c} y_j u_j(x)$. $= u_j(x)$. $(u_j \in m)$

By the choise of $\{u_j\}$, $\{\partial_j f\}$ span m/Δ. By 6.3 $m = \Delta + V_f = \Delta + V_f + m^{k+1} \forall k > 0$. Now apply Lemma 6.4.

Lemma 6.6. Let η have finite determinacy, with a universal unfolding (r,f). Then (r,f) is k-transversal $\forall\, k > 0$ and $r \geq \text{cod}\,\eta$.

Proof. Let $c = \text{cod}\,\eta$ and (c,g) be the unfolding of Corollary 6.5, which is k-transversal $\forall\, k > 0$. By the definition of universality $\exists$ a morphism $(\phi, \bar{\phi}, \varepsilon): (c,g) \to (r,f)$. So $g(x,y) = f(\phi(x,y)) + \varepsilon(y)$ where $(x,y) \in \mathbb{R}^n \times \mathbb{R}^c$, by (6.1).
$$= f(\phi^y x, \bar{\phi} y) + \varepsilon(y) \text{ with } \phi^y x = \pi_{x'}(\phi(x,y)),$$
choosing $x_1', \ldots, x_n'$ and $y_1', \ldots, y_r'$ as coordinates for $\mathbb{R}^{n+r}$. Now we have
$$\frac{\partial g}{\partial y_j}(x,0) = \sum_i \frac{\partial f}{\partial x_i'}(\phi^0 x, \bar{\phi} 0)\frac{\partial \phi_i^y}{\partial y_j}(x) + \sum_h \frac{\partial f}{\partial y_h'}(\phi^0 x, \bar{\phi} 0)\frac{\partial \bar{\phi}_h}{\partial y_j}(0) + \frac{\partial \varepsilon}{\partial y_j}(0).$$
$\phi^0 = \phi|\mathbb{R}^n \times 0 = 1$ and $\bar{\phi} 0 = 0$ by 6.1. Also $\frac{\partial \phi_i^y}{\partial y_j} \in E$ and $\frac{\partial \bar{\phi}_h}{\partial y_j}(0) \in \mathbb{R}$. So the first sum is in Δ, as $\frac{\partial f}{\partial x_i'}(x,0) = \frac{\partial \eta}{\partial x_i}(x)$, and the hth term in the second sum is $\frac{\partial f}{\partial y_h'}(x,0) \times$ constant. Remember $\partial_h f = \frac{\partial f}{\partial y_h'}(x,0) - \frac{\partial f}{\partial y_h'}(0,0) \in V_f$. So $V_g \subset \Delta + V_f$.

Now $m = \Delta + V_g$ $\forall\, k > 0$ by Lemma 6.4.

So $m \subset \Delta + V_f$ $\forall\, k > 0$, i.e. (r,f) is k-transversal $\forall\, k > 0$ by Lemma 6.4, $(\Delta, V_f \subset m)$. Also $r \geq \dim V_f \geq \dim m/\Delta = c$, follows at once.

Lemma 6.7. If η is k-determinate and if (r,f) and (r,g) are k-transversal unfoldings of η, then they are isomorphic.

Proof. (r,f) is k-transversal $\Rightarrow m = \Delta + V_f + m^{k+1}$ (Lemma 6.4) $\left.\begin{array}{l}\end{array}\right\}$ So $m = \Delta + V_f$.
η is k-determinate $\Rightarrow m^{k+1} \subset m\Delta \subset \Delta$ (Theorem 2.9) (6.8)

Let $\overline{\partial_j f}$ denote the image of $\partial_j f$ in m/Δ. Then (r,f) k-transversal means $\overline{\partial_j f}$ spans m/Δ. (r,f) and (r,g) are isomorphic if $\exists$ a morphism $(\phi, \bar{\phi}, \varepsilon): (r,f) \to (r,g)$ where $\phi, \bar{\phi}$ are diffeomorphisms. We write $f \cong g$.

Lemma 1. It suffices to prove Lemma 6.7 in the special case $\overline{\partial_j f} = \overline{\partial_j g}$ $\forall j$.

Proof. We introduce a standard unfolding (r,h) and show that $\exists\, h' \cong h$ such that $\overline{\partial_j h'} = \overline{\partial_j f}$, $j = 1, \ldots, r$. By symmetry $\exists$ also $h'' \cong h$ such that $\overline{\partial_j h''} = \overline{\partial_j g}$, $1 \leq j \leq r$. Assuming the special case of Lemma 6.7, $f \cong h' \cong h \cong h'' \cong g$.

Choose $u_1, \ldots, u_c \in m$ such that $\bar{u}_1, \ldots, \bar{u}_c$ form a base for m/Δ, where $c = \text{cod } \eta$, finite since $\det \eta$ is finite. Define $h: \mathbb{R}^n \times \mathbb{R}^c \times \mathbb{R}^{r-c} \times \mathbb{R}$

$$(x,v,w) \to \eta(x) + \sum_{j=1}^{c} v_j u_j(x) = \eta + vu,$$

where $v = (v_1 \ldots v_c)$, $u = \begin{pmatrix} u_1 \\ \vdots \\ u_c \end{pmatrix}$. $(w_1, \ldots, w_{r-c}$ are disconnected control coordinates, see below.)

Now $\overline{\partial_j f} = \sum_{h=1}^{r} a_{jh} \bar{u}_h$, $a_{jh} \in \mathbb{R}$. Denote the matrix (a_{jh}) by A. A has rank c since $\overline{\partial_j f}$ span m/Δ. Choose a matrix B such that AB is nonsingular, where AB is,

$$r \begin{array}{|c|c|} \hline c & r-c \\ \hline A & B \\ \hline \end{array} \quad \text{(not the matrix product)}$$

Define $\bar{\phi}: \mathbb{R}^r \to \mathbb{R}^c \times \mathbb{R}^{r-c}$, a linear isomorphism

$$y \mapsto (yA, yB).$$

This induces $h': \mathbb{R}^{n+r} \xrightarrow{\phi = 1 \times \bar{\phi}} \mathbb{R}^{n+r} \xrightarrow{h} \mathbb{R}$. $(1 \times \bar{\phi}, \bar{\phi}, 0): (r, h') \to (r, h)$ is

$$(x,y) \to (x, yA, yB) \mapsto \eta(x) + yAu$$

clearly an isomorphism, $\partial_j h = \begin{cases} u_j(x) & j \leq c \\ 0 & j > c \end{cases}$, $\partial_j h' = \sum_{h=1}^{r} a_{jh} u_h(x)$. So $\overline{\partial_j h'} = \sum_{h=1}^{r} a_{jh} \bar{u}_h(x) = \overline{\partial_j f}$.

Lemma 2. $m_s E_{n+s} = $ those germs in E_{n+s} vanishing on the $\mathbb{R}^n$-axis.

Proof. $\subseteq$: m_s is generated by $\{y_j\}$ which vanish on the $\mathbb{R}^n$-axis, where $x_1, \ldots, x_n$ are coordinates for $\mathbb{R}^n$ and $y_1, \ldots, y_s$ are coordinates for $\mathbb{R}^s$.

$\supseteq$: Suppose the function $\theta(x,y)$ vanishes on the $\mathbb{R}^n$-axis.

$$\theta(x,y) = [\theta(x,ty)]_0^1 = \int_0^1 \frac{\partial \theta}{\partial t}(x,ty)dt = \int_0^1 \sum_j \frac{\partial \theta}{\partial y_j}(x,ty) y_j dt$$

$$= \sum_j y_j \psi_j(x,y), \quad \psi_j \in E_{n+s}.$$

The continuing proof of Lemma 6.7 now mimics the first half of Theorem 2.9. Let $E^t = (1-t)f + tg$. Then assuming $\overline{\partial_j f} = \overline{\partial_j g}$,

$\overline{\partial_j E^t} = (1-t)\overline{\partial_j f} + t\overline{\partial_j g} = \overline{\partial_j f}$. So E^t is k-transversal. For $0 \leq t \leq 1$ we have a 1-parameter family of k-transversal unfoldings connecting f and g. Fix t_o, $0 \leq t_o \leq 1$.

<u>Lemma 3.</u> $\exists$ an isomorphism $(\phi^t, \bar{\phi}^t, \epsilon^t): (r, E^{t_o}) \to (r, E^t)$, $\forall t$ in some neighborhood of t_o.

This implies Lemma 6.7 by the compactness and connectedness of [0,1] (Cf. 2.9).

<u>Lemma 4.</u> $\exists$ a germ ϕ at $(0, t_o)$ of a map $R^{n+r} \times R$, $0 \times R \to R^{n+r}$, 0.
" " " $\bar{\phi}$ " " " " " $R^r \times R$, $0 \times R \to R^r$, 0.
" " " ϵ " " " " " function $R^r \times R$, $0 \times R \to R, 0$, such that

(1) $\phi^{t_o} = 1$ (so $\bar{\phi}^{t_o} = 1$), and $\epsilon^{t_o} = 0$, and $\forall t$ in a neighborhood of t_o,
(2) $\phi^t|R^n \times 0 = 1$; ϕ^t, $\bar{\phi}^t$ commute with $\pi: R^{n+r} \to R^r$, and (3) $E^t \phi^t + \epsilon^t \pi = E^{t_o}$.
(i.e. $E(x', y', t) + \epsilon(y, t) = E(x, y, t_o)$, where $\phi^t(x,y) = (x', y')$.

Lemma 4 = Lemma 3 because the set of diffeomorphisms is op±n in the space of maps. (See proof of 2.9)

<u>Lemma 5.</u> We can replace (3) by

(4) $\sum_i \frac{\partial E}{\partial x_i}(x', y', t) \frac{\partial x_i'}{\partial t}(x, y, t) + \sum_j \frac{\partial E}{\partial y_j'}(x', y', t) \frac{\partial y_j'}{\partial t}(y, t) + \frac{\partial E}{\partial t}(x', y', t) + \frac{\partial \epsilon}{\partial t}(y, t) = 0$.

Differentiation of (3) with respect to t gives (4). Integration with respect to t from t_o to t of (4) gives (3). (See 2.9)

<u>Lemma 6.</u> $\exists$ a germ X at $(0, t_o)$ of a map $R^{n+r} \times R$, $R^n \times 0 \times R \to R^n$, 0,
" " " Y " " " " " $R^r \times R$, $0 \times R \to R^r$, 0,
" " " Z " " " " " function $R^r \times R$, $0 \times R \to R, 0$ such that

(5) $\sum_i \frac{\partial E}{\partial x_i}(x, y, t) X_i(x, y, t) + \sum_j \frac{\partial E}{\partial y_j}(x, y, t) Y_j(y, t) + \frac{\partial E}{\partial t}(x, y, t) + Z(y, t) = 0$, $\forall$
(x, y, t) in a neighborhood of $(0, t_o)$. $(\frac{\partial E}{\partial x} \cdot X + \frac{\partial E}{\partial y} \cdot Y + \frac{\partial E}{\partial t} + Z = 0)$.

18. Classification of Elementary Catastrophes   543

Proof that Lemma 6 $\Rightarrow$ Lemma 5.

Let $(x',y') = \phi(x,y,t)$ be the unique solution of $\begin{cases} \dot{x}' = X(x',y',t), & x' = x \text{ at } t = t_0 \\ \dot{y}' = Y(y',t), & y' = y \text{ at } t = t_0 \end{cases}$.

Let $y' = \bar{\phi}(y,t)$ "   "   "   "   "

Let $\varepsilon(y,t) = \int_{t_0}^{t} Z(\bar{\phi}(y,\tau),\tau)\, d\tau$. So $\frac{\partial \varepsilon}{\partial t}(y,t) = Z(y',t)$. Substitute x', y', t for variables x, y, t in (5) and get (4).

$\phi^t|\mathbb{R}^n \times 0 = 1$ since $(x', y') = (x, 0)$ is a constant solution of
$\begin{cases} X(\mathbb{R}^n \times 0 \times \mathbb{R}) = 0 = \dot{x}', \\ Y(0 \times \mathbb{R}) = 0 = \dot{y}'. \end{cases}$

We now choose a mixture. Let A be a free E_{r+1}-module on $(r+1)$ variables (finitely generated), each $a = (Y_1,\ldots,Y_r,Z)$, some Y_j, $Z \in E_{r+1}$. Let B be a free E_{n+r+1}-module on n variables, each $b = X = (X_1,\ldots,X_n)$, some $X_i \in E_{n+r+1}$. Let C be E_{n+r+1} (finitely generated).

$\alpha: A \to C$ is given by $\alpha a = \frac{\partial E}{\partial y} \cdot Y + Z$; it is over π^* because it is linear in Y, Z. (π is projection $\mathbb{R}^{n+r+1} \to \mathbb{R}^{r+1}$)

$\beta: B \to C$ is given by $\beta X = \frac{\partial E}{\partial x} \cdot X$. (Recall mixture of Chapter 5).

$$\begin{array}{ccc} & & B \\ & & \downarrow \beta \\ A & \xrightarrow{\alpha} & C \\ \uparrow & & \uparrow \\ E_{r+1} & \xrightarrow{\pi^*} & E_{n+r+1} \end{array}$$

Lemma 7. $C = \alpha A + \beta B + (\pi^* m_{r+1})C$.

Proof that Lemma 7 $\Rightarrow$ Lemma 6. Apply Corollary 5.6 (to the Preparation Theorem) to give $C = \alpha A + \beta B$. Then $m_r C = \alpha(m_r A) + \beta(m_r B)$, where the E_r-module structures on C, A, B are induced by projection onto $\mathbb{R}^r$.

Now $\frac{\partial E}{\partial t} = g - f$. And $f|\mathbb{R}^n \times 0 = \eta = g|\mathbb{R}^n \times 0$ ($\forall t$). So $\frac{\partial E}{\partial t}$ vanishes on $\mathbb{R}^n \times 0 \times \mathbb{R}$ in $\mathbb{R}^{n+r+1}$. By Lemma 2 $\frac{\partial E}{\partial t} \in m_r C$, and so $\frac{\partial E}{\partial t} \in \alpha(m_r A) + \beta(m_r B)$, i.e. $\exists$ germs $X \in m_r B$, Y and $Z \in m_r A$ such that $-\frac{\partial E}{\partial t} = \frac{\partial E}{\partial x} \cdot X + \frac{\partial E}{\partial y} \cdot Y + Z$, as germs. Lemma 6 follows applying Lemma 2 a few times.

Proof of Lemma 7. (And hence of Lemma 6.7) As E^t is k-transversal $\forall t$, by (6.8) $m_n = \Delta + V_{E^{t_0}}$. So $E_n = \Delta + V_{E^{t_0}} + \mathbb{R}$. Let $\xi \in C$, and $\xi(x) = \xi(x, 0, t_0) \in E_n$. Then $\xi(x) = \sum_i \frac{\partial \eta}{\partial x_i} \cdot \underline{X_i} + \sum_j \partial_j E^{t_0} \cdot \underline{Y_j} + s$, where $\underline{X_i} \in E_n$, $\underline{Y_j} \in \mathbb{R}$ and $s \in \mathbb{R}$.

Let $\zeta(x,y,t) = \sum_i \frac{\partial E}{\partial y_i}(x,y,t)X_i(x,y,t) + \sum_j \frac{\partial E}{\partial y_j}(x,y,t)Y_j(y,t) -$

$\underbrace{\sum_j \frac{\partial E}{\partial y_j}(0,0,t_o)Y_j(0,t_o) + s}_{Z \in R}$. So $\zeta = \frac{\partial E}{\partial x}\cdot X + \frac{\partial E}{\partial y}\cdot Y + Z$

$\in \beta B + \alpha A$.

Now $\zeta(x,0,t_o) = \xi(x)$ because $E^{t_o}|R^n \times 0 = \eta$ and also

$\partial_j E^{t_o} = \frac{\partial E^{t_o}}{\partial y_j}|R^n \times 0 = \frac{\partial E^{t_o}}{\partial y_j}(0)$. So $\xi - \zeta$ vanishes on the fibre $R^n \times 0 \times t_o$.

By Lemma 2 $\xi - \zeta \in (\pi^* m_{r+1})C$. Hence $\xi \in \alpha A + \beta B + (\pi^* m_{r+1})C$, proving Lemma 7.

Given an unfolding of η, (r,f), $f: R^{n+r}, 0 \to R, 0$, we introduce d disconnected controls as follows. Let g be the composition,

$R^{n+r+d} = R^n \times R^r \times R^d \to R^{n+r} \to R$

$(x,y,w) \mapsto (x,y) \to f(x,y) = g(x,y,w)$.

We say $(r+d,g)$ is (r,f) with d <u>disconnected controls</u>. Using the morphisms $(1 \times \pi,\pi,0): (r+d,g) \to (r,f)$ and $(1 \times \iota,\iota,0): (r,f) \to (r+d,g)$, where ι is the injection map, we see that (r,f) is universal $\Leftrightarrow$ $(r+d,g)$ is universal. Clearly also if (r,f) is k-transversal so is $(r+d,g)$.

<u>Theorem 6.9</u>. If η has finite determinacy, and has (r,f) and (r,g) as universal unfoldings, then they are isomorphic.

<u>Proof</u>. By Lemma 6.6, (r,f) and (r,g) are both k-transversal, $\forall k > 0$. Choose some k such that η is k-determinate. Then Lemma 6.7 provides an isomorphism.

<u>Theorem 6.10</u>. If η is k-determinate, then an unfolding (r,f) is universal $\Leftrightarrow$ it is k-transversal.

<u>Proof</u>. $\Leftarrow$ is Lemma 6.6.

Given a k-transversal unfolding (r,f) we must show that for any unfolding (s,g) (also of η), $\exists$ a morphism $(s,g) \to (r,f)$. If $c = \text{cod } \eta$, choose $u_1, \ldots, u_c$ spanning m/Δ as in Corollary 6.5. Let h be the map $R^n \times R^{s+c} \to R$

$(x,y,v) \mapsto g(x,y) + \sum_{j=1}^{c} v_j u_j(x)$

so that $(s+c,h)$ is a k-transversal unfolding of η by Corollary 6.5.

Let $s + c + d = r + d'$, i.e. choose such integers d, d' (one can be zero). Let $(s+c+d,h')$ be $(s+c,h)$ with d disconnected controls, and $(r+d',f')$ be (r,f) with d' disconnected controls. Both will be k-transversal (as noted above), and we can apply Lemma 6.7 to show the existence of an isomorphism $(\phi, \bar\phi, \varepsilon)$. We now have, $(s,g) \xrightarrow{1 \times j_1, j_1, 0} (s+c, h) \xrightarrow{1 \times j_2, j_2, 0} (s+c+d, h') \xrightarrow{\phi, \bar\phi, \varepsilon} (r+d', f') \xrightarrow{1 \times \pi_r, \pi_r, 0} (r, f)$, with j_1, j_2 obvious injections, π_r a projection. This is the required morphism.

<u>Theorem 6.11</u>. If η has finite determinacy, it has a universal unfolding (c,f) where $c = \text{cod } \eta$, and moreover c is the minimum dimension of any universal unfolding of η.

<u>Proof</u>. By Corollary 6.5 a k-transversal unfolding (c,f) exists with $k \geq \det \eta$. (c,f) is universal by Theorem 6.10. Now use Lemma 6.6. for minimality.

CHAPTER 7. CATASTROPHE GERMS.

Let $\eta \in m^2$, and suppose η has an unfolding $f: \mathbb{R}^{n+r}, 0 \to \mathbb{R}, 0$. Represent f by a function $\tilde{f}: \mathbb{R}^{n+r}, 0 \to \mathbb{R}, 0$ and define M_f to be the subset of $\mathbb{R}^{n+r}$ on which $\frac{\partial \tilde{f}}{\partial x_1} = \ldots = \frac{\partial \tilde{f}}{\partial x_n} = 0$. Let the function $\tilde{X}_f$ be the composition $M_f \subset \mathbb{R}^{n+r} \xrightarrow{\pi_r} \mathbb{R}^r$. Observe that $0 \in M_f$ because $\eta \in m^2$. So we can define X_f to be the germ at 0 of $\tilde{X}_f$. X_f is called the <u>catastrophe germ</u> of f.

<u>Lemma 7.1</u>. Let $\eta \in m^3$ and cod $\eta = c$. Then $\exists$ a universal unfolding (c,f) such that M_f is diffeomorphic to $\mathbb{R}^c$. Then X_f is a germ at 0 of a map $\mathbb{R}^c, 0 \to \mathbb{R}^c, 0$.

<u>Proof</u>. $\eta \in m^3 \Rightarrow \Delta \subset m^2$. And so when choosing a base $u_1, \ldots, u_c$ for m/Δ, we can demand that $u_j(x) = \begin{cases} x_j & \text{if } j \leq n \\ \text{a monomial of degree} \geq 2, & \text{if } n < j \leq c. \end{cases}$

Let $f(x,y) = \eta(x) + \sum_{j=1}^{c} y_j u_j(x)$; (c,f) is k-transversal $\forall k > 0$, and so is universal using Theorem 6.10 with $k \geq \det \eta$. $\frac{\partial f}{\partial x_i} = \frac{\partial \eta}{\partial x_i} + y_i + \sum_{j=n+1}^{c} y_j \frac{\partial u_j}{\partial x_i} = 0 \equiv M_f$, i.e. M_f is the subset of $\mathbb{R}^{n+c}$ where $y_i = \psi_i(x_1, \ldots, x_n, y_{n+1}, \ldots, y_c)$ $\forall i = 1, \ldots, n$. So ψ is a map $\mathbb{R}^n_x \times \mathbb{R}^{c-n}_y \to \mathbb{R}^n_y$. The graph of such a polynomial map is diffeomorphic to its source, and M_f = graph of $\psi \subset \mathbb{R}^n_x \times \mathbb{R}^c_y = \mathbb{R}^n_x \times \mathbb{R}^{c-n}_y \times \mathbb{R}^n_y$, so $M_f \cong \mathbb{R}^c$.

We remark that M_f is not a manifold in general. E.g. $\eta = x^5$, $f = \frac{x^5}{5} + \frac{ax^3}{3}$, $\frac{\partial f}{\partial x} = x^4 + ax^2$, and for $(x,a) \in \mathbb{R}^2$, M_f looks like:

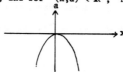

Lemma 7.2. Suppose η has finite determinacy, and $\eta = q + p$, where $q = x_1^2 + \ldots - x_\rho^2$ and p is a polynomial in $x_{\rho+1}, \ldots, x_n$ only, consisting of monomials of degree ≥ 3. Suppose (r,f) is a universal unfolding of p. Then if $g = q + f$, (r,g) is a universal unfolding of η and $X_f = X_g$.

Proof. By Lemma 6.6 (r,f) is k-transversal $\forall k > 0$, and in particular for $k \geq \det p = \det \eta$, Lemma 6.4 gives $m_\lambda = \Delta(p) + V_f + m^{k+1}$ which, with $m_\lambda^{k+1} \subset \Delta(p)$ (Theorem 2.9) gives $m_\lambda = \Delta(p) + V_f$. Here $\lambda = n - \rho$, and m_λ is the ideal of E_λ generated by $x_{\rho+1}, \ldots, x_n$. Similarly m_ρ is the ideal of E_ρ generated by $x_1, \ldots, x_\rho$. m and E denote m_n and E_n. Then $m_\rho E + m_\lambda E = m_\rho E + \Delta(p)E + V_f$.

Now $m = m_\rho E + m_\lambda E$ and $V_f = V_g$. Also $\Delta(\eta) = (x_1, \ldots, x_\rho, \frac{\partial f}{\partial x_{\rho+1}}, \ldots, \frac{\partial f}{\partial x_n})$
$= m_\rho E + \Delta(p)E$.

So $m = \Delta(\eta) + V_g = \Delta(\eta) + V_g + m^{k+1}$ for $k \geq \det \eta$ and so by Lemma 6.4 and Theorem 6.10, (r,g) is universal.

If $i \leq \rho$, $\frac{\partial g}{\partial x_i} = 2x_i$ ($= 0$ for M_g)
If $i > \rho$, $\frac{\partial g}{\partial x_i} = \frac{\partial f}{\partial x_i}$ ($= 0$ for M_g) $\Bigg\} \Rightarrow M_g = 0 \times M_f$.

We have $X_f: M_f \subset 0 \times \mathbb{R}^{r+\lambda} \xrightarrow{\pi_r} \mathbb{R}^r$

$X_g: M_g \subset \mathbb{R}^\rho \times \mathbb{R}^{r+\lambda} \xrightarrow{\pi_r} \mathbb{R}^r$ $\therefore X_f = X_g$.

<u>Lemma 7.3.</u> Suppose (r,f) and (s,g) are 2 unfoldings of η, and $\exists$ a morphism $(\phi, \bar{\phi}, \varepsilon): (s,g) \to (r,f)$. Then $M_g = \phi^{-1} M_f$, and X_g is the pullback of X_f under $\phi, \bar{\phi}$.

<u>Proof.</u> We have

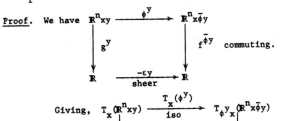

commuting.

Giving,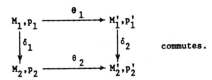

$\phi^0 = 1$, so ϕ^y is a diffeomorphism for small y, and $T_x(\phi^y)$ is an isomorphism for small y. $(x,y) \in M_g \Leftrightarrow T_x(g^y) = 0$ (definition of M_g)

$\Leftrightarrow T_{\phi^y x}(f^{\bar{\phi}y}) = 0$ (diagram commutes)

$\Leftrightarrow (\phi^y x, \bar{\phi} y) \in M_f$ (definition of M_f)

$\Leftrightarrow \phi(x,y) \in M_f$, i.e. $M_g = \phi^{-1} M_f$.

We have that

$$\begin{array}{ccc} \phi^{-1} M_f & \xrightarrow{\phi} & M_f \\ \downarrow X_g & & \downarrow X_f \\ \mathbb{R}^s & \xrightarrow{\bar{\phi}} & \mathbb{R}^r \end{array}$$

commutes, completing the lemma.

Recall that if θ_i is a germ $M_i, p_i \to M_i', p_i'$ where M_i, M_i' are C^∞ manifolds, $i = 1, 2$, then $\theta_1 \sim \theta_2 \Leftrightarrow \exists$ diffeomorphism-germs δ_1, δ_2 such that

$$\begin{array}{ccc} M_1, p_1 & \xrightarrow{\theta_1} & M_1', p_1' \\ \downarrow \delta_1 & & \downarrow \delta_2 \\ M_2, p_2 & \xrightarrow{\theta_2} & M_2', p_2' \end{array}$$

commutes.

Corollary 7.4. If $(\phi,\bar{\phi},\epsilon)$ is an isomorphism, $X_g \sim X_f$.

Proof. $\phi, \bar{\phi}$ will be diffeomorphism-germs; the requisite diagram is at the end of Lemma 7.3.

Lemma 7.5. If (r,g) and (r,f) are universal unfoldings of an η, of finite determinacy, then $X_f \sim X_g$.

Proof. This follows from Theorem 6.9 and Corollary 7.4.

Lemma 7.6. If η has finite determinacy and (s,g), (r,f) are universal unfoldings of η with $s > r$, then $X_g \sim X_f \times 1^{s-r}$.

Proof. Let (s,f') be (r,f) with $s-r$ disconnected controls. Then (s,f') is universal, so that $X_{f'} \sim X_g$ by Lemma 7.5. Also $M_{f'} = M_f \times R^{s-r}$,

$$\begin{array}{ccc} M_{f'} & = & M_f \times R^{s-r} \\ \downarrow X_{f'} & & \downarrow X_f \quad \downarrow 1^{s-r} \\ R^s & = & R^r \times R^{s-r} \end{array}$$

i.e. $X_{f'} = X_f \times 1^{s-r}$.

Lemma 7.7. If η has finite determinacy and is right equivalent to η', and if (r,f) and (r,f') are respective universal unfoldings, then $X_f \sim X_{f'}$.

Proof. We have $\eta' = \eta\gamma$ where $\gamma \in G$. Let $g = f(\gamma \times 1)$:
$$R^{n+r} \xrightarrow{\gamma \times 1} R^{n+r} \xrightarrow{f} R$$
$$\downarrow \pi_r \qquad \downarrow \pi_r$$
$$R^r \xrightarrow{1} R^r$$

This induces
$$\begin{array}{ccc} M_g & \xrightarrow{*} & M_f \\ \downarrow X_g & & \downarrow X_f \\ R^r & \longrightarrow & R^r \end{array}$$
$*$ is a diffeomorphism because γ is. And so $X_f \sim X_g$.

Now $g|R^n \times 0 = f\gamma|R^n \times 0 = \eta\gamma|R^n \times 0 = \eta'|R^n \times 0$. So (r,g) unfolds η', and (r,g) is a universal unfolding because (r,f) is, clearly. By Lemma 7.5, $X_g \sim X_{f'}$. Hence $X_f \sim X_{f'}$.

Theorem 7.8. If $\eta \in m^2$ of finite determinacy has a catastrophe germ X_f, then the equivalence class of X_f depends only upon the equivalence class of η. Moreover it is uniquely determined by the essential coordinates of η.

Proof. Denote the equivalence class of X_f by $[X_f]$. $[X_f]$ is independent of the choices of: n by Lemma 7.2, universal unfolding f by Lemma 7.5, r by Lemma 7.6, and of η by Lemma 7.7. Lemma 7.2 shows that $[X_f]$ is uniquely determined by the essential coordinates (of η).

Corollary 7.9. $\exists$ only 11 catastrophe germs if we restrict to those η of codimension ≤ 5.

Proof. If there are more than 2 essential coordinates of η, i.e. rank $\eta \leq n - 3$, then Lemma 4.11 shows $\text{cod } \eta > 5$. So restrict to $n \leq 2$. $+\eta$ and $-\eta$ give the same M_f and hence the same X_f. So the (distinct) essential coordinates giving distinct $[X_f]$'s are: x^3, x^4, x^5, x^6, x^7, $x^3 + xy^2$, $x^3 - xy^2$, $x^2y + y^4$, $x^3 + y^4$, $x^2y + y^5$, $x^2y - y^5$. These are the 11.

Definition. If $[X_f]$ is one of the 11 of Corollary 7.9 then $[X_f]$ is called an **elementary catastrophe**.

Corollary 7.10. If η has finite determinacy and (r,f) is a universal unfolding of η, where $r \leq 5$, then $[X_f]$ is an elementary catastrophe.

Proof. By Corollary 4.7 and the Reduction Lemma 4.9, $\eta \sim q + p$ and $p \in m^3$. Also Lemma 6.6 tells us that $r \geq c = \text{cod } \eta$, so that $c \leq 5$ and p is one of the germs written out in the proof of Corollary 7.9 (cod $p \leq 5$ and consult Diagram 4.1). By Lemma 7.1 applied to p $\exists$ a standard universal unfolding (c,g) of p such that X_g is a germ $\mathbb{R}^c, 0 \to \mathbb{R}^c, 0$. Now use Lemma 7.2 to provide a universal unfolding (c, f') of η such that $X_{f'} = X_g$. By Lemma 7.6 $X_f \sim X_{f'} \times 1^{r-c} = X_g \times 1^{r-c} : \mathbb{R}^r, 0 \to \mathbb{R}^r, 0$. Now $[X_g]$ is an elementary catastrophe by choice, and so in a certain obvious sense $[X_f]$ is an elementary catastrophe too. This is the same sense in which we said that "$[X_f]$ is independent of the choice of r by Lemma 7.6" in Theorem 7.8.

CHAPTER 8. GLOBALISATION.

We shall first define the Whitney C^∞ topology on the space of C^∞ functions $\mathbb{R}^{n+r} \to \mathbb{R}$, denoted by F.

Given $f: \mathbb{R}^{n+r} \to \mathbb{R}$ define a map $f^k: \mathbb{R}^{n+r} \to J^k_{n+r}$ (where, recall, $J^k_n = E_n/m_n^{k+1}$) which sends $p \in \mathbb{R}^{n+r}$ to the k-jet at 0 of the function $\mathbb{R}^{n+r} \to \mathbb{R}$

$w \mapsto f(p+w)$.

Then given a function $\mu: \mathbb{R}^{n+r} \to \mathbb{R}_+$ we define a basic neighborhood of 0 as $V^k_\mu = \{f \in F: \forall\, p \in \mathbb{R}^{n+r}, |f^k p| < \mu p\}$. For $f \in F$, $V^k_\mu(f) = \{g \in F: \forall\, p \in \mathbb{R}^{n+r}, |f^k p - g^k p| < \mu p\}$ is a basic open neighborhood of f. These form a base for a topology, called the Whitney C^k-topology. The topology with a base of all such $V^k_\mu(f)$, $\forall\, k \geq 0$, is called the <u>Whitney C^∞ topology</u>. F will be assumed to have this topology.

<u>Theorem 8.1.</u> If $r \leq 5$, then $\exists$ an open dense set $F_* \subset F$ such that if $f \in F_*$, then $\tilde{X}_f$ has only elementary catastrophes as singularities (and these are already classified), and M_f is an r-manifold.

We shall need several lemmas to prove the theorem.

Given $f \in F$, $\varepsilon > 0$, and $X \subset \mathbb{R}^{n+r}$, define an open set $V^k_{\varepsilon,X}(f)$ as $\{g \in F: \forall\, p \in X, |f^k p - g^k p| < \varepsilon\}$, so that ε controls all partial derivatives of order $\leq k$ on X. It is open because it is the union of all $V^k_\mu(f)$ for $\mu: \mathbb{R}^{n+r}, X \to \mathbb{R}_+, (0,\varepsilon)$.

<u>Definition.</u> Let J be a manifold. A <u>stratification</u> Q of J is a decomposition into a finite number of submanifolds $\{Q_i\}$ such that,

(1) $\partial Q_i = \bar{Q}_i - Q_i$ = the union of Q_j of lower dimension.

(2) If $z \in Q_j \subset \partial Q_i$ and a submanifold S of J is transverse to Q_j at z, then S is transverse to Q_i in a neighborhood of z. (8.2)

Following the construction of the k-jet prolongation of an unfolding (r,f) in Chapter 6, given $f \in F$ we let F be the induced map

$$\mathbf{R}^{n+r} \to J^k$$
$p = (x,y) \mapsto$ k-jet at 0 of the function $\mathbf{R}^n, 0 \to \mathbf{R}, 0$
$$x' \mapsto f(x+x',y) - f(x,y).$$

Given $X \subset \mathbf{R}^{n+r}$ we let $F^X = \{f \in F : \forall\, p \in X,\ F$ is transversal to Q at $p\}$, where Q is either a submanifold or a stratification of J^k.

<u>Open Lemma 1.</u> (OL1) If $X \subset \mathbf{R}^{n+r}$ is compact and $f \in F^X$, then $\exists$ a neighborhood $V^{k+1}_{\varepsilon,X}(f) \subset F^X$. (i.e. F^X is C^{k+1}-open.)

<u>Proof.</u> Given $p \in X$, F is transversal to Q at p. By continuity and (8.2) (if appropriate), F is transversal to Q in a neighborhood of p, in particular in a compact neighborhood N of p. This remains true for all sufficiently small changes of F and TF on N, and so for all sufficiently small changes in f^{k+1} on N. Because N is compact, $\exists\, \varepsilon > 0$ such that $V^{k+1}_{\varepsilon,N}(f) \subset F^N$. Cover compact X by a finite number of such N_i, and let $\varepsilon = \min \varepsilon_i$. Then
$$V^{k+1}_{\varepsilon,X}(f) = \bigcap_i V^{k+1}_{\varepsilon,N_i}(f) \qquad (\cup_i N_i = X)$$
$$\subset \bigcap_i V^{k+1}_{\varepsilon_i,N_i}(f) \qquad \text{relaxing controls}$$
$$\subset \bigcap_i F^{N_i} = F^X.$$

<u>Open Lemma 2.</u> (OL2) Let $X = \cup X_i$, a countable union of disjoint compact X_i with neighborhoods Y_i. Then F^X is C^{k+1}-open.

<u>Proof.</u> Choose a C^∞ bump function $\beta_i : \mathbf{R}^{n+r} \to [0,1]$, which takes values 1 on X_i and 0 outside Y_i, for each i. Let $\beta_0 = 1 - \sum_{i=1}^\infty \beta_i$. Given $f \in F^X$, then $f \in F^{X_i}$. So $\exists\, \varepsilon_i > 0$ such that $V^{k+1}_{\varepsilon_i,X_i}(f) \subset F^{X_i}$ \hfill (OL1)

Let $\mu = \beta_0 + \sum_{i=1}^\infty \varepsilon_i \beta_i$. Then $V^{k+1}_\mu(f) \subset \bigcap_{i=1}^\infty V^{k+1}_{\varepsilon_i,X_i}(f) \qquad (\mu = \varepsilon_i$ on $X_i)$
$$\subset \bigcap F^{X_i} = F^X.$$

<u>Density Lemma 3.</u> (DL3) $\forall\, p \in \mathbf{R}^{n+r}$ and $\forall\, f \in F$, $\exists$ a compact neighborhood N of p in $\mathbf{R}^{n+r}$ and $\exists$ a neighborhood V of $f \in F$ such that F^N is C^∞-dense in V.

<u>Proof.</u> Having chosen N and V we must show that $\forall\, g \in V$, $\exists$ an arbitrarily

C^∞-close $h \in F^N$. Now $F^N = \{f \in F: F$ is transversal to Q in $N\}$, where Q is (first) a submanifold of J^k. Given f let $z = F(0,0)$ and w.l.o.g. $p = (0,0)$.

Case 1. $z \in Q$. This is hard.

Case 2. $z \in \overline{Q} - Q$. This does not occur if Q is closed, but we need this case where Q is one stratum of a stratification.

Case 3. $z \notin \overline{Q}$. This is trivial.

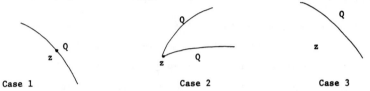

Case 1   Case 2   Case 3

Case 3. Pick N such that $FN \pitchfork Q$, and V such that $\forall g \in V$, $GN \pitchfork Q$. Then $g \in F^N$, trivially. So $V \subset F^N$, and $h = g$ will do.

Case 1. Let q be the codimension of Q in J^k. Choose a product neighborhood B of z in J^k and a projection $\theta: B \to \mathbf{R}^q$ such that $\theta^{-1}0 = B \cap Q$. Now J^k is spanned by monomials in $x_1, \ldots, x_n$. Of these choose $u_1, \ldots, u_q$ spanning the q-plane transverse to Q at z. Let e_w be the function $\mathbf{R}^n \to \mathbf{R}$

$$x \mapsto \sum_{i=1}^{q} w_i u_i(x), \text{ where } w_i \in \mathbf{R} \text{ form } w \in \mathbf{R}^q,$$

and so $e: \mathbf{R}^q \times \mathbf{R}^n \to \mathbf{R}$. As usual e induces $E: \mathbf{R}^q \times \mathbf{R}^n, 0 \to J^k, 0$

$(w,x) \mapsto$ k-jet of the function $\mathbf{R}^n, 0 \to \mathbf{R}, 0$. Then $(F+E): \mathbf{R}^q \times \mathbf{R}^{n+r}, 0 \to J^k, z$

$x' \mapsto e_w(x+x') - e_w(x)$. $(w,x,y) \mapsto F(x,y) + E(w,x)$,

is convenient notation. Now choose a compact neighborhood $W \times N$ of 0 in $\mathbf{R}^q \times \mathbf{R}^{n+r}$ such that $(F+E)(W \times N) \subset B$.

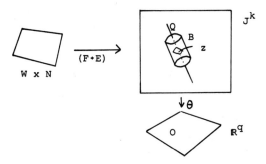

Choose a neighborhood V of f in F such that $\forall g \in V$, $(G+E)(W \times N) \subset B$. This is possible because $W \times N$ is compact and B is open.

Sublemma 1. The matrix of partial derivatives with respect to W at 0 of the composite map $W \times N, 0 \xrightarrow{(F+E)} B, z \to \mathbb{R}^q, 0$ is a nonsingular matrix.

Proof. $(F+E)(w,0,0) = F(0) + E(w,0) = z + E(w,0)$.
$E(w,0)$ is the k-jet at 0 of $\mathbb{R}^n \to \mathbb{R}$
$$x' \to e_w(x') - e_w(0) = \sum_{i=1}^{q} w_i u_i(x').$$
So $(F+E)(w,0,0) = z + \sum_{i=1}^{q} w_i u_i$, which is in the q-plane transverse to Q at z by construction. Hence $\theta(F+E)$ is transversal to 0 in $\mathbb{R}^q$.

Corollary. By choosing W, N, V sufficiently small, the matrix of partial derivatives of the composition map $\phi: W \times N \xrightarrow{(G+E)} B \to \mathbb{R}^q$ with respect to W is nonsingular at (w,p) $\forall (w,p) \in W \times N$, $\forall g \in V$.

Proof. By continuity from Sublemma 1.

Sublemma 2. (Implicit Function Theorem) Given $W^q \times N^{n+r} \to \mathbb{R}^q$ with the matrix of partial derivatives of ϕ with respect to W nonsingular $\forall (w,p) \in W \times N$, then $\exists$ a unique C^∞ map $\psi: N^{n+r} \to W^q$ such that $\phi^{-1} 0 = \text{graph } \psi$.

By Sard's Theorem choose a regular value w^* of ψ, arbitrarily small. Let ϕ^* be the map: $N^{n+r} \to \mathbb{R}^q$
$$p \mapsto \phi(w^*, p).$$

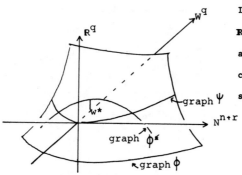

If N = east, W = north, R^q = vertical, ϕ = height above sea level, and ψ = coastline, then Sublemma 2 says that '∃ a coastline.

<u>Sublemma 3.</u> $\phi*$ is transversal to 0.

<u>Proof</u>. Suppose $\phi*p = 0$. Let $v = (w*,p)$, $\in$ graph $\psi \subset W \times N$ as $\phi v = 0$.
Consider $T_v(W\ N) \xrightarrow{T_v\phi} T_0 R^q$. $T_v\phi$ is surjective by the Corollary to Sublemma 1.
$R_w^q \times R^{n+r}$ R^q

Let K be the kernel of $T_v\phi$, $K = (T_v\phi)^{-1}0$. Dim $K = (q+n+r) - q = n+r$, by surjectivity.

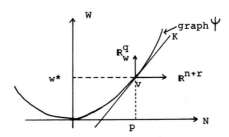

Because $w*$ is a regular value of ψ, the map $K^{n+r} \subseteq R_w^q \times R^{n+r} \xrightarrow{\pi_q} R_w^q$ is surjective. So K^{n+r} meets R^{n+r} transversely; dim $K^{n+r} \cap R^{n+r} = (n+r) + (n+r) - (n+r+q) = n + r - q$. Consider $T_p(N^{n+r}) \xrightarrow{T_p\phi*} T_0(R^q)$.
R^{n+r} R^q

Kernel of $T_p\phi*$ = kernel of $T_v\phi \cap R^{n+r} = K^{n+r} \cap R^{n+r}$, and so is of dimension $n + r - q$. Hence $T_p\phi*$ is surjective, and p is a regular point of $\phi*$. Thus $\phi*$ is transversal to 0.

We have chosen N and V. Choose now a bump function $\beta: \mathbf{R}^{n+r} \to [0,1]$ such that $\beta = 1$ on N, $\beta = 0$ outside a compact neighborhood of N. Given $g \in V$, choose w^* (dependent upon g), a regular value of ψ, w^* arbitrarily small. Define $h: \mathbf{R}^{n+r} \to \mathbf{R}$ by $h(x,y) = g(x,y) + \sum_{i=1}^{q} w_i^* u_i(x) \beta(x,y)$. Then by Sublemma 3, $\theta H = \phi^*$ is transversal to 0 on N. So H is transversal to Q, and $h \in F^N$. Given an arbitrary C^∞-neighborhood $V_\mu^\ell(g)$, we can reduce the partial derivatives of $w^* u \beta$ of order $\leq \ell$, below μ, on a compact neighborhood of N by making w^* sufficiently small. So $h \in V_\mu^\ell(g)$. h is arbitrarily C^∞-close to g.

This completes Case 1 of DL3.

Case 2. $z \in Q \subset \partial Q' = \overline{Q}' - Q'$ where $\{Q\}$ form a stratification. Given $g \in V$ we must show $\exists\, h$ such that H is transversal to both Q and Q' (and any other incident strata) at the same time, on N. Given g, find h as in Case 1 arbitrarily C^∞-close such that H is transversal to Q on N. Automatically by (8.2) H is transversal to Q' at all points in a compact neighborhood L of z in B.

Choose a product neighborhood B' of $Q' \cap (B-L)$, and a map $\theta': B' \to \mathbf{R}^{q'}$ such that $\theta'^{-1} 0 = Q' \cap (B-L)$, where q' is the codimension of Q' in J^k. Find now h' arbitrarily C^∞-close to h so that

(a) $\theta H'$ remains transversal to all points of θL, and

(b) $\theta' H'$ becomes transversal to 0 in $\mathbf{R}^{q'}$ by Case 1 for Q'.

Then H' is transversal to Q and Q' on N. By induction, $H^{(s)}$ is transversal to the stratification because there are only a finite number ($s+1$, say) of strata through z, by (8.2). Then $h^{(s)} \in F^N$ is arbitrarily C^∞-close to $g \in V$.

Density Lemma 4. (DL4) If $X \subset \mathbf{R}^{n+r}$ is compact, then F^X is C^∞-dense in F.

Proof. Given $f \in F$, cover X by a finite number of N_i given by DL3. Let $V = \cap_i V_i$. Then F^{N_i} is C^∞-open by OL1 (because C^{k+1}-open) and is C^∞-dense (by DL3) in V_i. So F^{N_i} is C^∞ open dense in V. Now $F^X \supset F^{\cup N_i} = \cap F^{N_i}$ is C^∞

open dense in V. So F^X is dense in V, i.e. $\forall\, f \in F,\ \exists\, V$ such that F^X is dense in V.

Therefore F^X is dense.

Density Lemma 5. (DL5) Let $X = \cup\, X_i$ as in OL2, then F^X is C^∞-dense.

Proof. Given $f \in F$ and given a basic C^∞-neighbourhood $V^\ell_\mu(f)$, we want $g \in V^\ell_\mu(f) \cap F^X$. Let $\{\beta_i\}$ be as in OL2. For each i choose $\varepsilon_i > 0$ such that $h \in V^\ell_{\varepsilon_i, Y_i} \Rightarrow \beta_i h \in V^\ell_\mu$. (This is possible by the boundedness of the derivatives of order $\leq \ell$ of β_i on Y_i). By DL4, choose $f_i \in V^\ell_{\varepsilon_i, Y_i}(f) \cap F^{X_i}$. Define $g = \beta_0 f + \sum_{i=1}^{\infty} \beta_i f_i$. Then $g = f$ outside $\cup\, Y_i$. On Y_i, $g = (1-\beta_i)f + \beta_i f_i = f + \beta_i(f_i - f)$. Now $f_i - f \in V^\ell_{\varepsilon_i, Y_i}$ by choice, so $\beta_i(f_i - f) \in V^\ell_\mu$. Meanwhile $g = f_i$ on X_i. But F_i is transversal to Q on X_i, and so G is also transversal to Q on X_i. Therefore $g \in \cap\, F^{X_i} = F^X$.

So $g \in V^\ell_\mu(f) \cap F^X$ as required.

The result of DL5 can also be proved by showing that F with the Whitney C^∞ topology is a Baire space, but the proof is longer.

Lemma 6. $F^{R^{n+r}}$ is C^{k+1}-open and C^∞-dense in F.

Proof. Choose $X,\ X'$ each as in OL2 such that $R^{n+r} = X \cup X'$. Then $F^{R^{n+r}} = F^X \cap F^{X'}$, each C^{k+1}-open and C^∞-dense, by OL2 and DL5 respectively.

Proof of Theorem 8.1. We describe the stratification Q of J^7 resulting from the classification of orbits in I^7 in Chapter 4.

(a) the open subspace $J^7 - I^7$,

(b) $n + 1$ orbits of jets of stable germs in m^2 of codimension 0 in I^7,

(c) the orbits of jets of germs in m^2 of codimension 1, 2, 3, 4 and 5 in I^7,

(d) the strata of the algebraic variety of jets of germs in m^2 of codimension ≥ 6 in I^7.

These come directly from Diagram 4.1.

Because Σ^7_6, i.e. (d), is of codimension $n + 6$ and we are not interested in its internal structure, we shall let Q be the stratification (a), (b), (c) of $J^7 - \Sigma_6^7$. The **strata** are the Γ_c^7 for $c = 0, 1, 2, 3, 4$ and 5, together with $J^1 - 0$ (this last making $J^7 - \Sigma_6^7$, rather than $I^7 - \Sigma_6^7$).

Lemma 7. Q satisfies (8.2) (and hence is a stratification).

Let $F_o = \{f \in F : F \text{ misses } \Sigma_6^7\}$, i.e. where F is transversal to Σ_6^7 if $r \leq 5$ (F maps $\mathbb{R}^{n+r}$ into J^7). By general position, F_o is C^0-open (and hence C^8-open) and C^∞-dense. Let $F_* = F_o \cap F^{\mathbb{R}^{n+r}}$, then $F_* \{f \in F : F$ is transversal to Q and $\Sigma_6^7\}$, and is C^8-open and C^∞-dense, using Lemma 6.

Suppose $f \in F_*$. Then F is transversal to $m^2/m^8 = I^7$, since I^7 is the union of strata of Q and Σ_6^7. So $F^{-1}(I^7)$ is of codimension n, and of dimension r. (I^7 is of codimension n in J^7). Now $F^{-1}(I^7)$ is the set of points (x,y) in $\mathbb{R}^{n+r}$ such that the 1-jet of $x' \to f(x+x',y) - f(x,y)$ is zero, i.e. such that $\frac{\partial f}{\partial x_1}(x,y) = \ldots = \frac{\partial f}{\partial x_n}(x,y) = 0$. So $F^{-1}(I^7)$ is precisely M_f and M_f is an r-manifold. Suppose that $\tilde{X}_f : M_f \to \mathbb{R}^r$ has a singularity at (x,y). Let η be the germ at (x,y) of $f|\mathbb{R}^n \times y$. W.l.o.g. $(x,y) = (0,0)$, so $\eta \in m^2$. The germ of f at $(0,0)$ is a 7-transversal unfolding of η, because $f \in F_*$ and so F is transversal to the orbit $(j^7 \eta)G^7$, contained in some stratum.

Lemma 8. If (r,f) is a 7-transversal unfolding of $\eta \in m^2$, and $r \leq 5$, then (r,f) is a universal unfolding.

Proof. By Lemma 6.4, $m = \Delta + V_f + m^8$. ($\Delta = \Delta(\eta)$). So dim $m/(\Delta + m^8) \leq$ dim $V_f \leq r \leq 5$, using (6.3). In the notation of Theorem 3.3, $\tau(j^8 \eta) \leq 5$. But cod $\eta = \tau(j^8 \eta) \leq 5$, by (3.5), and so by Lemma 3.1, det $\eta \leq 7$, and we can apply Theorem 6.10 to show that (r,f) is universal.

By Corollary 7.10 we now know that if X_f is the germ at $(0,0)$ of $\tilde{X}_f$, then $[X_f]$ is an elementary catastrophe.

So the only singularities of $\tilde{X}_f$ are elementary catastrophes.

Proof of Lemma 7. (Which we have used to complete Theorem 8.1). Q has a finite number of strata, each of which is a submanifold by Corollary 4.3. (There are in fact 7 strata.) Condition (1) of (8.2) follows from Corollary 3.6 since each Σ_c^7 is closed (Theorem 3.3). Note that $\bar{\Gamma}_c^7$ now refers to the closure in $J^7 - \Sigma_6^7$.

Condition (2): Let Q_1, Q_2 be strata, $z_1 \in Q_1 \subset \partial Q_2$, and S a submanifold of $J^7 - \Sigma_6^7$ transverse to Q_1 at z_1. Then S is transverse to $z_1 G^7$ at z_1.

Write α for the C^∞ map $J^7 \to C^\infty(G^7, J^7)$. $\alpha(z_1)$ is $z \to$ the map taking γ to $z \circ \gamma$.
now transversal to S in a neighborhood U of the identity e. Spanning, and hence transversality, is an open property, so $\exists$ an open neighborhood V of $\alpha(z_1)$ in $C(G^7, J^7)$ and a neighborhood U_1 of e (perhaps smaller than U) so that $\beta \in V$ implies β is transversal to S in U_1. $\alpha^{-1}(V)$ is open and contains z_1, and if $z \in \alpha^{-1}(V)$, $\alpha(z)$ is transversal to S in U_1; in particular zG^7 is transverse to S at z. But Q_2 is the finite union of such orbits zG^7. Hence S is transverse to Q_2 in $\alpha^{-1}(V)$, a neighborhood of z_1.

Thus condition (2) is satisfied, completing the proof of Lemma 7.

CHAPTER 9. STABILITY.

Given $f \in F_*$, let $X_f: M_f \to \mathbb{R}^r$ be induced by projection. (See Chapter 1) We have to show that X_f is locally stable at all points of M_f.

Definition. X_f is locally stable at $(x_o, y_o) \in M_f$ if given a neighborhood N of (x_o, y_o) in $\mathbb{R}^{n+r}$, $\exists$ a neighborhood V of f in F_*, such that given $g \in V$, $\exists (x_1, y_1)$ in $N \cap M_g$ such that X_f at (x_o, y_o) is locally equivalent to X_g at (x_1, y_1).

Let $\hat{f}$, $X_{\hat{f}}$ denote the germs of f, X_f at (x_o, y_o) and $\hat{g}$, $X_{\hat{g}}$ the germs of g, X_g at (x_1, y_1). Then $X_{\hat{f}}$, $X_{\hat{g}}$ agrees with the notation in Chapter 7, and we also have that

(9.1) $X_{\hat{f}} \sim X_{\hat{g}} \Leftrightarrow X_f$ at (x_o, y_o) is locally equivalent to X_g at (x_1, y_1).

Theorem 9.2. If $r \leq 5$ and $f \in F_*$, then X_f is locally stable at each point of M_f.

Proof. f induces $F: \mathbb{R}^{n+r} \to J^7$ as at the beginning of Chapter 8. Let (x_o, y_o) be in M_f, and $F(x_o, y_o) = z_o$. We suppose we are given a neighborhood N of (x_o, y_o). Since $f \in F_*$, F is transversal to $z_o G^7$ at z_o; hence we can choose a disc D^q with centre (x_o, y_o) contained in N, where q is the codimension of $z_o G^7$ in J^7, whose image under F intersects $z_o G^7$ transversely at z_o, and so that $F|_{D^q}$ is an embedding. $F(D^q)$ will then have intersection number 1 with $z_o G^7$. If F is perturbed slightly to G, $G(D^q)$ will still be a q-disc whose intersection number with $z_o G^7$ is still 1. I.e. $\exists$ an open neighborhood V_o of f in F with this property for $g \in V_o$. Write $V = V_o \cap F_*$. Given $g \in V$, G is transversal to $z_o G^7$ and we may choose $(x_1, y_1) \in D^q$ such that $G(x_1, y_1) = z_1 = G(D^q) \cap z_o G^7$. Then z_1 and z_o are in the same orbit and are right equivalent as germs $\mathbb{R}^n, 0 \to \mathbb{R}, 0$.

Let $f_o(x,y) = f(x_o+x, y_o+y) - f(x_o, y_o)$ and $g_1(x,y) = g(x_1+x, y_1+y) - g(x_1, y_1)$ define f_o and $g_1: \mathbb{R}^{n+r}, 0 \to \mathbb{R}, 0$. Then $z_o = j^7(f_o | \mathbb{R}^n \times 0)$ and $z_1 = j^7(g_1 | \mathbb{R}^n \times 0)$. Note that $F(\mathbb{R}^{n+r})$ is the same point-set as $F_o(\mathbb{R}^{n+r})$ and so F_o is transversal to $z_o G^7$ and $(r, \hat{f}_o)$ is a k-transversal unfolding of the germ z_o: so we can apply Lemma 8 in Chapter 8 (similarly for $\hat{g}_1$). As $r \leq 5$ the proof of this lemma gives that z_o (and so also z_1) is finitely determined as a germ. The result of the same lemma tells us that $\hat{f}_o$ and $\hat{g}_1$ are also universal unfoldings of germs z_o, z_1 respectively. Now apply Lemma 7.7 which says $X_{\hat{f}_o} \sim X_{\hat{g}_1}$ (germs at $(0,0)$ of X_{f_o}, X_{g_1}).

Now M_f is merely a translate of M_{f_o}: $M_f = M_{f_o} + (x_o, y_o)$. And so

$$X_f(x,y) = X_{f_o}(x-x_o, y-y_o) + y_o.$$

Then

$$\begin{array}{ccc} M_f, (x_o, y_o) & \xrightarrow{-(x_o, y_o)} & M_{f_o}, (0,0) \\ \downarrow X_f & & \downarrow X_{f_o} \\ \mathbb{R}^r, y_o & \xrightarrow{-y_o} & \mathbb{R}^r, 0 \end{array}$$

commutes, so that $X_{\hat{f}} \sim X_{\hat{f}_o}$ (by (9.1)).

Similarly $X_{\hat{g}} \sim X_{\hat{g}_1}$.

Hence $X_{\hat{f}} \sim X_{\hat{f}_o} \sim X_{\hat{g}_1} \sim X_{\hat{g}}$. This completes Theorem 9.2.
(Observe that $(x_o, y_o) \in M_f$ and $M_f = F^{-1}(I^7)$ so that z_o and $z_o G^7 \subset I^7$. Then $z_1 \in I^7$ and $(x_1, y_1) \in M_g = G^{-1}(I^7)$, i.e. $(x_1, y_1) \in N \cap M_g$ as required.)

Remark. This is a result about local stability. It would be interesting and useful to have a similar global stability result.

REFERENCES

1. M. Golubitsky & V. Guillemin, Stable Mappings and their singularities, Grad. Texts in Math., 14, Springer Verlag, New York, 1974.

2. H. I. Levine, Singularities of differentiable mappings. <u>Liverpool Symp. on Singularities</u>, (Springer Lecture Notes 192 (1971)), 1-89.

3. B. Malgrange, Ideals of differentiable functions, (Oxford Univ. Press., 1966).

4. J. N. Mather, Stability of C^∞ mappings I: The division theorem, <u>Annals of Math</u>. 87 (1968), 89-104.

5. J. N. Mather, Stability of C^∞ mappings III: Finitely determined map-germs, Publ. Math. I.H.E.S. 35 (1968), 127-156.

6. J. N. Mather, Right Equivalence (Warwick preprint, 1969).

7. J. N. Mather, On Nirenberg's proof of Malgrange's preparation theorem, Liverpool Symp. on Singularities (Springer Lecture Notes 192, (1971)) 116-120.

8. L. Nirenberg, A proof of the Malgrange preparation theorem, <u>Liverpool Symp. on Singularities</u> (Springer Lecture Notes 192 (1971), 97-105.

9. R. Thom, Les singularités des applications différentiables, Ann. Inst. Fourier (Grenoble) (1956), 17-86.

10. R. Thom, <u>Stabilité structurelle et morphogénèse</u>, Benjamin, (1972).

11. H. Whitney, Mappings of the plane into the plane, <u>Annals of Math</u>. 62 (1955), 374-470.

12. E. C. Zeeman, Applications of catastrophe theory, <u>Tokyo Int. Conf. on Manifolds</u>, April 1973.

19 THE UMBILIC BRACELET AND THE DOUBLE-CUSP CATASTROPHE

INTRODUCTION

Our objective is to understand the geometry of the double-cusp catastrophe, or in other words the 8-dimensional unfolding of the germ $f:R^2 \to R$ given by $f = x^4+y^4$. Now 8 dimensions are difficult to visualise and we only partially achieve this objective. So the question arises, why bother with this particular germ? There are several reasons both mathematical and scientific, as follows.

(i) <u>Modality.</u> The double-cusp is the simplest non-simple germ. More precisely any germ in two variables of codimension less than 8 is simple in the sense of Arnold [2,3], but the double-cusp is unimodal. Therefore a study of its geometry will help to give insight into the phenomenon of modality.

(ii) <u>Compactness.</u> The double-cusp is <u>compact</u>, in the sense that the sets $f \leq$ constant are compact. In Arnold's notation [3,4], the double cusp belongs to the family X_9, and in this family there are three real types of germ, according as to whether the germ has 0, 2 or 4 real roots. For example representatives of the three types are x^4+y^4, x^4-y^4 and $x^4+y^4 - 6x^2y^2$, respectively, and only the first of these is compact.

Compact germs play an important role in applications [9], because any perturbation of a compact germ has a minimum; therefore if minima

Published in Structural stability, the theory of catastrophes and applications in the sciences, Springer Lecture Notes in Mathematics, Volume 525, 1976, pages 328-366.

represent the stable equilibria of some system, then for each point of the unfolding space there exists a stable state of the system. By contrast, consider the fold-catastrophe x^3, which is not compact; this tends to be an incomplete model of any system, because at the fold point where the equilibrium breaks down, there is a catastrophic jump, but the model does not tell us where the system will jump to. The way to compactify the fold is to add a term x^4; in other words the fold-catastrophe can be regarded as a section of the cusp-catastrophe, which is compact. In this sense we may call the cusp the compactification of the fold. Similarly the double-cusp is important because it is the compactification of each of the three umbilics, the hyperbolic x^3+y^3, the elliptic x^3-3xy^2, and the parabolic x^2y+y^4.

(iii) Coupling. The commonest catastrophe in applications is the cusp, and in some applications two cusps appear, both depending upon the same parameters. In such cases the double-cusp (or one of its non-compact partners) describes the generic way that the two cusps can be coupled together, or can interfere with one another. A study of the geometry is necessary to give a full understanding of such coupling and interference.

(iv) Applications. Samples of the types of application in which the double-cusp appears are as follows. In economics [8] growth and inflation can each be modelled by a cusp, depending upon the same policy parameters such as devaluation, deflation, etc., and the problem is to see how they are coupled, so that one can be cured without harming the other. In linguistics Thom [11,12] uses a compact unfolding of the parabolic umbilic to model basic sentences, and is therefore implicitly using the double-cusp; the four nouns of a basic sentence are represented by the maximal set of 4 minima appearing in the unfolding. In brain-modelling [18] compact germs in 2 variables may be important because the cortex is a 2-dimensional sheet.

In developmental biology if an umbilic appears in the interior of an embryo then, since the embryo continues to exist, the compactification is implicit, and so there should be an accompanying sequence of catastrophes governed by a section of the double-cusp.

In structural engineering [13] the coalescence of two stable post-buckling modes, each governed by a cusp, can generate a highly unstable compound buckling and associated imperfection-sensitivity, governed by a double-cusp. For example this happens in a model due to Augusti* [5,13 Figure 100], consisting of a loaded vertical strut supported at its pinned end by two rotational springs at right-angles, when the strengths of the springs is allowed to coincide. Here the double-cusp is the non-compact $x^4+y^4-6x^2y^2$, with the boundary of stable equilibria representing the failure locus.

CONTENTS

The paper is divided into three sections :

1. The umbilic bracelet.
2. Catastrophe theory.
3. The double-cusp.

In Section 1 we describe the geometry of the discriminant of the real cubic. In Section 2 we establish a new form for the catastrophe map associated with a germ, and show how its singularities refine the canonical stratification of a jet space, which is independent of the unfolding. The new form yields new equations for the cuspoids and umbilics, which help to give further insight into the relationship between their geometries. In Section 3 we apply the results of the two previous sections to explore the geometry of the double-cusp. Other mathematical references containing information about the double-cusp are [1,7,10,17].

---

*I am indebted to Michael Thompson for drawing my attention to Augusti's example, and to Tim Poston for pointing out that it was a double-cusp.

SECTION 1 : THE UMBILIC BRACELET.

Since the double-cusp is a quartic form, its unfolding involves the umbilics, namely the cubic forms. Therefore we begin by studying the stratification of the space R^4 of real cubic forms in 2 variables. The point $(a,b,c,d) \in R^4$ corresponds to the form

$$f = ax^3 + bx^2y + cxy^2 + dy^3.$$

The stratification is given by general linear actions as follows. Let $G = GL(2,R)$ be the general linear group of real invertible 2×2 matrices. The left-action of G on the variables by matrix multiplication induces a right-action of G on R^4, as follows : given $f \in R^4$, $g \in G$, define fg by

$$(fg)v = f(gv), \text{ where } v = \begin{pmatrix} x \\ y \end{pmatrix}.$$

Define the <u>stratum</u> containing f to be the G-orbit, fG. The following lemma is classical.

<u>Lemma 1.</u> There are 5 strata in R^4, characterised by the type of roots.

Stratum	Dim	Example	Type of roots.
H, hyperbolic umbilics	4	$x^3 + y^3$	2 complex, 1 real
E, elliptic umbilics	4	$x^3 - 3xy^3$	3 real distinct
P, parabolic umbilics	3	x^2y	3 real, 2 equal
X, exceptional	2	x^3	3 real equal
O, the origin	0	0	indeterminate.

<u>Proof.</u> Real linear action preserves the type of roots. Conversely if f,f' have roots of the same type then there is a real projective map sending roots of f into f', and hence $g \in G$ such that f' = fg.

19. Umbilic Bracelet and Double Cusp Catastrophe

Remark. We call X the <u>exceptional</u> stratum because it underlies the exceptional singularities E_6, E_7, E_8 in Arnold's notation [2]. See also Lemma 13 below.

Discriminant. Define the discriminant $D = P \cup X \cup 0$, the union of the non-open strata. The equation of D is given by eliminating x, y from $f = \partial f/\partial x = 0$:

$$4(ac^3 + b^3 d) + 27a^2 d^2 - b^2 c^2 - 18abcd = 0.$$

However this is not much help in understanding the geometry, and so we shall pursue a different tack.

<u>Lemma 2.</u> The stratification of R^4 is conical with vertex 0.

<u>Proof.</u> If g = scalar multiplication by λ, then $fg = \lambda^3 f$. Hence the ray through f is contained in the stratum fG.

<u>Remark.</u> The importance of Lemma 2 is that to describe the stratification of R^4 it suffices to describe the induced stratification on the unit sphere $S^3 \subset R^4$, and then take the cone on the latter. We could, further, identify S^3 antipodally and describe the induced stratification of projective space, but we do not do this for two reasons. Firstly, when we come to apply the results to catastrophe theory, antipodal identification confuses maxima and minima, which are important to distinguish. Secondly our immediate aim is to visualise the stratification, and although the projective language is attractive (see Lemma 16), it is slanted towards the algebraic rather than the topological point of view, and consequently can hide some of the geometry. Therefore we shall consider the stratification of S^3, and visualise it in R^3 by removing a point "at infinity."

Figure 1.

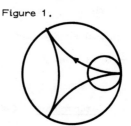

Recall that a <u>triangular hypocycloid</u> is the locus of a point on a circle of radius 1 rolling inside a circle of radius 3 (see Figure 1). It has 3 cusps and 3 concave sides.

<u>Theorem 1.</u> The induced stratification of S^3 is in the shape of a bracelet*, with a triangular hypocycloid section that rotates ⅓ twist going once round the bracelet. The strata H, E, P, X meet S^3 in the outside, inside, surface, and cusped edge of the bracelet, respectively.

Figure 2.

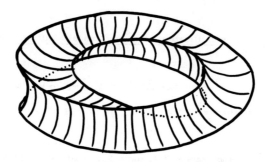

<u>Remark 1.</u> Since the discriminant is classical, this picture was probably known in the last century, but I have not found a reference to it. In Figure 2 we have sketched the bracelet in R^3, assuming that the point at infinity which has been removed from S^3 was a hyperbolic point. If an elliptic point had been removed the cusped edge would point inwards rather than outwards. The simplest way to project S^3 minus a point onto R^3 is by stereographic projection, but this is geometrically very distorting and in particular badly distorts the hypocycloidal sections. Hence in Figure 2 we

---

*The name "bracelet" arose when explaining the shape to my wife, who is a jeweller. Subsequently Tim Poston carved beautiful wooden bracelets of this shape.

have drawn a differentiably equivalent image, that preserves the concave curvature of the sections. The geometry is clarified by Lemma 6 below.

Remark 2. In his book [11, p99] Thom suggests that elliptic states are more fragile than hyperbolic (and deduces that males are more fragile than females), because elliptic states are limited and always followed by hyperbolic breaking. His arguement depends upon the stratification of real quadratic forms in 2 variables (see Lemma 10), and the observation that in the real projective plane elliptic forms correspond to the interior of a conic, and hence contain no projective lines. However the applications refer to the umbilics, in other words to cubic forms rather than quadratic forms; and we show in Lemma 5 that there are circles (corresponding to projective lines) both inside and outside the bracelet. Therefore from the qualitative point of view elliptic states are as robust as hyperbolic states, and any comparison between their fragilities would have to be quantitative depending upon some measure of the strata.

The circle group. To prove Theorem 1 it is convenient to use the circle group (which is the maximal torus of G) namely
$S^1 = SO(2) = \{g_\theta; 0 \le \theta < 2\pi\}$, where

$$g_\theta = \begin{pmatrix} \cos\theta & -\sin\theta \\ \sin\theta & \cos\theta \end{pmatrix}.$$

In turn, the circle group suggests the convenience of a complex variable $z = x + iy$, because then $g_\theta(z) = e^{i\theta}z$.

Lemma 3. With complex coefficients $(\alpha,\beta) \in C^2$, the generic real cubic form can be written

$$f = \mathcal{R}(\alpha z^3 + \beta z^2 \bar{z}).$$

Proof. Writing $\alpha = \alpha_1 + i\alpha_2$, $\beta = \beta_1 + i\beta_2$, then

$$f = \alpha_1(x^3 - 3xy^2) + \alpha_2(-3x^2y + y^3) + \beta_1(x^3 + xy^2) + \beta_2(-x^2y - y^3)$$
$$= (\alpha_1 + \beta_1)x^3 + (-3\alpha_2 - \beta_2)x^2y + (-3\alpha_1 + \beta_1)xy^2 + (\alpha_2 - \beta_2)y^3,$$

which is a permissible change of coordinates for R^4 from (a,b,c,d) because the matrix

$$\begin{pmatrix} 1 & 0 & 1 & 0 \\ 0 & -3 & 0 & -1 \\ -3 & 0 & 1 & 0 \\ 0 & 1 & 0 & -1 \end{pmatrix}$$

is non-singular.

Notation.

A denotes the α-plane, given by $\beta = 0$.

B denotes the β-plane, given by $\alpha = 0$.

A_0 denotes the unit circle in A, given by $|\alpha| = 1$, $\beta = 0$.

B_0 denotes the unit circle in B, given by $\alpha = 0$, $|\beta| = 1$.

We may write $R^4 = C^2 = A \times B$.

Lemma 4. S^1 acts orthogonally on $A \times B$ by rotating A thrice and B once.

Proof. $(fg_\theta)z = f(g_\theta z) = \mathcal{R}(\alpha e^{3i\theta}z^3 + \beta e^{i\theta}z^2 z)$. Therefore $(\alpha,\beta)g_\theta = (\alpha e^{3i\theta}, \beta e^{i\theta})$.

Lemma 5. $A_0 \subset E$, $B_0 \subset H$.

Proof. By Lemma 4 A_0, B_0 are S^1-orbits, and therefore contained in

G-orbits. A_0 contains the point $(\alpha, \beta) = (1,0)$, which corresponds to the form $x^3 - 3xy^2$, which is in E, and therefore $A_0 \subset E$. Similary B_0 contains $(0,1)$, corresponding to $x^3 + xy^2$ in H.

In Figure 2 A_0 is the horizontal core of the bracelet, and B_0 is the vertical axis of the bracelet (together with the point at infinity). Therefore A_0, B_0 represent projective lines in E, H confirming Remark 2 above.

Let $T = A_0 \times B$, the solid torus given by $|\alpha| = 1$. Radial projection from the origin gives a diffeomorphism $S^3 - B_0 \to T$. This is illustrated in Figure 3, where B is drawn symbolically as 1-dimensional rather than 2-dimensional, and B_0 as a point-pair rather than a circle. By Lemma 5 the bracelet does not meet B_0, and so is projected diffeomorphically into T. Therefore to prove the theorem it suffices to prove the existence of the bracelet in T, rather than S^3.

Figure 3.

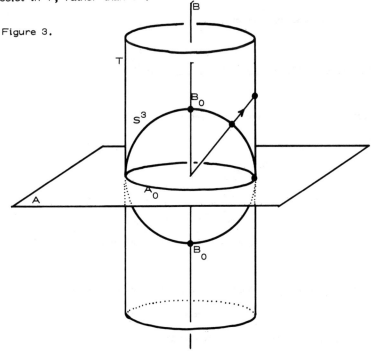

Lemma 6. D meets the plane $\alpha = 1$ in the triangular hypocycloid $\beta = 2e^{i\varphi} + e^{-2i\varphi}$, $0 \leq \varphi < 2\pi$ (see Figure 1).

Proof. If $(1,\beta) \in D$, then $f = \mathcal{R}(z^3 + \beta z^2 \bar{z})$ has a double root in $x:y$. Putting $x = e^{i\theta}$, then

$$f = \tfrac{1}{2}(e^{3i\theta} + e^{-3i\theta} + \beta e^{i\theta} + \bar\beta e^{-i\theta})$$

has a double root in θ. Multiplying by $e^{i\theta}$,

$$\tfrac{1}{2}(e^{4i\theta} + e^{-2i\theta} + \beta e^{2i\theta} + \bar\beta)$$

also has a double root in θ. Therefore the derivative vanishes,

$$2ie^{4i\theta} - ie^{-2i\theta} + i\beta e^{2i\theta} = 0.$$

$$\beta = -2e^{2i\theta} + e^{-4i\theta}.$$

Putting $2\theta = \varphi - \pi$ gives the required formula. Geometrically the formula represents the locus of a point on a circle of radius 1 rolling inside a circle of radius 3, namely the hypocycloid.

Lemma 7. X meets $\alpha = 1$ in the 3 cusp points.

Proof. With a triple root, the second derivative also vanishes. Therefore

$$\frac{d\beta}{d\varphi} = 2ie^{i\varphi} - 2ie^{-2i\varphi} = 0.$$

$$e^{3i\varphi} = 0.$$

$$e^{i\varphi} = 1, w, w^2, \text{ the cube roots of 1}.$$

$$\beta = 3, 3w, 3w^2, \text{ the three cusp points}.$$

Proof of Theorem 1. Apply the first third of the circle group S^1, for values $0 \leq \theta \leq \frac{2\pi}{3}$. By Lemma 4 this rotates the circle A_0 once, and gives the plane B a $\tfrac{1}{3}$-twist. Therefore it isotops the plane $\alpha = 1$ once round the torus T, and back onto itself with a ⅓-twist. Therefore it isotops the hypocycloid once round and back onto itself with a ⅓-twist, to form the bracelet shown in Figure 2. Hence the strata P and X meet T in the surface and cusped edge of the bracelet. Meanwhile the strata H and E meet in T in the exterior and interior by Lemma 5. This completes the proof of Theorem 1.

Corollary. The stratification of R^4 can be written parametrically as follows. Let

$$(\alpha,\beta) = \lambda(e^{3i\theta}, \mu e^{i\theta}(2e^{i\varphi} + e^{-2i\varphi})), \quad \lambda,\mu \geq 0, \quad 0 \leq \theta, \varphi < 2\pi .$$

Then the strata are given by

$$H : \lambda > 0, \mu > 1 .$$
$$E : \lambda > 0, 0 \leq \mu < 1 .$$
$$P : \lambda > 0, \mu = 1, \varphi \neq 0 .$$
$$X : \lambda > 0, \mu = 1, \varphi = 0 .$$
$$O : \lambda = 0.$$

Remark 3. Let G_+ denote the subgroup of G of index 2 consisting of matrices with positive determinant (G is the identity component). Then G_+-orbits in R^4 equal the G-orbits. In particular G_+ acts freely on H, and with index 3 on E. This is related to the geometric fact that there is only one tangent to the hypocycloid from an exterior point, but 3 from an interior point. It also underlies some of the qualitative differences between hyperbolic and elliptic umbilics, for example the bifurcation set of the former has one cusped edge, and the latter three (see Figures 6, 7, 8, 10 and Lemmas 12, 17).

SECTION 2 : CATASTROPHE THEORY.

We recall the construction of the catastrophe map associated with a determinate germ of a function [11,16]. Let $\mathcal{E}$ be the ring of germs of C^∞-functions $R^n \to R$, and m the maximal ideal. Given $f \in \mathcal{E}$, define the **Jacobian ideal** J of f by $J = (f_1, \ldots, f_n)_\mathcal{E}$, where $f_i = \partial f / \partial x_i$ and $(x_1, \ldots, x_n)$ are coordinates for R^n. Note that J is independent of choice of coordinates. Call f **determinate** if $J \supset M^q$, some q.

Let f be a fixed determinate germ, and suppose $f \in m^k$, $k \geq 3$. We shall mostly assume $f \notin m^{k+1}$, but this is not necessary. Define the *__unfolding space__ of f to be m/J, and define the __codimension__ of f to be $\dim(m/J)$. Choose a right inverse $\varepsilon: m/J \to m$ of the projection $m \to m/J$. Define the *__unfolding__ F of f associated with ε to be the map germ

$$F: R^n \times m/J \to R, \text{ given by}$$

$$F(x,c) = fx + (\varepsilon c)x, \quad x \in R^n, \ c \in m/J .$$

Note that the unfolding is not unique, since it depends upon the choice of ε, but is uniquely determined by ε in a coordinate-free way. Define the __catastrophe manifold__ $M \subset R^n \times m/J$ by the equations $F_1 = \ldots = F_n = 0$. Note that the determinacy of f ensures that these equations are independent, and so M is a manifold (or more precisely the germ of a manifold) of the same dimension as m/J. Define the __catastrophe map__

$$\chi_f : M \to m/J$$

to be the map germ induced by the projection $R^n \times m/J \to m/J$. Let sing χ_f denote the set of singularities of χ_f, and define the __bifurcation set__ to be

---

* By Mather's theory [16] our definition of unfolding is universal, but not minimal if $f \notin J$. However the particular germs that we shall be considering here will be homogeneous or quasi-homogeneous, in which case $f \in J$, and so our unfoldings are both universal and minimal.

χ_f(sing χ_f). Let strat χ_f denote the stratification of M induced by sing χ_f. We give a precise definition of strat χ_f below in terms of orbits. For the moment observe that strat χ_f is simpler than the bifurcation set, because the former does not contain self-intersections whereas the latter does (see Figure 6 for example).

For applications it is important to understand the geometry of χ_f, and in particular the geometry of the bifurcation set. This is what we should like to know in the case of the double-cusp, but as yet this problem is unsolved. The problem is made additionally awkward by the non-uniqueness of the bifurcation set, since it depends upon the choice of ε, and is unique only up to diffeomorphism. Since this problem is unsolved, we tackle here the simpler problem of studying sing χ_f and strat χ_f. Here the geometry is made slightly awkward by the fact that M is a non-linear manifold.

Now in applications the non-linearity of M is important, because M frequently represents a graph between cause and effect, and the very essence of catastrophe theory is the multivaluedness of this graph over the unfolding space, together with the catastrophic jumps that occur parallel to R^n, from fold points of M into other sheets of M. However, if we are to try and get an initial grip upon the geometry of strat χ_f, it is useful to replace M by a linear manifold. This is one of the purposes of Theorem 2. The theorem also shows that strat χ_f is a substratification of a canonical stratification, which is, unlike the bifurcation set, independent of choice of unfolding. Surprisingly the canonical stratification is even independent of f, and depends only upon the pair of integers n,k, as follows.

<u>Definition of canonical stratification.</u> Let $\mathcal{G}$ be the group of germs at 0 of C^∞-diffeomorphisms $R^n, 0 \to R^n, 0$. Then $\mathcal{G}$ acts on the right of $\mathcal{E}$

by composition, leaving m invariant, and hence induces actions upon the powers of the maximal ideal m^k, and the jet spaces m^j/m^k, for $j < k$. Define the <u>canonical stratifications</u> N^k of m^k and $N^{j,k}$ of m^j/m^k to be the sets of $\mathcal{G}$-orbits.

<u>Definition of strat χ_f.</u> Define a map germ $\varphi: R^n \times m/J \to m$ by

$$\varphi(x,c)\xi = F(x+\xi, c) - F(x,c)$$

where $x, \xi \in R^n$, $c \in m/J$ and F is the unfolding of f. Let $\varphi^{-1}N^1$ denote the pull-back under φ of the canonical stratification N^1 of m. Define

$$\text{strat } \chi_f = M \cap \varphi^{-1}N^1 .$$

Note that although N^1 is a global stratification of m, φ is only a map-germ, and so strat χ_f is only a stratification-germ of the manifold-germ M. We verify that this is a reasonable definition by the following lemma.

<u>Lemma 8.</u>

(i) $M = \varphi^{-1}m^2$

(ii) <u>Strat $\chi_f = \varphi^{-1}N^2$</u>

(iii) <u>Sing χ_f is given by the vanishing of the Hessian of F.</u>

(iv) <u>Singularities in the same stratum are equivalent.</u>

<u>Proof.</u> By Taylor expansion

$$\varphi(x,c)\xi = \xi F'(x,c) + \tfrac{1}{2}\xi^2 F''(x,c) + \ldots$$

where primes denote (in tensor notation) the derivatives with respect to x. Therefore

$$\varphi(x,c) \in m^2 \iff \text{coefficient } F' \text{ of } \xi \text{ vanishes}$$
$$\iff F_1 = \ldots = F_n = 0$$
$$\iff (x,c) \in M.$$

Hence $M = \varphi^{-1} m^2$, and strat χ_f is the pull-back of the canonical stratification N^2 of m^2. The Hessian H of F is given by $H = \det F'' = |F_{ij}|$. Let M_i be given by $F_i = 0$. Then $M = M_1 \cap \ldots \cap M_n$. The normal to M_i is

$$(F_{i1}, \ldots, F_{in}, \frac{\partial F_i}{\partial c_1}, \ldots, \frac{\partial F_i}{\partial c_p})$$

where $(c_1, \ldots, c_p)$ are coordinates for m/J. Therefore

$(x,c) \in \text{sing } \chi_f \iff \exists$ a tangent of M killed by $T\chi_f$

$\iff \exists v \neq 0, v \in R^n$, such that $\binom{v}{0} \subset TM$

$\iff \exists v \neq 0, \forall i, \binom{v}{0} \subset TM_i$

$\iff \exists v \neq 0, \forall i, \binom{v}{0} \perp$ normal M_i

$\iff \exists v \neq 0, F''v = 0$

$\iff H = 0$.

Finally suppose $(x,c) \in \text{sing } \chi_f$. Then $\varphi(x,c)$ is the local germ of F at (x,c) in the R^n direction, and by Mather's theory [16, Chapter 7] the stratum of this germ determines the equivalence class of the singularity of χ_f at (x,c). This completes the proof of Lemma 8.

Remark. Note that the converse to (iv) is not true: equivalent singularities do not necessarily lie in the same stratum. For example generic maxima and minima of F lie in distinct open strata of N^2, and hence pull back into distinct open strata of M, although, as regular points of χ_f, they are trivially equivalent. It is important for applications to keep maxima and minima distinct.

We are now ready to state the theorem. Recall our original assumption $f \in M^k$, $k \geq 3$. Therefore $J \subset m^{k-1}$, and $mJ \subset m^k \subset m^2$. Let π denote the projection

$$\pi : m^2/mJ \to m^2/m^k,$$

let $N = N^{2,k}$ denote the canonical stratification of m^2/m^k, and let $\pi^{-1}N$ denote the pull-back of N under π.

Theorem 2. The catastrophe map χ_f is equivalent to a map
$$\chi: m^2/mJ \to m/J$$
such that strat χ refines $\pi^{-1}N$.

Here <u>refines</u> means that strat χ is a substratification of $\pi^{-1}N$, in other words each stratum of strat χ is contained in a stratum of $\pi^{-1}N$. Note that N is independent of f, but the refinement in general depends upon both f and the unfolding F (although up to diffeomorphism it is independent of F). The simplest example of refinement can be seen below in the case of the hyperbolic and elliptic umbilics : here $\pi^{-1}N$ is given by a cone in R^3 (see Lemma 10), and the refinements are given by adding respectively one or three generators of the cone (see Examples 3 and 4).

Proof of theorem. Let θ denote the composition
$$R^n \times m/J \xrightarrow{\varphi} m \xrightarrow{\pi_1} m/mJ$$
where π_1 denotes projection. We shall show that θ is a diffeomorphism germ by proving the derivative $T\theta$ is an isomorphism, as follows. From the definition of φ and F,
$$\varphi(x,0)\xi = F(x+\xi,0) - F(x,0)$$
$$= f(x+\xi) - fx$$
$$= \xi f'x + \tfrac{1}{2}\xi^2 f''x + \ldots$$
in Taylor expansion. Therefore $T\varphi$ maps $R^n \times 0$ onto the subspace of m spanned by $f_1, \ldots, f_n$. Now the determinacy of f ensures that $f_1, \ldots, f_n$ are linearly independent modulo mJ, and, furthermore, span J/mJ [13, Lemma 3.8].

Therefore $T\theta$ maps $R^n \times 0$ isomorphically onto J/mJ. Meanwhile

$$\varphi(0,c) = f + \varepsilon c.$$

Therefore

$$T\varphi(0 \times m/J) = T\varepsilon(m/J).$$

Now $T\varepsilon$ maps m/J isomorphically onto a complement of J in m, because ε is a right inverse of the projection $m \to m/J$. Therefore $T\varphi$ maps $0 \times m/J$ isomorphically onto this same complement, and hence $T\theta$ maps $0 \times m/J$ isomorphically onto a complement of J/mJ in m/mJ. We have shown that $T\theta$ maps $R^n \times 0$, $0 \times m/J$ isomorphically onto complementary subspaces of m/mJ. Hence $T\theta$ is an isomorphism. Hence θ is a diffeomorphism germ. By Lemma 8, $M = \varphi^{-1}m^2 = \theta^{-1}(m^2/mJ)$. Therefore $\theta | M: M \to m^2/mJ$ is a diffeomorphism germ, and the map required by the theorem is given by composition $\chi = \chi_f(\theta|M)^{-1}$:

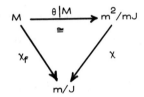

Let $\pi_2 = \pi\pi_1$. Then we have compositions

$$M \xrightarrow{\varphi} m^2 \xrightarrow{\pi_1} m^2/mJ \xrightarrow{\pi} m^2/m^k$$

with θ the composition $M \to m^2/mJ$ and π_2 the composition $m^2 \to m^2/m^k$.

By Lemma 8,

$$\text{strat } \chi_f = \varphi^{-1}N^2 = \varphi^{-1}(\varphi M \cap N^2).$$

Therefore

$$\text{strat } \chi = \theta(\text{strat } \chi_f) = \theta\varphi^{-1}(\varphi M \cap N^2) = \pi_1(\varphi M \cap N^2),$$

because $\pi_1|\varphi M$ is a diffeomorphism germ. Now π_2 commutes with the action of $\mathcal{G}$, and so N^2 refines $\pi_2^{-1}N$. Therefore $\varphi M \cap N^2$ refines $\pi_2^{-1}N$. Therefore

$\pi_1(\varphi M \cap N^2)$ refines $\pi_1 \pi_2^{-1} N = \pi^{-1} N$. This completes the proof of the theorem.

Indeterminate strata. Call a stratum of N <u>determinate</u> if the germs in the jets of that stratum are determinate, and call it <u>indeterminate</u> otherwise. Let N^* denote the subspace of indeterminate strata. Then N^* has codimension $k - 2$ in m^2/m^k. In the theorem the refinement is limited to $\pi^{-1} N^*$, and so

$$\text{strat } \chi = \pi^{-1} N, \text{ modulo codim } k - 2.$$

Consequently there is no refinement with the cuspoids ($n = 1$), but there is with the umbilics ($n = 2$) — see below.

<u>Definition.</u> Let $\sigma = \dim(m^k/mJ)$. If $\sigma = 0$ then $\pi = 1$, and $\pi^{-1} N = N$, as in the case of the cuspoids and elliptic and hyperbolic umbilics. If $\sigma > 0$ then $\pi^{-1} N \cong N \times \mathbb{R}^\sigma$. There are two possible reasons for $\sigma > 0$:

(i) f not homogeneous; for example $\sigma = 1$ for the parabolic umbilic $x^2 y + y^4$, because it is not 3-determinate.

(ii) f has modality > 0 (see [3,4]); for example $\sigma = 1$ for double-cusp $x^4 + y^4$, because $x^2 y^2 \notin mJ$.

<u>Lemma 9.</u> Codim $f = \dfrac{(n+k-1)!}{n!(k-1)!} - (n+1) + \sigma$.

<u>Proof.</u> Codim $f = \dim(m^2/mJ)$, by determinacy of f,

$$= \dim(\mathcal{E}/m^k) - \dim(\mathcal{E}/m^2) + \dim(m^k/mJ),$$

because $\mathcal{E} \supset m^2 \supset m^k \supset mJ$, which gives the required formula by counting monomials.

Example 1. The cusp, A_3.

Since this is a familiar example, we give the formulae in detail, in order to illustrate the theorem. The cusp catastrophe has symbol A_3 in Arnold's notation [2] and germ x^4, or the equivalent; for convenience of computation we choose the germ $f = \frac{1}{4}x^4$. Here $n = 1$, $k = 4$, $J = m^3$, and therefore $mJ = m^4$, $\sigma = 0$, codim $f = 2$. Choose for the unfolding space m/J the base $\{\frac{1}{2}x^2, x\}$ (regarded as 2-jets) and coordinates (u,v). Therefore a point $c \in m/J$ can be written

$$c = (u,v) = \tfrac{1}{2}ux^2 + vx.$$

Choose $\varepsilon: m/J \to m$ by reinterpreting x^2, x as germs rather than 2-jets. Then the unfolding $F: R \times m/J \to R$ is given by

$$F(x;u,v) = \tfrac{1}{4}x^4 + \tfrac{1}{2}ux^2 + vx.$$

The induced map $\varphi: R \times m/J \to m$ is given by

$$\varphi(x;u,v)\xi = \xi F' + \tfrac{1}{2}\xi^2 F'' + \ldots$$
$$= \xi(x^3 + ux + v) + \tfrac{1}{2}\xi^2(3x^2 + u) + \xi^3 x + \tfrac{1}{4}\xi^4.$$

The composition $\theta = R \times m/J \to m/mJ$ is given by

$$\theta(x,u,v)\xi = \xi(x^3 + ux + v) + \tfrac{1}{2}\xi^2(3x^2 + u) + \xi^3 x,$$

where ξ, ξ^2, ξ^3 are reinterpreted as 3-jets rather than germs. The restriction $\theta|M: M \to m^2/mJ$ is given by

$$(\theta|M)(x;u,v)\xi = \tfrac{1}{2}\xi^2(3x^2 + u) + \xi^3 x.$$

Choose for m^2/mJ the base $\{\xi^2, \xi^3\}$ and coordinates (a,b). Then $\theta|M$ is given by

$$\begin{cases} a = \tfrac{1}{2}(3x^2 + u) \\ b = x \end{cases}$$

Therefore $(\theta|M)^{-1}$ is given by

$$\begin{cases} x = b \\ u = 2a - 3x^2 = 2a - 3b^2 \\ v = -ux - x^3 = -2ab + 2b^3 \end{cases}$$

19. Umbilic Bracelet and Double Cusp Catastrophe

Therefore the catastrophe map $\chi: m^2/mJ \to m/J$ is given by

$$\begin{cases} u = 2a - 3b^2 \\ v = -2ab + 2b^3 \end{cases}.$$

This has Jacobian

$$\frac{\partial(u,v)}{\partial(a,b)} = \begin{vmatrix} 2 & -6b \\ -2b & -2a+6b^2 \end{vmatrix} = -4a.$$

Therefore sing χ is the b-axis, $a = 0$. The canonical stratification N of m^2/m^4 comprises 4 strata :

minima	$a > 0$
maxima	$a < 0$
fold	$a = 0, b \neq 0$
cusp*	$a = b = 0$.

The only indeterminate stratum is the last (which is why we have starred the word cusp), because codim $N^* = k - 2 = 2$, and so dim $N^* = 0$. Therefore since $\pi^{-1}N = N$, no refinement is possible. Therefore strat $\chi = N$. In Figure 4 the stratifications are shown by thick lines.

Figure 4.

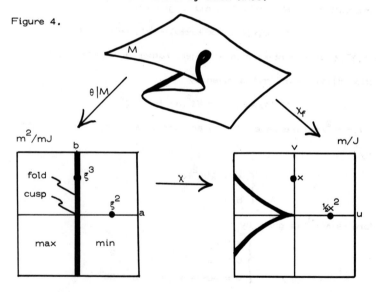

Example 2. The cuspoid, A_k.

In Arnold's notation [2] the kth cuspoid is A_{k+1}, with germ x^{k+2}, and codimension k. As in the cusp, strat $\chi = N$ and the stratification is given by a flag of linear subspaces, with one stratum in each odd, and two strata in each even codimension (except when k is even the origin is the only stratum of codimension k). This completes the case $n = 1$, and we now pass to $n > 1$.

Homogeneous forms. The jet space m^k/m^{k+1} can be identified with the space of real homogeneous forms of degree k in n variables. The $\mathcal{G}$-action in this case reduces G-action, where $G = GL(n,R)$ denotes the general linear group, because the non-linear action of $\mathcal{G}$ is quotiented out. Therefore the canonical stratification of m^k/m^{k+1} coincides with that induced by G. In particular when $n = 2$, $k = 3$ it is determined by the umbilic bracelet — which is why we proved Theorem 1. We now turn to the simpler case $n = 2$, $k = 2$ of quadratic forms in two variables :

$$q = ax^2 + 2bxy + cy^2.$$

Real quadratic forms are classified linearly by rank and signature, and therefore the stratification is determined by the discriminant cone C, given by $ac = b^2$. We can therefore state without proof :

Lemma 10. When $n = 2$ the canonical stratification N of quadratic forms m^2/m^3 has 6 strata, with indeterminate subspace N^* equal to the discriminant cone C.

Corank	Name		Dim	Example	Formula
0	minima		3	$x^2 + y^2$	$ac > b^2$, $a > 0$
	maxima		3	$-x^2 - y^2$	$ac > b^2$, $a < 0$
	saddles		3	$x^2 - y^2$	$ac < b^2$
1	folds*	C_+	2	x^2	$ab = b^2$, $a + c > 0$
		C_-	2	$-x^2$	$ac = b^2$, $a + c < 0$
2	umbilic*		0	0	$a = b = c = 0$

Figure 5.

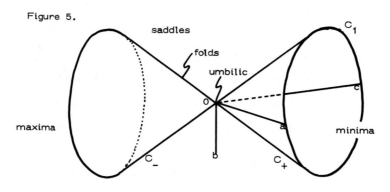

Definition of attaching map. Subsequently we shall need to describe how strata are attached to one another, and Figure 5 presents an opportunity to introduce and illustrate a useful definition. Suppose X, Y are strata (or disjoint unions of strata) in a manifold Z, such that $X \supset Y$. We say the map $\Psi: R \times M \to Z$ <u>attaches</u> X onto Y if $\Psi(0 \times M) = Y$ and Ψ maps the complement diffeomorphically onto X. Let $M_\delta = \Psi(\delta \times M)$. Intuitively we think of M_δ moving isotopically through X, tracing out the whole of X, as δ runs through non-zero values of R, and then crushing down onto Y as $\delta \to 0$. Given an open subset $V \subset Y$, we say Ψ <u>covers</u> V n times if $\Psi | \Psi^{-1}V : \Psi^{-1}V \to V$ is an n-fold covering. We say Ψ has <u>singularities at</u> $\Psi(\text{sing } \Psi)$, which, from the definition, is a closed subset of Y.

Example. In Figure 5 let C_1 denote the ellipse given by the intersection of C with the plane $a + c = 2$. Define $\Psi: R \times C_1 \to R^3$ by $\Psi(\delta, q) = \delta q$. Then Ψ attaches the fold strata $C_+ \cup C_-$ onto 0, with a singularity at 0. Other examples are given in Lemma 12 and Theorem 3 below.

Unfolding the umbilics. The stratification N in Lemma 10 and Figure 5 is the one that is refined by the hyperbolic, elliptic and parabolic umbilics (and also indirectly by the double-cusp) as we shall show below. In each case the refinement is non-trivial. Now there are many possible choices of unfolding, and different applications give rise to different choices and different (although diffeomorphic) bifurcation sets (see for example [14, Figure 6]). In order to best compare our formulae with those of Thom, we use his choice of unfolding for the parabolic umbilic [11, page 84]. Then, in order to best reveal the relationship between that and the other two we choose germs and unfoldings for the latter that are different to his. These give slightly different bifurcation sets, but yield simple formulae for χ.

Example 3. The hyperbolic umbilic D_4^+.

Choose the germ $f = x^2 y + \frac{1}{3} y^3$. Here $n = 2$, $k = 3$, $mJ = m^3$, and therefore $\sigma = 0$, codim $f = 3$. Choose for m/J the base $\{x^2, -x, -y\}$ and coordinates (t, u, v). Choose ε by reinterpreting the base jets as germs. Therefore the unfolding is

$$F = x^2 y + \tfrac{1}{3} y^3 + tx^2 - ux - vy.$$

Therefore
$$(\theta | M)(x,y;t,u,v)(\xi,\eta) = \xi^2(y+t) + 2\xi\eta x + \eta^2 y .$$

Choose for m^2/mJ the base $\{\xi^2, 2\xi\eta, \eta^2\}$ and coordinates (a,b,c). Therefore χ is given by

$$\begin{cases} t = a - c \\ u = 2ab \\ v = b^2 + c^2 \end{cases}$$

This has Jacobian $4(ac - b^2)$, confirming that

$$\text{sing } \chi = C,$$

the discriminant cone of Lemma 10. We now want to compute how strat χ refines N. Since $N^* = C$, the only strata to be refined are the two fold strata C_+ and C_-. The singularities of $\chi | C$ are given by $a^2 + 3b^2 = 0$. (This can be found by substituting $ac = b^2$ and computing where the Jacobian matrix drops in rank, or by Lagrange's method of undetermined multipliers). Hence $a = b = 0$. Therefore $\chi | C$ is singular along the c-axis, which is a generator of the cone C. This generator is separated by the origin into two half-lines. Therefore each of the two indeterminate strata, C_+ and C_-, is refined into two substrata, one substratum comprising a half-line of cusps, and the other comprising the complementary surface of folds. Therefore altogether strat χ has 8 strata.

The generator is mapped by χ into the parabola $u = 0$, $v = t^2$ which is the cusped edge of the bifurcation set.
Figure 6.

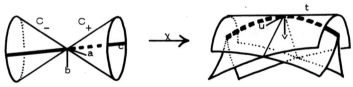

Example 4. The elliptic umbilic D_4^-.

Choose the germ $f = x^2y - \tfrac{1}{3}y^3$, and, apart from this one change of sign, exactly the same unfolding as the previous example:

$$F = x^2y - \tfrac{1}{3}y^3 + tx^2 - ux - vy .$$

Then χ is given by

$$\begin{cases} t = a + c \\ u = 2ab \\ v = b^2 - c^2 \end{cases}.$$

This time the singularities of $\chi|C$ are given by $a^2 - 3b^2 = 0$, which gives 3 generators of C, namely the c-axis and the lines with direction ratios $(a,b,c) = (3, \pm\sqrt{3}, 1)$. Therefore each of the two indeterminate strata is refined into two substrata, one substratum comprising 3 half-lines of cusps, and the other comprising the complementary 3 components of folds. Again strat χ has altogether 8 strata (only this time they are not all connected).

Each of the 3 generators is mapped by χ into a parabola touching the t-axis, and the sections of the bifurcation set perpendicular to the t-axis are triangular hypocycloids.

Figure 7.

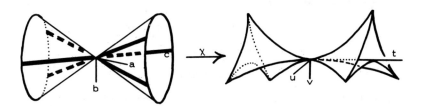

Example 5. The parabolic umbilic D_5.

Choose the germ $f = x^2y + \frac{1}{4}y^4$. Here $n = 2$, $k = 3$, $\sigma = 1$ because $y^3 \notin mJ$. Therefore codim $f = 4$ and $\pi^{-1}N = N \times R$. Following Thom [11, p. 84] choose for m/J the base $\{x^2, y^2, -x, -y\}$ and coordinates (t,w,u,v). Therefore the unfolding is

$$F = x^2y + \tfrac{1}{4}y^4 + tx^2 + wy^2 - ux - vy.$$

Therefore

$$(\theta|M)(x,y;t,u,v,w)(\xi,\eta) = \xi^2(y+t) + 2\xi\eta x + \eta^2(\tfrac{3}{2}y^2+w) + \eta^3 y.$$

Choose for m^2/mJ the base $\{\xi^2, 2\xi\eta, \eta^2, \eta^3\}$ and coordinates (a,b,c,d). Then χ is given by

$$\begin{cases} t = a - d \\ u = 2ab \\ v = b^2 + 2cd - 2d^3 \\ w = c - \tfrac{3}{2}d^2 \end{cases}.$$

Again this has Jacobian $4(ac-b^2)$, confirming that

$$\mathrm{sing}\ \chi = C \times R$$

where C is the discriminant cone of Lemma 10, and 0×R is the d-axis. This time there are 3 strata of N×R to be refined, namely 0×R, $C_+ \times R$, $C_- \times R$. The umbilic stratum 0×R is refined in 3 substrata

hyperbolic umbilics	$d > 0$
parabolic umbilic	$d = 0$
elliptic umbilics	$d < 0$.

Meanwhile the orther two strats $C_{\pm} \times R$ are refined by the formulae (which can be found by computing successive singularities of $\chi|C\times R$) :

folds	$ac = b^2 \neq -a^2 d$
cusps	$ac = b^2 = -a^2 d$, $a^2 \neq 4c$
swallowtails	$ac = b^2 = -a^2 d$, $a^2 = 4c$, $a \neq 0$.

We can draw pictures of the refinements by squashing each end of the cone flat; more precisely the projection $R^4 \to R^3$ given by $(a,b,c,d) \to (e,b,d)$, where $e = \frac{1}{2}(a-c)$, maps each of $C_{\pm} \times R$ diffeomorphically into R^3. Figure 8 shows the images of the refinements.

Figure 8.

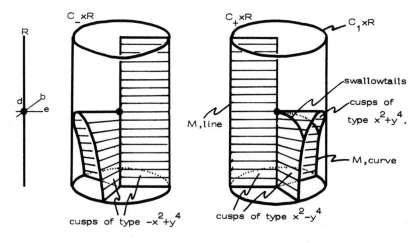

Note that the swallowtails only appear in $C_{+} \times R$ because $a^3 = 4ac = 4b^2 > 0$, and therefore $a > 0$. Therefore $C_{-} \times R$ is refined into only 2 substrata, namely one 2-dimensional substratum of cusps of type $-x^2 + y^4$, with 2 components given by

$$a = b = 0, \ c < 0$$
$$ac = b^2 = -a^2 d, \ a < 0;$$

and the complementary 3-dimensional substratum of folds of type $-x^2 + y^3$ (also with 2 components). Meanwhile $C_{+} \times R$ is refined into 4 substrata, namely one 1-dimensional substratum of swallowtails of type $x^2 + y^5$ (with 2 components); one connected 2-dimensional substratum of cusps of type $x^2 + y^4$, given by $a^2 > 4c$; another 2-dimensional substratum of (dual)-cusps of type $x^2 - y^4$, given by $a^2 < 4c$ (with 3 components); and the complementary

3-dimensional substratum of folds of type $-x^2 + y^3$ (with 2 components).

Summarising, we have shown :

Lemma 11. The catastrophe map of the parabolic umbilic has 12 strata as follows :

NxR	strata χ
minima	minima
maxima	maxima
saddles	saddles
C_+xR	folds, cusps, dual-cusps, swallowtails
C_-xR	folds, cusps
0xR	hyperbolic, elliptic, parabolic umbilics.

Note that the bifurcation set is much more complicated because of self-intersections (see [6,11,17]).

Before leaving the parabolic umbilic we relate it to the previous examples. Notice that Cxd meets the cusp strata in 1, 2 or 3 generators according as $d \gtreqless 0$, as follows :

$d > 0$: 1 generator with direction ratios (0,0,1,0)

$d = 0$: 2 generators with direction ratios (0,0,1,0),(1,0,0,0).

$d < 0$: 3 generators with direction ratios (0,0,1,0),(1,$\pm\sqrt{-d}$,-d,0) .

These correspond to Figures 6 and 7 because points on 0xR represent hyperbolic or elliptic umbilics according as $d \gtreqless 0$.

In Figure 8, 0xR is projected onto the vertical axis of each cylinder, and Cxd is projected onto the two horizontal sections at level d. Figure 8 illustrates how the hyperbolic stratum d > 0 lies locally in the closure of 1 sheet of cusps, whereas the elliptic stratum d < 0 lies in 3 sheets. Figure 8 also shows how the 3 generators merge smoothly into 1 by coalescing the other 2 at the parabolic point. We shall now rephrase these observations in terms of an attaching map, in order to furnish intuition for the analogous 7-dimensional result for the double-cusp (see Theorem 3 below).

Let $C_1 \times R$, M denote the intersection of the plane $a + c = 2$ with $C_+ \times R$, the cusp and swallowtail strata. In Figure 8, $C_1 \times R$ projects onto the cylinder of radius 1, and therefore M consists of a line and a curve; the line has equations $e + 1 = b = 0$, and the curve can be written parametrically $(e,b,d) = (\cos\theta, \sin\theta, -\tan^2\theta/2)$, $-\pi < \theta < \pi$. Define $\Psi: R \times M \to R^4$ by $\Psi(\delta,(q,d)) = (\delta q, d)$.

Lemma 12. <u>Ψ attaches the cusp and swallowtail strata onto the umbilic strata, covering the hyperbolic stratum once, the elliptic stratum thrice, with a singularity at the parabolic point.</u>

Proof. The line maps diffeomorphically onto $0 \times R$, and the curve is folded onto the elliptic stratum.

SECTION 3. THE DOUBLE CUSP.

The double cusp has germ $f = x^4 + y^4$, and belongs to the family X_9 in Arnold's notation [3,4]. Here $n = 2$, $k = 4$ and therefore $\dim(m^2/m^4) = 7$. Meanwhile $\sigma = 1$ because $x^2 y^2 \notin mJ$. Therefore codim $f = 8$. Therefore the problem of finding strat χ is reduced to finding

(i) the 7-dimensional stratification N, and

(ii) the 8-dimensional refinement of N×R.

However Looijenga [8] has shown that the last factor is trivial in the sense that there is a homeomorphism

$$\text{strat } \chi \cong (\text{strat } \chi') \times R$$

where χ' is the semi-universal unfolding

$$\chi' : m^2/m^4 \to m/J+m^4$$

defined in the same way as χ, only with mJ, J replaced by $m^4, J+m^4$. Therefore problem (ii) is reduced to finding

(iii) the 7-dimensional refinement strat χ' of N.

It is possible to list the strata and their incidence relations, and to write down equations for them as in the above examples. But to my mind the problem is not satisfactorily "solved" until one achieves a more global geometric description of the way the strata fit together, that one can somehow "visualise". In this sense we shall give a solution to problem (i), but as yet I have not been able to solve problems (ii) and (iii). So let us tackle problem (i).

We can decompose m^2/m^4 by the $\mathcal{G}$-invariant short exact sequence

$$m^3/m^4 \xrightarrow{\subset} m^2/m^4 \xrightarrow{p} m^2/m^3 ,$$

where p is the projection. In the case $n = 1$ this is easy to visualise, because it is just the left-hand plane pictured in Figure 4, with

$$\text{b-axis} \xrightarrow{\subset} R^2 \xrightarrow{p} \text{a-axis}.$$

19. Umbilic Bracelet and Double Cusp Catastrophe

However in the case n = 2 we are considering, the dimensions make visualisation more difficult :

$$R^4 \xrightarrow{\text{C}} R^7 \xrightarrow{p} R^3$$

However we have already done the two ends, because the stratification $N^{3,4}$ of the left-hand end R^4 is given by the cubic discriminant D, or the umbilic bracelet of Theorem 1, while the stratification $N^{2,3}$ of the right hand end R^3 is given by the quadratic discriminant cone C of Lemma 10. Since p commutes with the action of $\mathcal{G}$, the stratification $N = N^{2,4}$ of the middle R^7 that we are seeking is a refinement of $p^{-1}N^{2,3}$.

Product structure. It is convenient to choose a product structure $R^7 = R^3 \times R^4$ compatible with p. The easiest way to do this is to choose coordinates x,y. Then a 3-jet $f \in R^7$ can be written $f = pf + y$ by Taylor expansion, where $pf \in R^3$ is the unique 2-jet, and $y \in R^4$ the third order term, which is determined by, but depends upon, the choice of coordinates. Although the product structure depends upon choice, the constructions that we make below are $\mathcal{G}$-invariant, and hence independent of choice. For example the stratification $N^{2,3} \times R^4 = p^{-1}N^{2,3}$ is $\mathcal{G}$-invariant.

Lemma 13. In the double-cusp N contains 12 strata as follows :

Corank	$N^{2,3} \times R^4$	$N = N^{2,4}$	dim	strat χ'
0	minima	minima	7	A_1
	maxima	maxima	7	A_1
	saddles	saddles	7	A_1

19. Umbilic Bracelet and Double Cusp Catastrophe

Corank	$N^{2,3} \times R^4$	$N = N^{2,4}$		dim	strat χ'
1	$C_+ \times R^4$	T_+^6,	folds	6	A_2
		M_+^5,	cusps*	5	A_3, A_4, A_5, A_6, A_7
	$C_- \times R^4$	T_-^6,	folds	6	A_2
		M_-^5,	cusps*	5	A_3, A_4, A_5, A_6, A_7
2	$0 \times R^4$	H,	hyperbolic	4	D_4
		E,	elliptic	4	D_4
		P,	parabolic*	3	D_5
		X,	exceptional*	2	E_6
		0,	double cusp*	0	X_9.

Remark. The asterisks denote the indeterminate strata. The notation for the strata of N refer to Lemma 1 above and Lemma 14 below. The last column lists in Arnold's notation [2,3,4] the substrata that occur in the refinement strat χ' of N. One can show that these, and only these, substrata occur by the methods of A'Campo [1]. The substrata occur with multiplicities; for example all four substrata of cusps occur, $\pm x^2 \pm y^4$, two in each A_3. Although N is independent of the germ $x^4 + y^4$, the list of substrata depends upon the germ. In the case of non-compact germs of X_9, namely $x^4 - y^4$ and $x^2 + y^2 - 6x^2y^2$, different substrata occur; for example A_7 disappears, while D_6 appears in P, and E_7 appears in X. However I do not know the geometry of all the substrata.

Proof of Lemma 13. As we have already observed, the $\mathcal{G}$-action on $0 \times R^4$ reduces to linear action, and hence the refinement in N is given by the umbilic bracelet. There remains to check $C_+ \times R^4$. Any 2-jet in C_+ is equivalent to x^2, and therefore any 3-jet in $C_+ \times R^4$ is equivalent to $x^2 + y^3$ or x^2, giving a determinate substratum T_+^6 of folds, and an indeterminate substratum M_+^5 of cusps (indeterminate, because the 3-jet cannot distinguish between cusps, swallowtails, etc.). Similarly for C_-. This completes the proof, and we now look at these two substrata more closely.

<u>Lemma 14.</u> M_+^5 is a 5-dimensional Möbius strip and T_+^6 is a 6-dimensional solid torus.

<u>Proof.</u> Let $q_0 = x^2 \in C_+$, and let $f \in q_0 \times R^4$. Then
$$f = x^2 + ax^3 + bx^2y + cxy^2 + dy^3$$
$$= (x + \tfrac{1}{2}(ax^2 + bxy + cy^2))^2 + dy^3 \quad (\text{modulo } m^4)$$
$$\sim x^2 + dy^3.$$
Therefore $f \in M_+^5$ if and only if $d = 0$. Therefore
$$M_+^5 \cap (q_0 \times R^4) = q_0 \times R_0^3$$
where $R_0^3 \subset R^4$ is the linear subspace given by $d = 0$.

Analogous to Lemma 3, we can write the generic quadratic form using the complex variable $z = x + iy$, one complex coefficient $\gamma = \gamma_1 + i\gamma_2$, and one real coefficient δ, as follows:

Figure 9.

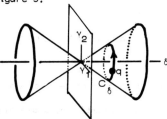

$$q = (\gamma, \delta) = \mathcal{R}(\gamma z^2 + \delta z \bar{z})$$
$$= \gamma_1(x^2 - y^2) - 2\gamma_2 xy + \delta(x^2 + y^2)$$
$$= (\delta + \gamma_1)x^2 - 2\gamma_2 xy + (\delta - \gamma_1)y^2$$

Therefore the discriminant cone C is given by $|\gamma| = |\delta|$, and $C_\pm$ given by $\delta \gtreqless 0$. Metrically the coordinates (γ, δ) have the attraction that C is now a circular cone with axis the δ-axis, whereas it was only an elliptical cone in the (a,b,c) coordinates of Lemma 10. Let C_δ denote the circle on C given by

δ = constant $\neq$ 0. As in Lemma 4 the action of the circle group S^1 is given by

$$qg_\theta = (\gamma,\delta)g_\theta = (\gamma e^{2i\theta},\delta).$$

In other words the circle group spins C twice around its axis, leaving the circles C_δ invariant. Now let

$$q = (\delta,\delta) = 2\delta x^2 \in C_\delta.$$

The first half of S^1 isotopes q once round C_δ back to itself. Meanwhile, by Lemma 4, S^1 acts on R^4 by $(\alpha,\beta)g_\theta = (\alpha e^{3i\theta},\beta e^{i\theta})$. Therefore when $\theta = \pi$, $g\pi$ maps R^4 antipodally, and hence maps R_0^3 onto itself reversing orientation. Therefore the first half of S^1 isotopes $q \times R_0^3$ over C_δ through $C_\delta \times R^4$ back onto itself with orientation reversal, tracing out a 4-dimensional Möbius strip M_δ^4. Therefore

$$M_+^5 \cap (C_\delta \times R^4) = M_\delta^4$$

because N is $\mathcal{G}$-invariant, and hence S^1-invariant. Finally we show M_+^5 is a Möbius strip by scalar multiplication, as follows. Let $R_\theta^3 = R_0^3 g_\theta$. Make a standard copy of the 4-dimensional Möbius strip by defining

$$M^4 = \bigcup \{e^{2i\theta} \times R_\theta^3; \ 0 \leq \theta < \pi\} \subset S^1 \times R^4.$$

Define

$$\Psi: R \times M^4 \to R^3 \times R^4$$
$$(\delta e^{2i\theta},y)) \to ((\delta e^{2i\theta},\delta),y).$$

Then $\Psi(\delta \times M^4) = M_\delta^4$, and Ψ maps $R_+ \times M^4$ diffeomorphically onto M_+^5, proving the latter is a 5-dimensional Möbius strip.

Meanwhile the complement of $q \times R_0^4$ in $q \times R^4$ is a pair of 4-cells, which the first half of S^1 isotops onto each other preserving orientation, and forming a 5-dimensional solid torus. Scalar multiplication by R_+ gives the 6-dimensional solid torus T_+^6. This completes the proof of Lemma 14.

To complete the description of N we need to show how the two Möbius strips are glued onto the umbilic bracelet.

Theorem 3. <u>Ψ attaches the cusp strata $M^5_{\pm}$ onto the umbilic strata $0 \times R^4$, covering the hyperbolic stratum H once, the elliptic stratum E thrice, and with singularities at D.</u>

<u>Remarks.</u> The wording of the theorem resembles that of Lemma 12, and so some intuition can be extracted from Figure 8. A corollary of the theorem is that the induced stratification on $R^3 xy$ is isomorphic to that of the hyperbolic umbilic in Figure 6 if $y \in H$, or to the elliptic umbilic in Figure 7 of $y \in E$. Consequently in the refinement strat χ' of N, the strat substrata A_4, A_5, A_6, A_7 of swallowtails, etc. do not neet the neighbourhood of H, E and only abut D, as in Figure 8.

<u>Proof of the theorem.</u> Let $\psi : M^4 \to R^4$ be induced by projection $S^1 \times R^4 \to R^4$. If we identify $M^4 = 0 \times M^4$, $R^4 = 0 \times R^4$ then $\psi = \Psi | M^4$. Since Ψ maps the complement of M^4 diffeomorphically onto $M^5_{\pm}$, the theorem reduces to showing ψ covers H once, E thrice, with singularities at D.

Define the <u>core</u> Q^3 of M^4 as follows. Recall that we defined $R^3_0 \subset R^4$ by $d = 0$; now define $R^2_0 \subset R^3_0$ by $c = d = 0$. Let $R^2_\theta = R^2_0 g_\theta$. Define
$$Q^3 = \bigcup_\theta \{e^{2i\theta} \times R^2_\theta ; \ 0 \leq \theta < \pi\} \subset M^4 .$$
Then Q^3 is a 3-dimensional solid torus, since the antipodal map of R^2_0 is orientation preserving.

<u>Lemma 15.</u> <u>Sing $\psi = Q^3$.</u>
<u>$\psi(\text{sing } \psi) = \psi Q^3 = \bigcup R^2_\theta = D.$</u>

Proof. Since ψ embeds each fibre, $\psi(e^{2i\theta} \times R^3_\theta) = R^3_\theta$,

$$\psi(\text{sing } \psi) \cap R^3_\theta = \lim_{\varphi \to \theta}(R^3_\varphi \cap R^3_\theta) = R^2_\theta .$$

Therefore $\psi(\text{sing } \psi) = \bigcup R^2_\theta$, and sing $\psi = Q^3$. The condition for f to lie in some R^2_θ is that f has at least two roots equal, which is the same condition for $f \in D$. This completes the proof of Lemma 15.

Projective notation. The projective point of view lends some insight, because it shows that the exceptional stratum determines both the bracelet and the Möbius strips. Let P^3 denote the 3-dimensional real projective space of lines through the origin in R^4. Then $R^2_\theta \subset R^3_\theta \subset R^4$ induce projective subspaces $P^1_\theta \subset P^2_\theta \subset P^3$. By Lemma 2, $X \subset D \subset R^4$ induces $\overline{X} \subset \overline{D} \subset P^3$.

Lemma 16. $\overline{X}$ *is a twisted cubic curve with tangents $\{P^1_\theta\}$ and osculating planes $\{P^2_\theta\}$. $\overline{D}$ is the ruled surface $\bigcup P^1_\theta$ and the envelope of $\{P^2_\theta\}$.*

Proof. The second sentence is a corollary of Lemma 15. $\overline{X}$ is a twisted cubic because it can be parametrised $[a,b,c,d] = [\lambda^3, 3\lambda^2\mu, 3\lambda\mu^2, \mu^3]$. The rest follows from the fact that the generators of a cubic developable are the tangents, and its tangent planes are the oscilating planes, of its edge of regression [15].

Having dealt with the singularities of ψ, to complete the proof of Theorem 3 we must now deal with its regularities.

Lemma 17. ψ *covers H once and E thrice.*

19. Umbilic Bracelet and Double Cusp Catastrophe

<u>Proof.</u> Given $\alpha \in C$, let $Y_\alpha \subset R^4$ denote the plane α = constant. Let $H_\alpha, E_\alpha, D_\alpha$ denote the intersections of Y_α with H, E, D, respectively. By Lemma 6, if $\alpha \neq 0$ then D_α is a triangular hypocycloid, with exterior H_α and interior E_α. If $\alpha = 0$ then $D_\alpha = 0$, with complement H_α, and E_α empty. Let M_α^2, Q_α^1 denote the intersection of $S^1 \times Y_\alpha$ with M^4, Q^3, respectively. Then M_α^2 is an ordinary 2-dimensional Möbius strip with core Q_α^1. Let ψ_α denote the restriction

$$\psi_\alpha = \psi | M_\alpha^2 : M_\alpha^2, Q_\alpha^1 \to Y_\alpha, D_\alpha .$$

Figure 10.

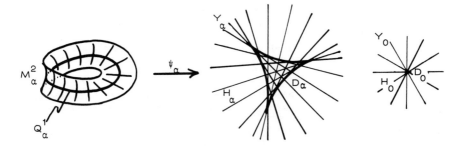

To complete the proof it suffices to show that ψ_α covers H_α once, E_α thrice, with singularities at D_α. Let $R_{\alpha,\theta}^1, R_{\alpha,\theta}^0$ denote the intersection of Y_α with R_θ^3, R_θ^2, respectively. By Lemma 16, R_θ^3 touches D along R_θ^2, and so $R_{\alpha,\theta}^1$ touches D_α at $R_{\alpha,\theta}^0$. Therefore ψ_α maps the fibres of M_α^2 to the tangents of D_α. The result follows from the fact that a triangular hypocycloid has only one tangent from an exterior point, but 3 from an interior point. This completes the proof of Lemma 17 and Theorem 3.

BIBLIOGRAPHY

1. N. A'Campo   Le groupe de monodromie du déploiement des singularités isolées de courbes planes I, Math. Ann. 213 (1975) 1-32.

2. V.I. Arnold,   Normal forms for functions near degenerate critical points, the Weyl Groups of A_k, D_k, E_k and Lagrangian singularities, Funk. Anal. i Ego Prilhozen 6, 4 (1972) 3-25; Eng. transl : Func. Anal. Appl. 6 (1973) 254-272.

3. V.I. Arnold,   Classification of unimodular critical points of functions, Funk. Anal. i Ego Prilhozen 7, 3 (1973) 75-76; Eng. transl\: Func. Anal. Appl. 7 (1973) 230-231.

4. V.I. Arnold,   Normal forms for functions in the neighbourhood of degenerate critical points, Uspehi Mat. Nauk. 29, 2 (1974) 11-49.

5. G. Augusti,   Stabilita' di strutture elastiche elementari in presenza di grandi spostamenti, Atti Accad. Sci. fis. math., Napoli, Serie 3^a, 4, No. 5 (1964).

6. A.N. Godwin,   Three dimensional pictures for Thom's parabolic umbilic, IHES, Publ. Math. 40 (1971) 117-138.

7. A.N. Godwin,   Topological bifurcation for the double cusp polynomial, Proc. Camb. Phil. Soc. 77 (1975) 293-312.

8. P.J. Harrison & E.C. Zeeman,   Applications of catastrophe theory to macroeconomics, Symp. Appl. Global Analysis, Utrecht Univ., 1973 (to appear).

9. C.A. Isnard & E.C. Zeeman,   Some models from catastrophe theory in the social sciences, Use of models in the Social Sciences (ed. L. Collins) Tavistock, London, 1976.

10. E. Looijenga,   On the semi-universal deformations of Arnold's unimodular singularities, Liverpool Univ. preprint, 1975.

11. R. Thom,   Structural stability and morphogenesis, (Eng. transl. by D.H. Fowler), Benjamin, New York, 1975.

12. R. Thom & E.C. Zeeman,   Catastrophe theory : its present state and future perspectives, Dynamical Systems - Warwick 1974, Springer Lecture Notes in Maths, Vol. 468 (1975), 366-401.

13. J.M.T. Thompson & G.W. Hunt, A general theory of elastic stability, Wiley, London, 1973.

14. J.M.T. Thompson, Experiments in catastrophe, Nature, 254, 5499 (1975) 392-395.

15. J.A. Todd, Projective and analytical geometry, Pitman, London, 1947.

16. D.J.A. Trotman & E.C. Zeeman, Classification of elementary catastrophes of codimension ≤ 5, this volume.

17. A.E.R. Woodcock & T. Poston, A geometrical study of the elementary catastrophes, Lecture Notes in Mathematics 373, Springer, Berlin, 1974.

18. E.C. Zeeman, Duffing's equation in brain modelling, Symp. for J.E. Littlewood's 90th birthday, 1975, Bull. Inst. Math. and Appl. (to appear).

DISCUSSION

DISCUSSION

Paper		Page
20.	Research ancient and modern.	605
21.	Catastrophe theory : its present state and future perspectives (with R. Thom).	615
22.	Afterthought.	651 – 656.

Paper 20 was written in a somewhat light-hearted vein for a conference on "Research in mathematics". The first half is not really relevant to our topic since it is about Greek mathematics; however the paper has been included in the book because the second half contains a discussion about the qualitative nature of catastrophe theory, and a non-trivial catastrophe model about education.

Paper 21 is a discussion between Thom and myself, written in 1973, on the development of catastrophe theory as we saw it. We discuss its various aspects in mathematics, physics, embryology, human sciences and linguistics. The bibliography, containing 98 references on mathematics and 82 references on applications, was added in 1974. In the subsequent three years to 1977 the subject has mushroomed, and the literature has expanded to such an extent that it would now be difficult to give a complete bibliography. The most striking advance, which neither Thom or I predicted in this discussion, has been the rapidity with which so many experimentalists in the various fields of applications have actively begun to use the method, assessing its value in comparison with other methods, fitting data, making their own models, and testing and verifying the resulting predictions.

Paper 22, the Afterthought, is the last paper, and the only one not published elsewhere. It contains a discussion of the philosophy and possible experimental testability of some of the more speculative psychological models, such as those described in Papers 1 and 9. It also touches upon Thom's predation loop.

20 Research, Ancient and Modern

Chapter 1. On government research establishments

THE main justification for mathematical research, and I think Professor Coulson* would have said the same, is that it is one of the oldest and most splendid endeavours of mankind. And that could well be the end of my talk.

But whether or not the government can afford to support vast numbers of us on fat salaries, to do what we enjoy, is another matter. The problem is as old as history. Recall that apocryphal story about Euclid told by Stobaeus[1]:

"Someone who had begun to read geometry with Euclid, when he had learnt the first theorem asked Euclid 'But what shall I get by learning these things?' Euclid called his slave and said 'Give him threepence, since he must make gain out of what he learns'."

Now you may laugh at Euclid's apparent gentle sarcasm, but I am not so sure. Judging from the humourlessness of Euclid's mathematical style, and remembering his position as a head of department of a new government research establishment, and recalling Professor Bondi's words about *his* experiences in a similar position, one could easily interpret Euclid's reply at face value. Anyway why did he offer threepence, when a penny would have made the point just as well? He probably had a research budget of £30, and being commissioned to produce a mammoth standard reference work on mathematics had, with an administrator's acumen, estimated it at about 1000 propositions, and was merely making use, like Bondi, of masses of cheap research student labour, as opposed to a few expensive professors. And that, alas, was probably the beginning of the bad effect of paying for, and promoting because of, research.

You may ask why do I describe the famous Mouseion at Alexandria as a government research establishment. In fact it was probably the first,[2] and it probably possessed all the attributes of a research establishment, tenured posts, excellent library and plenty of slave labour. One can easily conjecture the kind of conversation that must have once taken place.

One evening Alexander the Great as a youth comes up to his tutor and says:
Alexander: "I have a problem."
Aristotle (*who happened to be his tutor*): "Yes?"
Alexander: "In my plan to conquer the world it is obviously best to use a single well organised army. But as I capture each country, and then move on to the next, how do I keep control of the previous country?"
Aristotle (*after a pause, with a far seeing glint in his eye*): "Aha! I think I have the solution. You want to found a government research establishment. You could even name it after yourself. Then the sociology department (reference 2, p. 20) could manufacture suitable religions grafted on to the appropriate local beliefs that would keep the natives happy."

"As a matter of fact," and at this juncture Aristotle's tone of voice becomes noticeably casual, "as a matter of fact I have a very good student* who could do the architecture for you—he's eager to experiment with white marble—and another senior student† who would make a splendid first director of the place."

Aristotle's voice regains its normal timbre
"I suppose you'll have to have an arts man as first librarian—and there is an elderly Homer scholar‡ who would do—and he would have the advantage of being near retiring age so that as soon as he'd done the chore of setting up the catalogue system you could get rid of him and replace him by a proper scientist."

Aristotle's voice goes casual again. "And as a matter of fact I have just the man§ for the job, a student who is a brilliant all-rounder, interested in astronomy, geography, literature, the lot, but he needs a few more years of research before he takes on administrative chores. Oh yes—and I have another young student‖ who's a bit of a crank, but marvellous with his hands. His ambition is to build a giant lighthouse, but he can't get any funds. But in a government research establishment this would be well worth the cost, just from the prestige point of view alone, besides being actually quite a useful piece of equipment."

"I suppose you'll have to have a philosophy department, although to tell the truth the subject is a bit played out after Plato and myself, and most of my current students are rather second rate (reference 2, p. 40). On the other hand biology, psychology and medicine are really up and coming new subjects, and I have a splendid young man¶ who has done some fascinating work on the psychology of sex and nervous breakdowns, who would be ideal to head a research group."

"And let me see—you'll need a mathematician of course—and although I don't have any suitable students of my own available just at this moment, there is a young man** in Plato's academy. Not that he's very good at research, in fact I doubt he'll ever make his PhD, but he's quite a good scholar, and quite good at editing things. And although he's a bit humourless, he would make an excellent administrator, and so I'd recommend hiring him to set up the mathematics department."

"Oh—and another point—if I were you I would choose somewhere on the mediterranean coast, with a nice climate and a sandy beach with good bathing facilities, and not too far from the main shipping lanes.

* Dinocrates.
† Demetrius Phalerus.
‡ Zenodotus.
§ Eratosthenes.
‖ Sostratus.
¶ Erasistratus.
** Euclid.

---

* Professor Coulson would have been giving this talk, but for his untimely death.

Published in the Bulletin of the Institute of Mathematics and its Applications, Volume 10, 1974, pages 272-281.

As a matter of fact I had a vacation last year at just such a place, a little island called Ras-el-Tin (reference2, p.7). For that way you'll not only be able to attract some decent academics on to the staff, but you'll also guarantee a good flow of visitors each summer to keep the place academically alive. In fact it might even last a few centuries."

And that's exactly what Alexander did, in every detail, when he was 23.

Previous to Alexandria, there had only been private academies, which lacked permanence because they depended upon personalities, and tended to dissolve when the latter moved or died. The most famous example was Plato's Academy in Athens. Plato himself must have been one of the best, and stupidest, research supervisors in mathematics of all time. According to Tzetzes[3] he even had a notice written up over the porch of his academy: "Let no one unversed in geometry enter my doors," but when you come to see what he himself thought geometry was, you find a rather pedantic insistence upon ruler and compass constructions, which was stupid. It is a regrettable tendency amongst some mystics and philosophers to become so obsessed with some small facet of mathematics or science that happens to capture their imagination, that it biases their view of the whole.

The trouble was that Plato had so strong a personality, he almost managed to redirect the whole steam-roller of Greek mathematics in the wrong direction (as Cauchy later did succeed in doing, by discovering Cauchy's Theorem, and thereby redirecting the whole of nineteenth century mathematics away from real analysis into the jaws of complex analysis).

Chapter 2. Rediscovering Eudoxus of Cnidus

Luckily for the Greeks, Plato had at least one research student who was his match, and who was, in my opinion, the greatest of all the Greek mathematicians, Eudoxus. He was born in Cnidus about 408 BC, and Plato was about 20 years his senior (see reference 3 vol.1, pp. 320–334). Eudoxus was 23 when he became Plato's student, and so Plato must have been in his early forties, approaching the height of his powers. Meanwhile Cnidus was in what is now mainland Turkey, and so I suppose you could call Eudoxus something of a young Turk. One can imagine the opening conversation:

"Well young man, I have here 4 problems which you might like to try your hand at, left here by a fellow called Zeno."

One only has to look at Plato's Dialogues to know that he always went straight to the heart of the matter irrespective of whom he was talking to. And I believe this is a great virtue in a research supervisor: myself I much prefer Professor Atiyah's gold to Professor Roger's tin, especially during the first year of research.

But imagine Plato's astonishment when Eudoxus returns shortly with a closely written sheaf of papyrus claiming to have solved the lot. Then Plato's second virtue as a research supervisor comes out: insistence on clarity of communication.

"My dear young man," he says handing it straight back to the crestfallen Eudoxus, "you must *explain* the solution to me in words of one syllable, just as I explained the problem to you. We philosophers believe in the value of debate."

And as a matter of fact this used to be exactly Norman Steenrod's description of how to write a mathematical paper: "imagine you are going on a long walk with a friend, and you are telling him about the theorem—write the paper in *that* order."

But to return to Eudoxus' predicament of having to do mathematical research in an academy of philosophy. In effect he had to reduce the proof to as short a time as those argumentative philosophers would allow him to hold the floor. And being the greatest of all the Greek mathematicians, he meets this challenge. In effect he reduces the proof to one line, Definition 5 in Euclid Book V (see reference 1, vol. 2, p. 114, 120–129).

The problem was to define *ratio* between incommensurable magnitudes, when there was no definition of real numbers, nor any definition of how to add or multiply irrationals. His solution was to define the correct equivalence relation between pairs as follows. Let N denote the positive integers.

Eudoxus' Definition of Equivalence

$(a, b) \sim (a', b')$ if, for all $m, n \in N$, $ma \gtreqless nb$ as $ma' \gtreqless nb'$. The equivalence class of the pair (a, b) is called the *ratio* ($\lambda \acute{o} \gamma o s$) and is denoted $a\!:\!b$. Thus

$$a\!:\!b = a'\!:\!b'.$$

Now the beauty of the definition is its generality, because a, b, a', b' can be any kind of magnitudes, space intervals, time intervals, areas, volumes, musical notes, integers, rationals, irrationals, etc.—in fact the elements of any ordered set on which N operates. In the special case that they are real numbers (which *we* know to exist by Weierstrass and Dedekind, but which the Greeks did not), we can divide, and so the condition reduces to

$$\frac{a}{b} \gtreqless \frac{n}{m} \text{ as } \frac{a'}{b'} \gtreqless \frac{n}{m}.$$

Therefore $a/b = a'/b'$ because they determine the same Dedekind cut of the rationals.[4] Hence we can identify $a\!:\!b = a/b$. But Eudoxus' definition is much more general* than merely referring to the reals—it is the beginning of abstract algebra, and I shall develop arguments to suggest that he was by far the greatest algebraist of the Greeks, as well as being by far their greatest analyst, and in the very top rank both as geometer and astronomer.

Observe, in passing, that one can trace a direct line from Eudoxus to the discovery of the reals.

Eudoxus–Euclid–Bolzano–Weierstrass–Dedekind. The key link in this case is Bolzano†. He in his autobiography[5] confesses that the one work that really switched him on was the beauty of Euclid Book V. It was Bolzano who first introduced the ϵ, δ technique, and gave a rigorous definition of continuity. Bolzano never uses infinitesimals, while Cauchy at the same time was still using them freely. I think one can trace Bolzano's inspiration for this technique directly to Euclid Book V.

Now according to Proclus[1,3] Euclid Book V summarised Eudoxus' theory of proportion, which was traditionally recognised by the Greeks as being the "crown of Greek

---

* According to Proclus[3]: "Eudoxus . . . was the first to increase the number of so called general theorems."
† I am indebted to David Fowler for not only introducing me to the early nineteenth century, but also for a great deal of enjoyment of mathematics.

mathematics."[6] Personally, although I too admire the beauty of the book, and its rigour, and recognise it as the earliest surviving work on modern abstract algebra, nevertheless I suspect that it is a travesty. I suspect that Euclid ruined Eudoxus' theory by first misunderstanding it and then reporting only a fragment of it in the wrong order, so that he may have successfully prevented it from even being fully rediscovered. Because, alas, no actual words of Eudoxus survive today—the originals were probably burnt when the library at Alexandria was destroyed by Theophilus in 391 AD (reference 2, p. 55).

But first let me explain how Euclid used the little bit of Eudoxus that he did report. Before Eudoxus the Greeks were unable to state rigorously any similarity theorems. For example in Fig. 1 the equality $a:b = a':b'$ can only be stated either if one has the real numbers and a definition of division (neither of which the Greeks had) or if one has Eudoxus' definition of ratio. That is why Euclid had to postpone all similarity theorems to Book VI after Book V.

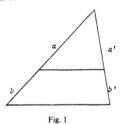

Fig. 1

That is also why Professor Penrose, as a boy, was astonished to discover the delightful similarity proof of Pythagoras' theorem:

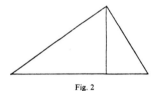

Fig. 2

For thanks to the dreadful influence of Euclid's pedantry upon schoolmasters for 2000 years this particular jewel of a proof was officially suppressed, because, in the interests of rigour*, it should properly be postponed until after either the proof of the existence of the real numbers, or Euclid Book V, neither of which is accessible to school mathematics. As a result school children were denied the benefit of intuition of the reals and forced to swallow the 47 propositions of Euclid Book I, in order to reach Euclid's ingenious but lumpish proof, independently of the reals. And of course what they gained on the swings they lost on the roundabouts, because they also had to swallow the rather shady axiom system of Book I.

Today we get exactly the same phenomenon, but *worse*. Because, in the interest of rigour*, and thanks to Euclid's modern counterpart Bourbaki, poor young French children have the axiom system for the reals forced down their unwilling gullets.

Myself I prefer the beautiful Chinese proof of Pythagoras' Theorem:

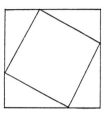

Fig. 3

For this has not only a pleasing symmetry, but the superior advantage of being provable either way, either with the reals, or by the chopping up methods of Euclid Book I.

Another essential place where Euclid has to use Eudoxus' definition of ratio is in Book XII. Here $a:b$ is the ratio of the volumes of two cones of the same height, and $a':b'$ is the ratio of the areas of their bases. This time the proof $a:b = a':b'$ (again without the reals and hence without integral calculus) is a real *tour de force*, also due to Eudoxus, for which he invented the theory of exhaustion (assuming false, and proving a contradiction by a finite inductive chopping up process). From this he deduces:

volume of cone = $\frac{1}{3}$ base × height.

There was no other rigorous proof of this simple fact until Dedekind[4] discovered the reals in 1854, thereby endorsing the use of integral calculus. And eventually in 1900 Dehn's solution[7] of Hilbert's 3rd problem[8] showed that, without the reals, this was the only way Eudoxus could have done it. This is an example of Eudoxus the analyst—not the only example because for instance he also invented the hippopede to describe planetary motion.[9]

But I am more interested in Eudoxus the algebraist, and so now let us turn to the more serious crimes of Euclid, the crimes of omission rather than commission. Euclid is the classic example of the dangers of putting a "good scholar" in charge of a university, or research institute, or curriculum reform, as opposed to the "researcher." *For good scholars tend to compartmentalise knowledge, while researchers try to synthesise it.* So Euclid put pure mathematics in the mathematics department, and applied mathematics in the applied departments, in the astronomy and geography departments at Alexandria, insisting upon separation. And in Book V he suppressed the marvellous stroke of genius of Eudoxus, that his definition of equivalence also suffices to define *velocity*.

One can even hear the humourless Euclid enunciating: "Velocity is done by the astronomy department."

Historically it is difficult to find any rigorous evidence for or against the hypothesis that Eudoxus solved Zeno's problems.[6] But if we look at the problem *intrinsically* from the content of the mathematics itself as opposed to *extrinsically* from the chance observations of a philosopher like Proclus, writing seven centuries after the event, then I put it to you that the evidence becomes overwhelming. For, at the time of Eudoxus, faced with Zeno's paradoxes,

---

* Rigor mortis.

the problem of defining velocity to the satisfaction of the philosophers must have been as formidable a task as the understanding of limits, and remember that to satisfy the philosophers was Eudoxus' particular predicament, because he was a research student in Plato's academy.

"If you can't even show us how to divide incommensurables" the philosophers would pityingly smile "how absurd to suggest that you can divide space by time."

Without the real numbers, and knowing the Greek repugnance for units—they would be most reluctant to base so fundamental a definition as velocity upon the arbitrariness of choice of unit—the problem must have seemed insoluble.

Surely this must have been the main content of Zeno's third paradox, The Arrow[3,6]:

A moving arrow at any instant is either at rest or not at rest, that is, moving. If the instant is indivisible, the arrow cannot move, for if it did the instant would immediately be divided. But time is made up of instants. As the arrow cannot move in any one instant, it cannot move in any time. Hence it always remains at rest.

The first step in resolving the paradox is to present a satisfactory rigorous definition of velocity, so that at any rate we can replace the word "moving" at the end of the first sentence by "it has a velocity." And Eudoxus can easily provide the required definition of velocity by a minor modification of his famous definition of equivalence above. For take a, b to be lengths, and a', b' to be time intervals, and define

$(a, a') \sim (b, b')$ *if the same condition holds.*

Define the *velocity* of an arrow travelling length a in time a' to be the equivalence class of (a, a'), which we again denote $a:a'$. There is plenty of evidence that Eudoxus was in the habit of switching terms like this, because in the special case that all four magnitudes are of the same type we find in Euclid Book V Proposition 16, the Alternando,

$a:a' = b:b' \Leftrightarrow a:b = a':b'.$

Again there is plenty of evidence that he was used to situations in which two of the magnitudes were of one kind, and the other two were of another kind—witness the result from Book XII where a, b are volumes and a', b' areas. Indeed in the famous definition he explicitly omits the word ὁμογενῶν (homogeneous), which is what he would have used had they been all of the same kind.

I have said that Eudoxus could easily have provided the rigorous definition of velocity. Of course we can never know since apparently no works of his survive. All we have are the beautiful fragments in Euclid Books V, X and XII. But I put it to you, given the environment, given the problems of the time, given that famous definition, and given how easily it solves both problems of defining ratios of incommensurables *and* the definition of velocity, and given that Eudoxus' theory of proportion was known as the crown of Greek mathematics, is it not irresistible to conclude that he must have intended it to do both? One can even conjecture the very words that he would have used:

(3α) Λόγος ἐστὶ δύο μεγεθῶν ὁμογενῶν ἡ κατὰ πηλικότητα ποιὰ σχέσις.

(3β) Τάχος ἐστι δύο μεγεθῶν, πρῶτον μὲν τοῦ μήκους δεύτερον δὲ τοῦ χρόνου, ἡ κατὰ πηλικότητα ποιὰ σχέσις.

The first occurs as Definition 3 of Euclid Book V (see reference 1, vol.2, p. 116), and the second is my invention* to match. My translations are

(3a) *Ratio is an equivalence class, with respect to size, of a pair of magnitudes of the same kind.*

(3b) *Velocity is an equivalence class, with respect to size, of a pair of magnitudes, the first a length and the second a time interval.*

Let us briefly summarise the first 5 definitions of Euclid Book V.

    Definition 1: Multiple
    Definition 2: Submultiple
    Definition 3: Ratio
    Definition 4: The Archimedean axiom†
    Definition 5: The equivalence relation

Beautiful. Absolutely minimal. But of course I conjecture that Eudoxus had (3α) *and* (3β). And as Euclid was copying out the list for his lectures on pure mathematics his scholarly compartmentalist approach induced him to snip out (3β) and send it along with a memo to the astronomy department. And of course the head of the astronomy department, like any sensible busy experimentalist of today, flicked it straight into the wastepapyrus basket. And so it was dropped from the syllabus. And then forgotten. And sometime later the original was burnt. And so irretrievably lost. Or perhaps not quite? Moral: Never entrust the safe-keeping of research to government research establishments, but rather to many individually opinionated academics, each of whom will have his own particular prejudice of what is important. Never trust mammoth coverages of all mathematics, like Bourbaki, but carefully keep original papers, especially the collected works of great mathematicians.

Returning to Euclid, we now see a familiar phenomenon take place. Whenever you introduce a subtle new concept in a lecture course (like equivalence class), and you only introduce one example, and that the most obvious (like ratio), then your audience will understand neither the subtlety of the new concept nor why you are making such a fuss. And that is exactly what happened to (3α). It has baffled all the translators of Euclid, from Barrow to de Morgan, from Heiberg to Heath. In the latest authoritative translation into English,[1] you will find a baffled essay by Heath. And if you look at Heath's translation of (3α) it is

(3a′) *A ratio is a sort of relation in respect to size between two magnitudes of the same kind.*

He has interpreted it as the teacher's reassuring fatherly pat to the 5-year old learning fractions for the first time "you see a fraction is a sort of relation between numerator and denominator." But nowhere else in the whole of Euclid do you ever find a reassuring fatherly pat. In fact I even suspect that he may have been neither a father, nor a person, but a Bourbaki‡. All the other Greeks did at least have a birthplace, and began their papers with an introduction, as we do today, but Euclid

---

* Thanks to assistance from Hugh Dickinson over the grammar.
† Eudoxus is of course 100 years before Archimedes.
‡ Bourbaki,[10] on the other hand, if not exactly a father figure, does at least adopt the rôle of traffic cop on dangerous bends, and does at least provide an introduction. But Bourbaki clearly identifies himself with Euclid by his choice of title, and he evidently finds Euclid *très sympathiques* because in the second sentence of that Introduction to Bourbaki Book 1 Part 1 Chapter 1 we find "...what constituted a proof for Euclid is still a proof for us..."

begins his:

"Definition 1. A point..."

The difference between our two translations all hinges upon the use of the phrase "ποιὰ σχέσις." Heath translates it as the warm fatherly pat "a sort of relation" whereas I translate it as a cold definition "an equivalence class." To the modern mathematical ear, one translation sounds vague and the other precise, but this is merely a question of conventional jargon. If you ask 99 per cent. of people—and I tried it out on my wife—they say both phrases sound fairly vague. Similarly 99 per cent. of ancient Greeks, and probably 100 per cent. of modern Greek scholars, will say that ποιὰ σχέσις is vague. But Eudoxus, and the algebra school of 370 BC, may have chosen that particular phrase as their particular jargon for equivalence class. Of course what needs to be done is a careful study of Greek papers in *abstract algebra* of that period, of which probably none survives. It may not be any good examining contemporary papers in geometry, because they may not have needed to use the jargon, nor papers of 100 years later in applied mathematics, like those of Archimedes. Nor will it be much use examining the later algebra papers of the Alexandrian school, because by that time Euclid will have fouled up the whole thing as we now unfold.

The next step in our detective story concerns Zeno's 4th paradox,[3,6] the Stadium:

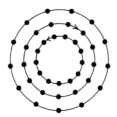

Fig. 4

Consider three rows of bodies, of which A is at rest while the other two B, C are moving with equal velocities in opposite directions. By the time they are all in the same part of the course B will have passed twice as many of the bodies in C as in A. Therefore the time it takes to pass A is twice as long as the time it takes to pass C. But the time which B and C take to reach the position of A is the same. Therefore double the time is equal to half the time.

To sort this out, all Eudoxus would have needed was the theorem

$$(a:a') : (b:b') = (a:b) : (a':b').$$

Then he would have completed the solution of all 4 of Zeno's paradoxes, the first two by ratios and the second two by velocity. But when we come to look for this theorem in Euclid Book V we find to our surprise that it is missing. We look a little closer and are shocked to find that *Euclid forgot to define the ratio of two ratios*, the kind of basic definition that anyone with an ounce of sympathy for algebra would have felt obliged to mention. Although it is needed for ratio of velocities, apparently it happened not to be needed for the geometry*, all that Euclid really cared about, and so perhaps that was why

* Mo Hirsch points out that it is needed for cross-ratios, which perhaps explains why Euclid never achieved projective geometry.

he omitted it. Then a little closer, to discover with the horror that he *couldn't* define the ratio of two ratios, because he had so messed up the order of the propositions, that he had made it impossible for himself. And at last we discover the cause of the mess—none other than the *Euclidean algorithm*, which properly lives in Book VII Proposition 2, where it is used to find the greatest common divisor of two integers. This little trick, that was obviously Euclid's only piece of research as a research student, he is eagerly awaiting to include in his lectures at the earliest possible opportunity, and he suddenly spots an opening to use a similar technique in Book V Proposition 8. But this use of the additive structure, and especially the use of *subtraction*, is unnecessary at this stage, and far too early for the theory of proportion. And putting it here wrecks the whole of Eudoxus' careful plan. Moreover it is clear that Euclid does not fully understand the theory of proportion, because he is so nervous of it—otherwise why, when he gets to his own familiar little back-garden in Book VII amongst the integers, why does he dare not use it?—and, instead, unnecessarily repeats a homely watered down version of it, that is useless for anything else.

The only way to disentangle Euclid's mess is to rewrite Book V in modern category theory.[11] And then the exquisite delicacy of Eudoxus' famous definition really becomes apparent for the first time.

In effect Euclid uses Eudoxus' definition as a functor

$$\mathscr{A} \xrightarrow{f} \mathscr{C}$$

from the category $\mathscr{A}$ of sets of magnitudes to the category $\mathscr{C}$ of sets of ratios. An object $A \in \mathscr{A}$ is an ordered set with additive structure. An object $C \in \mathscr{C}$ is an ordered set with unit, inverses and Q-action, where Q denotes the positive rationals. And of course there is no feedback functor $\mathscr{C} \to \mathscr{A}$, so Euclid could not define ratios of ratios.

However now look again at Eudoxus' definition: it is really a functor defined on a much more delicate category $\mathscr{B}$, where an object $B \in \mathscr{B}$ is just an ordered set with N-action. And this category sits neatly in between $\mathscr{A}$ and $\mathscr{C}$ with forgetful functors φ, ψ feeding into it from either side. Therefore Euclid's functor can be factored, $f = g\varphi$, where g denotes the more delicate Eudoxus functor. Therefore we have the diagram

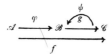

Now there is a feedback $g\psi$ enabling us to define ratios of ratios as desired.

I am not claiming that Eudoxus invented category theory, but I am convinced that he was aware of the essential mathematical content of the above diagram, otherwise why would he have chosen exactly so delicate a weapon?

If Euclid had had the same feeling for algebra that he had for geometry, then he would have given the additive structure in Book V the same royal treatment that he gave to the parallel postulate in Book I. He would have written the first half of Book V in the category $\mathscr{B}$, and the second half in the category $\mathscr{A}$. He would have postponed the crucial Proposition 8:

$$a < b \Rightarrow a{:}c < b{:}c,$$

upon which so many of the key results at the end of the Book depend, until the second half, because it belongs in the category $\mathscr{A}$, and is actually *false* in the category $\mathscr{B}$. By proving it too early, in his eagerness to use his algorithm techniques, Euclid is forced to introduce the additive structure, and particularly subtraction, too early, and therefore, in effect, to trample the delicate category $\mathscr{B}$ out of existence.

Back to Eudoxus: of course once he had ratios of ratios he would easily have been able to reiterate the process and finish up with a most interesting algebraic object, an ordered Archimedean group, that was commutative, but *not necessarily associative*. The multiplication is defined by

$$ab = (a{:}1){:}(1{:}b).$$

If I have persuaded you that Eudoxus knew how to take ratios of ratios, then it follows inexorably that he must have possessed a form of group theory. But not the same as ours, because ours is multiplicative, whereas his was dividative*, and they are subtly different.[11] For example for us associativity is natural and commutativity exceptional, whereas with him it is the other way round. For him commutativity follows from

$$(1{:}b){:}(a{:}1) = (1{:}a){:}(b{:}1)$$

whereas there is no natural expression of associativity in terms of division. Of course it never occurs to modern group theorists to study non-associative groups, nor to look at group theory from the dividative point of view, but this is only because they absorbed multiplication and inverses with their mothers' milk, and look upon division as a secondary operation.

"But multiplication is more natural" they will insist, "because it represents composition of maps." However that is only a pure mathematician's point of view. Looking at nature, the applied mathematician is always comparing lengths, temperatures, musical notes, etc., and so, as Eudoxus would have said, perhaps ratios are more natural than products. There is a moral here, due to René Thom: just as a baby babbles in the phonemes of all the languages of the world, but after listening to its mother's replies, soon learns to babble in only the phonemes of her language, so we mathematicians, babbling in all the phonemes of mathematics should perhaps cock an ear now and then towards mother nature. More of this in Chapter 4 below.

The lost category of Eudoxus, $\mathscr{B}$, contains several interesting objects, which he may or may not have known about, such as space-time and the tangent-bundle of the reals.[11] If we apply this functor, g, to either of these, we obtain another interesting object, a non-associative extension, G, of the multiplicative group of positive reals, R_+. One can construct G from R_+ by adding on either side of each rational, r, a predecessor r_- and a successor r_+. It is as if Eudoxus is saying reassuringly to Pythagoras and Zeno: "Don't worry, Achilles can safely approach either side of a rational tortoise, but at the same time you were quite right in feeling that irrationals were less approachable."

---

* My research supervisor, Shaun Wylie (who is both mathematician and classical scholar) says "dividative" ought to be "divisive," but somehow I feel this conjures up the wrong overtones. Anyway he also says that bicycles ought to be called dicycles.

Conjecture: G is the unique maximal group in $g\mathscr{B}$.

The smallest non-associative group in $\mathscr{C}$ is of order 3, and is the quintessence of dividative group theory, in the same way that the smallest non-abelian group of order 6 is the quintessence of multiplicative group theory, and I am tempted to name this little group the *eudoxan, E*. The eudoxan appears as the group of ratios of any ordered set of more than one element, on which N acts trivially. Consequently E has a natural notation and multiplication table:

	<	=	>
<	<	<	=
=	<	=	>
>	=	>	>

There is also an additive form $\{-, 0, +\}$ of the eudoxan, because it appears as the fibre over Q of the orientation-bundle plus zero-section of the tangent bundle of any ordered set containing Q. The exponential map induces an isomorphism:

$$\{-, 0, +\} \xrightarrow{\exp} \{<, =, >\}.$$

One can argue that in the additive eudoxan *two minuses make a plus*, because the minus can be interpreted as the reversal of orientation of the underlying set. On the other hand one can argue that in the multiplicative eudoxan *two minuses make a minus*, from the multiplication table. If Eudoxus ever got an inkling of *that*, I can imagine him delightedly putting it as a paradox to all his colleagues, especially to tease the philosophers, because there is plenty of evidence that paradoxes were fashionable in those days. But alas, they probably couldn't understand, and so the only legacy that he was able to donate to his successors may have been a thorough nervousness about the minus sign, and an instinct never to touch it with a barge pole unless it was absolutely necessary. Perhaps that is why the Greeks were so chary of using the minus sign. All except Euclid, that is, who triumphantly needed it for his algorithm, and proceeded to trample with it all over Eudoxus' delicate and beautiful theory of proportion. I hope that Euclid is turning in his grave (or their graves).

I hope you will forgive my rather longwinded story of Eudoxus, which I have told for several reasons. Firstly, as you will have guessed I am a Eudoxus fan and hopeful that I might persuade one or two of you, if you have not already tried it, to blow the dust off your copies of Euclid, and follow Bolzano into the beauties of Book V. Secondly I wanted to bring to life my opening remark about mathematics being one of the oldest endeavours of man, and one through which our colleagues of yesterday can still speak to us with the freshness of today. Greek mathematics can still stimulate current research problems.

Thirdly I wanted to make a point about grand treatises like Euclid and Bourbaki, or centralised curriculum reforms. They are splendid in some places, but cannot help being biased in others. Euclid was splendid on geometry, but poor in algebra and applied mathematics. Bourbaki is splendid in algebra and analysis, but poor in geometrical thinking and applied mathematics. For example Dr. Howlett would have been as astonished by the position of Cauchy's Theorem in Bourbaki, as was Professor Penrose by that of Pythagoras' Theorem in Euclid. These treatises are trying to impose an artificial unity from without, that is liable to stifle growth, whereas allowing free expression to the opinions of many individual mathematicians enables the subject to evolve its own unity from within. More of this in Chapter 4. Meanwhile these grand treatises can cause mathematical loss, not only as we have seen in classical times, but even today. For compare the cautionary tales set out in Table I.

Table I

The year	The mind	The Greek	The modern
0	The discoverer	c.540 BC Pythagoras discovers that $\sqrt{2}$ is irrational.	Newton and Leibniz discover calculus, and in 1686 Newton publishes his "Principia."[12]
80	The iconoclast	c.460. Zeno poses his Paradoxes.[3,6]	In 1734 Bishop Berkeley publishes "The Analyst."[13]
160	The resolver	c.380. Eudoxus creates the theory of proportion.	Weierstrass c. 1850 and Dedekind on November 24th, 1854, create the real numbers.[4]
240	The expositor	c.300. Euclid writes his Elements.[1]	In 1939 Bourbaki begins publishing his "Éléments."[10]
320	The loser	By 220 the bias towards geometry has caused half the theory to be lost.	By 2020 the bias towards algebra will have caused the other half to be lost.

"How can we lose mathematics today," you may ask, and I will tell you. By the year 2010 the exponential communication explosion will probably have pushed most books off the library shelves on to tape. Soon after that, or maybe earlier, computer control will be introduced increasingly into libraries, with the automatic WC-instruction, which says: *if a tape is neither consulted, nor cited, for 10 consecutive years, then Wipe Clean.* So by 2020, sure enough, much of our mathematics will be wiped clean. And then, for the next 2000 years, our children at school will have their horizons bounded by *Bourbaki Book I, just as for the last 2000 our forefathers had theirs bounded by Euclid Book I.

Chapter 3. Teaching and research today

Turning to the present day, I should like, if I may, to put forward some tentative personal opinions. People often speak of the conflict between teaching and research, but I find the reverse. Of course there is the natural conflict with the myriad other interesting things to do in life, because the day only has 24 hours. But apart from that I find much of my teaching stimulates research, and much of my research is oriented towards teaching. For example my interest in the Greeks arose a couple of years ago, because I have to give an annual presidential address to the Polygons, our local sixth-formers association, and in order to keep the interest of the staff I search around for a new topic each year. On balance I agree with UGC policy to pay only for teaching: lecturers should be paid to lecture to the students, readers to read to them, and professors to profess to them. How well we can do this depends largely upon the staff/student ratio, and ultimately upon the richness of the country. Some universities in poor countries have to cope with ratios of 1:100 or more, but, as well as teaching, the staff there still seem to manage to do a little research in the crevices of time, in the early mornings, or late at night, or on Sundays, because they deeply enjoy research—and that is the secret. We will always do research because we love it, and we will always promote people for it because they win our admiration.

I find myself very much in two minds about research institutes: it is clear that the IAS at Princeton has benefited US mathematics enormously, but then the US is a rich country. In a relatively poor country like India it could be argued that the Tata Institute may have done as much harm as good, by taking many of the best mathematicians out of the universities and largely away from teaching. Of course government research establishments are a different matter, and it is clear that governments should support the development of socially useful projects like controlled fusion. But as I have tried to explain with my story of Eudoxus it would be a fundamental mistake to try and separate teaching from research, assigning teaching to the universities and research to research institutes and research establishments, as some advisers would seem to have us do in the future.

Another of my hobby-horses is the advantage of ignorance, in that it encourages creativity, both in the young and the old. May I tell a story of my first few diffident steps as a young research supervisor? I dutifully started running a seminar for my students on manifolds in about 1958, and Professor Penrose started coming along. I explained the embedding $M^n \subset R^{2n-1}$ by general position, and then conjectured that $M^n \subset R^{2n}$ because we couldn't find any counterexamples. In my ignorance I did not know that Whitney[14] had proved it 14 years previously, and that it was a well known result. The normal well educated thing would have been to equip my students with the techniques of Whitney's proof. Instead we all had a go at proving it ourselves. Roger Penrose said:

"Well if it only crosses itself in isolated double-points, why couldn't we eliminate each one, by putting a loop going off on one sheet and coming back on the other, and then putting a cone on the loop, which wouldn't meet M again if $n > 2$, and then we could slip the double-point off the top of the cone."

And I replied: "But if the loop was sort of knotted up with M, then M would get entangled with the knot as we slipped it off." And there we stuck until one day I mentioned it to Henry Whitehead over a beer. He said "That's OK by regular neighbourhoods—there's an old 1939 paper of mine that nobody ever read because* of the war." And so Penrose and I had the honour of being joint authors of Whitehead's last paper.[16] For the result of those two brief conversations was a proof, different from Whitney's, that gave, instead of one result, a whole cascade of results that reopened a chapter of geometric topology.

Another point I would like to make is that even administration can sometimes help research, not only the research of those administered to, but also that of the administrator himself—although it probably requires pretty strong self-discipline to survive as a researcher for more than 5 years of heavy duty administration. I remember when the Warwick Research Centre was being

---

* "Théorie des ensembles."[10]

* Actually nobody read it because it is practically unreadable.[15]

set up, its Advisory Board thought that in addition to running symposia* in the subjects of Warwick expertise, it ought also to run one now and then in fields that were booming internationally, but by accident happened to be underrepresented in this country. Examples of such symposia were 1966/7 Harmonic analysis, 1968/9 Qualitative theory of differential equations, 1971/2 Algebraic geometry. For the hope was that we might thereby stimulate a few British mathematicians to enter those fields. And what with the administrative business of finding out what those fields actually were, and whom we ought to invite, I found myself hoist by my own petard. I became a victim of my own administrative policy.

Chapter 4. Qualitative developments in science

I was particularly intrigued by the meaning of the word "qualitative." One might define a *qualitative property* to be a diffeomorphism invariant, as opposed to a *quantitative property* which is an affine invariant. In a science in which different experimenters may use non-linearly related scales to measure the same data, only qualitative conclusions can be deduced from the resulting experimental graphs. Thus the laws of those sciences must be expressed in qualitative language. This is particularly true of the social sciences.

Now, although diffeomorphism invariants include topological invariants, many of which have been known for a long time, such as Betti numbers and homology groups, very few have in fact proved useful in describing experimental graphs. The qualitative language available for describing graphs has been of such poverty, until recently, that its existence was barely recognised. It was limited to a few words such as "increasing" or "single-valued," and as a result two things occurred. Firstly any qualitative mathematical statement about a graph could be translated so easily into everyday language, that it was not recognised as being mathematical by scientists. Secondly such statements were so obvious that they were too simple to be classed as laws, and were accused of being both trivially true, and trivially false. Let me give an example:

Hypothesis 1.
Mathematical enjoyment is an increasing function of creativity.

Or, better still, the translation into English: "the more creative the more enjoyable." Most people would admit that this is trivially true, but sometimes it can become trivially false, if one happens to be enjoying reading somebody else's work, for instance, rather than busy creating one's own.

The trouble is that the mind hops on to the statement too quickly and hops off again to consider all the exceptions. If, on the other hand, the statement has the power to arrest the mind for a while, and to synthesise a variety of phenomena, one is more ready to accept the statement as a first approximation to the truth, and even perhaps to call it a law. One is more ready to forgive the law for not being quite true in all circumstances, and rather than admit that those circumstances actually disprove the law, one is inclined to take the more lenient view that perhaps the law needs modifying occasionally. This is certainly the case with the great laws of physics such as Newton's Law of gravity or Boyle's Law for

---

* Symposia are named after Plato's original idea of the Symposium, which was great conversation over drinks.

---

gases, both of which are false. Nevertheless we are still quite happy to call them laws, and to have them existing alongside modifications such as relativity and Van der Waals' equation. I suspect that philosophers are inclined to be a little too overimpressed by the so called laws of physics, and that social scientists a little too overawed, and I would hope that the latter should begin to approach the whole matter of laws in a more adventurous spirit. If I may be forgiven, may I quote from a recent paper with Carlos Isnard[17]:

"*A scientific law is an intellectual resting point*. It is a landing, that needs must be approached by a staircase, upon which the mind can pause, before climbing further to seek modifications."

Of course there may be more than one staircase rising from that landing. And so a science begins to grow and fork like the trunk and branches of a tree, with the forks representing the laws of that science, synthesising the ideas below them. Each science is like a grove of trees, and the most delicate features of that grove will be the blossoms and leaves, opening fresh each spring, and giving it its shape as seen from a distance. The blossoms represent the conversations between the scientists involved, and the leaves their experiments and research papers. The blossoms allow for cross-fertilisation, and the leaves provide the wherewithal to help new twigs to swell into new strong branches. And each autumn the leaves fall, providing humus to feed and strengthen the main trunks. Some trees may rot and fall, but the grove is refreshed and sustained by the appearance of new saplings, just as a science is refreshed and sustained by the appearance of new paradigms.[18] This view of science frees us from vain attempts to impose artificial unity from without, and allows us to admit to an open ended concept of unity evolving from within.

"It's all very well waxing lyrical" the social scientist will reply "but the poverty of the qualitative language offered to us by the mathematicians makes it very difficult to get any kind of tree off the ground, let alone a grove." But that is just where the social scientists might be wrong, because there is a new fertiliser on the market, called catastrophe theory,[17,19] ideally suited for stimulating the growth of new paradigms.

Catastrophe theory substantially enriches the language of diffeomorphism-invariants with words that are more subtle, neither trivial, nor so easily translatable into non-mathematical language. And they have impressive power of synthesis. In fact if I had to select one extra tool out of the whole of mathematics to add to the Hammersley tool-kit, I would choose the cusp catastrophe, pictured in Fig. 5. The cusp catastrophe describes the canonical way that two control factors can interfere with one another when influencing the same behaviour mode.

Chapter 5. Why mathematics is sometimes exciting and sometimes dreary

To conclude the paper let me suggest one application of the cusp catastrophe that synthesises much of the intuition about the teaching of research students described by previous speakers. This example is an extension of Hypothesis 1 above, but instead of being trivial it fulfills the requirement of giving the mind pause. It is a light hearted example, with no claim to be a law, and mathematically it is not as serious as the social science

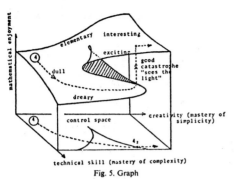

Fig. 5. Graph

models in reference 17, because there is no dynamic maximising the probability, nor suggested design of experiment*. Therefore the surface pictured above represents only the most probable behaviour rather than the actual behaviour. The purpose of the example is to summarise, to give insight, and hopefully to contain a few germs of truth.

Hypothesis 2.
Creativity (mastering of simplicity) is a normal factor, and technical skill (mastering of complexity) is a splitting factor, influencing mathematical enjoyment.

The definition of normal and splitting factors are given in reference 17, but for our purposes it is sufficient to say that the hypothesis means that the graph of enjoyment, as a function of creativity and technique, looks like Fig. 5. In particular in elementary mathematics, where little technique is needed, Hypothesis 1 becomes a special case of Hypothesis 2, and so the latter is a generalisation of the former. Meanwhile the acquisition of technique has a splitting effect, causing the function to become double-valued, and the graph to become double-sheeted into exciting and dreary mathematics. The middle shaded piece of the graph represents least probable situations, and so is irrelevant—we only use the upper and lower sheets.

We now view teaching techniques at various levels by the paths shown in the control space Fig. 6. In each case the resulting enjoyment of the students can be traced by lifting the path to the surface in Fig. 5. For example the path (4) is shown lifted, with a "good" catastrophe occurring above the point 4_2.

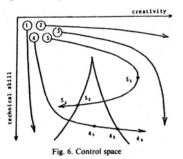

Fig. 6. Control space

* Leo Rogers points out that Kurt Lewin's[20] work may provide experimental support.

(a) School children learning tables*

Path (1) represents the old-fashioned drilling technique, which became very dreary. Path (2) represents the opposite approach of allowing children only to play, which was quite enjoyable, but did not give much skill. Path (3) represents the better modern approach, which allows children first to discover multiplication for themselves, by playing with stones in egg containers for instance, and then encouraging them to fill excitedly exercise books, mastering the technique. At least in early primary education (and before the prison doors of Bourbaki's set theory have had time to clang shut in secondary education) we seem at least to be returning a little towards the spirit of the pre-Euclidean Greeks, who believed that education should be enjoyable. For instance Plato writing in the Laws,[3] says:

"First there should be calculations specially devised as suitable for boys, which they should learn with amusement and pleasure, for example, distributions of apples or garlands where the same number is divided among more or fewer boys, . . ."

The three basic ingredients of school mathematics should be geometric intuition, physical intuition, and a sense of fun. Then the fourth ingredient, a sense of rigour, will grow of its own accord.

(b) Undergraduate lecture courses

Path (1) represents a bad lecturer giving bad material, which the students find first dull and then dreary. Path (3) represents a good lecturer giving good material, which the students find first interesting and then exciting. Path (4) represents a bad lecturer with good material—the sort of lecturer who spends most of the course setting up the machinery without giving any motivation, and then brings it all together to prove the major theorems in the last few lectures. The good students suddenly see the light at point 4_2 and jump catastrophically from the lower sheet to the upper sheet, from the dreariness of the machinery, 4_1, to the excitement of the theorem, 4_3. But, alas, the bottom half of the class, who generally fall behind towards the end of a lecture course, get stuck at the dreary point 4_1, and never see the light.

Finally path (5) represents the good lecturer giving bad material. The students enjoy the lectures at the time, 5_1, but when they come to revise it, at 5_2, they suddenly realise the material is pretty dull, and jump catastrophically from the upper sheet to the lower sheet, finishing somewhat disillusioned at 5_3.

(c) Research students and research supervisors

Path (1) represents a poor student with a poor supervisor, writing a dreary thesis. Path (3) represents a good student with a good supervisor writing an exciting thesis. Path (4) represents a good student with a poor supervisor, who writes a dull thesis at 4_1, made of tin, but once he gets free of his supervisor, at 4_2, suddenly blossoms into doing exciting research at 4_3. Path (5) represents a poor student with a good supervisor, who imparts a spurious creativity to the student during his PhD years, causing him to write an interesting thesis at 5_1, made of gold, but once he gets free of his supervisor, at 5_2, he collapses into writing dull papers using the same old techniques.

(d) Mathematicians in a rut

A mathematician may have bravely started on path (3),

* I am indebted to Ruth Rees for this example.

and written several good papers, but the path may then bend round into (5) as he reaches the limit of his creativity in that field, or gets bored with the subject, or gets imprisoned in the very techniques that he himself has created. Once he appreciates that he has suffered the bad catastrophe at 5_2, and is now in a rut at 5_3, there is only one thing to do: change fields. This is what Professor Bondi was recommending even as early as immediately after the PhD. Since our mathematician has no technique in the new field, he jumps straight back to the beginning of, we hope, path (3). It is very important that he soon tries his hand at a little creativity in the new field, in order to round the right hand side of the cusp, before learning all the new techniques. Otherwise those techniques will prove not interesting, but increasingly dreary as he follows path (4), perhaps to get stuck at 4_1. The more technical skill he acquires in the new field, the further away recedes the point 4_2, because the cusp is ever widening, and so the more creative he is required to become before he can ever have a chance of doing anything exciting. This was the serious intent behind my apparently flippant remark above, in favour of ignorance, and why I told the story of Whitehead's last paper.

References

1. Euclid, "Elements," Heath, T.L., Translation and commentary, Dover, New York, reprinted 1956.
2. Forster, E. M., "Alexandria, a history and a guide," Anchor, Doubleday, New York, 1961.
3. Heath, T.L., "A history of Greek mathematics," Clarendon Press, Oxford, 1921, reprinted 1960.
4. Dedekind, R., "*Stetigkeit und irrationale Zahlen*," Braunschweig, 1872; translation, Benman, W. W., "Essays on theory of numbers," Chicago, Open Court, 1901, 1–27.
5. Bolzano, B. P. J. N., "Paradoxes of the Infinite," *in* Přithonský, F., *Editor*, 1851; translation Steele, D., Routledge & Kegan Paul, London 1950.
6. Bell, E. T., "Men of mathematics," Vol. I, Penguin, London 1953.
7. Dehn, M., "*Ueber raumgleiche Polyeder*," Nachr. Akad. Wiss. Göttingen, 1900; "*Über den Rauminhalt*," Math. Annalen, 1902, **55**, 465–478.
8. Hilbert, D., "Mathematical Problems," Int. Congress of Math. 1900, *Nachr. Akad. Wiss. Göttingen*, 1900, 253–297; translation, Newson, M. W., *Bull. Amer. Math. Soc.*, 1902, 437–479.
9. Neugebauer, O., "On the 'Hippopede' of Eudoxus," *Scripta Mathematica*, 1953, **19**, 225–229.
10. Bourbaki, N., "*Elements de mathematiques*," Livres 1, 2, ..., Hermann, Paris, 1939; translation, Addison Wesley, USA, 1968.
11. Zeeman, E. C., "The lost group theory of Eudoxus," *in the press*.
12. Newton, I., "*Philosophiæ naturalis Principia mathematica*," Pepys Præses, London, 1686.
13. Berkeley, G., "The Analyst, or A Discourse Addressed to an Infidel Mathematician," London and Dublin, 1734; Luce, A. A., and Jessop, T. E., "Works," Vol. 4, Nelson, London, 1951, 53–102; Extract, Newman, J.R., "World of Mathematics," Vol. 4, Allen & Unwin, London, 1956, 286–293.
14. Whitney, H., "The self-intersection of a smooth n-manifold in $2n$-space," *Annals of Math.*, 1944, **45**, 220–246.
15. Whitehead, J. H. C., "Simplicial spaces, nuclei, and m-groups," *Proc. London Math. Soc.*, 1939, **45**, 245–327.
16. Penrose, R., Whitehead, J. H. C., and Zeeman, E. C., "Imbedding of manifolds in Euclidean space," *Annals of Math.*, 1961, **73**, 613–623.
17. Isnard, C. A. and Zeeman, E. C., "Some models from catastrophe theory in the social sciences," Edinburgh Conference, 1972, *in* Collins, L., *Editor*, "The use of models in the social sciences," Tavistock, London, 1974.
18. Kuhn, T. S., "The structure of scientific revolutions," Int. Encycl. United Science, Vol.2, No. 2, Chicago University Press, 1962, enlarged 1970.
19. Thom, R., "*Stabilite structurelle et morphogenes*," Benjamin, New York, 1972.
20. Lewin, K., "The conceptual representation and the measurement of psychological forces," Duke University Press, 1938; Johnson Reprint, USA, 1968.

21 CATASTROPHE THEORY : ITS PRESENT STATE AND FUTURE PERSPECTIVES.

We reprint here from Manifold, a student Mathematical magazine at Warwick University, a survey by René Thom of catastrophe theory and a reply by Christopher Zeeman. They are followed by a new response by Thom to Zeeman's criticisms. All references are to the large bibliography that follows the three articles, and should prove valuable to anyone wanting to explore this fascinating new subject created by Thom.

Anthony Manning.

La Théorie des Catastrophes : Etat Présent et Perspectives.   René Thom.

Existe-t-il, à proprement parler, une "théorie des catastrophes"? Dans les applications (Physique, Biologie, Sciences humaines), on ne peut considérer la théorie des catastrophes comme une théorie scientifique au sens usuel du terme, c'est à dire un ensemble d'hypothèses dont on peut déduire des conséquences nouvelles vérifiables expérimentalement. Dans ces domaines, le modèle des catastrophes est à la fois beaucoup moins, et beaucoup plus qu'une théorie scientifique; on doit le considérer comme un langage, une méthode, qui permet de classifier, de systématiser les données empiriques, et qui offre à ces phénomènes un début d'explication qui les rende intelligibles. En fait, n'importe quelle phénoménologie peut être expliquée par un modèle convenable de la théorie des catastrophes. Et, comme me l'a fait remarquer, très pertinemment, le biologiste anglais L. Wolpert, une théorie qui explique tout n'explique rien. Ceci montre simplement qu'on ne doit pas attendre du modèle le même usage que d'une loi quantitative de la Physique, ni d'un fait d'expérience à la manière de la méthode expérimentale de Claude Bernard, en Biologie. Nous essaierons de préciser ci-dessous, pour chaque discipline particulière, les services particuliers qu'on peut en attendre, et les bénéfices qu'on peut raisonnablement espérer tirer de son usage.

S'il est donc clair qu'il n'y a pas de théorie des catastrophes dans les domaines appliqués, existe-t-il alors une "Théorie des Catastrophes" en Mathématique Pure? Ici encore, on peut en douter. En effet, là où la théorie a pu acquérir la rigueur proprement mathématique, elle s'est fragmentée en autant de sujets distincts relevant de branches mathématiques particulières (Systèmes Dynamiques, théorie des singularités d'applications différentiables, équation aux Dérivées Partielles, Actions de groupes ... etc.), ne conservant

---

Published in Dynamical Systems, Warwick 1974, (Edited A.K. Manning) Springer Lecture Notes in Mathematics, Volume 468, pages 366-401.

du modèle original que des idées très générales, comme généricité, transversalité, déploiement universel ... Voici, je crois, comment se présente la situation actuellement.

La théorie des catastrophes en Mathématique.

Le modèle initial, celui du champ "métabolique" de dynamiques locales, fait appel de manière essentielle à la theorie de la bifurcation des systemes dynamiques. Mais les résultats récents de cette théorie manifestent une pathologie initialement insoupçonné: non densité des flots structurellement stables, existence générique d'une infinité d'attracteurs dans une variété compacte (contre-exemple de Newhouse) et instabilité topologique des dits attracteurs, jets de singularités de champs de vecteurs non stabilisables à partir de la codimension trois (F. Takens); tous ces résultats indiquent à quel point les fondements mathématiques du modèle sont précaires. Cependant, une analyse plus approfondie laisse quelques raisons d'espérer. On sait qu'à tout attracteur est associée une fonction de Liapunov locale (qui joue le rôle d'une entropie locale). Il est raisonnable de penser que les seuls attracteurs assez stables pour engendrer une morphologie empirique sont ceux dont la fonction de Liapunov n'a pas de jets trop dégénérés. Par exemple, dans l'exemple de Newhouse, où il y a une infinité d'attracteurs, la fonction de Liapunov globale correspondante a nécessairement des jets plats. (L'idée récente d' associer la stabilité structurelle à une filtration sur la variété correspond à l'existence d'une fonction de Liapunov discrète ...) De même, dans le formalisme hamiltonien qui est celui de la Mécanique Quantique, les seuls "états stationnaires" décelables expérimentalement sont ceux pourvus d'un "hamiltonien local" de caractère central, à jet non trop dégénéré. En un certain sens, l'attracteur n'existe que grâce à sa fonction de Liapunov, et on peut aussi exiger que si le système dynamique est perturbé, si l'attracteur disparait par bifurcation, alors la fonction de Liapunov locale subit elle-même une bifurcation <u>pas trop dégénéré</u> Imposer aux fonctions de Liapunov de n'avoir que des points singuliers algébriquement isolés, c'est faire la théorie des catastrophes élémentaires. Que cette théorie soit insuffisante n'est que trop évident, comme le montre l'exemple de la bifurcation de Hopf d'un attracteur ponctuel dans le plan. Tout le problème revient donc à évaluer le caractère "pas trop dégénéré" d'une bifurcation. De ce point de vue, l'apparition par bifurcation de nouveaux groupes de symétries (comme le groupe S^1 dans le cas de Hopf) semble un phénomène encore bien mal compris. Peut-être faudra-t-il le rattacher au rôle si mystérieux des groupes de Lie associés aux singularités d'hypersurfaces complexes dont on parlera ci-dessous.

Parmi les problèmes particulièrement urgents que pose en mathématique la théorie des catastrophes, citons :

(i) La théorie du delploiement universel d'une singularité d'applications composées.
(ii) Le deploiement universel d'un germe d'action d'un groupe de Lie G dans l'espace euclidien.
(iii) La bifurcation des singularités de fonctions G-invariantes.

Ces problèmes paraissent en particulier importants pour la Physique. (Les questions (i) et (ii) pour la Mécanique Quantique : cf. la thèse de Pham; la question (iii) pour les transitions de phase.)

Enfin, en dépit de progrès récents, il reste de nombreux problèmes ouverts dans l'étude des singularités algébriques ou analytiques complexes (a fortiori réelles). Outre les problèmes toujours ouverts liés à la classification, l'équisingularité, attend toujours sa définition algébrique. On a pu classifier les singularités de fonctions complexes jusqu'en codimension ≤ 8 (Arnol'd Siersma). C'est là qu'Arnol'd a remarqué que tant qu'aucun module n'apparait dans la singularité on peut associer à celle-ci un groupe de Lie classique G comme suit : (1) En deux variables, la désingulatisation (à la Hironaka) de la singularité introduit un graphe en droites projectives; ce graphe est alors isomorphe au schéma de Dynkin du groupe G. (2) Si on considère la variété discriminant D dans le déploiment universel U de f; alors le complémentaire U-D est topologiquement un $K(\pi,1)$, où π est le groupe des tresses du groupe de Weyl du groupe G. Enfin, dans la théorie différentiable, il reste à décrire les généralisations du symbole de Boardman nécessaires pour définir la stratification minimale d'un ensemble (ou d'un morphisme) analytique. Aucun progrès n'a été enregistré sur ce problème depuis les tentatives de B. Morin.

Physique (et Chimie).

De nombreux phénomènes, en Physico-Chimie, sont justiciables des modèles de la théorie des catastrophes. Mais il est un problème qui domine tous les autres; c'est celui des <u>transitions de phases</u>. En Mécanique Statistique, on ne reconnait la notion de phases que sur l'espace infini, parce qu'il s'agit d'un état invariant par translation (ou par un sous-groupe des translations à domaine fondamental compact). Il faut donc d'abord localiser cette notion, grâce à la notion de pseudo-groupe. Puis on s'efforce de préciser la nature des singularités qui peuvent se presenter 'génériquement' pour les surfaces limitant ces phases locales. Si G désigne le groupe d'isomorphisme local d'un pseudo-groupe K d'une phase locale, alors, dans le modèle des catastrophes, le groupe G opère dans l'espace des variables internes, et le minimum associé à la phase est décrit par une fonction G-invariante. C'est donc la bifurcation des fonctions G-invariantes (problème (iii) plus haut) qui va décrire la transformation de la phase K en une autre phase K'. Très souvent K' est un sous-pseudo-groupe de K, ce qui correspond à ce que les physiciens appellent une cassure de symétries (breaking of symmetry). J'ai proposé d'interpréter la dualité onde-corpuscule de la Mécanique Quantique d'une manière analogue : un champ serait un milieu qui peut se présenter sous deux phases (locales) : une phase homogène, invariante par le groupe D des déplacements; une phase corpusculaire, invariante par le sous-groupe SO(a), si la particule est localisee en a. Une telle manière de voir a l'avantage de réduire le formalisme quantique à celui de la mécanique (statistique) classique. Il est d'ailleurs douteux que cette simplification conceptuelle puisse offrir des perspectives de solution aux difficultés de la physique théorique actuelle, difficultés dues au caractère fondamentalement quantitatif de cette discipline.

Dans la transition Liquide-Gaz, le modèle de van der Waals suggère un potentiel en v^4 sur l'espace (p,T) des variables 'de controle'. Mais, on le sait, ce modèle est inexact au voisinage du point critique. Ceci peut être dû, soit au fait qu'il y a plus d'une variable interne, soit, comme je l'ai proposé, que l'on doive user d'un modèle métabolique au lieu d'un modèle 'statique'.

Parmi les sujets d'étude qui semblent relever des catastrophes, citons : les configurations complexes d'ondes de choc (réflexion de Mach), les dislocations de réseaux cristallins, et des cristaux liquides; le géomorphologie et la morphologie des objets célestes (galaxies, éruptions solaires ...). Les équilibres chimiques complexes et la cinétique chimique rapide.

Les Physiciens adressent à l'emploi du modèle des catastrophes deux objections, d'ailleurs liées : la première est l'objection quantitative classique : il n'y a physique que s'il y a loi exprimable en équations, et de ce fait contrôlables par l'expérience. La seconde est : la Nature n'est pas 'générique', comme le montre l'exactitude déraisonnable des lois physiques (selon l'expression si juste d'E. Wigner). On répondra d'abord que, ne serait-ce que pour interpréter physiquement les grandeurs qui figurent dans les équations, certaines considérations qualitatives sont indispensables en Physique (comme ailleurs). Par ailleurs, les équations exprimant les lois physiques doivent être indépendantes des unités qui servent à mesurer ces grandeurs. Il en résulte que toute loi physique quantitative est nécessairement liée à un groupe d'homothéties sur les variables de base, donc fait appel nécessairement, au caractère localement affiné de l'espace-temps. Or il n'y a aucune raison de penser que les variables externes qui déploient une catastrophe admettent localement un tel groupe local d'homothéties. On peut d'ailleurs parfois définir un tel groupe d'homéomorphismes à un paramètre : c'est le principe de la 'scaling hypothesis' en théorie du point critique, par exemple. Mais ce simple fait de la dimensionnalité des grandeurs physiques montre que des lois quantitatives précises ne sont possibles que dans la mesure où le phénomène étudié est solidaire de la géométrie de l'espace-temps; c'est pourquoi les seules lois physiques rigoureuses, celles de la gravitation, de l'électromagnétisme classique sont liées à la géométrie de l'espace-temps, comme l'exprime la relativité générale. (La Mécanique Quantique elle-même dans la mesure où elle est quantitativement rigoureuse, exprime sans doute certaines régularités dans la régulation, métrique ou topologique, de l'espace-temps.) Exiger que tout phénomène naturel soit régi par une loi quantitative, c'est en fait exiger que tout phénomène soit réductible à la géométrie de l'espace-temps. J'ai beau être un géomètre professionnel, je n'en trouve pas moins ce postulat quelque peu exorbitant.

Biologie.

C'est l'étude du développement embryologique qui a conduit à la création de la théorie des catastrophes. Il ne semble pas, cependent, que ces idées aient fait de grands progrès dans les milieux de la recherche biologique. A cela, une raison majeure : l'abime psychologique qui sépare la démarche biologique actuelle de toute pensée théorique. Le biologiste expérimentateur n'a nulle besoin de théorie pour trouver des faits : n'importe quel matériel peut donner lieu à une suite pratiquement infinie d'expériences. De la composition chimique des gaz intestinaux du Cobaye à l'ultrastructure du centriole, de la croissance des racines d'Arum à la teneur en ACTH dans les membres de l'embryon d'Axolotl, tout donne lieu à expérimentation, à publication dans une revue spécialisée. La seule partie un peu théorique de la Biologie, à savoir la Génétique, s'est trouvée ramenée par le 'dogme central' à l'étude d'une morphologie particulière, la composition chimique de l'ADN. D'où la croyance universellement répandue que la seule analyse biochimique va suffire, via le 'code génétique', à élucider toute l'évolution des formes vivantes. A l'heure actuelle, la Biologie n'est qu'un immense cimetière de faits, vaguement synthétisés par un petit nombre de formules creuses, comme : 'information codée dans l'ADN', 'stimulation différentielle des gènes' ... etc. Certes, l'apport de la Biologie Moléculaire a été considérable; mais cette discipline a eu l'effet psychologique désastreux de favoriser l' état d'esprit biochimique : il consiste à chercher, pour tout phénomène de la vie, un agent matériel spécifique responsable (acide nucléique, enzyme, substance inductrice ou répresseur), puis, une fois l'agent trouvé et isolé (les candidats, en général, ne manquent pas) à se reposer sur ses lauriers, sans se préoccuper des mécanismes qui, lorsque c'est nécessaire, provoquent l'apparition ou la disparition de la dite substance ou ses changements morphologiques. C'est que la description globale d'un schéma de régulation impliquant un assez grand nombre de paramètres exige une figuration multidimensionnelle qui, évidemment, n'est pas dans l'équipment conceptuel du biologiste contemporain. Le théoricien des catastrophes, qui s'intéresse avant tout à l'évolution spatio-temporelle de la forme embryonnaire, sans trop s'occuper de sa composition biochimique, a peu en commun avec la biochimiste, dont les intérêts sont exactement opposés. Et un vrai dialogue de sourds s'instaure.

L'expérimentateur : Si vos modèles sont bons à quelque chose, ils doivent prévoir des faits nouveaux, et je ne demande pas mieux que de vous faire les expériences correspondantes.

Le théoricien : Avant de prévoir des faits nouveaux, j'ai besoin de systématiser, pour les comprendre, la masse des faits déjà connus. Rien ne sert d'ajouter à l'acquit expérimental, déjà énorme, si l'on n'a pas d'abord une théorie qui explique les faits connus – et surtout les plus classiques d'entre eux, ceux qui sont dans tous les manuels élémentaires.

L'expérimentateur : Mais alors votre construction théorique peut-elle avoir une utilité, un quelconque rapport avec le concret?

Le théoricien : Cela sert à comprendre ce qui se passe.

L'expérimentateur : Comprendre ne m'intéresse pas, si je ne peux en tirer une idée d'expérience ...

Le théoricien : Il faut vous convaincre que les progrès de la Biologie dépendent moins d'un enrichissement de données expérimentales que d'un élargissement des capacités de simulation mentale des faits biologiques, de la création d'une nouvelle 'intelligence' chez le Biologiste. Il y faudra sans doute quelque temps, une génération peut-être ...

La raison majeure qui rend la théorie des catastrophes si peu accessible à l'expérience est notre ignorance de la nature des paramètres qui déploient les grandes catastrophes de l'embryologie, les fameux 'gradients épigénétiques'. Déjà, dans le cas du déferlement hydrodynamique, les paramètres pertinents ne peuvent être directement explicités, car ils sont de nature banale et dépendent étroitement des conditions initiales qui ont un effet 'focalisant' ('focussing') à l'avenir. Il en va probablement de même en Embryologie, où ces paramètres peuvent, de plus, être de nature cinétique, et par suite échapper aux techniques d'analyse biochimique. Pour toutes ces raisons, il y a tout lieu de croire que le gouffre entre théorie et expérience ne sera pas comblé de sitôt. Pendant de longues années, le 'modélisme' théorisant va fleurir, de manière pratiquement indépendante de la recherche en laboratoire.

On peut regretter cette situation. Car privé de tout appui concret, le modélisme pourra s'embarquer dans des constructions artificielles inutilement compliquées. La théorie des catastrophes, avons-nous dit, est un language; comme la langue d'Esope, elle peut exprimer le meilleur, ou le pire, et comme il n'y aura pas de contrôle expérimental, seul un sentiment d'esthétique, d'économie intellectuelle permettra de trier le bon grain de l'ivraie. En dépit de la venue inévitable de tels déchets, il faut poursuivre la modélisation de l'embryologie, de la morphogénèse en général. Et ceci moins dans le but immédiat d'y intéresser le biologiste (cela prendra du temps) que pour perfectionner la théorie elle-même. Déjà, en Physiologie, le besoin de modèles est plus fortement ressenti. Certes, là aussi, l'esprit biochimique, avec ses enzymes et ses agents spécifiques, exerce ses ravages; mais les physiologistes ont plus conscience de leur misère théorique que leurs collègues de Biologie Moléculaire, et pour eux le problème de la régulation ne peut être éludé par l'emploi d'un verbalisme trompeur. Aussi, je ne serais pas étonné si les premiers succès marquants de la théorie des catastrophes apparaîtront dans la description des grandes régulations organiques. De ce point de vue, la théorie de E.C. Zeeman sur l'activité cardiaque est des plus prometteuses.

Mais, avons-nous dit, la raison la plus forte pour poursuivre la modélisation de la vie est dans le perfectionnement de la théorie de la régulation, et de la reproduction. Nos idées actuelles sur l'interconnexion des catastrophes, la constitution globale des figures de régulation multi-dimensionnelles (ce que j'ai proposé d'appeler les 'logoi') sont encore extrêmement rudimentaires. Il n'y a pas de doute qu'en embryologie, par exemple, la dynamique vitale utilise des procédés qu'il y aurait tout intérêt à bien comprendre, à bien expliciter. Car ces mêmes procédés peuvent également jouer, sous une forme moins visible, en d'autres circonstances, par exemple en Physique fondamentale. L'hypothèse réductionniste, qui sait, devra peut-être un jour être retournée : ce sont les phénomènes vitaux qui pourront nous expliquer certaines énigmes sur la structure de la matière ou de l'énergie. Après tout, n'oublions pas que le principe de la conservation de l'énergie a été exprimé pour la première fois par von Mayer, un médecin ...

Science humaines : psychologie, sociologie.

De la physiologie, il n'y a qu'un pas jusqu'à l'éthologie, science des comportements animaux, et pour l'homme, à la psychologie. Là aussi, les perspectives d'applications de la théorie sont considérables. On est moins tenté, dans un esprit 'réductionniste', d'exiger un support matériel à des facteurs psychologiques généraux comme l'agressivité, l'attention, ... Par ailleurs, la formalisation des comportements en champs morphogénétiques, en 'chréodes', est chose assez immédiate, car cette morphologie se situe précisément entre la morphologie organique, décrite en Biologie, et la morphologie de la description verbale, qu'on peut faire pour l'homme et les animaux supérieurs. Aussi beaucoup de ces questions sont mûres pour une approche 'catastrophique'. Evidemment il n'en faudrait pas conclure que ces modèles pourront ipso facto avoir une utilisation pratique (cf. notre conclusion). En sociologie les perspectives sont moins bonnes, parce que l'espace substrat, la morphologie sociale proprement dite, n'est pas encore clairement explicitée.

Linguistique, sémantique, philosophie.

La théorie des catastrophes jette une certaine lumière sur la nature du langage (considéré comme une morphologie d'origine organique simulant la réalité extérieure). Elle explique ainsi les structures syntaxiques, la nature du verbe en tant que catégorie grammaticale. Dans la mesure même où on comprendra mieux la 'figure de régulation' (le 'logos') des êtres extérieurs, vivants ou non vivants, on comprendra mieux la nature des concepts qui y réfèrent, et qui en sont des structures analogiques simplifiées. On pourra alors explorer le monde du substantif, du lexique, qui est encore la grande terre inexplorée de la Sémantique. Lâchons un peu la bride à la spéculation : la théorie des catastrophes nous laisse entrevoir la possibilité d'un langage multidimensionnel, aux possibilités syntaxiques infiniment plus complexes que la langue ordinaire, où une bonne part du raisonnement pourrait se formaliser, comme un calcul. Bien mieux, on pourrait concevoir une mathématique nouvelle, où la démarche du mathématicien serait décrite par un chemin continu, les 'étapes' du raisonnement correspondant seulement a des variations qualitatives catastrophiques sur cet espace multidimensionnel.

Sur le plan de la philosophie proprement dite, de la métaphysique, la théorie des catastrophes ne peut certes apporter aucune réponse aux grands problèmes qui tourmentent l'homme. Mais elle favorise une vision dialectique, héraclitéenne de l'univers, d'un monde qui est le théâtre continuel de la lutte entre 'logoi', entre archétypes. C'est à une vision fondamentalement polythéiste qu'elle nous conduit : en toutes choses, il faut savoir reconnaitre la main des Dieux. Et c'est peut-être là aussi qu'elle trouvera les limites inéluctables de son efficacité pratique. Elle connaitra peut-être le même sort que la psychanalyse. Il ne fait guere de doute que l'essentiel des découvertes de Freud en psychologie ne soit vrai. Et cependant, la connaissance même de ces faits n'a eu que très peu d'efficacité sur le plan pratique (pour la cure des troubles mentaux, notamment). De même que le héros de l'Iliade ne pouvait s'opposer à la volonté d'un Dieu, tel Poséidon, qu'en invoquant le pouvoir d'une divinité opposée, telle Athéna, de même nous ne pourrons restreindre l'action d'un archétype qu'en lui opposant un archétype antagoniste, en une lutte ambigue au résultat incertain. Les raisons mêmes qui nous permettent d'étendre nos possibilités d'action en certain cas nous condamneront à l'impuissance en d'autres. On pourra peut-être démontrer le caractère inéluctable de certaines catastrophes, comme la maladie ou la mort. La connaissance ne sera plus nécessairement une promesse de réussite, ou de survie; elle pourra être tout aussi bien la certitude de notre échec, de notre fin.

Catastrophe theory : A reply to Thom.   E.C. Zeeman.

René Thom's article on the present and future perspectives of catastrophe theory in the previous issue of Manifold was very much in his own inimitable style : a fascinating mixture of tantalising hints and deeply profound remarks about mathematics and science, spiced with a few provocative cracks at the experimentalists, and garnished with some fairly wild speculations. In a sense Thom was forced to invent catastrophe theory [148] in order to provide himself with a canvas large enough to display the diversity of his interests. Ever since the disappearance of natural philosophy from our universities and the fragmentation of mathematicians into pure and applied, our canvases have steadily been growing smaller and smaller. At least catastrophe theory marks a revival of natural philosophy, to be enjoyed once again for a while at any rate. One could wish that more mathematicians should enliven our literature by writing in this vein, were it not for the fact that the speculation by lesser men often leads to nonsense. In fact it makes an amusing little application of the cusp catastrophe.

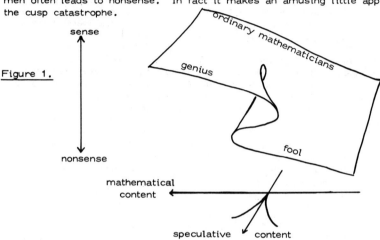

Figure 1.

Thom certainly puts himself out of the ordinary by his courageous speculative ventures, but however close he sails to the edge, he somehow always manages to stay on the upper surface.

Nevertheless I must confess that I often find his writing obscure and difficult to understand, and occasionally I have to fill in 99 lines of my own between each 2 of his before I am convinced. Of course sometimes this is just due to sheer laziness on his part over mathematical details, but at other times the obscurity is the reverse side of a much more important coin : in order to create profound new ideas, profound because they can be developed a long way with immense consequences, it is necessary to invent a personal shorthand for one's own thinking. The further the development, the more subtle must be the shorthand, until eventually the shorthand becomes part of the paradigm. But, until it does, the shorthand needs decoding. Meanwhile Thom has thought ahead for so many years that now, when he speaks to us, he often uses his shorthand and forgets to decode it. Maybe this is because the IHES has no undergraduates. When I get stuck at some point in his writing, and happen to ask him, his replies generally reveal a vast new unsuspected goldmine of ideas. In trying to "trier le bon grain de l'ivraie" I discover "plus du bon grain". Therefore in this spirit let me return to the topics in his article.

Mathematics.

It is in mathematics itself, as Thom modestly omits to mention, that catastrophe theory has already made its greatest contribution to date. I agree that there is strictly speaking no "catastrophe theory", but then this is more or less true for any non-axiomatic theory in mathematics that attempts to describe nature. For instance the "theory of differential equations" is not well defined : it uses odd bits of analysis, topology and algebra in its foundations, and then proliferates into a ragbag of techniques. This is why differential equations are perpetually awkward to fit into any undergraduate syllabus. It is only those pieces of mathematics that are far from nature that make tidy theories, because all the messy interrelationships with other branches of mathematics can be artifically ruled out of the game by judicious choice of axioms. The comparison between differential equations and catastrophe theory is an interesting one : Newton invented differential equations in order to describe smooth phenomena in nature, and this in turn forced the development of calculus, analysis, Taylor series etc. Similarly Thom invented catastrophe theory in order to describe discontinuous phenomena in nature, and this in turn forced major developments in the theories of singularities, unfolding, stratifications, the preparation theorem etc.

Let us take one example, the preparation theorem. In a sense this is more fundamental than the Taylor series, and no doubt will slowly transform the face of applied mathematics of the future. For up till now applied mathematicians, in using Taylor expansions, have implicitly had to artificially restrict themselves to analytic functions in order that the series should converge, which is a very severe straightjacket due to the uniqueness of analytic continuation. Now, with Malgrange's preparation theorem, they have the sudden freedom and flexibility to use C^∞-functions; there is no longer any need for the series to converge, only for the jet to be determinate. Whereas before, the tail of the Taylor series wagged the dog, in future it can be amputated with impunity, because, by the uniqueness of unfoldings, germs can be replaced by jets, and so the ∞-dimensional problem in analysis can be replaced by a finite dimensional problem in algebraic geometry. It was in struggling to prove the uniqueness of unfoldings (which is the heart of the classification theorem) that Thom narrowed the gap in the proof down to the preparation theorem, and so persuaded Malgrange firstly (against his will) that it was true and secondly to prove it [30]. In this sense catastrophe theory is a driving force determining mainstream direction of research within mathematics.

I do not know whether Thom has ever written down that delightful analogy he once gave in a lecture on mathematical education at Warwick; it runs as follows : Just as, when learning to speak, a baby babbles in all the phonemes of all the languages of the world, but after listening to its mother's replies, soon learns to babble in only the phonemes of its mother's language, so we mathematicians babble in all the possible branches of mathematics, and ought to listen to mother nature in order to find out which branches of mathematics are natural.

Physics.

I find the application of catastrophe theory to phase transition very difficult, and do not fully understand it yet [114,133-6,148,150]. Van der Waals' equation for liquid-gas is easy enough, and gives a beautiful canonical cusp catastrophe surface, but what is strikingly absent is any dynamic minimising the potential. Nor can there be, because if there were, then boiling and condensation would obey the delay rule rather than Maxwell's rule. It is true that they can be exceptionally delayed in states of supersaturation and superevaporation, but normally boiling point equals condensation point, and so Maxwell's rule prevails. Out of the hundred or more applications of catastrophe theory in several different fields this is the only one I know that unambiguously obeys Maxwell's rule. Now there must be a mathematical reason underlying Maxwell's rule. By this I mean Maxwell's rule must be a theorem rather than a hypothesis, in the same way that the delay rule is a theorem based on the hypothesis of the existence of an underlying dynamic minimising the potential. The question is : what is the corresponding hypothesis that would lead to Maxwell's rule ? L. Schulman [133] has pointed out that the answer must lie in statistical mechanics, with the internal variables in a Hilbert space of states. Now there is no rigorous treatment yet of catastrophe theory for an infinite dimensional state-space. But suppose there were : then the free energy expressed as an integral over the state-space is dominated by, and therefore approximated by, the state with minimum energy — hence Maxwell's rule. Perhaps an analysis of this approximation will reveal why the elementary model is inexact at the critical point. A full understanding may entail a rewriting of the foundations of statistical mechanics.

To my mind the other outstanding catastrophe theory problem in physics is the breaking of waves [148,166]. Although I agree that the hyperbolic umbilic seems to be diffeomorphic to the shape of a wave breaking on the sea-shore, I do not yet see how to identify the catastrophe variables with the classical variables of hydrodynamics. Such a programme is ambitious in the sense that it implies that both water and air are obeying the same differential equation. This observation gives insight that the programme may be too naive, because there does not seem to be any variable in water that falls off by the square-root of the distance from the surface, as does one of the internal variables of the hyperbolic umbilic. It is possible that the breaking wave is not the hyperbolic umbilic after all, but a 3-dimensional Maxwell section of the double-cusp that happens to be diffeomorphic to the hyperbolic umbilic, just as phase transition is a 1-dimensional Maxwell section of the cusp diffeomorphic to two folds. In which case the breaking wave is more complicated than phase-transition and for a full understanding must involve the statistical mechanics underlying hydrodynamics.

Another possible application of catastrophe theory suggested by T. Poston [130] is to soap bubbles, but this again, as in most applications in physics, requires an infinite dimensional state-space, as well as hard geometric analysis. In engineering there are several potential applications including structural stresses, non-linear oscillations, cybernetics, and various types of regulators. Perhaps the richest application of the umbilics so far have been in light caustics [117,119,125,144] and elasticity [157-9].

Biology.

I think Thom is a little hard on the biologists in his Manifold article. But I understand his impatience, because it is over 5 years since he first explained the idea to them [141-4,147-9]. And it is such a magnificent idea - the first rational explanation of how the local genetic coding could possibly cause the global unfolding of the embryo. However it is a very difficult application to understand, because at first sight the only observable feature is part of the bifurcation set in the space of external variables, space-time. The space of internal variables must be of so high a dimension that it has to remain implicit. And the potential is probably only a Liapunov function, in other words is a purely mathematical construction, one step even further removed from the concrete. By the time we have used the classification theorem in order to reduce the dimensions to those of the useable models of the elementary catastrophes, even if we manage to achieve an interpretation and identify the 1 or 2 internal variables of the model with some elusive morphogens, the potential of the model will almost certainly be biologically meaningless. Only the bifurcation set of the model will retain its marvellous clarity of meaning.

The biologists can hardly be blamed for their despair of understanding the mathematical subtleties of what can be explicit, what must remain implicit, what can be meaningful, and what must perforce be biologically meaningless. No wonder they fall back upon the defence of "how can I test this model against other models?" This point of view is in effect a simple insurance policy, because if they can dispose of a theory by proving it wrong experimentally, it saves the time and effort of having to wade through all that formidable looking mathematics. What they do not, and must find difficult to, appreciate is the infinite and all-embracing variety of models that the theory automatically encompasses and classifies. Eventually the only way to fully appreciate this fact is to go through all the details of the proof of the classification theorem [84]; only then does one feel that true weight of mathematical power behind the few elementary models. I must confess it took me several years to achieve this objective myself, and it is only the very exceptional biologist who would have the expertise, time and inclination to follow suit. On the other hand a slow migration is beginning of mathematical students into experimental biology, and it is through them that I anticipate communication will eventually take place. At present the leading biologists freely admit the void of explanation in developmental biology, and would eagerly welcome a theory. But, and here I differ from Thom in emp**hasis**, any theory must face up to the classical scientific method of prediction, experiment and verification. I see no reason why his theories should be sacrosanct on the grounds of being qualitative rather than quantitative. There are plenty of qualitative predictions in science, and plenty of quantitative experiments in which the quantities depend upon the individual, but the quality is common to all individuals.

Thom has already shown how several morphologies in embryology are geometrically similar to elementary catastrophes [148,149]; what is now needed is a closer identification of the catastrophe variables involved with space-time variables and morphogens. Better still if there can be alternative identifications, i.e. alternative models within catastrophe theory itself, which the experimenters can test between. Towards this end Thom himself is at times unfortunately counterproductive, because each time he writes he tends to embroider upon his previous models with interesting new ideas, but without distinguishing clearly between those features that can be deduced from the original catastrophe theory, and those that are part of the embroidery; this is particularly confusing when he justifies the embroidery only by analogy, rather than by also basing it upon clearly stated additional mathematical hypotheses.

A glaring example is his interpretation* of the middle unstable sheet of the cusp catastrophe as the formation of mesoderm in amphibian gastrulation [149]. He is tempted into this pitfall by the fact that mesoderm forms as an intermediary layer between ectoderm and endoderm, analogous to the accidental topological situation occuring in the canonical model of the cusp catastrophe, where since the space of internal variables happens to be 1-dimensional, the unstable sheet happens to lie in between the two stable sheets representing ectoderm and endoderm. Now, except in applications such as light caustics, where the geodesic path is not necessarily the shortest, the saddles and maxima in catastrophe theory play a totally different role to the minima, and it is not only mathematically contradictory to mix the two, but also very confusing to the biologist who is trying to master the theory. In this particular case there are perfectly good alternative models of mesoderm within the theory, using either the butterfly catastrophe, or a primary wave switching ectoderm into mesoderm [174].

The same fault implicitly occurs when Thom refers to the mushroom shape of the parabolic umbilic, [148 p. 102] because, although a mushroom does occur as a section of the bifurcation set, the stalk of the mushroom bounds a region of minima while the head of the mushroom bounds a region of saddles, which are quite different. I confess that I do not yet fully understand the embryological applications of the elliptic and parabolic umbilics; for me they do not yet have the beautiful translucence of the applications of the cusp and swallow-tail to gastrulation.

Another example where Thom's use of analogy is misleading is in his discussion of cliff regulation [149]. To explain the formation of a regulator from a potential well he appeals to the analogy of perturbations having some

Figure 2.

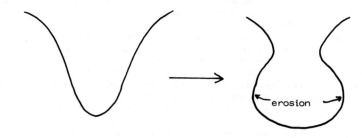

---

* Thom replies that this piece of embroidery is the mysterious phenomenon of "threshold stabilisation", apparently well known to physicists. In support he appeals to (1) the analogy of wet sand clinging to a maximum (which implicitly involves more mathematical hypotheses) or (2) the maximum reached by the Liouville measure of the energy-level of a saddle (but this argument only works for 2-dimensions).

eroding effect upon the substrate. Here I think this is a fallacious appeal to physical intuition, because the curves are only mathematically meaningful in the sense that they represent the stationary values of the catastrophe potential. To appeal to the concept of "erosion" is implicitly to attach a mathematical meaning to the difference between the insides and outsides of the curves. However the fact that the curves are of codimension 1 is again an accident arising from the use of the canonical model with only 1 internal variable. As soon as one uses 2 or more internal variables, which must certainly be the case if one appeals to the "substrate", then the curves are of codimension 2 or more, and consequently no longer have insides or outsides to be eroded. Nevertheless the concept of cliff regulation is a splendid notion, as Thom says obviously fundamental to the future understanding of physiology, and there are alternative ways of introducing it mathematically into the theory. For example given a dynamical system on R^n, and given an attractor point, having a 1-dimensional slow manifold with bounded basin of attraction thereon, then homeostasis implies a cliff regulator (see [168]).

Summarising the situation in biology : Thom is fully justified in his impatience that this magnificent theory has not yet had greater impact upon developmental biology, but I think he should turn some of his criticism away from the biologists and redirect it towards us, his fellow mathematicians, who are far too ignorant of embryology, and have done far too little to analyse and develop the models of specific morphologies. Another promising area for catastrophe theory in biology, as yet practically unexplored, is evolution[112,162].

<u>Human sciences.</u>

Most of my own contributions to catastrophe theory have been in the human sciences, biology, psychology, sociology and economics.
In psychology, as Thom points out, one is studying the regulators in the brain underlying behavioural patterns. There has been plenty of research in both laboratory and clinic upon the behavioural changes caused by dosing the brain with chemicals, but as yet very little has been achieved on the connection between behaviour and the electrochemical activity of the normal undosed brain. It is an exciting prospect that catastrophe theory may provide one of the first systematic links between the psychology and neurology of a normal brain [170].

But just as frutiful areas for applications are the social sciences, where many individuals are involved instead of one. Economists are already demanding models that can allow for catastrophic changes and divergent effects. And I believe that sociology may well be one of the first fields to feel the full impact of this new type of applied mathematics, in spite of the prevailing mood at Princeton, and in spite of Thom's own doubts about the social morphologies not being yet sufficiently explicit. It is true that in sociology there is less likely to be a general theory so much as a variety of particular models to describe the divisions and swings of opinion, emergence of compromise, voting habits, social habits, social changes, effects of stress, effects of overpopulation and pollution, policy changes, political moves, emergence of classes, divergence of taste, evolution of laws, etc. Moreover this type of individual model will in general be much easier to understand than those in physics and biology because the internal variables tend to be explicit and few in number. The external and internal variables tend to play the role of cause and effect, the former representing control factors influencing the latter, which represent the resulting behaviour. The potential is often best understood as a probability function, and the dynamic as a sociological or psychological pressure. Let us give a simple example (see also [122-4,160-61,177-9]).

Example : strength of opinion.

Consider the opinions held by the individuals of a population on some issue. For simplicity suppose there are just two possible opinions, called left and right. Let the behaviour variable x measure the strength of the opinion :

```
x ↑
  ┼ strongly held left-wing opinion
  ┼ weakly held left-wing opinion
0 ┼ neutral
  ┼ weakly held right-wing opinion
  ┼ strongly held right-wing opinion
```

The two main control factors c_1, c_2 influencing opinion are bias and involvement. The bias to the left or right may be due to self-interest, heredity, environment, political persuasion, information, ignorance or prejudice. The involvement may be voluntary or involuntary. The potential function $P_c(x)$ is the probability of opinion x given control factors $c = (c_1, c_2)$. In the case of probability functions the maxima are important, rather than the minima. Bias will be a normal factor. We take as hypothesis that involvement is a splitting factor, in other words the more involved he is, the more strongly the individual is likely to adhere to his chosen opinion, and the less likely he is to be neutral even though he may be relatively unbiased. Therefore $\partial P/\partial x = 0$ gives as model the cusp catastrophe surface. So far everything we have described is explicit, and

Figure 3.

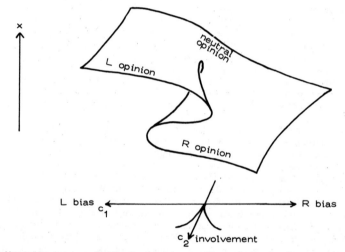

possibly collectable by a suitably designed questionnaire : for instance the individual might be asked to position himself or herself on three continuous scales indicating political point of view, involvement and strength of opinion (the word "bias" has perhaps the wrong overtones for soliciting the desired information from a questionnaire).

What is implicit in the model is some underlying dynamic representing the influence of communication on people as they make up their minds, pushing them towards the most probable opinion. The peaks of probability represent not only most probable behaviour, but also the asymptotic behaviour. Here of course the model can only be a first approximation to the truth, due to the random elements in communication, and the irrational elements in human nature, and this may be the basis of Thom's doubts. Nevertheless it is often only a first such approximation that the sociologist themselves are trying to capture, while fully admitting to the unpredictability of the free choice of individuals, and I suggest even this crude model may give some qualitative insight as follows.

We may regard the individuals as a cloud of points clustered in the neighbourhood of the catastrophe surface. If the issue is such that more and more people become involved, for instance as in the Dreyfuss affair or the Watergate affair, we can envisage this as a slow drift of the points along the surface in the direction of the c_2-axis. Unbiased individuals find themselves caught into taking sides, and even families are liable to be split. Those most involved find themselves sharply divided in opinion along the c_2-axis; at the same time there is a continuous change of opinion along a path going round the top of the cusp through the less involved, and a slight overlap amongst the more involved due to individuals near the centre who may have changed their bias yet paradoxically remain entrenched in the old opinion. Both latter features are common to "polarised" populations, but seldom exhibited so clearly in a model.

Suppose that we now change the bias of individuals by propaganda and persuasion, moving the points parallel to the c_1-axis. The uninvolved will hardly register any change of opinion, the slightly involved will change their minds smoothly, and the more involved will tend to suddenly switch opinion after some delay, not uncommonly to the surprise of both friend and foe, while the fanatics will be very hard to change, but once persuaded, will tend to become fanatical and irreversible converts.

The whole model can be elaborated to include the emergence of a compromise opinion by using the butterfly catastrophe [124]. So much for this elementary example illustrating the type of model possible in sociology; let us now return to the general discussion.

At present there seem to be two types of sociologist, the majority approaching the subject from the point of view of the humanities, and the minority approaching from the sciences. The latter tend to use statistics as their main tool, and are often accused by the former of missing the real point. In turn the latter accuse the former of basing their theories upon intuition rather than upon scientifically collected evidence. Nevertheless it may well be that the former have a better understanding of that underlying social morphologies, and are justifiably distressed by the way the certain quantitative analyses may seem to miss the point. One of the main benefits of catastrophe theory to sociology may well be to reinforce some of the theories of the non-mathematical sociologists. For, by providing models in which continuous causes can produce discontinuous and divergent effects, catastrophe theory may enable them to retain, indeed confirm and develop, theories which at present are being thrown into doubt by misinterpretation of quantitative data.

Summarising : the two ways in which catastrophe theory may alter the face of sociology are in the design of experiment, and the synthesis of data. In future the sociologist may redesign his experiment with not only the objective of drawing a smooth curve to illustrate the trend, but also the aim of detecting those critical points where the curve, or its derivative, may be discontinuous, and hence revealing the social morphology that is taking place.

Linguistics.

Thom's application of catastrophe theory to linguistics [145,151-4] is another extremely exciting possibility, because this is the first coherent attempt to explain the brain activity behind language. Linguists make little attempt to link neurology and linguistics, and even Chomsky falls back on the suggestion that ability to appreciate the deep structure of language must be hereditary, without indicating how the genes inside each cell could possibly store such an ability. This blind faith in heredity is one step even further removed from credibility than the biochemists' euphemism about developmental instructions being "coded in the genes".

By contrast Thom suggests that the deep structure of language is yet another aspect of universal morphologies, and his approach would at the same time explain how animals, or children before they have learnt to speak, can reason logically (a simple observation all too often overlooked by linguists and philosophers). His main idea is that a basic sentence begins as a single thought, represented by a bifurcation of a dynamical system describing the neurological activity, with the attractors of the system representing the nouns, and the surfaces separating their basins of attraction representing the verb. Speech is a mechanism that subsequently lists the component parts of the bifurcation, and speech-recognition is the reverse mechanism that synthesises a duplicate model of the same bifurcation, and thereby simulates another single thought analogous to the original thought. The simplest bifurcations are the elementary catastrophes, and Thom suggests that these give rise to the basic types of spacio-temporal sentences, which are the foundation stone of any language. I find this idea very convincing. However when Thom gets down to the business of formulating the relationship between the mathematics and the neurology I find him less convincing, and possibly open to improvement, as follows.

He rests his model [148,p.336 and 145,p.232] on a fibering $f : X \to R^4$, from a manifold X representing the relevant brain-states, to R^4 representing conceptual space-time, which he suggests arises from our early awareness of space-time. The synaptic connections in the brain, he goes on to say, determine a dynamic on the fibre F, parametrised by R^4. The basic sentences are represented by bifurcations over paths in R^4, and these are classified by the elementary catastrophes with control-space R^4 and state-space F. The fallacy is that at any given moment the brain state $x \in X$ can only lie in one attractor (or in the basin of one attractor), of the dynamic on F. Therefore the brain can only think of one actor, or one noun, at a time, whereas what Thom really wants in his model is for the brain to think of the whole sentence simultaneously.

After discussions with P. Winbourne and M. Godwin, I should like to propose an alternative formulation as follows. We begin with an analogy of visual perception. Let $C_+^\infty(D^2)$ denote the space of non-negative C^∞-functions on a disk D^2. A function $p \in C_+^\infty(D^2)$ represents a picture in D^2, with $p(y)$ representing the light intensity at $y \in D^2$. The maxima of p represent the brightest spots. If D^2 now denotes the visual field, then the visual mechanism gives a map $C_+^\infty(D^2) \to X$, from pictures to brain-states. Meanwhile the faculty of visual perception must imply the existence (to within some tolerance) of an inverse map $f: X \to C_+^\infty(D^2)$, where D^2 is now the conceptual (as opposed to the external) visual field, otherwise the mind could not make head nor tail of the resulting jumble of brain-states. When the brain-state is at x the mind-state or perceived picture is fx.

We now return to Thom's suggestion of our early awareness of space-time. Our early experience of space is primarily an awareness of matter. If D^3 denotes a region of space, the most direct mathematical description of matter is a (possibly discontinuous) density distribution $m: D^3 \to R_+$, where $m(y)$ denotes the density of matter at $y \in D^3$. However from the psychological point of view of awareness this direct description is inadequate for two reasons. Firstly we cannot see or touch inside solid objects to tell how dense they are. Secondly animate objects (including ourselves) tend to have a nest of significant neighbourhoods around them, of which we are aware. For instance the insides of a person are more vulnerable than his skin, his 10-centimetre neighbourhood is a territory that he has a strong instinct to defend if invaded, his 1-metre neighbourhood lies within his reach, his 2-metre neighbourhood lies within striking distance, and outside his 10-metre neighbourhood is outside his immediate striking area, unless he has a gun*. Both these inadequacies are met to some extent if we replace m by some smoothed density distribution $s \in C_+^\infty(D^3)$; for example we might define $s: D^3 \to R_+$ by the transform

$$s(y) = \int_{D^3} e^{-|y-z|^2} m(z) \, dz$$

Then the nests of neighbourhoods are given by the level surfaces of s. Summarising: awareness of space can be formally represented by smooth density distributions.

We now take as our <u>main hypothesis</u> that space-time awareness is represented by a brain $\to$ mind map

$$f: X \to C_+^\infty(D^3 \times T),$$

where X is a manifold representing the brain-states underlying spacio-temporal thoughts, D^3 is a conceptual region of space, and T a conceptual interval of time. The justification for this hypothesis is twofold, firstly the analogue above, implied by visual perception, and secondly the representation of space-awareness by the smooth density distributions. We now show how this hypothesis leads to a type of catastrophe theory that is simpler than Thom's model, and subtly different. Since $C_+^\infty(D^3 \times T) = C^\infty(T, C_+^\infty(D^3))$, given a brain-state $x \in X$, then the resulting mind-state $s = fx$ is a conceptual time-path s_t, $t \in T$, of smooth density distributions $s_t: D^3 \to R_+$. Mathematically we can regard s as a catastrophe potential, with control-space T and state-space D^3,

* Notice the slight out-of-context jump in the mind at the word "gun". A larger catastrophic jump occurs in the perceived neighbourhoods of a person if he pulls out a gun,

which is much simpler than Thom's model. The maxima of s_t will represent the "centres" of solid objects in D^3, in other words the actors of Thom, or the nouns of the corresponding basic sentence. However this set-up differs from the usual catastrophe theory application, because there is no dynamic* maximising s_t. Therefore in this case the maxima themselves are less important then their nests of neighbourhoods given by the level surfaces of s_t.

Consider a particular example, the message sentence "A gives B to C" - see [145]. At any given time t, it is important to know whether AB together form a "closer" subset than, for example, BC. Psychologically we would recognise this by observing that the matter (or rather the smoothed distribution) in between A and B is denser than that between B and C. Mathematically we can detect this by checking that the saddle between A and B is higher than that between AB and C   (Figure 4a).

Figure 4.   (a)   (b)

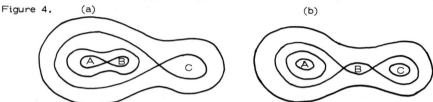

A catastrophe happens if the two saddles are at the same level, (Figure 4b), and semantically this occurs at the moment that the message B leaves A's hand and enters C's hand. Therefore we might call it a <u>transfer of proximity</u> catastrophe. These catastrophes are characterised by the Maxwell sets between saddles of index 1 lying on the same component of level surface. I think that this formulation leads to   mathematics that is much closer to Thom's original conception.

Summarising : in this application to linguistics there are three types of 1-dimensional catastrophe, entrances and exits, represented by the familiar fold together with an orientation, and the above transfer of proximity. What needs to be studied is the various sequences of folds and transfers that can occur along paths in higher dimensional control spaces, near the organising centres of higher catastrophes. When classifying the higher catastrophes the same restriction appears as for probability functions [124], because distribution functions are positive, and therefore bounded below. This restriction eliminates all the original 7 elementary catastrophes except for the cusp, x^4, and the butterfly, x^6. The classification up to dimension 8 contains only two more cuspoids, x^8 and x^{10}, and the <u>double cusp</u>, $x^4 + y^4$. The two cuspoids would represent sentences whose 4 or 5 actors must all be in single file, such as the messenger sentence "A sends B by C to D". The most interesting key to linguistics, therefore, seems to lie in the study of paths in the double cusp, and the associated sequences of entrances, exits and transfers between the 4 actors involved, and the comparison of these paths with Thom's original classification [145] of basic sentences.

---

* In the notation of [177] is an application at Level 1 rather than Level 2.

Answer to Christopher Zeeman's reply.   René Thom.

    The reply by Christopher to my Manifold article was for me something of a surprise; I did not expect that my article deserved a reply, nor did I feel that I had been in this article unnecessarily provocative. Now his reply contains, with a good deal of eulogy, fairly severe criticisms. I shall not dwell on the eulogy, be it merited or not, but I will concentrate on the criticisms, as these are worthy of discussion, and they may throw some light on our basic differences regarding the scientific status of Catastrophe Theory.

    I will not discuss the question of my book's obscurities; they are obvious; but consider only that, if one puts into a work the content of almost one full life of scientific thought, then some right to imprecision, or even to error, may be granted to the author, especially in fields of such complete novelty. I think that Christopher's criticisms arise basically from a fairly strict, dogmatic view of catastrophe theory (CT), which he* identifies with "elementary catastrophe theory" (ECT); i.e. the theory of catastrophes induced by a field of gradient dynamics. I strongly believe that CT has to be considered as a theory of general morphology, hence it may be necessary for us to use all kinds of catastrophes (generalised catastrophes, composed maps catastrophes, G-invariant catastrophes and so on); the aim of CT is to find the syntax describing the aggregation of such catastrophes. Now the elementary catastrophes of ECT, although undoubtedly fundamental, are nothing but the simplest constituents of such a syntax. To take the linguistic analogy, elementary catastrophes are the phonemes of the text composed by the morphogenetic fields; ECT is to CT no more than phonology is to grammar.

    If you accept this general view of CT, then a fundamental problem arises : given a sequence of morphological events, is there a _unique_ model explaining it? More precisely, given two models M_1 and M_2 explaining the morphology, does there exist a covering model M such that M_1, M_2 are specialisation of M ? Unless we restrict our class of models to a very narrow class (like ECT), there is no such uniqueness theorem. Hence, when confronted with two distinct models (M_1), (M_2) for the same morphology, the mathematics, by itself, would not give enough to decide between them. Possibly experiment might do it; if it is impossible to imagine an experimental criterion between the models, then only a subjective feeling of elegance, of mathematical or conceptual economy may decide. In general, it makes no sense to claim that a model is false; we may only say that one model is preferable to another, with a convenient justification. This kind of vagueness for the choice of models is felt by scientists of strict positivist or Popperian opinion (and they are prevalent among scientists in UK or US; particularly physicists and biologists) as an overwhelming objection against the scientific claims of CT. Needless to say, I do not share this prejudice : for me, the scientific status of CT is founded on its internal, mathematical consistency, which allows making deductions, generating

---

* Footnote reply by Zeeman : Of course I agree with René here that ECT ≠ CT. My emphasis on ECT has been mainly because of its usefulness in applications. By contrast CT is not yet sufficiently developed, and as yet has been less useful for applications, but is more interesting mathematically.

new forms from another set of forms, thus allowing in some favourable cases qualitative predictions, and in general realising a considerable "reduction of arbitrariness" in the description. This fundamental point will be discussed later, when dealing with CT met in the Social Sciences. Let us now review the list of difficult points in Christopher's reply.

Physics.

Phase transitions. The situation here is quite unsatisfactory, as well from the CT viewpoint as from the standard molecular interaction point of view. The results obtained by physicists in the theory of phase transitions are not convincing for two basic reasons :

i) they always start with a cubic lattice in Euclidean space which means that they start "a priori" with an orderly situation (cf. the Ising model),

ii) they blindly use the Gibbsian formalism in the study of phase transition; but the only motivation for justifying Gibbs formalism is ergodicity of the local dynamic of interacting molecules; but, precisely, the existence of different phases seems to show that this ergodicity breaks down.

From the experimental point of view, I wonder to what extent the notion of "characteristic exponent" is a valid notion. If, as they say, the position of the critical point itself is not well defined, how could the exponent α of a curve of type $y = \exp \alpha x$ arriving at this point be well defined?

Hydrodynamic breaking.

In general the temporal evolution of the surface limiting two phases (for instance, liquid water from air) is given locally by an Hamilton-Jacobi equation, hence one should expect the appearance of all elementary catastrophes, in particular elliptic and hyperbolic umbilics. (Note that the local potential, defined on the space of normalised covectors, is defined by the initial data, and not intrinsically by thermodynamic parameters.) But the specific property of these phases may affect this Hamilton-Jacobi evolution; for instance, the surface-tension phenomena, which tend to minimise locally the area of the surface limiting liquid water, prohibit the appearance of the elliptic umbilic. If the situation starts from the hyperbolic umbilic (giving standard shallow-water breaking) and tends toward the elliptic side, then this evolution is stopped at the parabolic point, which becomes stabilised, and unfolds transversally, thus giving rise to the mushroom type of curve : it is true that the curve bounding the hat of the mushroom limits the domain of a saddle; I think that this saddle undergoes threshold stabilisation through coupling with an external variable, so that the global evolution describes the emission of a drop at the end of a (breaking) cylindrical jet. Here again, we cannot claim the exact metrical validity of this model. This description may seem unduly complicated, but there is little doubt that the theory of breaking phenomena is indeed quite complicated ...
It is true that, when I used the umbilics to describe some biological phenomena, like phagocytosis, ropheocytosis, ... I had not a clear idea of this complexity. Much work has to be done in this field before the situation may be satisfactorily clarified.

Biology, in particular Embryology.

Let me first discuss a specific point, namely mesoderm stabilisation. I attribute that to "threshold stabilisation" of the unstable sheet of the cusp catastrophe. More precisely, the attractor associated to mesoderm is the attracting limit cycle exhibited by the flow defined by the gradient of the Riemann-Hugoniot potential :

$$V = x^4/4 + u\, x^2/2 + v\, x$$

for $u = -1$, with respect to the hyperbolic metric $ds^2 = dx^2 - dv^2$.

This construction suggests that the v unfolding variable, after some time, becomes an internal variable, with a damping effect on the original catastrophe expressed by the $-$ sign of dv^2 in ds^2. I believe this model better than the butterfly model, because of the fact that the attractor cycle has a limit, when the hyperbolic metric degenerates to dv^2, an hysteresis cycle (as in the Van der Pol equation), and this hysteresis cycle has a biological significance related to mesoderm's embryological vocation : pumping energy from endoderm (intestinal mucosa and liver), to capture the prey and to bring it to the mouth. In fact, in vertebrates, this hysteresis cycle is realised organically by blood circulation. Here you see that global considerations on development have a bearing on the choice of a local model (namely for gastrulation).

More on threshold stabilisation : coupling with a "plastic dynamic" like putting a lump of clay inside a potential well, or applying some general erosion-generating dynamic, like in Geomorphology, are the most obvious mechanisms which may generate these effects. But there may be many more : coupling with a high frequency source may stabilise a maximum of potential like in the Kapitza pendulum; more generally, given an infinite dimensional linear oscillator, like a vibrating string, with evolution defined by a differential equation : $dX/dt = H(X)$, where X is a vector in Hilbert space, H a self-adjoint operator, the spectrum of this operator is nothing but the set of singular values of the distance function to the origin : $d^2 = |H|^2$, restricted to the "ellipsoid" defined by $\langle X, H(X) \rangle = 1$. These values correspond to critical points where only the lowest critical value, the "ground value" is associated to a minimum of d^2; all other points are saddles, and correspond to stationary "excited" states of the oscillator. Now it happens that in most situations (in classical or quantum mechanics) these excited states are fairly easy to realise, more or less as if they were, under certain circumstances, "attractors" of the dynamic. (For instance a horn player may extract out of his horn, just by strengthening or relaxing his lips, several harmonics above the ground frequency.) It seems that some phenomenon of "self-coupling" is the cause of such attracting effects, as the eigenfrequencies are the only ones which may be excited through resonance. More generally, if we start with a gradient dynamic, and if we allow adding some kinetic energy to the potential, we get a blurring of the original catastrophe scheme : it would be worthwhile to have a good theory of such kinetic effects ...

Coming back to Biology, I think we have to be open-minded and try to test any kind of model which may be proposed to explain development. Your idea of a "secondary wave" of spatial changes following a "primary wave" of metabolic catastrophe seems to be interesting; but I am extremely suspicious of all kinds of very concrete morphological modelling involving such usual

language verbs like : tearing, folding, stripping, etc ... I believe that
mechanical properties of cells are very poorly known. (Do not forget that the
"phase" of living matter is not defined : solid, liquid ? As hydrodynamic
breaking is still so mysterious, I wonder how you could state anything very
serious in that matter ...) Moreover, this strikes me as a paradigmatic
inconsistency : why use first all the abstract refinements of ECT, and finally
end in this intuitive cell-interaction modelling, the precision of which is illusory?
You may leave that to professional biologists, like Wolpert, who are unable to
concieve anything else. It is giving up all the progress made by replacing the
anthropocentric notion of Force by the mathematical notion of potential.

   I use here this opportunity to give a word of caution. Many people,
understandably eager to find for Catastrophe theory an experimental
confirmation (?), may embark into precarious quantitative modelling, where
explicit observable interpretations are given to unfolding parameters (even to
internal ones ...). Needless to say, many (if not all) of these interpretations
will break down. This may cause - among positivist-minded Scientists - a
"backlash" reaction against Catastrophe theory, a reaction already noticeable
among some scientists in UK. (In France and in the US, Catastrophe theory
is still too ignored to have provoked such a reaction.) In the same line, I
would also like to add a didactic warning : when presenting CT to people, one
should never state that, due to such and such a theorem, such and such a
morphology is going unavoidably to appear. In no case has mathematics any
right to dictate anything to reality. The only thing one might say is that, due
to such and such a theorem, one has to expect that the empirical morphology
will take such and such a form. If reality does not obey the theorem - that
may happen - this proves that some unexpected constraints cause some lack of
transversality, which makes the situation all the more interesting.

   As a general conclusion about CT in Biology, I feel that we should not
hurry for any "experimental confirmation"; I think that a lot of theoretical
thinking, of speculative modelling, has to be done before one might really
start to experiment to make a choice between models. Even so, it is doubtful
that these experiments would interest very much present-day biologists, as they
would be unable to understand their motivation. Quite likely, there is very
little which can be done about the present situation : I agree with P. Antonelli
[101], when he states that theoretical biology should be done in Mathematical
Departments; we have to let biologists busy themselves with their very concrete
- but almost meaningless - experiments; in developmental Biology, how could
they hope to solve a problem they cannot even formulate ?

Social Sciences.

   Let us start with Linguistics. I think that the "connectedness objection"
is taken care of in my Bahia paper, Langage et catastrophes [151], through
confusion of actants, the unicity of the attractor associated to the global meaning
of a sentence is assured. Today, I would be less certain that the full set of
"archetypal morphologies" described by kernel sentences can be derived in a
strict formal way from ECT. I think that linguistic constraints, like persistence
of the subject, play a fundamental role in determining these morphologies. But
it may be worthwhile to look at the double cusp from that point of view.

   The model proposed by Christopher (that the internal space of
linguistic dynamics is the space of density distributions on space-time) is

certainly an interesting idea, but I wonder whether he would consider*, for instance, that the subject is more dense than the object, that is : has the density distribution a semantic meaning, or is it just the representation of density of matter ?

In social sciences, still more than in exact sciences, the hope of finding quantitative modelling of catastrophes is very slight. Granted that CT leads to basically qualitative modelling, what may be the interest of such models? Certainly not experimental confirmation, which would not be at all surprising, since the model is constructed precisely to generate the given morphology. A first answer, I think, is as follows : CT is - quite likely - the first coherent attempt (since Aristotelian Logic) to give a theory on analogy. When narrow-minded scientists object to CT that it gives no more than analogies, or metaphors, they do not realise that they are stating the proper aim of CT, which is to classify all possible types of analogous situations. In that respect, Christopher's models for agression, for paradox, for strength of opinion etc..., are quite illuminating. Now the positivist objection may be rephrased as follows : whereas quantitative modelling allows us to use computation, and therefore is more powerful than ordinary common sense intuition, how could qualitative modelling be stronger than usual, ordinary language deduction? How can a qualitative model be something more than an idle, superfluous geometric picture of common sense intuition? This objection, I believe, has some validity. But it will lose its strength, precisely in so far as a complete CT will be constructed, which will allow formal deduction, and combinatorial generation of new forms from a set of given forms. In as much as CT develops into a formal syntax of (pluri-dimensional) catastrophes, we will be able to go from a purely verbal description to an abstract, topological morphology which we will be able to handle with purely formal, algebraic tools. Hence we might put into connection apparently disjoint facts, predict unexpected situations, or, at least, reduce the arbitrariness of the description. As I said earlier, reducing the arbitrariness of the description really is the proper definition of scientific explanation. This definition is rejected by some scientists (Mitchison, private correspondance), because, as they say, it appeals to the subjective feeling of the observer : mystical or magical explanations do achieve this reduction of arbitrariness, and they are opposite to scientific method. For instance, you could explain all natural phenomena by saying that they appear as a result of God's will. Now it happens that all these magical explanations imply the use of non-formalisable concepts, like : God, entelechy, order, complexity, programme, force, message, information, meaning, spirit, randomness, life, etc ... All these concepts have morphologically the common feature of being transpatial (non-local) : they prescribe long range order, long range constraints in the morphology to which they apply. (Think for instance of the concept of meaning for linguistic morphology.) Now scientific thinking is basically formal thinking : it is based on spatial local concatenation of forms, which excludes any long range manipulation of symbols (or basic forms, or morphogenetic fields in CT terminology). Quite frequently, in many sciences, people use freely these "magical" concepts, without being aware of their magical character. Molecular Biology, for instance, with its intemperate use of : "Message, Information, Code", clearly exhibits its fundamental impotence to deal with the spatio-temporal ordering of events in living matter. I am afraid the same is true in many social sciences, where one meets with such

---

* Footnote reply by Zeeman : Yes indeed. Since the 3-dimensional space involved is conceptual, the density distribution therein most definitely has a semantic meaning.

concepts as : information, authority, collectivity, sense of history, conflict, consciousness, etc... All these concepts have an illusory explanatory power. It is perhaps the major interest of CT to clear all sciences of these old, biologically deeply inrooted concepts, and to replace their fallacious explanatory power by the explicit geometric manipulation of morphogenetic fields. In all sciences, CT calls for the same cleaning of intuition, as Hilbert advocated in his "Grundlagen der Geometrie" for the foundations of Geometry : Eliminate the "obvious" meaning, and replace it by the purely abstract geometrical manipulation of forms. The only possible theoretisation is Mathematical.

Bibliography on Catastrophe Theory.

The bibliography is by no means exhaustive, but is given as a background for the Thom-Zeeman debate, and as a source for further reading. It is divided into two parts, mathematics and applications. The mathematics papers are primarily concerned with singularities of maps and their classification, which in particular includes the elementary catastrophes, and which further requires the preparation theorem and stratifications. Also included are papers on bifurcations, which form the beginning of a generalised catastrophe theory for dynamical systems. Nearly all the applications so far involve the elementary catastrophes, but a few papers are included which implicitly involve other types of generic singularities.

Several of the references are to be found in the volumes, which have been abbreviated as follows :

[T] : <u>Towards a Theoretical Biology</u>, Proceedings of Serbelloni Conferences 1967-1970, Ed. C.H. Waddington, Edinburgh University Press, Volumes 1 - 4, 1968-1972.

[L] : <u>Proceedings of Liverpool Singularities Symposium I</u>, Ed. C.T.C. Wall, Springer Lecture Notes, 192, Springer-Verlag, Berlin, 1971.

[D] : <u>Dynamical Systems</u>, Proceedings of Symposium at Salvador, Brazil, 1971, Ed. M.M. Peixoto, Academic Press, New York, 1973.

MATHEMATICS.

1. V.I. Arnol'd, Singularities of smooth maps, Uspehi Mat. Nauk, 23, 1 (1968) 3-44; (Eng. Transl.) Russian Math. Surveys, 23, 1 (1968) 1-43.

2. V.I. Arnol'd, On braids of algebraic functions and the cohomology of swallowtails, Uspehi Mat. Nauk, 23, 4 (1968) 247-248.

3. V.I. Arnol'd, On matrices depending on parameters, Uspehi Mat. Nauk, 26, 2 (1971) 101-114; (Eng. transl.) Russian Math. Surveys, 26, 2 (1971) 29-43.

4. V.I. Arnol'd, Lectures on bifurcations and versal families, Uspehi Mat. Nauk, 27, 5 (1972) 119-184; (Eng. transl.) Russian Math. Surveys, 27, 5 (1972) 54-123.

5. V.I. Arnol'd, Integrals of rapidly oscillating functions and singularities of projections of Lagrangian manifolds, Funktsional. Anal. i Prilozhen., 6, 3 (1972) 61-62; (Eng. transl.) Functional Anal. Appl., 6 (1973) 222-224.

6. V.I. Arnol'd, Normal forms for functions near degenerate critical points, the Weyl Groups of A_k, D_k, E_k and Lagrangian singularities, Funktsional. Anal. i Prilozhen., 6, 4 (1972) 3-25; (Eng. transl.) Functional Anal. Appl., 6 (1973) 254-272.

7. V.I. Arnol'd, Classification of unimodal critical points of functions, Funktsional. Anal. i Prilozhen., 7, 3 (1973) 75-76; (Eng. transl.) Functional Anal. Appl., 7 (1973) 230-231.

8. V.I. Arnol'd, Normal forms for functions in the neighbourhood of degenerate critical points, Uspehi Mat. Nauk, 29, 2 (1974) 11-49.

9. V.I. Arnol'd, Singularities of differentiable functions, Invited address, Int. Congress of Math., (Univ. Br. Columbia, Vancouver, 1974).

10. T.F. Banchoff, Polyhedral catastrophe theory I : Maps of the line to the line, [D] 7-22.

11. J.M. Boardman, Singularities of differentiable mappings, IHES Publ., Math., 33 (1967) 21-57.

12. T. Bröcker, Differentierbäre Abbildungen, (Regensburg Lecture Notes 1973).

13. P. Brunovsky, One parameter families of diffeomorphisms, Warwick Symposium on Differential Equations (Ed. D. Chillingworth, Springer Lecture Notes 206, Berlin, 1971) 29-33; & Comment. Math. Univ. Cardinae, 11 (1970) 559-582.

14. M.S.B. de Carvalho, Liapunov functions for diffeomorphisms, Thesis, Warwick University, 1973.

15. F. Dumortier, Singularities of vector fields on the plane, (Thesis, Brussels, 1974).

16. A.N. Godwin, Three dimensional pictures for Thom's parabolic umbilic, IHES, Publ. Math., 40 (1971) 117-138.

17. A.N. Godwin, Methods for Maxwell sets of cuspoid catastrophes, (Lanchester Polytechnic preprint, Rugby, 1974).

18. M. Golubitsky & V. Guillemin, Stable mappings and their singularities, Grad. Texts in Math., 14 (Springer, N.Y., 1974).

19. M. Golubitsky, Contact equivalence for Lagrangian submanifolds, these Proceedings.

20. J. Guckenheimer, Bifurcation and catastrophe, [D] 95-110.

21. J. Guckenheimer, Catastrophes and partial differential equations, Ann. Inst. Fourier (Grenoble), 23 (1973) 31-59.

22. J. Guckenheimer, Solving a single conservation law, these Proceedings.

23. B. Heatley, Local stability theories equivalent to catastrophe theory, Thesis, Warwick University, 1974.

24. E. Hopf, Abzweigung einer periodischen Lösung von einer stationären Losung einer Differentialsystems, Ber. Verh. Sächs, Akad. Wiss. Leipzig. Math. Phys., 95 (1943) 3-22.

25. T-C. Kuo, On C^0-sufficiency of jets, Topology, 8 (1969) 167-171.

26. T-C. Kuo, A complete determination of C^0-sufficiency in $J^r(2,1)$, Inv. Math., 8 (1969) 225-235.

27. T-C. Kuo, The ratio test for analytic Whitney stratifications, [L] 141-149.

28. T-C. Kuo, The jet space $J^r(n,1)$, [L] 169-177.

29. H.I. Levine, Singularities of differentiable mappings, (Notes of Lectures by R. Thom, Bonn, 1959) [L], 1-89.

30. B. Malgrange, Ideals of differentiable functions, (Oxford U.P., 1966).

31. L. Markus, Dynamical Systems - five years after, these Proceedings.

32. J. Martinet, Lectures on singularity theory, to be published in French by PUC, Rio de Janeiro, 1974.

33. J. Mather, Stability of C^∞-mappings :
    I The division theorem, Ann. Math., 87 (1968) 89-104.
    II Infinitessimal stability implies stability, Ann. Math., 89 (1969) 254-291.
    III Finitely determined map germs, IHES Publ. Math., 35 (1968) 127-156.
    IV Classification of stable germs by R-algebras, IHES Publ. Math., 37 (1969) 223-248.
    V Transversality, Adv. in Math., 4 (1970) 301-336.
    VI The nice dimensions, [L] 207-253.

34. J. Mather, Right equivalence, (Warwick preprint, 1969).

35. J. Mather, Notes on topological stability, (Harvard preprint, 1970).

36. J. Mather, On Nirenberg's proof of Malgrange's preparation theorem, [L] 116-120.

37. J. Mather, Stratifications and mappings, [D] 195-232.

38. K.R. Meyer, Generic bifurcations of periodic points, Trans. Amer. Math. Soc., 149 (1970) 95-107.

39. S. Newhouse, On simple arcs between structurally stable flows, these Proceedings.

40. S. Newhouse & J. Palis, Bifurcations of Morse-Smale dynamical systems, [D] 303-366.

41. L. Nirenberg, A proof of the Malgrange preparation theorem, [L] 97-105.

42. J. Palis, Arcs of dynamical systems : bifurcations and stability, these Proceedings.

43. F. Pham, Introduction à l'étude topologique des singularités de Landau, (Gauthier-Villars, Paris, 1967).

44. F. Pham, Remarque sur l'équisingularité universelle, (Nice preprint, 1970).

45. V. Poenaru, Zakalyukin's proof of the (uni)versal unfolding theorem, these Proceedings.

46. V. Poenaru, The Maslov index for Lagrangian manifolds, these Proceedings.

47. V. Poenaru, Analyse Différentielle, Lecture Notes in Mathematics 371, Springer, 1974.

48. I.R. Porteous, Geometric differentiation - a Thomist view of differential geometry, [L] Volume 2, Lecture Notes in Mathematics, 209, Springer, 1971, 121-127.

49. I.R. Porteous, The normal singularities of a submanifold, Jour. Diff. Geom., 5 (1971) 543-564.

50. F. Sergeraert, La stratification naturelle de $C^\infty(M)$, (Thesis, Orsay, 1971).

51. M. Shub, Structurally stable diffeomorphisms are dense, Bull. A.M.S. 78 (1972) 817-818.

52. D. Siersma, Singularities of C^∞ functions of right-codimension smaller or equal than eight, Indag. Math. 25 (1973) 31-37.

53. S. Smale, On gradient dynamical systems, Ann. of Math (2) 74 (1961) 199-206.

54. S. Smale, Differentiable dynamical systems, Bull. A.M.S. 73 (1967) 747-817.

55. S. Smale, Stability and isotopy in discrete dynamical systems [D], 527-530.

56. J. Sotomayor, Generic one-parameter families of vector fields, Bull. A.M.S., 74 (1968) 722-726; & IHES Publ. Math., 43 (1973) 5-46.

57. J. Sotomayor, Structural stability and bifurcation theory, [D] 549-560.

58. J. Sotomayor, Generic bifurcations of dynamical systems, [D] 561-582.

59. P. Stefan, A remark on right k-determinacy, Bangor preprint, 1974.

60. F. Takens, A note on sufficiency of jets, Inv. Math., 13 (1971) 225-231.

61. F. Takens, Singularities of functions and vector fields, Nieuw. Arch. Wisk, (3), 20 (1972) 107-130.

62. F. Takens, Introduction to global analysis, (Math. Inst. Utrecht Univ. 1973).

63. F. Takens, A nonstabilisable jet of a singularity of a vector field, [D] 583-597.

64. F. Takens, Integral curves near mildly degenerate singular points of vector fields, [D] 599-617.

65. F. Takens, Singularities of vector fields, IHES Publ. Math., 43 (1973) 47-100.

66. F. Takens, Unfoldings of certain singularities of vector fields : generalised Hopf bifurcations, J. Diff. Equations, 14 (1973) 476-493.

67. F. Takens, Constrained differential equations, these Proceedings.

68. R. Thom, Une lemme sur les applications différentiables, Bol. Soc. Mat. Mexicana, (2) 1 (1956) 59-71.

69. R. Thom, Les singularités des applications différentiables, Ann. Inst. Fourier (Grenoble), 6 (1956) 43-87.

70. R. Thom, La stabilité topologique des applications polynomiales, L'Enseignement Mathématique, 8 (1962) 24-33.

71. R. Thom, Sur la théorie des enveloppes, J. Math. Pures Appl. (9) 41 (1962) 177-192.

72. R. Thom, L'équivalence d'une fonction différentiable et d'un polynome, Topology, 3 (1965) 297-307.

73. R. Thom, On some ideals of differentiable functions, J. Math. Soc. Japan, 19 (1967) 255-259.

74. R. Thom, Ensembles et morphismes stratifiés, Bull. A.M.S., 75 (1969) 240-284.

75. R. Thom, Sur les variétés d'ordre fini, Global Analysis (Papers in honour of K. Kodaira) Tokyo, 1969, 397-401.

76. R. Thom, The bifurcation subset of a space of maps, Manifolds Amsterdam 1970, Lecture Notes in Mathematics 197, Springer 1971, 202-208.

77. R. Thom, Singularities of differentiable mappings (notes by H.I. Levine), [L] 1-89.

78. R. Thom, Stratified sets and morphisms : Local models, [L] 153-164.

79. R. Thom & M. Sebastiani, Un résultat sur la monodromie, Invent. Math., 13 (1971) 90-96.

80. R. Thom, Sur le cut-locus d'une variété plongée, J. Diff. Geom. (Dedicated to S.S. Chern & D.C. Spencer), 6 (1972), 577-586.

81. R. Thom, Méthodes Mathématiques de la Morphogénèse, Edition 10-18, U.G.E. Paris, Octobre 1974.

82. R. Thom, On singularities of foliations, Intern. Conf. on Manifolds (Tokyo University, 1973).

83. J-C. Tougeron, Idéaux de fonctions différentiables, (Springer-Verlag, 72, Berlin, 1972).

18 84. D.J.A. Trotman & E.C. Zeeman, Classification of elementary catastrophes of codimension ≤ 5, (Warwick Lecture Notes, 1974).

85. C.T.C. Wall, Introduction to the preparation theorem, [L] 90-96.

86. C.T.C. Wall, Stratified sets : a survey, [L] 133-140.

87. C.T.C. Wall, Lectures on C^∞-stability and classification, [L] 178-206.

88. C.T.C. Wall, Regular stratifications, these Proceedings.

89. G. Wasserman, Stability of unfoldings, Lecture Notes in Mathematics 393, Springer, Berlin, 1974.

90. G. Wasserman, (r,s)-Stability of unfoldings, (Regensburg preprint, 1974).

91. H. Whitney, Mappings of the plane into the plane, Annals of Math. 62 (1955) 374-470.

92. H. Whitney, Singularities of mappings of Euclidean spaces, Symp. Internat. Topologia Algebraica, Univ. Nacional Autonoma de Mexico, Mexico City, 1958, 285-301.

93. H. Whitney, Tangents to an analytic variety, Annals of Math., 81 (1965) 496-549.

94. F. Wesley Wilson, Smoothing derivatives of functions and applications, Trans. Amer. Math. Soc. 139 (1969) 413-428.

95. A.E.R. Woodcock & T. Poston, A geometrical study of the elementary catastrophes, Lecture Notes in Mathematics 373, Springer, Berlin, 1974.

96. E.C. Zeeman, C^0-density of stable diffeomorphisms and flows, Proc. Symp. Dyn. Systems, Southampton University, 1972.

19 97. E.C. Zeeman, The umbilic bracelet, to appear.

Numbering in this book.

98. E.C. Zeeman, Words in catastrophes, in preparation.

APPLICATIONS.

99. R. Abraham, Introduction to morphology, Quatrième Rencontre entre Math. et Phys. (1972) Vol. 4 Fasc. 1, Dept. Math. de l'Univ de Lyon, Tome 9 (1972) 38-114.

100. J. Amson, Equilibrium and catastrophic modes of urban growth, London Papers in Regional Science Vol. 4, Space-time concepts in urban and regional models, 291-306.

101. P. Antonelli, Transplanting a pure mathematician into theoretical biology, Proc. Conference on Mathematics, Statistics and the Environment, Ottawa 1974.

102. N.A. Baas, On the models of Thom in biology and morphogenesis, (Univ. Virginia preprint, 1972).

103. C.P. Bruter, Secondes remarques sur la percepto-linguistique, Document 6, Centre Intern. Sémantique, Urbino (série A, 1971), 1-7.

104. C.P. Bruter, Quelques aspects de la percepto-linguistique, TA Informations, 2 (1972) 15-19.

105. C.P. Bruter, Sur la nature des mathématiques (Collection Discours de la Méthode) Gauthier-Villars, Paris, 1973.

106. C.P. Bruter, Topologie et perception, Tome I : Bases Mathématiques et Philosophiques (Collection Interdisciplinaire) Maloine-Doin, Paris, 1974. Tome II : Aspects neurophysiologiques (in preparation).

107. G.A. Carpenter, Travelling wave solutions of nerve impulse equations, (Thesis, Wisconsin, 1974).

108. D.R.J. Chillingworth, Elementary catastrophe theory, IMA Bulletin, in press.

109. D.R.J. Chillingworth, The catastrophe of a buckling beam, these Proceedings.

110. D.R.J. Chillingworth & P. Furness, Reversals of the earth's magnetic field, these Proceedings.

111. C.T.J. Dodson & M.M. Dodson, Simple non-linear systems and the cusp catastrophe, York University preprint, 1974.

112. M.M. Dodson, Darwin's law of natural selection and Thom's theory of catastrophes, Math. Biosciences (to appear).

113. M.M. Dodson & E.C. Zeeman, A topological model for evolution (in preparation).

114. D.H. Fowler, The Riemann-Hugoniot catastrophe and van der Waals' equation, [T] 4, 1-7.

115. N. Furutani, A new approach to traffic behaviour, Tokyo Preprint, 1974.

116. B. Goodwin, Review of Thom's book, Nature, vol. 242, 207-208, (16th March 1973).

117. J. Guckenheimer, Caustics, Proc. UNESCO Summer School, Trieste; 1972, to be published by the International Atomic Energy Authority, Vienna.

118. J. Guckenheimer, Review of Thom's book, Bull. A.M.S. 79 (1973) 878-890.

119. J. Guckenheimer, Caustics and non-degenerate Hamiltonians, Topology, 13 (1974) 127-133.

120. J. Guckenheimer, Shocks and rarefactions in two space dimensions, Arch. for Rational Mechanics and Analysis (to appear).

121. J. Guckenheimer, Isochrons and phaseless sets, Jour. Math. Biology, (to appear).

13 122. C. Hall, P.J. Harrison, H. Marriage, P. Shapland & E.C. Zeeman, A model for prison riots, to appear.

123. P.J. Harrison & E.C. Zeeman, Applications of catastrophe theory to macroeconomics, (to appear in Symp. Appl. Global Analysis, Utrecht Univ., 1973).

10 124. C.A. Isnard & E.C. Zeeman, Some models from catastrophe theory in the social sciences (Edinburgh conference 1972), in Use of models in the Social Sciences (Ed. L. Collins, Tavistock, London, 1974).

125. K. Jänich, Caustics and catastrophes, these Proceedings.

126. C.W. Kilmister, The concept of catastrophe (review of Thom's book), Times Higher Educ. Supplement, (30th November 1973), 15.

127. J.J. Kozak & C.J. Benham, Denaturation; an example of a catastrophe, Proc. Nat. Acad. Sci. U.S.A., 71 (1974) 1977-1981.

— Numbering in this book.

128. G. Mitchison, Topological models in biology : an Art or a Science? (M.R.C. Molecular Biology Unit, Cambridge, preprint).

129. H. Noguchi & E.C. Zeeman, Applied catastrophe theory (in Japanese), Bluebacks, Kodansha, Tokyo, 1974.

130. T. Poston, The Plateau problem, Summer College on Global Analysis, ICTP, Trieste, 1972.

131. T. Poston & A.E.R. Woodcock, On Zeeman's catastrophe machine, Proc. Camb. Phil. Soc., 74 (1973) 217-226.

132. D. Ruelle & F. Takens, On the nature of turbulence, Comm. Math. Phys., 20 (1971) 167-192.

133. L.S. Schulman & M. Revzen, Phase transitions as catastrophes, Collective Phenomena, 1 (1972) 43-47.

134. L.S. Schulman, Tricritical points and type three phase transitions, Phys. Rev., Series B, 7 (1973) 1960-1967.

135. L.S. Schulman, Phase transitions as catastrophes, these Proceedings.

136. L.S. Schulman, Stable generation of simple forms, Indiana Univ. preprint, 1974.

137. S. Smale, On the mathematical foundations of electrical circuit theory, J. Diff. Geometry, 7 (1972) 193-210.

138. S. Smale, Global analysis and economics :

   I : Pareto optimum and a generalisation of Morse theory, [D] 531-544.
   IIA : Extension of a theorem of Debreu, J. Math. Econ., 1 (1974) 1-14.

   III : Pareto optima and price equilibria, (to appear).

139. F. Takens, Geometric aspects of non-linear R.L.C. networks, these Proceedings.

140. R. Thom, Topologie et signification, L'Age de la Science, 4 (1968) 219-242.

141. R. Thom, Comments on C.H. Waddington : The basic ideas of biology, [T] 1, 32-41.

142. R. Thom, Une théorie dynamique de la morphogénèse, [T] 1, 152-179.

143. R. Thom, A mathematical approach to Morphogenesis : Archetypal morphologies, Wistar Inst. Symp. Monograph 9. Heterospecific Genome Interaction, Wistar Inst. Press, 1969.

144. R. Thom, Topological models in biology, Topology, 8 (1969) 313-335, & [T] 3, 89-116.

145. R. Thom, Topologie et Linguistique, Essays on Topology and related topics (ded. G. de Rham; ed. A. Haefliger & R. Narasimhan) Springer, 1970, 226-248.

146. R. Thom, Les symmetries brisées en physique macroscopique et la mécanique quantique, CRNS., RCP 10 (1970).

147. R. Thom, Structuralism and biology, [T] 4, 68-82.

148. R. Thom, Stabilité structurelle et morphogénèse, Benjamin, New York, 1972; English translation by D.H. Fowler, Benjamin-Addison Wesley, New York, 1975.

149. R. Thom, A global dynamical scheme for vertebrate embryology, (AAAS, 1971, Some Math. Questions in Biology VI), Lectures on Maths. in the Life Sciences, 5 (A.M.S., Providence, 1973) 3-45.

150. R. Thom, Phase-transitions as catastrophes, (Conference on Statistical Mechanics, Chicago, 1971).

151. R. Thom, Langage et catastrophes : Eléments pour une Sémantique Topologique, [D] 619-654.

152. R. Thom, De l'icône au symbole; Esquisse d'une théorie du symbolisme, Cahiers Internationaux de Symbolisme, 22-23 (1973) 85-106.

153. R. Thom, Sur la typologie des langues naturelle : essai d'interprétation psycho-linguistique, in Formal Analysis of Natural languages, ed. Moutin, 1973.

154. R. Thom, La linguistique, discipline morphologique exemplaire, Critique, 322 (March 1974) 235-245.

155. R. Thom, Gradients in biology and mathematics, and their competition, (AAAS, 1974, Some Mathematical Questions in Biology VII), Lectures on Mathematics in the Life Sciences, 6 (A.M.S. Providence, U.S.A., 1975), in press.

156. R. Thom, D'un modèle de la Science à une science des modèles, to appear.

157. J.M.T. Thompson, Instabilities, bifurcations and catastrophes, Physics Letters A, (to appear).

158. J.M.T. Thompson & G.W. Hunt, A general theory of elastic stability, Wiley, London, 1973.

159. J.M.T. Thompson & G.W. Hunt, Towards a unified bifurcation theory, University College, London, preprint, 1974.

160. M. Thompson, The geometry of confidence : An analysis of the Enga te and Hagen moka; a complex system of ceremonial pig-giving in the New Guinea Highlands, (Portsmouth polytechnic preprint, 1973), to appear in Rubbish Theory, Paladin.

161. M. Thompson, Class, caste, the curriculum cycle and the cusp catastrophe, to appear in Rubbish Theory, Paladin.

162. C.H. Waddington, A catastrophe theory of evolution, Annals N.Y. Acad. Sci., 231 (1974) 32-42.

163. A.T. Winfree, Spatial and temporal organisation in the Zhabotinsky reaction, Aahron Katchalsky Memorial Symp. (Berkeley 1973).

164. A.T. Winfree, Rotating chemical reactions, Scientific American, 230, 6 (June 1974) 82-95.

165. A.E.R. Woodcock & T. Poston, A higher catastrophe machine, Williams College preprint, 1974.

166. E.C. Zeeman, Breaking of Waves, Warwick Symp. Dyn. Systems, (Ed. D.R.J. Chillingworth) Lecture Notes in Mathematics, 206, Springer, 1971, 2-6.

167. E.C. Zeeman, The Geometry of catastrophe, Times Lit. Supp., (December 10th, 1971) 1556-7.

3 168. E.C. Zeeman, Differential equations for the heartbeat and nerve impulse, [T] 4, 8-67, & [D] 683-741.

15 169. E.C. Zeeman, A catastrophe machine, [T] 4, 276-282.

170. E.C. Zeeman, Catastrophe theory in brain modelling, Intern. J. Neuroscience, 6 (1973) 39-41.

171. E.C. Zeeman, Applications of catastrophe theory, Intern. Conf. on Manifolds, (Tokyo University, 1973).

11 172. E.C. Zeeman, On the unstable behaviour of stock exchanges, J. Math. Economics, 1 (1974) 39-49.

20 173. E.C. Zeeman, Research ancient and modern, Bull. Inst. Math. and Appl., 10, 7 (1974) 272-281.

4 174. E.C. Zeeman, Primary and secondary waves in developmental biology, (AAAS, 1974, Some Mathematical Questions in Biology, VIII), Lectures on Maths in the Life Sciences, 7 (A.M.S., Providence, USA, 1974), 69-161.

175. E.C. Zeeman, Differentiation and pattern formation, (Appendix to J. Cooke, Some current theories of the emergence and regulation of spatial organisation in early animal development) Annual Rev. of Biophys. and Bioengineering, 1975, in press.

176. E.C. Zeeman, Catastrophe theory in biology, these Proceedings.

2 177. E.C. Zeeman, Levels of structure in catastrophe theory, Proc. Int. Congress of Math. (Vancouver, 1974).

Numbering in this book.

12 178. E.C. Zeeman, Conflicting judgements caused by stress, (to appear).

179. E.C. Zeeman, Applications de la théorie des catastrophes à l'étude du comportement humain, (to appear).

9 180. E.C. Zeeman, Duffing's equation in brain modelling, (in preparation).

↑ — *Numbering in this book.*

22 AFTERTHOUGHT

READER. After reading your dialogue about embryology [19], I'd like to hear what you have to say about some of the more speculative applications of catastrophe theory, such as your psychological models.

MATHEMATICIAN. Since they are more speculative, anything I say must necessarily be somewhat tentative, and we may want to modify them later as we gain more knowledge.

R. Fair enough; I won't hold your words against you. Let's look at some specific models such as the one about rage and fear, and the one about anorexia nervosa [14,15]. Are you going to deduce those from hypotheses, as you did for the primary and secondary waves [19]?

M. Those are still as yet in what I called the descriptive stage [19], and we may not be able to push them back into an explanatory stage until we know more about the brain.

R. And have you been successful in predicting experiments?

M. A number of qualitative predictions have already been confirmed. For instance, while working on the anorexia model I found that the mathematics suggested a number of predictions that I thought at first were wrong, but which to my surprise my coworker Hevesi subsequently confirmed from observations of his patients [15]. The rage and fear model gave insight into how a loss of temper can have a cathartic effect upon a mood of self-pity [14,15], which I have subsequently often observed in both children and adults, and in both real life and fiction. Other models [18] have given insights into other aspects of human behaviour, and help to explain how moods persist and delay before changing, and why when the change comes, it is liable to be sudden. Such psychological models can on occasions be tested by introspection. I realise that introspection is sometimes seen as "subjective" rather than "objective", but in several cases it is the subjective feelings that are the important objective elements (the observables).

R. But can you call such models scientific? Even if they do contain germs of truth, are they not more like proverbs?

M. They are scientific in the sense of satisfying Thom's criterion [12] of "reducing the arbitrariness of description". The elementary catastrophes are profoundly simple concepts, higher dimensional analogues of maxima, minima and thresholds. (Do not confuse their simplicity with the complexity of the proof of their simplicity.) They provide a natural language for synthesising many ideas. Hevesi said of the anorexia model that it permitted him to give a coherent synthesis of a large number of observations of his patients that would otherwise have appeared disconnected [15].

R. That is an interesting point, but surely to study thresholds you only need elementary mathematics.

M. I agree, and that is the same elementary mathematics that you need for catastrophe theory with only one parameter (or one causal factor). But we are discussing the type of phenomenon that only sometimes has a threshold, and at other times changes smoothly. Therefore we need to consider the threshold of the appearance of a threshold; and that is a cusp catastrophe. And for the cusp catastrophe you need the full works. As soon as you have more than one parameter (or more than one causal factor) then you need the full complexity of the mathematics. This is why bifurcation theory has been confined to only one parameter for so long, and is only now becoming more multi-parameter.

R. Let's get back to this business of prediction. If a model cannot give predictions that you can test quantitatively can you call it scientific?

M. Are you saying that meteorology and economics are unscientific because they fail that test? But leaving this objection aside, I am personally very much in sympathy with you on this question. I have suggested procedures for quantitative testing of my psychological models [16,17], and discussed the possibilities with doctors, psychologists and animal behaviourists; some of them have even offered me facilities for experimentation. However the logistics of setting up such experiments are formidable, and I do not have the competence myself, so we must wait and see if any experimentalists are sufficiently interested in spending the time, energy and resources.

R. I still don't see how you are going to make quantitative predictions that fit all individuals.

M. The predictions would only be quantitative for one individual in one experiment. Each experiment would have its own quantitative thresholds, but the global picture extracted from the data should be qualitatively equivalent (in a precise sense [16]) to that arising from other experiments and other individuals.

R. That is very different from the classical paradigms of science, for instance those of physics.

M. Physicists do their experiments in laboratories, and are lucky in that the observable phenomena are very similar to those that occur outside the laboratory; this fact of course justifies their isolation in the laboratory. However in the behavioural sciences the phenomena must be observed much more in vivo. Although the qualitative patterns corresponding to various phenomena (psychological, sociological, physiological, neurological, etc.) can be recognised and measured and modelled in retrospect, prediction may be unreliable because of the interference between these patterns. The next step is to recognise the patterns of interference (between patterns), and this will lead to hierarchies of patterns. That is one of the reasons why Thom is so keen on linguistics, and upon the language of catastrophe theory, in which the elementary catastrophes are only some of the phonemes [7,9,10,11]. This is part of the natural language of dynamical systems, and if you believe that our brains can be modelled by dynamical systems, then this language must underlie our thought processes. I have always insisted that my catastrophe models in psychology are really models of the brain [15,16].

R. That surprises me, because I always thought you were a structuralist rather than a reductionist.

M. Well of course it is very important to be both. And there are theories of the brain that are not only reductionist in the sense of being based on the local neural network of nerve cells and synaptic connections, but are also structural in the sense of correlating psychological structures with global anatomical and dynamic structures of the brain. For instance I find Paul MacLean's concept of the triune brain very appealing, and I consider his accumulation of experimental and medical evidence very impressive [1,2,3,4,5,6]. I tend to think of my psychological models as descriptions of the limbic brain, where, according to MacLean and others, emotions and moods are probably generated. Once the mood is determined, then the detailed behaviour is selected by the neocortex within the framework of that mood.

R. But surely it's unwise and unnecessary to base psychological models upon so vague, and possibly controversial, a foundation. Wouldn't it be better to stay within psychology, and restrict yourself to psychological indices that can be measured directly?

M. There is one very important point. If you confine yourself within psychology, what is the analogue of homeostasis? What is the dynamic that holds the mood in equilibrium? And when that equilibrium breaks down, what is the dynamic that causes the catastrophic jump into another mood? The only way of explaining the dynamic is to go back to the neurology [16].

R. But I doubt if you could measure the full dynamic. And anyway, since most neurological measurements appear to be unrelated, or only very tenuously related, to psychological behaviour, I think most psychologists would question the usefulness of going back to the neurology.

M. That may be true, but there are also an increasing number of experiments in which the two appear to be related [2-6,17].

Furthermore the relationship might be increased by using different mathematical techniques for handling the data [16]. Also the mere consideration of the relationship may be a useful theoretical exercise.

R. Can you give me an example of where that has been useful?

M. For instance, when working on the anorexia model, Hevesi and I forced ourselves to give a neurological interpretation of the behaviour in terms of inputs to the limbic brain. The model then suggested that he must be doing somthing during his therapy to reduce those inputs in the brains of his patients. So he went back and watched himself with new eyes, and discovered what he now believes to be the operative suggestions of his therapy [15]. As a result he is now in a better position to communicate his techniques to other psychotherapists. For instance one psychotherapist reported shortly after reading about our model [14] that he "had applied the technique in one case of anorexia nervosa with startling success" [13].

R. I agree that sounds impressive. But I would like to see a more careful study done before we draw any firm conclusions. I must say, compared with Thom, you seem to be much more interested in practical matters. His approach seems to be more philosophical, and less concerned with prediction. As a reader I found his book difficult to read. Take his predation loop, for instance [7,8]; what precisely does he mean by his assertion that "the hungry predator is his prey"? Isn't his predation loop just a metaphor?

M. I think his reply might be "yes", but that would be because he believes that metaphors can be very profound [9,10,11]. And I would agree that his predation loop is profound, so let me try and put into my own words some of the points that I think he is making.
   The predation loop is about many things. On the surface it is about a simulator in the brain of a primitive predator. The assertion means that during the search for prey, the simulator contains only one image, namely that of the prey. After perception of the prey, the simulator has to contain two images, both the predator and the prey, in order to make the "computations" about distance and velocity necessary to effect a capture. After eating the prey, the simulator contains only one image again, that of the predator. As a matter of fact there is no need for the simulator to contain any images during the digestion of the prey, and Thom draws attention to this by his second assertion "sleep is the revenge of the prey upon the predator", implying, incidentally, that this may be the origin of sleep. And so back to the beginning of the loop again.

R. Even so, I'd hesitate to call that profound.

M. But on a deeper level Thom is suggesting that those primitive predators whose central nervous systems were able to store and simulate the predation loop would have had an evolutionary advantage over their competitors. Next, those predators whose actual physical shape and movements began to mimic the abstract dynamics stored inside the simulator would have had a further evolutionary advantage, because they would be more efficient. Thus, he argues, there could be evolutionary pressure, firstly for the brain to acquire, and secondly for the embryological morphogenesis to mimic, properties of abstract dynamics. He then goes on to suggest how such mimicry might have led to the evolution of the three germ layers, and subsequently to that of the perceptual, locomotory, circulatory and digestive systems of vertebrates [7]. He has similar ideas about reproduction.

R. That sounds highly speculative to me.

M. Since we know so little about all the underlying mechanisms involved, it may take several years before it is possible to formulate precise questions and test experimentally the cascade of ideas that must spring from so revolutionary an approach. This is why many of Thom's ideas are only descriptive, and must necessarily remain non-predictive for some time. But I think his suggestion of how function might have caused the evolution of embryological development through the dynamics of the brain is very interesting.

R. I always thought it would have been the other way round.

M. Exactly, and that is just the point. If function is to influence form, then it must do so through the intermediary of some storage system, and Thom is suggesting that the natural language for such a storage system is catastrophe theory [7,8,9,10,11]. At the first of Waddington's conferences at Serbelloni on theoretical biology Crick suggested that the three great unsolved problems of biology were evolution, development and the brain, and now Thom has reposed them as a single problem, in a rather unexpected form.

R. Well, I will go and have another look at the predation loop. By the way, if the elementary catastrophes are the phonemes of the language, tell me a word.

M.

References.

1. P.D. MacLean, Psychosomatic disease and the "Visceral Brain", Recent developments bearing on the Papez theory of emotion. Psychosom. Med. 11 (1949) 338-353.
2. -------- Man and his animal brains, Modern Medicine, 32 (1964) 95-106.
3. -------- The triune brain, emotion, and scientific bias, The Neurosciences Second Study Program (Ed. F.O. Schmitt) Rockfeller Univ. Press, New York (1970) 336-349.
4. -------- The limbic brain in relation to the psychoses, Physiological correlates of emotion, Academic Press Inc. N.Y., (1970) 129-146.
5. -------- The triune concept of the brain and behaviour. I : Man's reptilian and limbic inheritance. II : Man's limbic brain and the psychoses. III : New trends in man's evolution. Hinks Memorial Lectures, Univ. Toronto Press, 1973.
6. -------- The imitative-creative interplay of our three mentalities, Astride the two cultures : Arthur Koestler at 70 (Ed. H. Harris), Random House, New York (1976) 187-213.

7. R. Thom, Structural stability and morphogenesis, Benjamin, New York, (French edition 1972, English translation, by D.H. Fowler 1975).
8. -------- A global dynamical scheme for vertebrate embryology, Lectures on Maths in the Life Sciences, Amer. Math. Soc., Providence, Rhode Island, 5 (1973) 3-45.
9. -------- Modèles mathematiques de la morphogénèse, Union Generale d'Editions, Paris, 1974.
10. -------- La lingistique, discipline morphologique exemplaire, Critique, 322 (1974) 235-245.
11. -------- D'un modèle de la science à une science des modèles, Synthese 31 (1975) 359-374.
12. R. Thom & E.C. Zeeman, Catastrophe theory : its present state and future perspectives, Dynamical Systems, Warwick 1974, Springer Lecture Notes in Maths, 468, 366-401.

13. S.F. Wallace, letter (1976). Euclid Guidance Associates, Cleveland, Ohio.

14. E.C. Zeeman, Catastrophe theory, Sc. Am. 234 (1976) 65-83.
15. -------- Draft of Scientific American article, (this book §1).
16. -------- Brain modelling, Structural stability, the theory of Catastrophes, and Applications to the Sciences, Springer Lecture Notes in Maths. 525 (1976) 328-366.
17. -------- Duffing's Equation in brain modelling, Bull. Inst. Math. & Appl. 12 (1976) 207-214.
18. -------- Catastrophe theory, Proc. Roy. Instn. Gt. Br. 49 (1976) 77-92.
19. -------- A dialogue between a biologist and a mathematician, Biosciences Communications, (1977).

Numbering in this book.

INDEX

INDEX

abnormality, *see* anorexia nervosa
abstract dynamics, 655
abstraction, 17
A'Campo,N. 595
access
  denial of, 15,34,69
  regaining, 44
adhesiveness of cells, 143,200,
    240-242,247,250,258,261,263,
    270,272,282
  photographs, 264-266
aggression, 3-8,12-17,67,335
Airy pattern, 57-58
alcohol, 373-374,379-381
alienation, 75,387,395-396,404-405
amphibian embryology, 169-215,
    235-258
analogy, 17,637
anger, 16(fig.)
anorexia nervosa, 33-52,298,651,654
  abnormality, 34(fig.),35,37,41,47,49
    of attitude, 39,46-49
    of behaviour, 34,35,37,39,41-42
    escalation of, 44-46
    measurement of, 44-49
  bias factor, 36-37,41
  brain dynamics, 34,44,52,292,654
  bulimia, 34
  butterfly factor, 41-42
  cure, 33-35,44,48-49,51,654
  deception, 45
  dieting, 33,37,44-45,49
  drugs, 48
  fasting, 33-34,37,42,51
  gorging (bingeing, stuffing) 34,
    37,42
  Hevesi, J. 33,35,51,298,651
  hunger, 34-37,41
  hypnosis, 51
  hysteresis cycle, 34,37
  inaccessibility, 34,44
  insecurity (anxiety), 39,45,47-49
  insight, 50
  irreversibility, 44,46,48
  knock-out, 37(fig.),38,42
  let-go, 37(fig.),38,42
  limbic brain, 35-36,48
  loss of self-control, 36-37,41
  malnutrition, 33,48
  monster inside, 38,51
  obesity, 45
  operative suggestions, 40,51-52,
    298,654

anorexia nervosa *continued*
  personality
    complete, 33
    dissociated, 43,50
  phases, first and second, 34,37
  prohibitions, 43,51
  puberty, 45
  purging, 37-38,51
  qualitative experiments, 33,51-52,651
  quantitative testing, 33,52,298
  reassurance, 41,43,48-49
  rebirth, 44
  reconciliation, 43,51
  rigidity of neurosis, 44,46,48
  sleep, 36,39,298
  slimming, 44,49
  therapy, 33,40-44,49,52,654
  trance, 33,40-44,48,51
  Wallace,S.F. 654,656
Anorexic Aid Society, 33
Antonelli,P. 636
anxiety, 15,20,45,49
argument, 15
Arnold,V.I. 63,563,617,639
  notation, 563,567,581,583,585-589,595
arthropod segmentation, 236
attaching maps, 584,592,598
attack, 6(fig.),29(fig.),67
attractor (attractors)
  in brain, 12,39,69,266,296
  cycle (stable closed orbit), 62,76,
    92,293-296,446,451,480,635
  infinity of, 616
  point (stable equilibrium) 69,76,91
  slow manifold, 91-92,95,97,101-102,295
  strange, 275
  surface, 101-102,295,363,365,367
  Van der Pol, 77,295,635
Augusti,G. 565

Balinsky,B.I. 169,170,171,175,176,182,201
Barnaby,K.C. 452,492
basin of attraction, 39,478,480
behaviour (state), 5,306
  abnormal, 34,37,39,48,50
  aggressive, 3-8,12-17
  axis, 6(fig.),15
  bimodal, 5-7,18,29,34
  of brain, 287
  catastrophic change of, 7-8,13-17,
    69,312-313,335
  continuous scale of, 350
  divergent, 17-18,69,71,335-336

behaviour (state) *continued*
  effect of bias on, 37,339-341
  graphs, 4,314-315,324,328
  models, 5-7,15-16,34,298,651
  most and least likely, 5-7
  space, 23,315
    dimension of, 24,332
    see also state space
  of stock exchanges, 361-371
  surface, 6-7
    sections of, 31
  trimodal, 29
behavioural sciences, 652
Bellairs,R. 258,263-266,282
Berry,M.V. 57-58,64
bias factor, 30-32,36-37,337,339-341,
    468
  see also factors
bifurcation
  of brain dynamics, 12,17,34,293-300
  of dynamical systems, 616
  Hopf, 24,77,138,294(fig.),616
  non-elementary (generalised),
    294-296,616,633
  of non-linear oscillators, 294-295
  theory, 652
bifurcation set in
  biology, 72-73,162-163,625
  buckling, 59-61,419,428,435,437
  light caustics, 55-58
  ship dynamics, 442,460-464,469,476,
    487
bifurcation set of catastrophes, 7,
    574
  butterfly, 31,337
  cusp, 7,329,582
  double cusp, 575
  elliptic umbilic, 26,587
  hyperbolic umbilic, 26,56,61,586
  parabolic umbilic, 28,590
  swallowtail, 26
bimodal distribution, 5,7,379
bimodality, 18-19,20,23,29,46,70
biological clocks, 136-137, 202-204,
    207-208,235,246-252,258,275,
    277-278,281,298-300
biologist, dialogue, 267-286,619
biology
  brain, 12-13,35-36,287-300,655
  developmental (embryology)
    gastrulation, 141,145,169-215,258
    hypotheses, 143,151-152,268-277
    primary & secondary waves,
      71-73,141-286
    slime mold, 216-231

biology
  developmental (embryology) *continued*
    somites, 176,189,195-197,201-209,235-266
  ecology, 148-151,283
  evolution, 136-139,296,655
  experimental testing, 116,133-186,
    214-215,230-231,243-244,250-252,
    258,263-266,278-283,290-292,295,
    299-300,618-619,625,636,651-654
  heartbeat & nerve impulse, 73,81-140
  philosophy of modelling, 81,137,
    267-286,618-620,625-627,635-636,
    651-656
  see also each topic
boiling, 53
Bootle-Bumtrinket, 463
Bourbaki,N. 607,608,610,611
brain
  α-rhythm, 136,291,295
  alcoholic effects, 373-385
  anorexia, 34-36,40,52,298
  association, 297
  attractors, 12,39,295-296
  bifurcation, 12,17,34,298-299
  cerebral input, 35,40,52
  circadian rhythms, 298-300
  conceptual space-time, 631
  drugs, 48,299
  dynamics, 12,34,44,52,287-292,653-655
  EEG, 290,295-296
  estimating speed, 376-381
  experiments, 33,51-52,133-136,288,290,
    292,295,299-300,651-654
  language, 17,620-621,630-632,636-637,655
  limbic system, 12-13,35-36,48-50,52,295,
    298,653
  manic-depression, 298-300
  memory, 46,287,297
  mind, 17,297
  modelling, 287-300,564,653-655
  mood, 13,50,69,297,651,653
  neocortex, 12-13,35,295,653
  neurons, 12,82,287-288
    self-firing, 93,136
  non-verbal, 48-49
  oscillators, 12,293-300
  resonance, 296-299
  selectivity, 375,377,380,382
  sensory inputs, 50,296-297
  simulator, 288,374,654
  sleep, 35-36,39-40,136,299,654
  somatic input, 35,40,52
  trance, 40-44,48,51
  see also nerve impulse
breaking of waves, 624,634

buckling, 58-61,417-439,565
  arch, 418
  compound, 565
  constrained strut, 431
  cusp catastrophe, 59,419
  double cusp, 565
  dual cusp, 60,428,438
  Euler force, 59,418,421
  evolute of ellipse, 435-436
  free strut, 58-59,421-426
  global, 433-438
  harmonics, 59,422,429
  hyperbolic umbilic, 61
  imperfection sensitivity, 60,429
  load bearing capacity, 428
  panel, 60-61
  pinned strut, 60,427-431
buoyancy locus, 444-445,476
  complete, 487
  ellipse, 464
  ellipsoid, 482
  rectangle, 467
  wall-sided ship, 448
butterfly catastrophe, 29-32,336-344
  bifurcation set, 31,337,469
  dual, 469
  factors
    bias, 30-32,36-37,337,339-341,
      468
    butterfly, 30-32,41-42
    see also factors
  formulae, 27,30,337-341,468
  models
    anorexia, 36-42
    compromise, 29,336,344-345,629
    opposition, 345-349
    ship stability, 441,469
    war policy, 344-345
  pictures, 31,42,338,340,469-470
  pocket, 30-31,42,337,345
  trimodality, 29

canonical
  formulae for catastrophes, 27,326
  stratification of
    cubic forms, 566
    germs & jet spaces, 575-576
    quadratic forms, 583-584
  unfolding of umbilics, 585
capsizing, 441,455,458,489
  angles, 461-462,473,477,479,482
  bifurcation set, 464,469
Carpenter,G.A. 76
catastrophe (catastrophes)
  classification theorems, 22-25,
    66,332,343,497-504,519-520

catastrophe (catastrophes) *continued*
  compact (complete), 342-343,563,632
  cuspoids, 27,519-520,523,583
  dual, 12,60,332,339,357,415,428,438,
    462,469
  elementary, 1,25-27,326,497-499,651
  formulae, 27,519,581-589
  germ, 545,548
  machine, 8-12,409-415
  manifold, 574
  map, 476,502,574
  non-elementary (generalised), 24,
    289,294,633
  seven, 25-28
  space-time, 72-73,276
  transfer of proximity, 63
  umbilics, 25-28,519-520,524-527,
    585-592
  see also names of catastrophes :
    fold, cusp, swallowtail, butterfly,
    elliptic, hyperbolic and parabolic
    umbilics
catastrophe theory, 1-2,65,303-304,
    574-575,615,633
  in biology, 80,267-268,279-280,
    618-620,625,635-636,654-655
  future perspectives, 615-638
  as language, 612,615,620-621,630-633
  levels of structure, 65-78
  in mathematics, 66,496-500,574
    616-617,623
  name, 1,69,303
  in physics, 1,53-63,408,617-618,624
    634
  in social sciences, 302,319-324,612,
    620,627-630,636-638
catastrophic change (sudden jump), 1,
    10-12,18,69,103,303,312-313,575
  action potential, 73,83-84,94
  of amplitude, 62,294,297,299,454
  in anorexia, 37
  in attention, 296
  of behaviour, 7-8,13-17,69,312-313,
    335
  boiling, 53
  in brain dynamics, 13,291,295,297-299
  buckling, 59,418-419,432,438
  capsizing, 463
  in catastrophe machine, 10-12,410-412
  in cell development, 72,160-161,245,
    277,279
  in chemical reactions, 77
  cure, 48-49
  decensoring, 354,356
  declaration of war, 330,335
  disillusion, 20,613

catastrophic change (sudden jump) *continued*
  Duffing effect, 62,294,297,299,454
  dynamic cause, 10,24,69
  in dynamical systems, 93-94
  of estimated speed, 380
  heartbeat, 73,111-112
  heeling, 463
  in hysteresis cycle, 12,31
  inspiration, 20,613
  loss of temper, 16
  in manic-depression, 298
  of mood, 13,297-298,651,653
  phase-shift, 62,294,297,299,454
  of policy, 312-313,330
  political conversion, 629
  riots, 75,78,388,392,405
  shock-wave, 62
  stock market crash, 21,74,368
  threshold, 7,81,93-95,100,652
  in Van der Pol oscillator, 77,295
catharsis of self-pity, 15-16,651
cause-effect graph, 67,324,328,423,575
caustics, 24,54-58,497,626
cell (cells)
  adhesiveness, 143, 200,240-242,247,250,258,261,263-266,270,272,282
  catastrophic switch, 72,142,160-161,245,277,279
  continuity of
    development, 151,158,273
    dynamics, 157,274
    morphology, 151,159,276
    tissue, 151,158,272-273
    variables, 157,273-275
  curvature, 141,183-187,191-193,282
  determination, 141,153,270,279
  development path, 161-162,239,245
  differences, 143,155,263-266,269-270,282
  differentiation, 141,143,152-155,158,230,244,268-273
  flask, 182-183
  gene-systems, 137,142-143,161,275
  grafting, 156
  homeostasis, 71,89,136,143-144,151,156-157,273-275,653
    breakdown of, 161-162,204,277
  migration, 163,270-271
  mitosis, 207,241,275,278
  movement, 143,183,188-190,218-230,241-242,268-282
  muscle, 96
  nerve, 62

cell (cells) *continued*
  photographs, 264-266
  shapes, 241-243,252,263-266,282
  states, 71,157,245,273-275
  submerging, 143,145,179,181-183,218-224
censorship
  books & art, 349-355
  military, 355-356
central nervous system, *see* brain
characteristics, 61-62
chartist, 22,361
chemical reaction, 76-77
chick embryology, 258-266
Chillingworth,D.R.J. 424,645
chreod, 150
circadian rhythm, 299
classification
  cubic forms, 566-568
  cuspoids, 27,519,523
  elementary catastrophes, 26-27,497-561
  quadratic forms, 583-584
  space, time catastrophes, 73
  theorems, 22-25,66,332,343,497-504,519-520 (Note : different, but related versions are used in different papers.)
  umbilics, 27,519-520,524-527
cliff regulation, 626
clock & wavefront model, 235-256
  see also biological clocks
closed orbit (periodic solution, limit cycle), 62,76,92,293-296,446,451,480,635
cloud of points, 67
codimension, 513-518
  of germ (cod), 513,574
  of submanifold, 517
compact (= complete) singularity, 342-344,563,632
compromise, 29,336,344-345,629
configuration space, 474,480
conflicting factors, 19-20
  see also factors
conflicting judgements, 373
Conley,C. 77
constraint, 431
continuity
  in biology of
    cell development, 151,158,273
    dynamics of homeostasis, 157,274
    environment, 150
    morphology, 151,159,276
    tissue, 151,158,272-273
    variables, 157,273-275
  in sociological scales, 320,350

control (parameter)
  behaviour graph, 314-315,324,328
  factors, 19,313
    see also factors
  point, 8
  space, 8,23,25-27,69,100,314,473
Cooke,J. 202,235-256,258,277-280
cortex, see brain, neocortex
cost, 3,16,23,313-319,333-336
cotangent bundle
  in ships, 474,480
coupling
  of cusps, 564
  non-linear, 492
  of oscillators, 288,296
creativity, 613
Crisp,A.H. 50,64
critical exponents, 370
Cuban missile crisis, 17
cubic forms, 524,566,569
cure
  of anorexia, 33-35,44,48-49,
    51,654
  of manic-depression, 299
curvature, 436,444,447,472
  in embryology, 141,183-187,
    191-193
cusp, 7
  equation, 19
  in evolute, 435,437,458,466-467
  at metacentre, 458
cusp catastrophe, 6-7,18-19,103-105,
    329-333
  bifurcation set, 7,329
  canonical, 326,329
  conflicting factors, 19,332
    see also factors
  dual, 12,60,332,339,357,415,428,
    438,462
  feedback flow, 66,73-77,106,128,
    295,362,388,391
  fold curve, 7,329
  formula, 19,27,103,329,581-582
  normal & splitting factors, 19,331
    see also factors
  pictures, 6,19,et passim
  qualitative properties, 18-19,
    319-324,331-333,612
  stratification, 581-582
  surface, 6
  theorem, 23,331
  threshold of a threshold, 652
  unfolding, 581
cusp catastrophe models
  aggression, 6,13-17,67,69
  anorexic behaviour, 34

cusp catastrophe models *continued*
  argument, 15
  buckling, 59-60,419,428
  catastrophe machine, 8-11,412-413
  catharsis of self-pity, 16
  caustics, 54
  censorship, 352
  delinquency, 20
  driving under alcohol, 382
  Duffing's equation, 62-63,294
  economic growth, 71
  economic policy, 20
  emotional response, 20
  fight/flight, 6
  heartbeat, 73,111
  manic-depression, 298
  mathematical enjoyment, 613
  misestimation of speed, 382
  more haste less speed, 68,70
  nerve impulse, 73,102,128
  nonsense, 622
  opposition to government, 348
  phase-transition, 53
  primary wave, 71-72,160-162,253
  prison disturbances, 75,388,405
  ship stability, 460
  shock waves, 61-62
  stock exchanges, 21,73,368
  strength of opinion, 628
  territorial fish, 14
  war policy, 16-17,330
  Van der Pol equation, 295
  Van der Waals equation, 53
cusped edge, 25-26,568
cuspoids, 27,519-520,523,583

damped flow, 478
damping in
  catastrophe machine, 11,410
  non-linear oscillators, 62,293-294
  ships, 442,446,476
data fitting
  driving under alcohol, 382-384
  nerve impulse, 122-133
  prison disturbances, 394-397,405
delay
  before catastrophe, 318-319
  before mood changing, 13,653
  rule (convention), 117,161,306-310,
    413
  sociological justification, 310-313
delinquency, 20
denial of access, 15,34
density theorem, 550
description, 267,637-638,651,655

determinate, 507-508,512,574
  stratum, 580
development, 71-73
developmental
  biology see biology
  path, 160-162
  wave, see primary wave
dialogue with
  biologist, 267-286
  reader, 651-655
differential equations
  Duffing, 62-63,293-294
  fast & slow, 86-88,294
  heartbeat, 81-82,96-98,107-119
  Hodgkin-Huxley, 124-127,133-135
  nerve impulse, 81-85,105-107,
    119-136
  non-linear, 90-107,293-295
  propagation, 129-131
  quasi partial, 61
  reaction-diffusion, 76-77
  relaxation oscillations, 93
  ship stability, 446,451,457,476,
    478,486
  theory of, 623
  Van der Pol (Liénard), 81,91,
    294-295
differentiation of cells, 141,143,
    150,152,153-155,158,230,244,
    268-273
diffraction patterns, 57
diffusion
  chemical, 66,75-77
  developmental biology, 143-145,
    147,151,219-220,238-239
  ecology, 150
dimension of
  control (parameter, unfolding)
    space, 24,574
  state (behaviour) space, 25-27
diplomacy, 17,335
discriminant, 567,583
dissipative dynamics, 24,410,479
distributions, unimodal & bimodal,
  5,7,315
divergence, 17-18,23,69-71,335-336
division theorem, 527-528
DNA code, 274-275,618
dog, 4-8,17-18,69
double-cusp catastrophe, 343,563-565,
    593-600
  applications, 564-565,632
doves & hawks, 16,29(fig.)
    307-318,330
Drew,G.C. 373-374
Dreyfuss affair, 629

driving under alcohol, 373,379-381
dual catastrophe, 12,357
  cusp, 12,60,332,339,415,428,438,462
  butterfly, 469
Duffing's equation, 62-63,293-294
  in brain modelling, 288-289,293-300
  in ship rolling, 454
dummy variable, 522
Durrell,G. 463
dynamic (dynamics)
  of brain, 12,34,44,52,287-292,653-654
  capsizing angle, 479
  in catastrophe models, 8-12,66,69-74,
    473-474
  gradient-like (dissipative), 24,
    69,163,410,479
  modelling of, 81,287-288,441-442,
    491-492
  non-gradient, 24,294-296
  of ships
    Hamiltonian, 442,476,486
    linear, 444-455
    non-linear, 473-492
  of stock exchange, 362
dynamical systems, 85-107,288

eating, 34-37,42,300,654
ecology, 148-151,283
economics, 627
  growth, 70-71,564
  malaise, 347
  policy, 20
  stock exchanges, 21-22,361-371
education, 20,613-614
elasticity, modulus of, 418
elementary catastrophe, see catastrophe
elliptic umbilic catastrophe, 25-27,
    524,566,569,587
Elsdale,T. 258,280-287
embryology, see biology, developmental
emotion, 13,297,653
emotional response, 20
energy
  buckling, 419,422,426,430,435
  elastic, 10,410
  epidemic, 147
  graphs, 10-11,412
  Hamiltonian, 442,473-475,481,483,
    485-486
  kinetic, 474-483
  morphogenetic, 154-155,166-168
  potential, 410,474,483,485
equilibrium
  graph (surface, manifold), 10, 459,
    474,487
  stable & unstable, 10,81,363,459-460,
    469, et passim
    see also attractor

equivalent
  catastrophe germs, 548
  functions (potentials), 104
  germs, 506-507
  graphs, 324-327
  locally, 332,357,502
  maps, 502
  right, 506
  singularities, 23,66,576
  surfaces, 23,367,389
  (Note : different but related definitions are used in different papers.)
essence of germ, 513
essential variable, 522
estimating speed, 376-381
Euclid, 605-611
Eudoxus, 606-611
Euler,L. 58-61,417-439,444
evolute of
  buoyancy locus, 457,475
  ellipse, 434-437,466
  ellipsoid, 466,482
  parabola, 458
evolution, 23,29,136-139,296,655
exceptional singularities, 567
excess demand, 21-22,362
experiments
  fitted data
    alcohol, 382-384
    nerve impulse, 122-133
    prisons, 393-398
  methodology, 70,287-292,296,628, 630
  neuro-psychological, 295,653
  proposed
    alcohol, 354-355
    anorexia, 51-52
    brain, 296-300
    censorship, 350
    ecology, 283
    gastrulation, 214-215
    heartbeat, 116
    manic-depression, 299
    nerve-impulse, 133-136
    opposition, 3
    slime mold, 230-231,281
    somites, 250-252,278-282
    stock exchanges, 370
  philosophy of, 267-277,618-620, 625,635-638,651-655
  tested
    anorexia, 51-52,654
    caustics, 56-58
    ecology, 283
    heartbeat, 34
    somites, 243-244,258,262-266, 278-282

explanation, 267,637,651
explicit versus implicit, 24,65, 81,274,287-292,295-296,382, 389,653

facial expressions of dog, 4
factor analysis, 404
factors
  bias, 30-32,36-37,337,339-341,344
    failure of reform, 347
    invulnerability, 345
    lopsidedness, 468
    loss of self-control, 36-37,41
  butterfly, 30-31,42,337,344
    isotopy of hull, 468
    length of war, 344
    pressure of government, 348
    reassurance, 41-42
  conflicting, 19-20,332
    abnormality & insecurity, 47
    balance of payments & unemployment, 20
    devaluation & deflation, 71
    frequency & non-linearity, 63,294
    frustration & anxiety, 16
    haste & skill, 68
    rage & fear, 6
    size & closeness, 14
    space & time, 62,72,160
    temperature & pressure, 53
    threat & cost, 16-17
  control, 19,310
    centre of gravity, 474,485
    position, 8,410
  normal & splitting, 19-20,30,314, 331-332,337
    approach/avoidance & anxiety, 20
    creativity & skill, 613
    environment & abnormality, 298
    erotic content & aesthetic value, 352
    demand & speculation, 21,368
    hunger & abnormality, 34
    imperfection & load, 60,427-428
    load & compression, 59,419,425
    malaise & reform, 347
    mathematics & speculation, 622
    pacemaker & tension, 111
    politics & involvement, 628
    potential & potassium, 123
    reward & punishment, 20
    speed & selectivity, 382
    tension & alienation, 388
    threat & cost, 71
    tiresomeness & demandingness, 20
fast
  dynamic (flow), 69,73-77,363
  eigenvalue, 86
  equation, 86,294
  foliation, 87
  orbit, 86

fear, 3-8,14-15
feedback
  in anorexia, 49
  flow on cusp catastrophe, 66,
      73-77,295,362,388(fig.)391
fight/flight, 5-8,15
fish, 13-14
flow
  feedback, 66,73-77,295,362,
      388(fig.),391
  Hamiltonian, 442,477-478
fold
  catastrophe, 3,24,67
    boundary of cloud of points, 67
    chemical reactions, 77
    formula, 27
    rainbow, 55
    rolling of ship, 441,454
    see also catastrophic change
  curve, 7,23,26,100-101,329,460
  point, 26,325
  surface, 25
  threshold, 94, 100
foliation,
  fast, 87
  of space-time, 72,276
  of stratification, 66,72-73
forced oscillations, 62,293-300,
    449-455,480
forest frontier moving, 148-150,283
Fowler, D.H. 28,103,646
frame of mind, 7-8,15,69
friction, 11,410
  see also damping
frontier
  forest, 149,283
  movement of, 71-72,80,142,149,152,
      161-163,268-283
  stabilisation of, 152,163,166
  theorem, 71,152,157-164,276
frustration, 15-16
function
  generic, 23,66,104,157,314,356,
      502,550
  locally stable, 502,558
  smooth, 23,315,356
  (Note : different, but related,
  definitions of generic are used
  in different papers.)
fundamentalist, 22,361
funnel, 74

Gartree prison, 387,394,404
gas, 53
  dynamics, 61

gastrulation
  amphibia, 141,145,169-215,258,
      269-272
  birds, 258-262
  pictures, 169-174,178,182,195,260
generic
  function, 23,66,104,157,314,356,
      502,550
  system,surface, 367
  (Note : different, but related,
  definitions are used in different
  papers.)
genetic code, 268,274-275,618
geodesic
  distance, 24,55
  flow, 115
  spray, 475-476,487
germs, 504
  catastrophe, 545
    uniqueness of, 548
  classification of, 503,519-527
global
  buckling of strut, 433-438
  dynamics of heartbeat, 117-119
  dynamics of ship, 477-480,484-487
  metacentric locus, 464-471
globalisation from germs to functions,
      498-499,550-558
gradient
  dynamics, 103,157
  -like dynamics, 24,69,473
  non-, 24,275,291,296
gradients in biology, 239,244,246,
      253-254,619
graphs, 3-7,23,314-315,324-328
  cannonical models of, 326
  cause-effect, 67,324,328,423,575
  control-behaviour, 314-315,324,328
  of energy, 10-11
  of equilibria, 10,418,459
  equivalence of, 324-327
  qualitative properties, 322,326,612
  theorems about, 23,331,343
Greek mathematics, 605-611
growth, economic, 70-71

Hall,C.S. 387-401,404
Hamiltonian-Jacobi equation, 634
Hamiltonian dynamics, 442,475-486
hard spring, 63,294,454,472
harmonic
  of buckled beam, 59,422
  oscillator, 293
Harrison,P.J. 387-401,404
Hartley,P. 33

hawks, 16, 29(fig.), 307-318,330
heartbeat, 73,81-82,96-98,107-119
  anatomy, 108
  atrial & ventricular beat, 109,
      112-113,116-119
  cardiac failure, 110-113
  conclusions & experiments, 116-119
  contraction & relaxation, 96-99,
      111-112,115
  diastole & systole, 82,98,112,
      117-118
  ectopic pacemakers, 93
  electrocardiogram, 114
  global dynamics, 117-119
  local equations, 107-111
  membrane potential, 96-97,114
  pacemaker wave, 82,109-110,
      113-115
  Rybak experiment, 107-110
  Starling's Law, 109-110,113
heeling, 441
  angle, 461-462
  bifurcation set, 463-464,469
  when loading, 490-491
Hensen's node, 260-262,279,281
hermitian rank of germs & jets, 521
hessian, 67,577
Hevesi,J. 33,35,40,51-52,298,651
history, laws of, 320
Hodgkin,A.L. & Huxley,A.F. 81,83-85,
    107,120-135
homeostasis, 71,89,136,143-144,150,
    151,156-157,273-275,291,653
  breakdown of, 161-162,204,277
homoclinic orbit, 76-77
Hooke's law, 10,413
Hopf bifurcation, 24,77,138,
    294(fig.),616
human sciences, 620,627
hunger, 34-37,41,654
Hunt,G. 60,61,64,417
Huxley,A.F. *see* Hodgkin
hydrodynamic breaking, 624,634
hyperbolic umbilic catastrophe,
    25-27,585-586
  bifurcation set, pictures, 26,56,
      61,586
  formula, 27,585
  models
    breaking of waves, 624,634
    buckling of panel, 61
    caustics, 56-57
    evolute of ellipsoid, 466,482
    metacentric locus, 441,466,482
  unfolding & stratification, 585-586
hypnosis, 51

hypocycloid, 568
hypotheses in
  biology, 267-277
    ecology, 150
    gastrulation, 180-181,211
    primary waves, 143-151-152,
      157-159,267-277
    slime mold, 218
    somites, 202,235,277
  economics
    stock exchanges, 361,363-369
    modelling, 267-268,287-292,
      319-324,651-655
  physics, 267-268,617-618
    ships, 473-474
  psychology, 651
    alcohol, 375,382
  sociology
    censorship, 350,355
    opposition, 347
    prisons, 389-391,405
    war policy, 315-318,344
hysteresis, 11,18,23,70
  cycle, 11-12,34,37,295,635
hysteria, 15

imperfection sensitivity, 59-60
    427-429,463,565
implicit function theorem, 553
implicit versus explicit, 24,65,
    81,274,287-292,295-296,382,
    389,653
inaccessibility, 15,18,34
indeterminate stratum, 580
index of stock exchange, 21,361
induced rolling in ships, 451-452
inexact sciences, 18,304
insight, 50,65,323-324,492,651
institutional disturbances, *see*
    prison disturbances
interference patterns, 57
introvert/extrovert, 374,382
invariants, qualitative, 291,
    321-322
Isnard,C.A. 303-359

Jacobian ideal, 507(Δ),574(J)
Jänich,K. 28,64
jet (jet space), 506,576

Kapitzer pendulum, 635
kinematic wave, 151,246,277,
    280-281
kinetic energy, 474,483,485
knee, 287
Kopell,N. 76

language
  of catastrophe theory, 615,653,655
  nature of, 620,623,630-632,636-637
  qualitative, 322-324,326,612
laws, 320,323,333,476,612,615
least-likely behaviour, 7
light caustics, 24,54-58,497,626
likelihood, 5
limbic system, see brain
linear theory of rolling, 444-452
linguistics, 620-621,630-632,636,
    653
load bearing capacity, 427-428
local ring, 504
locally
  equivalent, 332,357,502
  stable, 502,558
Lorenz,K. 3,4(fig.),64,67
lose temper, 16(fig.),20
Lyapunov function, 24,69,164,292,
    616,625

machine, catastrophe, 8-12,409-415
MacLean,P. 13,35,48,52,64,297,
    653,655
magnetism, 11
Malgrange preparation theorem, 498,
    503,527-536
manic-depression, 298-300
manifold
  equilibrium, 487
  slow, 87,627
map
  attaching, 584,592,598
  catastrophe, 476,502,574
Markus,L. 293,641
Marriage,H. 373,387-401,404
mathematical enjoyment, 612-614
Mather,J. 332,498,574,577,641
maxima & minima, 10,23,577,626,651
maximal ideal, 504
Maxwell,J.C. 64,310
  convention (rule), 53,75,306-310,
    624
  line (set), 53(fig.),624,632
memory, 46,287,297
mesoderm, 173,179-180,186-215,626
    635
metacentre, 444
metacentric
  height, 444,447,449
  locus, 457,464-471,475-476,482,
    487-489
metaphor, 44,637,654
minima & maxima, 10,23,577,626,651
misestimations of speed, 373,376-381

mitosis, 207,241,275,278
mixture (mixed homomorphism), 536
modality, 66,503,563
  see also bimodality, trimodality
modelling, 137,267-268,287-292,
    633-638
models, see under names of catastrophes,
    cusp, butterfly & hyperbolic
    umbilic
Möbius strip, 596-600
molecular biology, 275,619-620
mood, 13,50,69,297-298,651,653
more haste less speed, 68,70
Morin,B. 25,617
morphogen, 145,164,238-239,280,625
morphogenesis, 141-146,183,268-272,497,655
  birds, 258-262
  definition, 269,272
  energy in, 154-155,166-168
  gastrulation, 169-173,178-184
  neurulation, 175-178,189-202
  slime mold, 216-231
  somites, 176,202-209,235-256,266
morphogenetic movement, 167,269-271
morphologies
  linguistic, 630-633,636-638
  social, 620,629
most-likely behaviour, 7
movement of frontier, 71-72,80,142,
    149,152,161-163,268-283

Nakayama's lemma, 508
neo-cortex, see brain
nerve impulse, 73,81-85,105-107,
    119-136,287
  action potential, 83-85,121,
    132-133
  conclusions, 133-136
  electric equations, 125
  Hodgkin-Huxley data, 85,121-126,133
    equations, 124-127,133-135
  local equations, 120,122-129
  neuron, 82,287,296
    self-firing, 93,136
  propagation wave, 120,129-132
  resting potential, 83,127
  sodium, 107,122-126,133,135
  synapses, 82-83
  velocity, 131-132
  voltage clamp, 122-124
neuro-psychology, 12-13,287-300,
    653-655
neurulation, 175-178,189-202,209-213
Newhouse,S. 616,392-393
Newton's law of motion, 11,446,476
noise, 74-75,370,391-392

non-linear
 changes of scale, 321
 coupling, 492
 differential equations, 90-107,293
 group-action, 583
 invariants, 291,321
 oscillations, 62-63,293-296
 resonance, 296-299,441
non-verbal thinking, 17,48
normal bundle, 434,475-476,487
normal & splitting factors, 19-20,30,
  331-332,337
 see also factors
notochord, 141,145,189-197,200-202,
  207,213,215,242,260-261,269-272
 pictures, 171,173-176,195,201,260

Occam's razor, 267
operative suggestion in anorexia, 40,
  51-52,654
opinion, 628-629
opposite pairs, 18,20
opposition to government, 345-349
oscillators, 62-63,293-296
 brain, 12,288,293-300
 cellular, 235
 Duffing, 62-63,293-294,454
 harmonic, 293
 resonance, 63,293,296-297,441,449,
  452-455,482
 ship rolling, 449-452
 Van der Pol, 77,92-93,293-295

pacemaker
 heartbeat, 82,109-110,113-115
 slime mold, 219
parabolic umbilic catastrophe, 25
 formulae, 27,526,589
 mushroom, 626
 pictures, 28,590
 unfolding & stratification, 589-591
paradigm, 612,622,652
paradox of naval architecture, 452
parameter (control) space, 8,23,
  25-27,69,100,314,441,473-474,
  485
pattern formation, 202-209,236-238,
  270
Pearcey pattern, 57-58
Pearson,M. 243,258,285
Peixoto,M.M. 297
pendulum of fashion, 349,354-355
Penrose,R. 607,611
periodic solution (closed orbit,
  limit cycle), 62,76,92,
  293-296,446,451,480,635

personality
 change of, 33,44,
 dissociated, 43,50
 of institution, 392-393
perturbation theory, 423
phase
 linked, 235,297
 portrait, 477-478
 shift, 62,294,297,299
 transition, 53,75,617,624,634
philosophy of
 catastrophe theory, 615-638
 modelling, 18,81,267-268,
  287-292,296,441-442,491-492
 reductionism & structuralism, 619
 science, 319-324,612,615-621,
  636-638
physics, 1,64,617-618,624,634,652
 breaking of waves, 624,634
 buckling, 58-61,417-439
 light caustics, 24,54-58,497,626
 oscillations, 62-63,293-296
 phase transition, 53,75,617,624,634
 shock waves, 61-62
physiology, 107-140,620
playing it cool, 387,400
pocket, 30,31(fig.),32,42(fig.)
positional information, 235,239-240,
  244-245,253-254,281
Poston,T. 9,63,415,568,624,647
potassium, 107,122-126,133
potential
 action, 83-85,121,132-133
 cardiac membrane, 114
 energy, 410,474,483
 function, 103-104,410
 resting, 83,127
prediction *see* experiments
preparation theorem, 498,503,
  527-536
primary waves, 71-73,141-154,268-276
 action potential, 120,147
 in amphibia, 179-181,209-215,
  244-247,280-281
 beginning, 180,218,259
 in birds, 257-262,281-282
 & clock, 202-209,235,244-250,277-282
 contraction, 115,147
 deepening, 165,283
 detection of, 155-156,250-252,278-283
 in embryology, 151-153
 experiments
  proposed, 214-215,230-231,250-252,262
  tested, 241-244,251-252,258,278-283
 forest frontier, 148-150,283
 hypotheses, 143,151-152,267-277

primary waves *continued*
  quantitative aspects, 164-168
  regulation, 208-209,219-220,
    236-240,247-250,277-278
  ripple ahead of, 168,283
  ripple on, 203
  in slime mold, 218-220
  spurts, 203-204,248-250,277-278
  stabilisation, 154,161-163,166,
    276,278,283
  theorem, 71-73,152,157-164,276
  timing, 209-213
prison disturbances, 74,387-406
probability distribution, 67,
    305-328
prolongation of germs, 537
proverbs, 67-68,323,651
psychological models
  aggression, 3-8,12-13,67-69
  anorexia, 33-52
  argument, 15
  association, 297
  attention, 296-297
  delinquency, 20
  emotional response, 20
  manic-depression, 298-300
  misestimation of speed, 373-385
  mood, 13,297-298,651-653
  more haste less speed, 68,70
  opinion, 628-629
  recall, 297
  self-pity, 15-16
psychology
  neuro-, 287-292,293,295,654-655
  philosophy of modelling, 620,
    627,651-655
public opinion, 304-318,350-355
pulse solution, 76

qualitative
  equivalence, 324-327
  experiments, 33,51-52
  meaning of, 319-324,612
  properties (features, invariants),
    18,23,70,291,321-322
  and quantitative, 82,145,363,403,
    441,615,625,636-637,652
quantitative
  aspects, 164-168,209-213,226-227,
    243-244
  data fitting, 122-133,182-384,
    393-398,404-406
  estimates, 187,191-193,252,446-456,
    489
quantum mechanics, 616-618

questionnaires, 346,350-351,628
queue d'aronde *see* swallowtail
    catastrophe

rage, 4,67
rainbow, 55
rank of germs & jets, 521
Rankine-Hugoniot condition, 61
reaction-diffusion equations, 76-77
reducing the arbitrariness of
    description, 51,637,651
reduction lemma, 522
reductionist philosophy, 619
regulation
  cliff, 626
  physiological, 147,620
  slime mold, 219-220
  somites, 208-209,236-240,247-250,
    277-278
repeatability, 143,150,152,157,159,
    275-276
repeller, 91-92,95,101-102,294-295,
    367
research, 605,611,613
resonance, 63,293,296-297,441,449
    452-455,482
Riemann-Hugoniot catastrophe *see*
    cusp catastrophe
right-equivalence, 506
ring of germs, 504
riots, 75,392(fig.),399
ripple
  ahead of wave, 168,283
  on wave, 203
rolling *see* ship stability
Rosemary, v,568,609
rule (convention)
  delay, 69,306-313,413-414
  Maxwell, 53,75,624,306-310
  voting, 309,311,345
Rybak,B. 107-110

saddles (unstable periodic solutions)
    62,294
Schmocker,M. 298
Schulman,L. 624
science, philosophy of, 18,51-52,
    267-268,319-324,612,615-638,
    651-655
secondary waves, 141-155,166-168,635
  adhesiveness, 143,200,240-242,247,
    250,258,261,263-266,270,272,282
  curvature, 141,183-187,191-193
  energy, 154-155,166-168

secondary waves *continued*
  experiments
    proposed, 214-215,230-231,
      250-252,262
    tested, 241-244,248,251-252,
      258,278-283
  gastrulation, 169-215,258,269-272
    pictures, 169-174,178,182,195,
      260
  mesoderm, 173,179-180,186-215,626,
    635
  morphogenesis, 141-146,153-155,
    268-272
  neurulation, 175-178,189-202,
    209-213
  notochord, 189-197,200-202,207,213,
    215,242,260-261,269-272
    pictures, 171,173-176,195,201,260
  recruitment, 154,280
  relaxation, 115,147
  repolarisation, 120,147
  slime mold culmination, 216-231
    see also somites
  submerging, 143,145,179,181-183,
    218-224
  wavefronts, 174,180,195-197,
    259-262,281
segmentation *see* somites
selectivity, 375,377,382
self-pity, 15-16
sensitive course, 453-454,482
sensitivity, imperfection, 60,429,
  463
sensory perception, 296
seven elementary catastrophes, 25-28
Shapland,P. 373,387-401,403
ship stability, 441-493
  buoyancy locus, 444-445,448,464,
    467,476,482,487
  capsizing, 441,455,458,464,469,489
    angle, 461-462,473,477,479,482
  catastrophe model, 473-474
  configuration space, 474,480,483,
    485
  cotangent bundle, 474,480,485
  dynamics, 441,477-480,492
  energy, 473-474,481,483,485
  equilibrium surface, 459-460,
    469,476
    manifold, 487
  Hamiltonian, 442,475,481,483,
    485-486
    flow, 476-480
  heaving, 456-457,482-484
  heeling, 441,461-464,469,490-491
  imperfection-sensitivity, 463

ship stability *continued*
  lever arm curve, 446,471-473
  linear theory, 444-446
  loading, 458,484-491
  metacentre (metacentric), 444
    height, 444,447,449
    locus, 437,464-471,475-476,
      488-489
  paradox, 452
  parameter space, 473-474,480,
    483,485
  phase-portrait, 478
  pitching, 455-456,480-482
  qualitative aspects, 441-443,
    491-492
  quantitative estimates, 446-452,
    489-490
  resonance, 449, 452-455,482
  righting couple, 445, 471
  rolling, 444-455,473-480,482
  seaway, 449-452,480
  sensitive cause, 453-454,482
  state space, 473-474,480,483,486
  statics, 441,459
  wall-sided, 448,458
  waves, 449-450,480
shock waves, 61-62
Siersma,D. 326,513
simplicity, 17,65,68,81,163,287,
  323,612,651
  hypothesis, 276
singular set, 574
singularity, 23,66,331
  compact (= complete), 342-344,
    563,632
  equivalent, 33,66,332,502,576
    locally equivalent, 332,357,558
    right equivalent, 506
  G-invariant, 616
  stable, 449
    locally stable, 502,558
  of vector field, 616
sleep, 35-36,39-40,136,299,654
slime mold, 141,145,216-231,247,281
slow
  eigenvalue, 86
  equation, 86,111,294
  feedback, 73,77,295,362,388
  flow, 105-106,128,295,363,369
  wavefront, 235
smooth (C^∞), 23(footnote),315,
  356,502
social sciences, 302,319-324,612,
  620-621,627-630,636-638,652-653
sociological models,
  censorship, 349-356

sociological models *continued*
  compromise, 29,629
  delinquency, 20
  education, 20,612-614
  opinion, 628-630
  opposition, 345-349
  war policy, 16-17,305-345
sodium, 107,122,126-127,133
soft spring, 63,294,454,472
somites, 80,202-209,235-266,
    277-283
  amphibian, 176,202-209,235-256
  Bellairs,R. 258,263-266,281-282
  cells,
    adhesiveness, 200,240-242,247,
      250,258,261,263-266,270,272,
      282
    catastrophe, 160,181-183,245,
      249,258,272,279
    compression, 261-262,263,282
    shape, 242,264-266,282
  chick, 257-266
  clock, 202-204,207,235,247-250
    & wavefront, 202-209,244-250,
      279-280
  Cooke,J. 202,235-256,278-280
  Elsdale,T. 258,280-281
  experiments
    proposed, 214-215,250-252,262
    tested, 241-244,248,251-252,
      258,278-282
  Hensen's node, 261-262,279,281
  mitosis, 207,278
  pattern formation, 202-209
  pictures
    anatomy, 176-177,201,242,260,266
    cells, 264-266
    data, 243
    fate maps, 174,195,260
    model, 203,249,253
  primary wave, 202-204,244-246,257,
    268-277
  regulation, 208-209,236-240,
    247-250,277-278
  secondary wave, 195-197,204-209,
    246-250,257
  segmentation, 204-209,241-244,249,
    261,277
  sizes, 252-254
  wavefronts, 174,195-197,235,253,
    257,259-262,281
space
  of C^{∞}-functions, 550
  control (parameter), 8,23,25-27,
    69,100,314,329,441,473-474,485
  of germs, 504

space *continued*
  of jets, 506
  state (behaviour), 23-24,69,315,473
    see also state space
  time
    in biology, 71,157,160-163,249,281
    catastrophes, 72-73,276
    conceptual, 630-631,636-639
    foliated by time, 72-73,276
  unfolding, 574
speech, 630
splitting factor, 19-20,30,331-332,337
  see also factors
springs, hard & soft, 63,294,454,472
stabilisation of frontier, 152,154,
    161-163,166,276,278,283
stability of ships, *see* ships
stable
  catastrophe map, 66,558-560
  closed orbit (limit cycle,periodic
    solution, attractor), 62,76,
    293-296,451,480
  equilibrium, 10,23,81,274,363,389,
    423,469-470
  function (potential), 104
  heeling angles, 461,469
  locally, 502,558
  singularities, 499-500
  under constraint, 431
state (behaviour) space, 23-24,69,315,473
  brain, 288,296
  cell, 157,245,273-275
  dimension of, 24,332
  institution, 389
  machine, 8,410
  ship, 473-474,480,486
  strut, 424
stochastic noise, 74-75,370,391-392
stock exchanges, 21-22,73-74,361-371
  chartists & fundamentalists, 22,362
  hypotheses, 361,363-364,368-369
  slope of recession, 370
stratification
  canonical, 576,583-584
  of catastrophe manifold, 575
  of cubic forms, 524,566,573
  of cusp catastrophe, 581-582
  definition, 550
  of double cusp, 593-595
  foliated, 66,73
  of germs, 576
  of jets, 557,575-576
  of quadratic forms, 569,583-584
  refinement, 578
  simple, 66
  of umbilics, 566-568,585-592

stress, 373
structure, levels of, 65-78
strut, buckling, 58-60,417-439
subjective feelings, 651
submerging, 143,145,179,181-183,
    218-224
sudden jump, *see* catastrophic change
support function, 314
surface
  behaviour-, 6,10
    sections of, 31
  cusp catastrophe, 6
  equilibrium-, 459,474,487
  fold-, 25
swallowtail catastrophe, 25-27,441,
    488-489
symbolic umbilic (=E_6), 524,527,
    566-567
symmetry of Hamiltonian, 485
synthesis, 17,33,51,333,367,612,651

Takens,F. 74,616,643
tears, 15
temper, loss of, 16(fig.),20
tension, 387,395-396,404-405
territory, 7,14
theoretical point of view, 619
therapy,
  anorexia, 33,40-41,44,49,52,654
  manic-depression, 298-299
Thom,R.
  book, ix,1,28,63
  catastrophe theory, 1-2,65,102,134,
    144,303,361,364,381,388,417,
    615-650
  caustics, 58
  classification theorems, 22,66,104,
    158,332,343,497-504
  elementary catastrophes, 1,27-28,81,
    103,137,336,367,569,585,633
  embryology, 618-620,625,635-636
  experiments, 619,625,633,637
  human sciences, 620,627,636-638
  linguistics, 564,620-621,630-632,
    653
  mathematics, 497-499,616,623
  papers, 643-644,647-648
  philosophy, 51,319,620-621,651
  physics, 617-618,624,634
  predation loop, 654-655
  transversality, 499,550-558
  unfoldings, 498,537-545
Thompson,J.M.T. 60,64,417
Thompson,M. 370
threat, 16,313-319,333-336
threshold
  of appearance of a threshold, 652

threshold *continued*
  for behavioural change, 7
  curve, 128
  frequency, 296-297
  model in embryology, 255
  point, 94,100
  stabilisation, 626,634-635
  for triggering an action, 81,
    93-95,100
Tolstoy,L. 320
traffic flow, 61
transfer of proximity, 63
transversal unfolding, 539
transversality lemmas, 504,
    539-540,551-556
trimodality, 29
Trotman,D.J.A. 497-561
Tvergaard,V. 60
twisted cubic curve, 599

umbilic (umbilics)
  bracelet, 563,568
  catastrophes, 25-28
  classification, 519-520,524-527
  stratification, 566,585-592
  unfolding, 27,585-589
  see also names, elliptic,
    hyperbolic & parabolic
unfolding (unfoldings)
  category of, 537
  compositions, 616
  cuspoids, 27,581-583
  existence lemma, 539
  G-actions, 616
  of germ, 537,574
  morphism of, 537
  space (control space), 574
  transversal, universal, 537
  umbilics, 27,585-590
  uniqueness theorem, 544
unimodal
  distribution, 5,7,305,315
  singularity, 66,563
unstable
  behaviour of stock exchanges,361-371
  capsizing angles, 461,469,479
  closed orbit (limit cycle, periodic
    solution, repellor or saddle),
    62,293-296
  constrained, 431
  equilibria, 11,23,423,460,469-470

Van der Pol equation, 77,92-93,
    293-295,635
Van der Waal's equation, 53,617,
    624

Vogt, W. 169, 171, 172, 173, 174, 181, 182, 187
voting rule, 309, 311, 345

Waddington, C.H. 137, 193, 195, 198, 201, 267, 268, 283
wall-sided ship, 448, 458
Wallace, S.F. 654, 656
war policy, 16-17, 29, 305-345.
  applications, 333-336, 344-345
  butterfly model, 336-338
  cusp model, 330
  delay rule, 306-313
  hypotheses, 315-318, 344
Wassermann, G. 72
wave (waves)
  breaking of, 624, 634
  kinematic, 151, 246, 277, 280-281
  ocean, 449-450, 480
  trains, 76-77
  see also primary & secondary waves
wavefronts, 174, 195-197, 235, 253, 257, 259-262, 281
Weiss, P. 207, 275
Wever, R. 299
Whitehead, J.H.C. 611, 614
Whitney, H. 101, 497, 499, 611
  C^∞-topology, 550
Winfree, A. 76-77, 252, 254
Wolpert, L. 183, 235, 239-240, 254
Woodcock, A.E.R. 9, 63, 129, 415
word, 655

Zeno, 606-611
Zhabotinsky reaction, 76

PUBLICATION DETAILS

The papers are reproduced by permission of the original copyright owners.

Paper

1. (Shortened version) Sc. Am. 234 (1976) 4, 65-83.
2. Proc. Int. Cong. Math. Vancouver (1974) 2, 533-546.
3. Towards a theoretical biology, (Ed. C.H. Waddington) Edinburgh Univ. Press 4 (1972), 8-67.
4. Lectures on Maths in the Life Sciences 7 (1974) Amer. Math. Soc., Rhode Island, 69-161.
5. J. Theoretical Biology 58 (1976), 455-476.
6. Structural Stability, the theory of catastrophes and applications in the sciences, Springer Lecture Notes in Maths 525 (1976), 396-401.
7. Biosciences Communications (1977).
8. Same as 6; 367-372.
9. Bull. Inst. Math & Appl. 12 (1976), 207-214.
10. Use of models in the social sciences (Ed. L. Collins), Tavistock, London (1975), 44-100.
11. J. Math. Economics 1 (1974), 39-49.
12. Br. J. maths & stats. Psychology 29 (1976), 19-31.
13. Same as 12; 66-80.
14. Same as 6; 402-406.
15. Same as 3; 276-282.
16. Same as 6; 373-395.
17. Proc. Esc. Lat. Am. Math. 3 (1976).
18. Same as 6; 263-327.
19. Same as 6; 328-366.
20. Bull. Inst. Math. & Appl. 10 (1974), 272-281.
21. Dynamical Systems, Warwick 1974 (Ed. A.K. Manning), Springer Lecture Notes in Math 468, 366-401.
22. ——.

The chronological order in which the papers were written was :

3,15,18,10,20,11,12,21,2,4,13,1,19,16,14,6,8,5,9,17,7,22.